Race, Class, and Gender

Intersections and Inequalities

TENTH EDITION

MARGARET L. ANDERSEN
University of Delaware

PATRICIA HILL COLLINS
University of Maryland

CENGAGE

Australia • Brazil • Mexico • Singapore • United Kingdom • United States

Race, Class, and Gender:
Intersections and Inequalities,
Tenth Edition
Margaret L. Andersen and
Patricia Hill Collins

Product Director: Thais Alencar

Product Manager: Ava Fruin

Product Assistant: Megan Nauer

Marketing Manager: Tricia Salata

Content Manager: Rita Jaramillo

Production Service:
Sharib Asrar - MPS Limited

Art Director: Marissa Falco

Cover Designer: Marissa Falco

Cover Image: Sam Rodriquez

Compositor: MPS Limited

For product information and technology assistance, contact us at **Cengage Customer & Sales Support, 1-800-354-9706** or **support.cengage.com**.

For permission to use material from this text or product, submit all requests online at **www.cengage.com/permissions**.

Library of Congress Control Number: 2019931157

ISBN: 978-1-337-68505-4

Cengage
20 Channel Center Street
Boston, MA 02210
USA

Cengage is a leading provider of customized learning solutions with employees residing in nearly 40 different countries and sales in more than 125 countries around the world. Find your local representative at **www.cengage.com**.

Cengage products are represented in Canada by Nelson Education, Ltd.

To learn more about Cengage platforms and services, register or access your online learning solution, or purchase materials for your course, visit **www.cengage.com**.

Printed in the United States of America
Print Number: 01   Print Year: 2019

Contents

★ = new pieces

Preface

We write this preface at a time when the social dynamics of race, class, and gender have flared up in ways not even imagined when we wrote the previous edition of this book. When we published the ninth edition of this book a mere five years ago, the nation was in a more optimistic mood. The United States had elected its first Black president and many believed we had entered a "postracial" age. Women seemed to have it made by reaching highly visible positions of power. The nation was in the midst of a major recovery from an economic recession that had decimated many people's communities and individual resources. Many of the laws and policies that had historically restricted LGBTQ rights were being dismantled. Although many problems remained, there was a sense that we had turned a corner from our past.

Now, just as we go to press, the nation has witnessed the mass murder of several Jewish people who were quietly worshiping in their synagogue—the largest such slaughter in over sixty years. Police shootings of Black men and women are all too common and frequently reported in the national and local news media. Young immigrant children are being torn from their families as their mothers and fathers seek asylum in the United States from violence in their native countries. With the exception of Native people and Chicanas in the American southwest, whose land was appropriated in 1848 (as the result of The Treaty of Guadalupe Hidalgo) following the Mexican-American war), the United States has historically been a nation of immigrants. Now, fears about those defined as somehow "other" are being stoked for the benefit of people who feel that they are threatened by the changes that a more diverse U.S. population is bringing. And, many laws and policies that have loosened restrictions on women, people of color, and LGBTQ people are now jeopardized, with many such policies being reversed or eliminated. Many people are worried about America's future commitment to the rights of all. This makes the understanding of the interconnections among race, class, gender, sexuality, and nationality all the more critical.

Not everyone understands these events and changes in the same way. How are race, class, and gender systematically interrelated, and what is their relationship to other social factors? This is the theme of this book: how race, class, and gender simultaneously shape social issues and experiences in the United States. Race, class, and gender are interconnected, and they must be understood as operating together if one wants to understand the experiences of diverse groups and particular issues and events in society. This book will help students see how the lives of different groups develop in the context of their race, class, and gender location in society.

Since the publication of the first edition of this book, the study of race, class, and gender has become much more prevalent. Over the years that this book has been published, there has been enormous growth in the research scholarship that is anchored in an intersectional framework. Social justice advocates have also insisted on an "intersectional perspective" when addressing many of the social wrongs in our nation. Still, most people continue to treat race, class, and gender in isolation from each other. Some also treat race, class, and gender as if they were equivalent experiences, or as if one were more important than another. Although we see them as interrelated—and sometimes similar in how they work—we also understand that each has its own dynamic, but a dynamic that can only be truly understood in relationship to the others. With the growth of intersectional studies, we can also now better understand how other social factors, such as sexuality, nationality, age, and disability, are connected to the social structures of race, class, and gender. We hope that this book helps students understand how these structural phenomena—that is, the social forces of race, class, and gender and their connection to other social variables—are deeply embedded in the social reality of society.

This anthology is more than a collection of readings. It is strongly centered in an analytical framework about the intersections and inequalities brought on by the dynamic relationships among and between race, class, and gender. The organization of the book features this framework, including a major revision of the last section to reflect the growing activism of people who resist the many inequalities that race, class, and gender create. Our introductory essay distinguishes an intersectional framework from other models of studying "difference." The four parts of the book are intended to help students see the importance of this intersectional framework, to engage critically the core concepts on which the framework is based, and to analyze different social institutions and current social issues using this framework, including being able to apply it to understanding social change.

ORGANIZATION OF THE BOOK

The four major parts of the book reflect these goals. We introduce each part with an essay by us that analyzes the issues raised by the reading selections. These essays are an important part of this book, because they establish the conceptual foundation that we use to think about race, class, and gender.

As in past editions, we include essays in Part I ("Why Race, Class, and Gender Still Matter") that engage students in personal narratives, as a way of helping them step beyond their own social location and to see how race, class, gender, sexuality, and other social factors shape people's lives differently. We include here the now classic essay by Audre Lorde, who so beautifully laid out the vision for inclusive thinking in one of her early and influential essays. Other pieces in this section show students the very different experiences that anchor the study of race, class, and gender, including the feelings that such inequalities generate. The study of race, class, and gender is not just knowledge for knowledge's sake. As some of these introductory essays point out, the dynamics of race, class, and gender result in hurt and anger even while building the desire for change.

Part II, "Systems of Power and Inequality," provides the conceptual foundation for understanding how race, class, and gender are linked together and how they link with other systems of power and inequality, especially ethnicity and sexuality. Here we want students to understand that individual identities and experiences are structured by intersecting systems of power. The essays in Part II link ethnicity, nationality, and sexuality to the study of race, class, and gender. We treat each of these separately, not because we think they stand alone, but to show students how each operates so they can better see their interlocking nature. The introductory essay provides working definitions for these major concepts and presents some of the contemporary data that will help students see how race, class, and gender stratify contemporary U.S. society.

Part III, "Social Institutions and Social Issues," examines how intersecting systems of race, class, and gender shape the organization of social institutions and how, as a result, these institutions affect current social issues. Social scientists routinely document how Latinos, African Americans, women, workers, and other distinctive groups are affected by institutional structures. We know this is true but want to go beyond these analyses to scrutinize how institutions are themselves constructed through race, class, and gender relations. This section also showcases social issues that are very much on the public mind: jobs, work and the economy; families and social relationships; education and health care; citizenship and national identity; and violence and criminalization.

We have substantially revised Part IV, "Intersectionality and Social Change" to show students the contemporary ways that people are resisting the inequalities that this book addresses. Many anthologies use their final section to show how students can make a difference in society, once they understand the importance of race, class, and gender. We think this is a tall order for students who may have had only a few weeks to begin understanding how race, class, and gender matter—and matter together. By showing some examples of how people mobilize for social change, we hope to show students how an intersectional framework can shape one's action in both local, national, and global contexts. This section focuses particularly on the use of social media as a tool for challenging dominant social structures.

This book is grounded in a sociological perspective, although the articles come from different perspectives, disciplines, and experiences. Several articles provide a historical foundation for understanding how race, class, and gender have emerged. We would have liked to include articles that bring a global dimension to the study of race, class, and gender, because we know these are played out in different ways in different societies. We hope the analytical tools this book provides will help people question how race, class, and gender are structured in different contexts, but the pressing issues in the United States are more than enough to fill an anthology such as this.

As in earlier editions, we have selected articles based primarily on two criteria:

1. accessibility to undergraduate readers and the general public, not just highly trained specialists;
2. articles that are grounded in race *and* class *and* gender—in other words, intersectionality.

We try not to select articles that focus exclusively on one intersectional component while ignoring the others. In this regard, our book differs significantly from other anthologies on race, class, and gender that include many articles on each factor, but do less to show how they are connected. Luckily, the growth of intersectional scholarship in the years since this book was first published made finding such insightful articles easier than in the past.

We also distinguish our book from those that are centered in a multicultural perspective. Although multiculturalism is important, we think that race, class, and gender go beyond the appreciation of cultural differences. Rather, we see race, class, and gender as embedded in the structure of society and significantly influencing group cultures and opportunities. Race, class, and gender are structures of group opportunity, power, and privilege, not just cultural differences. We always search for articles that are conceptually and theoretically informed and at the same time accessible to undergraduate readers. Although it is important to think of race, class, and gender as analytical categories, we do not want to lose sight of how they affect human experiences and feelings; thus, we include personal narratives that are reflective and analytical. We think that personal accounts generate empathy and also help students connect personal experiences to social structural conditions.

We know that developing a complex understanding of the interrelationships between race, class, and gender is not easy and involves a long-term process engaging personal, intellectual, and political change. We do not claim to be models of perfection in this regard. We have been pleased by the strong response to the first nine editions of this book, and we are fascinated by how race, class, and gender studies have developed since the publication of our first edition in 1992. We know further work is needed. Our own teaching and thinking has been transformed by developing this book. We imagine many changes still to come.

NEW TO THE TENTH EDITION

We have made several changes in the tenth edition, including the following:

- 29 new readings;
- a completely revised final section that focuses on the use of social media in mobilizing for change;

- four revised introductions, one of the noted strengths of our book compared to others; and,

- new material on important current issues, including immigration and mass deportation, white nationalism, the criminalization of immigrants, growing inequality, transgender identities, undocumented students, and the changing composition of the U.S. population, among other topics.

PEDAGOGICAL FEATURES

We realize that the context in which you teach matters. If you teach in an institution where students are more likely to be working class, perhaps how the class system works will be more obvious to them than it is for students in a more privileged college environment. Many of those who use this book will be teaching in segregated environments, given the high degree of segregation in education. Thus, how one teaches this book should reflect the different environments where faculty work. Ideally, the material in this book should be discussed in a multiracial, multicultural atmosphere, but we realize that is not always the case. We hope that the content of the book and the pedagogical features that enhance it will help bring a more inclusive analysis to educational settings that might or might not be there to start with.

We see this book as more than just a collection of readings. The book has an analytical logic to its organization and content, and we think it can be used to format a course. Of course, some faculty will use the articles in an order different from how we present them, but we hope the four parts will help people develop the framework for their course. The introductions that we have written for each part also frame the readings in ways that will ground faculty and students in the thinking that undergirds the articles we have included.

We also provide pedagogical tools to expand teaching and learning beyond the pages of the book, including an instructor's manual and sample test questions. We include features with this edition that provide faculty with additional teaching tools. Including the following:

- *Instructor's manual.* This edition includes an instructor's manual with suggestions for classroom exercises, discussion and examination questions, and course assignments.

- *Index.* The index will help students and faculty locate particular topics in the book quickly and easily.

- *Cengage Learning Testing, powered by Cognero Instant Access.* This is a flexible, online system that allows you to author, edit, and manage test bank content from multiple Cengage Learning solutions; create multiple test versions in an instant; and deliver tests from your LMS, your classroom, or wherever you want.

A NOTE ON LANGUAGE

Reconstructing existing ways of thinking to be more inclusive requires many transformations. One transformation needed involves the language we use when referring to different groups. Language reflects many assumptions about race, class, and gender; and for that reason, language changes and evolves as knowledge changes. The term *minority,* for example, marginalizes groups, making them seem somehow outside the mainstream or dominant culture. Even worse, the phrase *non-White,* routinely used by social scientists, defines groups in terms of what they are not and assumes that Whites have the universal experiences against which the experiences of all other groups are measured. We have consciously avoided using both of these terms throughout this book, although this is sometimes unavoidable.

We have capitalized Black in our writing because of the specific historical experience, varied as it is, of African Americans in the United States. We also capitalize White when referring to a particular group experience; however, we recognize that White American is no more a uniform experience than is African American. We use *Latina/o* and Latinx interchangeably, though we recognize that is not how groups necessarily define themselves. When citing data from other sources (typically government documents), we use *Hispanic* because that is usually how such data are reported. We recognize that people feel strongly about the terms used to describe them, because the language we use to describe and categorize people is highly political. We hope language we use is respectful of the changing context of how people define themselves, even while discussing whole groups homogenizes the experiences of many.

Language becomes especially problematic when we talk about features of experience that different groups share. Using shortcut terms like Hispanic, Latina/o, Native American, and women of color homogenizes distinct historical experiences. Even the term *White* falsely unifies experiences across such factors as ethnicity, religion, class, and gender, to name a few. At times, though, we want to talk of common experiences across different groups, so we have used labels such as Latina/o, Asian American, Native American, LGBTQ, and women of color to do so. Unfortunately, describing groups in this way reinforces basic categories of oppression. We do not know how to resolve this problem but want readers to be aware of the limitations and significance of language as they try to think more inclusively about diverse group experiences.

ACKNOWLEDGMENTS

An anthology rests on the efforts of more people than the editors alone. This book has been inspired by our work with scholars and teachers from around the country who are working to make their teaching and writing more inclusive and sensitive to the experiences of all groups. Over the years of our own collaboration, we have each been enriched by the work of those trying to make higher education a more equitable and fair institution. In that time, our work has grown from many networks that have generated new race, class, and gender scholars. These associations continue to sustain us. Many people contributed to the development of this book. We especially thank Maxine Baca Zinn, Elizabeth Higginbotham, Valerie Hans, and the Boston Area Feminist Scholars Group for their inspiration, ideas, suggestions, and support.

We also thank the team at Cengage for their encouragement and support for this project. Most particularly, we thank Ava Fruin, Sarah Kaubisch, Samen Iqbal, Rita Jaramillo and Sharib Asrar for expertly overseeing the many plans and details for publishing a new edition. We also thank the anonymous reviewers who provided valuable commentary on the prior edition and thus helped enormously in the development of the tenth edition.

This book has evolved over many years, and through it all we have been lucky to have the love and support of Richard, Roger, Valerie, Lauren and Patrice. We thank them for the love and support that anchors our lives. And, with this edition, we give our love to Aden Jonathan Carcopo, Aubrey Emma Hanerfeld, Harrison Collins Pruitt and Grant Collins Pruitt with hopes that the worlds they encounter will be just and inclusive, so that they can thrive in whatever paths they take.

About the Editors

Margaret L. Andersen (B.A. Georgia State University; M.A., Ph.D., University of Massachusetts, Amherst) is the Edward F. and Elizabeth Goodman Rosenberg Professor Emerita at the University of Delaware where she has also served as the Vice Provost for Faculty Affairs and Diversity, Executive Director of the President's Diversity Initiative and Dean of the College of Arts and Sciences. She has received two teaching awards at the University of Delaware. She has published numerous books and articles, including *Race in Society: The Enduring American Dilemma* (Rowman and Littlefield, 2017); *Thinking about Women: Sociological Perspectives on Sex and Gender* (11th ed., Pearson, 2020); *Race and Ethnicity in Society: The Changing Landscape* (edited with Elizabeth Higginbotham, 4th ed., Cengage, 2016); *On Land and On Sea: A Century of Women in the Rosenfeld Collection* (Mystic Seaport Museum, 2007); *Living Art: The Life of Paul R. Jones, African American Art Collector* (University of Delaware Press, 2009); and *Sociology: The Essentials, 10th ed.* (coauthored with Howard F. Taylor; Cengage, 2020). She received the American Sociological Association's Jessie Bernard Award for expanding the horizons of sociology to include the study of women the Eastern Sociological Society's Merit Award, and Robin Williams Lecturer Award. She is a past vice president of the American Sociological Association and past president of the Eastern Sociological Society.

Patricia Hill Collins (B.A., Brandeis University; M.A.T., Harvard University; Ph.D., Brandeis University) is Distinguished University Professor Emerita of sociology at the University of Maryland, College Park, and Charles Phelps Taft Professor Emerita of African American Studies and Sociology at the University of Cincinnati. She is the author of numerous articles and books, including *On Intellectual Activism* (Temple University, 2013), *Another Kind of Public Education: Race, Schools, the Media and Democratic Possibilities* (Beacon, 2009), *From Black Power to Hip Hop: Racism, Nationalism and Feminism* (Temple University, 2006); *Black Sexual Politics: African Americans, Gender and the New Racism* (Routledge, 2004), which won the Distinguished Publication Award from the American Sociological Association; *Fighting Words* (University of Minnesota, 1998); and *Black Feminist Thought: Knowledge, Consciousness, and the Politics of Empowerment* (Routledge, 1990, 2000), which won the Jessie Bernard Award of the American Sociological Association and the C. Wright Mills Award of the Society for the Study of Social Problems. Her most recent books include *Intersectionality: Key Concepts* (Polity, 2016) with Sirma Bilge, and *Not Just Ideas: Intersectionality as Critical Social Theory* (Duke, 2019).

About the Contributors

Joan Acker was Professor Emerita of sociology at the University of Oregon. She founded and directed the University of Oregon's Center for the Study of Women in Society and was the recipient of the American Sociological Association's Career of Distinguished Scholarship Award as well as the Jessie Bernard Award for feminist scholarship. She authored *Class Questions, Feminist Answers*, as well as many other works in the areas of gender, institutions, and class.

Elizabeth A. Armstrong is Professor of Sociology and Organizational Studies at the University of Michigan, Ann Arbor. She has been a fellow at the Radcliffe Institute for Advanced Study at Harvard University and a recipient of a National Academy of Education/Spencer Postdoctoral Fellowship. She is the author (with Laura Hamilton) of *Paying for the Party: How College Maintains Inequality*.

Elizabeth M. Armstrong is Associate Professor of Sociology at Princeton University with joint affiliations in the Woodrow Wilson School and the Office of Population Research. Her research interests include public health, the history and sociology of medicine, risk in obstetrics, and medical ethics. She is the author of *Conceiving Risk, Bearing Responsibility: Fetal Alcohol Syndrome and the Diagnosis of Moral Disorder* and she was a Robert Wood Johnson Foundation Scholar in Health Policy Research at the University of Michigan.

Linda Bartolomei is the Director of the Centre for Refugee Research and the co-coordinator of the Master of Development Studies program in the School of Social Sciences at the University of New South Wales in Sydney, Australia. She is the co-author of *Improving Reponses to Refugees with Backgrounds of Multiple Trauma*.

Marianne Bertrand is Chris P. Dialynas Professor of Economics and Neubauer Family Faculty Fellow at Chicago Booth University School of Business. Her work has been published in the *Quarterly Journal of Economics*, the *Journal of Political Economy*, the *American Economic Review*, and the *Journal of Finance*, among others.

Eduardo Bonilla-Silva is Professor of Sociology at Duke University. He is the author of numerous books, including *Racism without Racists and Whiteout: The Continuing Significance of Racism* (co-authored with Woody Doane). He has served as President of the American Sociological Association and the Southern Sociological Society.

Hana Brown is Associate Professor of Sociology at Wake Forest University. She studies the relationship between politics, the state, and social inequality, as well as on racial divisions in policy outcomes. She has published works in numerous academic journals.

David J. Connor is Professor of Special Education and Learning Disabilities at Hunter College. He is the author of numerous books and articles, including among others: *Urban Narratives: Portraits-in-Progress–Life at the Intersections of Learning Disability, Race, and Social Class* and *DisCrit: Critical Conversations Across Race, Class, & Dis/Ability.*

Heidi M. Coronado is Assistant Professor of Counselor Education at California Lutheran University. Her work examines ethnic identity development, educational access, and resiliency in immigrant, first- and second-generation Latino/a and indigenous youth; indigenous epistemologies and wisdom traditions for youth empowerment; class, race, gender and ethnicity in education; and indigenous/Latino/a mental health and healing practices.

Richard Cortes is a counselor at Glendale Community College. His research focuses on social justice counseling and psychological factors for low-income minority and immigrant college students.

Rebecca Covarrubius is Assistant Professor of Psychology at the University of California Santa Cruz. Her research focuses on identity and educational access for underrepresented students.

Bonnie Thornton Dill is Professor of women's studies and dean of the College of Arts and Humanities at the University of Maryland, College Park. Her books include *Women of Color in U.S. Society,* co-edited with Maxine Baca Zinn, and *Across the Boundaries of Race and Class: An Exploration of Work and Family among Black Female Domestic Servants.*

Marlese Durr is Professor of sociology and anthropology at Wright State University. Among other works, she has published *The New Politics of Race: From Du Bois to the 21st Century* and *Race, Work, and Families in the Lives of African Americans.*

Roberta Espinoza is Associate Professor of Sociology at Pitzer College. Among her publications are two books: *Working-Class Minority Students' Routes to Higher Education* and *Pivotal Moments: How Educators Can Put All Students on the Path to College.* She is a former fellow of the American Sociological Association's Minority Fellowship Program.

Walter A. Ewing is an anthropologist and Senior Researcher at the American Immigration Council. His research interests include immigration policy, migration, and immigrant labor. He can be followed at walkerewing.com.

Joe Feagin is the Ella C. McFadden Professor of Sociology at Texas A&M University. His research interests include racial and ethnic studies, urban political economy, and gender relations. He is the author of numerous books, including *The White Racial Frame: Centuries of Racial Framing and Counter-Framing* and *How Blacks Built America*, among many others.

Abby L. Ferber is Professor of sociology at the University of Colorado at Colorado Springs. She is the author of numerous books, including *White Man Falling: Race, Gender and White Supremacy; Hate Crime in America: What Do We Know?*; and *Making a Difference: University Students of Color Speak Out*. She is co-author of *Sex, Gender, and Sexuality: The New Basics,* and co-editor of *Privilege: A Reader* with Michael S. Kimmel.

Staycie L. Flint is a pediatric chaplain in Illinois who works within the Advocate Health Care system to help the most vulnerable populations develop a spiritual life.

Stephanie A. Fryberg is the William P. and Ruth Gerberding Professor of Psychology at the University of Washington. Her research is on American Indian/American Indian self concepts and the experiences of first-generation college students.

Vivian L. Gadsden is the William T. Carter Professor of Child Development and Education at the Graduate School of Education at the University of Pennsylvania. She is also the Director of the National Center on Fathers and Families. Her research focuses on literacy and at-risk youth, as well as fathers and families, and intergenerational learning.

Charles A. Gallagher is the chair of the Department of Sociology at LaSalle University with research specialties in race and ethnic relations, urban sociology, and inequality. He has published several articles on subjects such as color-blind political narratives, racial categories within the context of interracial marriages, and perceptions of privilege based on ethnicity.

Herbert J. Gans has been a prolific and influential sociologist for more than fifty years. His published works on urban renewal and suburbanization are intertwined with his personal advocacy and participant observation, including a stint as consultant to the National Advisory Commission on Civil Disorder. He is the author of the classic *The Urban Villagers*.

Tonya Golash-Boza is Professor of Sociology at the University of California, Merced. She is the author of *Deported: Immigrant Policing, Disposable Labor and Global Capitalism, Race and Racisms: A Critical Approach,* and *Due Process Denied: Detentions and Deportations in the 21st Century*, among other publications.

Kishonna L. Gray is Assistant Professor in Communication and Women and Gender Studies at the University of Illinois at Chicago. Her interests are in race gender, identity, and digital media. Her two books include *Race, Gender, and Deviance in Xbox Live: Theoretical Perspectives from the Virtual Margins* and *Woke Gaming: Digital Challenges to Oppression and Social Injustice.*

Laura T. Hamilton is Associate Professor of Sociology at the University of California, Merced. Her research and teaching interests include gender, sexuality, family, education, and social class. She is the co-author (with Elizabeth A. Armstrong) of *Paying for the Party: How College Maintains Inequality.*

Jamie D. Hawley is a minister in the United Church of Christ, he also served as chaplain at the University of Michigan Health System where he worked to address health disparities in marginalized communities.

Aída Hurtado is Professor and Luis Leal Endowed Chair in Chicana Studies at the University of California, Santa Barbara. She studies equity issues in education, feminist studies, and images of ethnic and racial groups in the media. She is the author of numerous research articles.

Sarah J. Jackson is Associate Professor of Communication at Northeastern University. She is the author of two books, including *Hashtag Activism: Race and Gender in America's Networked Counterpublics* and *Black Celebrity, Racial Politics, and the Press.*

Sujatha Jesudason holds a Ph.D. in sociology from the University of California, Berkeley and is the Executive Director of the CoreAlign Team, an organization that builds a network of leaders working innovatively to change policies, culture, and conditions that support all people's sexual and reproductive decisions.

Denise L. Johnson is Senior Associate Professor and Program Chair at Bellevue College. She teaches courses on the social lives of children and youth, as well as on race and ethnicity, sociology of the body, sex and sexualities, and other subjects.

Jennifer A. Jones is Assistant Professor of Sociology at the University of Notre Dame. She studies migration, the growing multiracial population, and shifting social relations between different racial groups. She is the co-author of *Afro-Latinos in Movement: Critical Approaches to Blackness and Transnationalism in the Americas.*

Jonathan Ned Katz was the first tenured Professor of lesbian and gay studies in the United States (Department of Lesbian and Gay Studies, City College of San Francisco). He is the founder of the Queer Caucus of the College Art Association. He is also the co-founder of the activist group Queer Nation.

Haunani-Kay Trask is a Hawaiian scholar and poet and has been an indigenous rights activist for the Native Hawaiian community for over 25 years. She is a former Professor of Hawaiian Studies at the University of Hawaii at Manoa and is the author of several books of poetry and nonfiction.

Gloria Ladson-Billings is the Kellner Family Professor of Urban Education in the Department of Curriculum and Instruction at the University of Wisconsin–Madison. She is the author of *The Dreamkeepers: Successful Teachers of African-American Children* and former president of the American Educational Research Association.

Peter A. Leavitt is Visiting Assistant Professor of Psychology at Dickinson College where he teaches courses on how social identities, such as how race, social class, and gender shape how people behave and how they perceive the behavior of others.

Donovan Lessard is the Director of Research and Senior Data Analyst at the Physician Assistant Education Association where he evaluates various health prevention programs, including obesity, cancer, HIV/AIDS, and tobacco control, among others.

Nancy López is Professor of Sociology at the University of New Mexico where she founded the Institute for the Study of Race and Social Justice, and served as the statewide coordinator of New Mexico's Race, Gender Class Data Policy Consortium. Her book *Hopeful Girls, Troubled Boys: Race and Gender Disparity in Urban Education* focuses on the race-gender experiences of Dominicans, West Indians, and Haitians to explain why girls are succeeding at higher rates than boys. She is also the editor of *Mapping Race: Critical Approaches to Health Disparities Research.*

Audre Lorde was a poet, essayist, teacher, activist, and writer dedicated to confronting and addressing the injustices of racism, sexism, and homophobia. Her numerous writings include, among others: *Sister Outsider; The Cancer Journals; From a Land Where Other People Live;* and *The Black Unicorn.*

Gregory Mantsios is the Director and Founder of the Joseph S. Murphy Institute for Worker Education and Labor Studies at the City University of New York. This center also publishes the journal *New Labor Forum.*

Amy D. McDowell is Assistant Professor of Sociology at the University of Mississippi. Her research interests include religion, popular culture, gender, and race, especially in how young people use defiant music to contest dominant racial frames.

Peggy McIntosh is Senior Research Associate of the Wellesley Centers for Women. She is the founder of the National SEED Project on Inclusive Curriculum—a project that helps teachers make school climates fair and equitable. She is also the co-founder of the Rocky Mountain Women's Institute.

Mignon R. Moore is Associate Professor of Sociology at Barnard College. She is the author of *Invisible Families: Gay Identities, Relationships and Motherhood among Black Women*, as well as articles on LGBT families and religious life among LGBT people.

Alfonso Morales is Associate Professor in the Department of Urban and Regional Planning at the University of Wisconsin—Madison. His work examines how urban agriculture, food distribution, and community and economic development.

Sendhil Mullainathan is Professor of economics at Harvard University. He is the co-founder of the Abdul Latif Jameel Poverty Action Lab at MIT that uses randomized evaluations to study poverty alleviation. He is the co-author of *Scarcity: Why Having Too Little Means So Much.*

Kathyrn Norton-Smith is a graduate student in sociology at the University of Oregon. Her research focuses on health, political sociology, and the life course with a special interest in young adults with cancer.

Michael Omi is Associate Professor in Asian American and Asian Diaspora Studies at the University of California, Berkeley. He is the co-author of *Racial Formation in the United States* (with Howard Winant). He has received UC-Berkeley's Distinguished Teacher Award.

William Perez is Associate Professor of Education at the Claremont Graduate University. His research and teaching focus on the education of immigrant students, minority access to education, and prejudice and discrimination in educational settings.

Yvonne A. Perez is the Program Coordinator for International Development at Pima Community College.

Eileen Pittaway is the Director of the Venter for Refugee Research and lecturer in the School of Social Work at the University of New South Wales. She is the co-author of *Improving Reponses to Refugees with Backgrounds of Multiple Trauma.*

Karen D. Pyke is Professor and Distinguished Teaching Professor at the University of California, Riverside. Her research is on the institutional practices that undergird faculty gender inequity in academia, as well as research that examines the "normal American family" as a controlling ideology that informs how adult children of immigrants understand their family lives.

Karina Ramos a Senior Staff Psychologist at the University of California, Irvine Counseling Center. She is passionate about multicultural counseling and issues related to the educational experiences and the mental health and well-being of undocumented students, enjoys working with the diverse student population at UC Irvine, and also serves as the liaison for UCI Dreamers and the American Indian Resource Program.

Victor M. Rios is Professor of Sociology at the University of California, Santa Barbara. He is the author of numerous books and articles, including *Punished: Policing the Lives of Black and Latino Boys* and *Human Targets: Schools, Police, and the Criminalization of Latino Youth.*

Dorothy Roberts is the George A. Weiss University Professor of Law and Sociology and the Raymond Pace and Sadie Tanner Mossell Alexander Professor of Civil Rights at University of Pennsylvania Law School. She is an acclaimed scholar of race, gender, and the law, and author of *Killing the Black Body: Race, Reproduction and the Meaning of Liberty;* and *Fatal Intervention: How Science, Politics and Big Business Re-Create Race in the 21ˢᵗ Century,* among others.

Lillian B. Rubin was an internationally known lecturer and writer. Some of her books include *The Man with the Beautiful Voice, Tangled Lives, The Transcendent Child,* and *Intimate Strangers.*

Rubén G. Rumbaut is Distinguished Professor of Sociology at the University of California, Irvine. He is the author of *America's Languages: Investing in Language Education for the 21ˢᵗ Century* as well as numerous books, policy publications, and articles on immigrant experiences and policies.

Joelle Ruby Ryan is Senior Lecturer in Women's Studies at the University of New Hampshire. Her research and teaching interests are on gender, LGBT issues, film, social activism, and writing.

J. Lotus Seeley is Assistant Professor of Sociology at Florida Atlantic University. Her research and teaching interests are gender, sexuality, care work, and emotional labor, among others.

Thomas M. Shapiro is Pokross Professor of Law and Social Policy at the Heller Scholar for Social Policy and Management at Brandeis University. He also directs the Institute on Assets and Social Policy. His most recent book is *Toxic Inequality: How America's Wealth Gap Destroys Mobility, Deepens the Racial Divide and Threatens Our Future.* He is also the author of *The Hidden Cost of Being African American: How Wealth Perpetuates Inequality.*

Jennifer M. Silva is Assistant Professor of Sociology at Bucknell University. She is the author of *Coming Up Short: Working-Class Adulthood in an Age of Uncertainty.* Her research and teaching focus on social class and inequality, especially involving young people.

Mrinal Sinha is Assistant Professor of Psychology at California State University, Monterey Bay. His research is on the racialization of masculinities, Chicana Feminisms, and educational success with Latina/o students.

C. Matthew Snipp is the Burnet C. and Mildred Finley Wohlford Professor of Humanities and Sciences in the Department of Sociology at Stanford University where he founded the Center for Native American Excellence. He is the author of *American Indians: The First of This Land;* and *Public Policy Impacts on American Indian Economic Development.* His tribal heritage is Oklahoma Cherokee and Choctaw.

Natalie J. Sokoloff is Professor Emerita of sociology at the John Jay College of Criminal Justice, SUNY. She is author of numerous books and publications about intimate partner violence, and women and men's incarceration. She is the editor of *Domestic Violence at the Margins: Readings on Race, Class, Gender, and Culture.*

Michael Stambolis-Ruhstorfher is Assistant Professor of American and Gender Studies at the Université Bordeaux Montaigne. His research compares the United States and France as he focuses on sexuality, gender, and cultural studies.

Amy Steinbugler is Associate Professor and Chair of Sociology at Dickinson College. She is the author of *Beyond Loving: Intimate Racework in Lesbian, Gay and Straight Interracial Relationships* and writes about gender, race, and intimacy.

Veronica Terríquez is Associate Professor of Sociology at the University of California, Santa Cruz. Her research focuses on civic organizations and schools, especially in low-income, immigrant, and Latino communities. She is principal investigator for two large studies: the California Young Adult Study and the Youth Leadership and Health Study.

Bhoomi K. Thakore is a research associate at Northwestern University. She is the author of numerous papers on racial representations in the popular media.

Jeremiah Torres is a graduate of Stanford University, where he studied symbolic systems. His article "Label Us Angry" appeared in the book *Asian American X,* a collection of essays about the experiences of contemporary Asian Americans.

Jessica Vasquez-Tokos is Associate Professor of Sociology at the University of Oregon where she studies race, gender, class, and romantic unions. Her books include *Marriage Vows and Racial Choices* and *Mexican Americans across Generations: Immigrant Families, Racial Realities.*

Mary C. Waters is M.E. Zukerman Professor of Sociology and Harvard College Professor at Harvard University. She is the author of *Black Identities: West Indian Immigrant Dreams and American Realities; Ethnic Options: Choosing Identities in America;* and numerous articles on race, ethnicity, and immigration.

Sandra E. Weissinger is Assistant Professor of sociology at the Southern Illinois University Edwardsville. Her work focuses on intragroup marginalization, inequalities, community activism, and African American communities and institutions.

William Julius Wilson is the Lewis P and Linda L. Geyser University Professor at Harvard University. He is past president of the American Sociological Association and a MacArthur Prize Fellow. He is the author of numerous books, including *When Work Disappears, The Declining Significance of Race,* and *More Than Just Race: Being Black and Poor in the Inner City,* among many others.

Howard Winant is Professor of Sociology at the University of California, Santa Barbara where he is also affiliated with Black Studies, Chicano/a Studies, and Asian American Studies. He is the author of *Racial Formation* (with Michael Omi), as well as *The New Politics of Race: Globalism, Difference, Justice,* along with other books and articles.

Adia M. Harvey Wingfield is Professor of sociology at Washington University, St. Louis. She specializes in the study of race, class, gender, and work and is the author of several books, including *No More Invisible Man: Race and Gender in Men's Work, Framing and the 2008 Presidential Campaign* (with Joe Feagin), and *Doing Business with Beauty: Black Women, Hair Salons, and the Racial Enclave Economy.*

Min Zhou is Professor of sociology and Asian American studies, Walter and Shirley Wang Endowed Chair in U.S.–China Relations and Communications, and the founding chair of the Asian American Studies Department at UCLA. She is the author of *Chinatown: The Socioeconomic Potential of an Urban Enclave; The Transformation of Chinese America;* and *Contemporary Chinese America: Immigration, Ethnicity, and Community Transformation.*

PART I

Why Race, Class, and Gender Still Matter

MARGARET L. ANDERSEN
AND PATRICIA HILL COLLINS

The United States is a nation where people are supposed to be able to rise above their origins. Those who want to succeed, it is believed, can do so through hard work and solid effort. Although equality has historically been denied to many, there is now a legal framework in place that guarantees protection from discrimination and equal treatment to all citizens.

Historic social movements, such as the civil rights movement and the feminist movement, raised people's consciousness about the rights of African Americans and women. Moreover, these movements have generated new opportunities for multiple groups—African Americans; Latinos; disabled people; Asian Americans; and lesbian, gay, bisexual, transgendered, queer (LGBTQ) people; and, older—people to name some of those who have been beneficiaries of civil rights action and legislation.

African Americans have held distinguished elected offices, including the presidency. Gays and lesbians now have the right to same-sex marriage. Women, including one Latina, sit in very high places as Supreme Court justices and CEOs of major companies. Disabled people have rights of access to work and schools and are protected under federal laws. The vast majority of Americans, when asked, say that they support equal rights and nondiscrimination policies.

Yet, despite these accomplishments, race, class, and gender continue to structure society in ways that value some lives more than others. Currently, some groups have great opportunities and resources, while other groups struggle.

Race, class, and gender matter because they remain the foundations for systems of power and inequality that, despite our nation's diversity, continue to be among the most significant social facts of people's lives. Despite having removed formal barriers to opportunity, the United States is still highly unequal along lines of race, class, and gender.

In this book, we ask you to think about how race, class, and gender as *systems of power* produce social inequality. We also encourage you to imagine ways to change, rather than reproduce, existing social arrangements. This starts with shifting one's thinking so that groups who are so often silenced or ignored become heard. It also requires groups who are heard from most often to learn to listen. We all need to remember that, as individuals, we are each located in systems of power wherein our social location can shape what we know—and what others know about us. Yet, we are also part of social groups whose ideas carry different weight within systems of power. As a result, dominant forms of knowledge have been constructed largely from the experiences of the most powerful—that is, those who have the most access to high quality education and systems of communication. To acquire a more inclusive view—one that pays attention to group experiences that may differ from your own—requires that each of us develop a new frame of vision.

You can think of this as if you were taking a photograph. For years, people of color, poor people, and women were either outside the frame of vision of more powerful groups or distorted by the views of the powerful. If you move your angle of sight to include those whose experiences have been excluded or distorted, however, you might be surprised by how incomplete or just plain wrong your earlier view was. Completely new subjects come into view. This is more than a matter of sharpening your focus, although that is required for clarity. Instead, this new angle of vision means actually seeing things differently, perhaps even changing the lens you look through—thereby removing the filters (or stereotypes and misconceptions) that you bring to what you see and think.

WHY RACE, CLASS, AND GENDER MATTER

How might race, class, and gender matter in shaping everyone's experiences? Thinking from a perspective that engages race, class, and gender is not just about illuminating the experiences of oppressed groups. Instead, inclusive thinking changes how we understand groups who are on both sides of power and privilege. For example, the development of women's studies has changed what we know and how we think about women. At the same time, it has changed what we know and how we think about men. This does not mean that women's studies is about "male bashing." It means taking the experiences of women and men seriously and analyzing how race, class, and gender shape the experiences of both—in different, but interrelated, ways. Likewise, studying indigenous people, Blacks, Latinos/as, Asians and other racial and ethnic groups begins by learning their diverse histories and experiences. Doing so can shift our understanding of White people's experiences. Rethinking class means learning to think differently about privilege, disadvantage, and opportunity. The vastly different experiences of wealthy, middle-class, working-class, and poor people in the United States remain mysterious if we don't know the diverse histories and experiences of these groups. The exclusionary thinking that comes from past frames of vision simply does not reveal the intricate interconnections of gender, race, and class that link different groups within U.S. society.

We remind you that race, class, and gender have affected the experiences of all individuals and groups. For example, gender is not just about women. Nor is class only about the poor. We also stress that thinking about race, class, and gender is not just a matter of studying victims. Relying too heavily on the experiences of poor people, women, and people of color can erase our ability to see race, class, and gender as an integral part of everyone's experiences. Such a perspective focuses your attention on the dynamics of privilege, not just oppression.

So, you might ask, how does reconstructing knowledge work? To begin with, knowledge is not just some abstract thing—good to have, but not all that important in everyday life. There are real consequences to having partial or distorted knowledge. First, knowledge is not just about content and information; it provides an orientation to the world. What you know frames how you behave and how you think about yourself and others. If what you know is wrong because it is based on exclusionary thought, you are likely to act in discriminatory ways, thereby reproducing the racism, class oppression, sexism, and homophobia in society. This may not be because you are intentionally racist, elitist sexist, or homophobia. It can simply be because you do not know any better. But, you can learn. Challenging oppressive race, class, and gender relations requires reconstructing what you know so that you have a basis from which to change these damaging and dehumanizing systems of oppression.

Second, learning about other groups helps you realize the partiality of your own perspective. Further, this is true for both dominant and subordinate groups. Knowing only the history of Puerto Rican women, for example, or seeing their history only in single-minded terms will not reveal the historical linkages between the oppression of Puerto Rican women and the exclusionary and exploitative treatment of African Americans, working-class Whites, Asian American men, and similar groups. Moreover, both dominant and subordinated people can engage in exclusionary thinking about other groups. Because we each have different partial perspectives, we each have different things to learn about power relations of race, class, and gender.

Third, distorted or partial thinking can compromise your everyday interactions, such as when you say something out of ignorance that offends another person or group. Even if you say nothing when someone else makes an offensive comment, you are inadvertently colluding in maintaining a system of inequality.

Finally, having misleading and incorrect knowledge leads to the formation of bad social policy—policy that then reproduces, rather than solves, social problems. As an example, U.S. immigration policy has often taken a one-size-fits-all approach, failing to recognize that vast differences among groups coming to the United States privilege some and disadvantage others. Taking a broader view of social issues fosters more effective public policy.

INTERSECTIONAL FRAMEWORKS AND RACE, CLASS, AND GENDER

The fact that race, class, and gender shape the experiences of all people in the United States is widely documented in research and, to some extent, commonly understood. For years, social scientists have studied the consequences of race, class, and gender as *separate* forms of inequality for different groups in society. In the United States, race, class, and gender constitute fundamental systems of inequality and, as a result, we know a lot about them individually.

The framework of race, class, and gender studies presented here, however, explores how race, class, and gender operate *together* in people's lives. Fundamentally, race, class, and gender are *intersecting* categories of experience that affect all aspects of human life; they *simultaneously* structure the experiences of all people in this society. At any moment, race, class, or gender may feel more salient or meaningful in a given person's life, but they are overlapping and cumulative in their effects.

In this volume, we focus on several core features of this intersectional framework. First, we emphasize *social structure* in our efforts to conceptualize intersections of race, class, and gender. We use the matrix of domination approach to analyze race, class, and gender. A matrix of domination sees social structure as having multiple, interlocking levels of domination that stem from the societal configuration of race, class, and gender relations. This structural pattern affects individual consciousness, group interaction, and group access to institutional power and privileges (Collins 2000). Within this structural framework, we focus less on comparing race, class, and gender as separate systems of power than on investigating the structural patterns that join them. Even with their simultaneity, you can see the intersection of race, class, and gender in individual stories and experiences. In fact, much exciting work on the intersections of race, class, and gender appears in autobiographies, fiction, and personal essays. We do recognize the significance of these individual narratives and include many here, but we also emphasize social structures that provide the context for individual experiences.

Second, applying an intersectional framework to race, class, and gender within a context of social structures helps us understand how race, class, and gender are manifested differently, depending on their configuration with the others. Thus, one might say African American men are privileged *as men,* but this may not be true when both race and class are taken into account. Otherwise, how can we possibly explain the particular disadvantages African American men experience in the criminal justice system, in education, and in the labor market? Studying the connections among race, class, and gender reveals that divisions by race and by class and by gender are not as clear-cut as they may seem. White women, for example, may be disadvantaged because of gender but privileged by race and perhaps (but not necessarily) by class. Increasing class differentiation among White people reminds us, too, that race is not a monolithic category. Certainly, Whites are overrepresented among the most powerful members of society, but they also constitute the largest number of poor people in America.

Third, an intersectional framework for race, class, and gender studies is historically grounded in a particular matrix of domination. We emphasize the intersections of race, class, and gender because these systems of power have had and continue to have special impact in the United States. Yet race, class, and gender intersect with other categories of experience, such as sexuality, ethnicity, age, ability, religion, and nationality. We take up these other systems of power in this volume, but keep a trained eye on race, class, and gender because they are foundational within the United States.

Race, class, and gender constitute fundamental categories of analysis in the U.S. setting, so significant that in many ways they influence all other categories. Systems of race, class, and gender have been so consistently and deeply codified in U.S. laws that they have had intergenerational effects on economic, political, and social institutions. For example, the capitalist class relations that have characterized all phases of U.S. history have routinely privileged or penalized groups based on race, class, *and* gender. U.S. social institutions have reproduced economic equality for poor people, women, and people of color from one generation to the next. Thus, in the United States, race, class, and gender demonstrate visible, long-standing, material effects that in many ways foreshadow more recently visible categories of ethnicity, religion, age, ability, and/or sexuality.

The intersectional framework used here has grown tremendously since the first edition of this book. Race, class, and gender still constitute the bedrock of this book, but the type of thinking we have argued for across many editions of this book has shaped the contours of intersectionality (Collins and Bilge 2016). We see *intersectionality* as the current term that best encompasses the thinking we propose in this volume—that is, a focus on the structural foundations of race, class, and gender inequality. Intersectionality is now a term widely used both in activist projects and in scholarly work, but, because of its widespread usage, its specific meaning can become diluted and vague. We see a continued need for a structural analysis of the specific intersections of race, class, and gender in the United States. That is the focus of this book.

INTERSECTIONALITY, INEQUALITY, AND DIFFERENCE

How does intersectionality differ from other ways of conceptualizing race, class, and gender relationships? We think this can be best understood by contrasting *intersectional frameworks* on race, class, and gender studies to what might be called *difference frameworks*. Difference frameworks, though viewing some of the common processes in race, class, and gender relations, tend to focus on unique group experiences. Difference frameworks often approach race, class, and gender primarily at the level of culture, often minimizing how structures of power shape culture. *Culture* is traditionally defined as the "total way of life" of a group of people. It encompasses both material and symbolic components, and is an important dimension of understanding human life. Analysis of culture per se, however, tends to look at the group itself rather than at the broader conditions within which the group lives. Of course, as anthropologists know, a sound analysis of culture situates group experience within these social structural conditions.

Although we think understanding difference goes far in sensitizing people to other people's cultures, a difference perspective, taken in and of itself, misses the broader point of understanding how racism, class relations, and sexism have shaped one another. Imagine, for example, looking for the causes of poverty solely within the culture of poor people, as if patterns of unemployment, unexpected health care costs, or the high cost of living in so many urban areas have no effect on people's opportunities and life decisions. Or, imagine trying to study the oppression of LGBTQ people in terms of gay culture only. Obviously, doing this turns attention to the group itself and away from the dominant society. Likewise, studying race only in terms of Latinx culture, Asian American culture, or African American culture, or studying gender only by looking at women's culture, encourages thinking that blames the victims for their own oppression. Difference frameworks are useful because of how they document unique histories and cultural contributions.

You might think of the distinction between the difference frameworks and intersectional frameworks as one of *thinking comparatively* versus *thinking relationally*—the hallmark of an intersectional approach. For example, difference frameworks encourage individuals to compare their experiences with those supposedly unlike them. When you think comparatively, you might look at how different groups have, for example, encountered prejudice and discrimination or you might compare laws prohibiting interracial marriage to debates about same-sex marriage. These are important and interesting questions, but they are taken a step further when you think beyond comparison to the structural relationships between different group experiences.

In contrast, thinking relationally requires moving beyond comparing, for example, gender oppression with race oppression or the oppression of LGBTQ people with that of racial groups. Instead, intersectional thinking asks you to look for the connections or links between two or more entities. Recognizing how intersecting systems of power shape different groups' experiences also positions you to think about how race, class, and gender simultaneously shape a given social problem, or, more broadly, how the relationship among race, class, and gender operate together as a system of power.

We think that intersectionality is more analytical than the difference framework *because of its focus on structural systems of power and inequality as well as its embrace of relational thinking.* This means that race, class, and gender involve more than either comparing and adding up oppressions or appreciating cultural diversity. Intersectionality requires analysis and criticism of existing systems of power and privilege. An intersectional analysis examines the *interrelationships* among race, class, and gender within a given matrix of domination. Studying intersections of race, class, and gender as structural phenomena means recognizing and analyzing the hierarchies and systems of domination that permeate society and limit our ability to achieve true democracy and social justice.

RACE, CLASS, AND GENDER THROUGH
AN INTERSECTIONAL LENS

Once you understand that race, class, and gender are *simultaneous* and *intersecting* social structural systems of behavior and beliefs within particular historical contexts, you have the foundation for intersectional thinking. To emphasize the expansive nature of intersectional thinking, we open this anthology with a classic essay by noted feminist Audre Lorde, now deceased. Her essay, "Age, Race, Class and Sex: Women Redefining Difference," remains a critical and inspiring statement about the importance of thinking inclusively and across the differences that too often divide us.

Understanding the diverse histories, cultures, and experiences of groups who have been defined as marginal in society has been vital to the formation of our social institutions. Beyond this important fact is the reality that groups outside of the dominant culture have been silenced over the course of history, leading to distortions and incomplete knowledge of how society has been organized. Ignoring the experiences of those who have been on the margins results in a distorted view of how the nation has developed.

You might ask yourself how much you learned about the history of group oppression in your formal education. Maybe you touched briefly on topics such as the labor movement, slavery, women's suffrage, perhaps even the Holocaust, but most likely these were brief excursions from an otherwise dominant narrative that largely ignored the perspectives of working-class people, women, and people of color, along with others. How much of what you study now is centered in the experiences of the most dominant groups in society? Have classic social science studies been generated from research samples that have included only certain populations (such as middle-class college students in psychology studies, or mostly White people in other social science research)? Is the literature you are assigned or artistic creations that you study mostly the work of Europeans? Men? White people? How much has the creative work of Native Americans, Muslim Americans, new immigrant populations, Asian Americans, Latinos/as, African Americans, gays, lesbians, or women informed the study of "American" art and culture?

Minimizing the experiences and creations of these different groups suggests that their work and creativity is less important and less central to the development of culture than is the history of White American, elite men. What false or incomplete conclusions does this exclusionary thinking generate? When you learn, for example, that democracy and egalitarianism were central cultural beliefs in the early history of the United States, how do you explain the genocidal policies targeted toward indigenous peoples, the enslavement of millions of African Americans, the absence of laws against child labor, and the denial of the vote to anyone except propertied White men?

Haunani-Kay Trask ("From a Native Daughter") examines the gap between dominant cultural narratives and people's actual experiences. As she—a native Hawaiian—tells it, the official history she learned in schools was not what she was taught in her family and community. Dominant narratives can try to justify the oppression of different groups, but the unwritten, untold, subordinated truth can be a source for knowledge in pursuit of social justice. This book asks you to think more inclusively. Without doing so, you are prone to understand society, your own life within it, and the experiences of others through stereotypes and the misleading information that is all around you. What new experiences, under-standings, theories, histories, and analyses do these readings inspire? What does it take for a member of one group (say a Latino male) to be willing to learn from and value the experiences of another (for example, an Indian Muslim woman)? These essays show that, although we are caught in multiple systems, we can learn to see our connection to others.

Engaging oneself at the personal level is critical to this process of intersec-tional thinking about race, class, and gender. Changing one's mind is not just a matter of assessing facts and data, although that is important; it also requires examining one's feelings. We incorporate personal narratives that reflect the diverse experiences of race, class, gender, and/or sexual orientation into the opening section of this book to encourage you to think about your personal story. We intend for the personal nature of these accounts—especially those that describe what exclusion means and how it feels—to build empathy among groups. We think that empathy encourages an emotional stance that is critical to relational thinking and developing an inclusive perspective. As an example, Jeremiah Torres's essay ("Label Us Angry") describes how his seemingly trouble-free childhood changed overnight when he became the victim of a hate crime in a community known for its acceptance of multiculturalism. His narrative reminds us that learning about race, class, and gender is not just an intellectual exercise. Experiences with these forms of oppression are personal and often painful.

Likewise, Jamie D. Hawley and Staycie L. Flint ("'It Looks Like a Demon:' Black Masculinity and Spirituality in the age of Ferguson") recount the "soul murder" that comes from the ongoing demonization of African American men. One consequence, as argued by these two spiritual workers, is a deep and collec-tive grief within Black America.

We hope that understanding the significance of race, class, and gender will encourage readers to put the experiences of the United States itself into a broader context. Knowing how race, class, and gender operate within U.S. national borders should help you see beyond those borders. We hope that developing an awareness of how the increasingly global basis of society influences the configuration of race, class, and gender relationships in the United States will encourage readers to cast an increasingly inclusive perspective on the world itself.

REFERENCES

Collins, Patricia Hill. 2000. *Black Feminist Thought: Knowledge, Consciousness, and Empowerment*. New York: Routledge.

Collins, Patricia Hill and Sirma Bilge. 2016. *Intersectionality*. Cambridge, UK: Polity.

1

Age, Race, Class and Sex

Women Redefining Difference

AUDRE LORDE

Much of Western European history conditions us to see human differences in simplistic opposition to each other: dominant/subordinate, good/bad, up/down, superior/inferior. In a society where the good is defined in terms of profit rather than in terms of human need, there must always be some group of people who, through systematized oppression, can be made to feel surplus, to occupy the place of the dehumanized inferior. Within this society, that group is made up of Black and Third World people, working-class people, older people, and women.

As a forty-nine-year-old Black lesbian feminist socialist mother of two, including one boy, and a member of an interracial couple, I usually find myself a part of some group defined as other, deviant, inferior, or just plain wrong. Traditionally, in American society, it is the members of oppressed, objectified groups who are expected to stretch out and bridge the gap between the actualities of our lives and the consciousness of our oppressor. For in order to survive, those of us for whom oppression is as American as apple pie have always had to be watchers, to become familiar with the language and manners of the oppressor, even sometimes adopting them for some illusion of protection. Whenever the need for some pretense of communication arises, those who profit from our oppression call upon us to share our knowledge with them. In other words, it is the responsibility of the oppressed to teach the oppressors their mistakes. I am responsible for educating teachers who dismiss my children's culture in school. Black and Third World people are expected to educate white people as to our humanity. Women are expected to educate men. Lesbians and gay men are expected to educate the heterosexual world. The oppressors maintain their position and evade responsibility for their own actions. There is a constant drain of energy which might be better used in redefining ourselves and devising realistic scenarios for altering the present and constructing the future.

Institutionalized rejection of difference is an absolute necessity in a profit economy which needs outsiders as surplus people. As members of such an economy, we have *all* been programmed to respond to the human differences between us with fear and loathing and to handle that difference in one of three

SOURCE: Lorde, Audre. "Age, Race, Class, and Sex." *Sister Outsider*, published by Crossing Press, Random House Inc. Copyright © 1984, 2007 by Audre Lorde. Used herein by permission of the Charlotte Sheedy Literary Agency, Inc.

ways: ignore it, and if that is not possible, copy it if we think it is dominant, or destroy it if we think it is subordinate. But we have no patterns for relating across our human differences as equals. As a result, those differences have been mis-named and misused in the service of separation and confusion.

Certainly there are very real differences between us of race, age, and sex. But it is not those differences between us that are separating us. It is rather our refusal to recognize those differences, and to examine the distortions which result from our misnaming them and their effects upon human behavior and expectation.

Racism, the belief in the inherent superiority of one race over all others and thereby the right to dominance. Sexism, the belief in the inherent superiority of one sex over the other and thereby the right to dominance. Ageism. Heterosexism. Elitism. Classism.

It is a lifetime pursuit for each one of us to extract these distortions from our living at the same time as we recognize, reclaim, and define those differences upon which they are imposed. For we have all been raised in a society where those distortions were endemic within our living. Too often, we pour the energy needed for recognizing and exploring difference into pretending those differences are insurmountable barriers, or that they do not exist at all. This results in a vol-untary isolation, or false and treacherous connections. Either way, we do not develop tools for using human difference as a springboard for creative change within our lives. We speak not of human difference, but of human deviance.

Somewhere, on the edge of consciousness, there is what I call a *mythical* norm, which each one of us within our hearts knows "that is not me." In America, this norm is usually defined as white, thin, male, young, heterosexual, Christian, and financially secure. It is with this mythical norm that the trappings of power reside within this society. Those of us who stand outside that power often identify one way in which we are different, and we assume that to be the primary cause of all oppression, forgetting other distortions around difference, some of which we ourselves may be practising. By and large within the women's movement today, white women focus upon their oppression as women and ignore differences of race, sexual preference, class, and age. There is a pretense to a homogeneity of experience covered by the word *sisterhood* that does not in fact exist.

Unacknowledged class differences rob women of each others' energy and creative insight. Recently a women's magazine collective made the decision for one issue to print only prose, saying poetry was a less "rigorous" or "serious" art form. Yet even the form our creativity takes is often a class issue. Of all the art forms, poetry is the most economical. It is the one which is the most secret, which requires the least physical labor, the least material, and the one which can be done between shifts, in the hospital pantry, on the subway, and on scraps of surplus paper. Over the last few years, writing a novel on tight finances, I came to appreciate the enormous differences in the material demands between poetry and prose. As we reclaim our literature, poetry has been the major voice of poor, working-class, and Colored women. A room of one's own may be a necessity for writing prose, but so are reams of paper, a typewriter, and plenty of time. The actual requirements to produce the visual arts also help determine, along class lines, whose art is whose. In this day of inflated prices for material, who are our sculptors, our painters, our photographers? When we speak of a broadly based women's culture, we need to be aware of the effect of class and economic differences on the supplies available for producing art.

As we move toward creating a society within which we can each flourish, ageism is another distortion of relationship which interferes without vision. By ignoring the past, we are encouraged to repeat its mistakes. The "generation gap" is an important social tool for any repressive society. If the younger members of a community view the older members as contemptible or suspect or excess, they will never be able to join hands and examine the living memories of the community, nor ask the all-important question, "Why?" This gives rise to a historical amnesia that keeps us working to invent the wheel every time we have to go to the store for bread.

We find ourselves having to repeat and relearn the same old lessons over and over that our mothers did because we do not pass on what we have learned, or because we are unable to listen. For instance, how many times has this all been said before? For another, who would have believed that once again our daughters are allowing their bodies to be hampered and purgatoried by girdles and high heels and hobble skirts?

Ignoring the differences of race between women and the implications of those differences presents the most serious threat to the mobilization of women's joint power.

As white women ignore their built-in privilege of whiteness and define woman in terms of their own experience alone, then women of Color become "other," the outsider whose experience and tradition is too "alien" to comprehend. An example of this is the signal absence of the experience of women of Color as a resource for women's studies courses. The literature of women of Color is seldom included in women's literature courses and almost never in other literature courses, nor in women's studies as a whole. All too often, the excuse given is that the literatures of women of Color can only be taught by Colored women, or that they are too difficult to understand, or that classes cannot "get into" them because they come out of experiences that are "too different." I have heard this argument presented by white women of otherwise quite clear intelligence, women who seem to have no trouble at all teaching and reviewing work that comes out of the vastly different experiences of Shakespeare, Moliere, Dostoyevsky, and Aristophanes. Surely there must be some other explanation.

This is a very complex question, but I believe one of the reasons white women have such difficulty reading Black women's work is because of their reluctance to see Black women as women and different from themselves. To examine Black women's literature effectively requires that we be seen as whole people in our actual complexities—as individuals, as women, as human—rather than as one of those problematic but familiar stereotypes provided in this society in place of genuine images of Black women. And I believe this holds true for the literatures of other women of Color who are not Black.

The literatures of all women of Color re-create the textures of our lives, and many white women are heavily invested in ignoring the real differences. For as long as any difference between us means one of us must be inferior, then the recognition of any difference must be fraught with guilt. To allow women of Color to step out of stereotypes is too guilt provoking, for it threatens the complacency of those women who view oppression only in terms of sex.

Refusing to recognize difference makes it impossible to see the different problems and pitfalls facing us as women.

Thus, in a patriarchal power system where white-skin privilege is a major prop, the entrapments used to neutralize Black women and white women are not the same. For example, it is easy for Black women to be used by the power structure against Black men, not because they are men, but because they are Black. Therefore, for Black women, it is necessary at all times to separate the needs of the oppressor from our own legitimate conflicts within our communities. This same problem does not exist for white women. Black women and men have shared racist oppression and still share it, although in different ways. Out of that shared oppression we have developed joint defenses and joint vulnerabilities to each other that are not duplicated in the white community, with the exception of the relationship between Jewish women and Jewish men.

On the other hand, white women face the pitfall of being seduced into joining the oppressor under the pretense of sharing power. This possibility does not exist in the same way for women of Color. The tokenism that is sometimes extended to us is not an invitation to join power; our racial "otherness" is a visible reality that makes that quite clear. For white women there is a wider range of pretended choices and rewards for identifying with patriarchal power and its tools.

Today, with the defeat of ERA, the tightening economy, and increased conservatism, it is easier once again for white women to believe the dangerous fantasy that if you are good enough, pretty enough, sweet enough, quiet enough, teach the children to behave, hate the right people, and marry the right men, then you will be allowed to co-exist with patriarchy in relative peace, at least until a man needs your job or the neighborhood rapist happens along. And true, unless one lives and loves in the trenches it is difficult to remember that the war against dehumanization is ceaseless.

But Black women and our children know the fabric of our lives is stitched with violence and with hatred, that there is no rest. We do not deal with it only on the picket lines, or in dark midnight alleys, or in the places where we dare to verbalize our resistance. For us, increasingly, violence weaves through the daily tissues of our living—in the supermarket, in the classroom, in the elevator, in the clinic and the schoolyard, from the plumber, the baker, the saleswoman, the bus driver, the bank teller, the waitress who does not serve us.

Some problems we share as women, some we do not. You fear your children will grow up to join the patriarchy and testify against you, we fear our children will be dragged from a car and shot down in the street, and you will turn your backs upon the reasons they are dying.

The threat of difference has been no less blinding to people of Color. Those of us who are Black must see that the reality of our lives and our struggle does not make us immune to the errors of ignoring and misnaming difference. Within Black communities where racism is a living reality, differences among us often seem dangerous and suspect. The need for unity is often misnamed as a need for homogeneity, and a Black feminist vision mistaken for betrayal of our common interests as a people. Because of the continuous battle against racial erasure that Black women and Black men share, some Black women still refuse to recognize that we are also oppressed as women, and that sexual hostility against Black women is practiced not only by the white racist society, but implemented within our Black communities as well. It is a disease striking the heart of Black nation-hood, and silence will not make it disappear. Exacerbated by racism and the pressures of powerlessness, violence against Black women and children often becomes a standard within our communities, one by which manliness can be measured. But these woman-hating acts are rarely discussed as crimes against Black women.

As a group, women of Color are the lowest paid wage earners in America. We are the primary targets of abortion and sterilization abuse, here and abroad. In certain parts of Africa, small girls are still being sewed shut between their legs to keep them docile and for men's pleasure. This is known as female circumci-sion, and it is not a cultural affair as the late Jomo Kenyatta insisted, it is a crime against Black women.

Black women's literature is full of the pain of frequent assault, not only by a racist patriarchy, but also by Black men. Yet the necessity for and history of shared battle have made us, Black women, particularly vulnerable to the false accusation that anti-sexist is anti-Black. Meanwhile, woman hating as a recourse of the powerless is sapping strength from Black communities, and our very lives. Rape is on the increase, reported and unreported, and rape is not aggressive sex-uality, it is sexualized aggression. As Kalamu ya Salaam, a Black male writer points out, "As long as male domination exists, rape will exist. Only women revolting and men made conscious of their responsibility to fight sexism can col-lectively stop rape."

Differences between ourselves as Black women are also being misnamed and used to separate us from one another. As a Black lesbian feminist comfortable with the many different ingredients of my identity, and a woman committed to racial and sexual freedom from oppression, I find I am constantly being encour-aged to pluck out some one aspect of myself and present this as the meaningful whole, eclipsing or denying the other parts of self. But this is a destructive and fragmenting way to live. My fullest concentration of energy is available to me only when I integrate all the parts of who I am, openly, allowing power from particular sources of my living to flow back and forth freely through all my dif-ferent selves, without the restrictions of externally imposed definition. Only then can I bring myself and my energies as a whole to the service of those struggles which I embrace as part of my living.

A fear of lesbians, or of being accused of being a lesbian, has led many Black women into testifying against themselves. It has led some of us into destructive alli-ances, and others into despair and isolation. In the white women's communities, heterosexism is sometimes a result of identifying with the white patriarchy, a rejec-tion of that interdependence between women-identified women which allows the self to be, rather than to be used in the service of men. Sometimes it reflects a die-hard belief in the protective coloration of heterosexual relationships, sometimes a self-hate which all women have to fight against, taught us from birth.

Although elements of these attitudes exist for all women, there are particular resonances of heterosexism and homophobia among Black women. Despite the fact that woman-bonding has a long and honorable history in the African and African American communities, and despite the knowledge and accomplishments of many strong and creative women-identified Black women in the political, social and cultural fields, heterosexual Black women often tend to ignore or discount the existence and work of Black lesbians. Part of this attitude has come from an understandable terror of Black male attack within the close confines of Black society, where the punishment for any female self-assertion is still to be accused of being a lesbian and therefore unworthy of the attention or support of the scarce Black male. But part of this need to misname and ignore Black lesbians comes from a very real fear that openly women-identified Black women who are no longer dependent upon men for their self-definition may well reorder our whole concept of social relationships.

Black women who once insisted that lesbianism was a white woman's problem now insist that Black lesbians are a threat to Black nationhood, are consorting with the enemy, are basically un-Black. These accusations, coming from the very women to whom we look for deep and real understanding, have served to keep many Black lesbians in hiding, caught between the racism of white women and the homophobia of their sisters. Often, their work has been ignored, trivialized, or misnamed, as with the work of Angelina Grimke, Alice Dunbar-Nelson, Lorraine Hansberry. Yet women-bonded women have always been some part of the power of Black communities, from our unmarried aunts to the amazons of Dahomey.

And it is certainly not Black lesbians who are assaulting women and raping children and grandmothers on the streets of our communities.

Across this country, as in Boston during the spring of 1979 following the unsolved murders of twelve Black women, Black lesbians are spearheading movements against violence against Black women.

What are the particular details within each of our lives that can be scrutinized and altered to help bring about change? How do we redefine difference for all women? It is not our differences which separate women, but our reluctance to recognize those differences and to deal effectively with the distortions which have resulted from the ignoring and misnaming of those differences.

As a tool of social control, women have been encouraged to recognize only one area of human difference as legitimate, those differences which exist between women and men. And we have learned to deal across those differences with the urgency of all oppressed subordinates. All of us have had to learn to live or work or coexist with men, from our fathers on. We have recognized and negotiated these differences, even when this recognition only continued the old dominant/subordinate mode of human relationship; where the oppressed must recognize the masters' difference in order to survive.

But our future survival is predicated upon our ability to relate within equality. As women, we must root out internalized patterns of oppression within ourselves if we are to move beyond the most superficial aspects of social change. Now we must recognize differences among women who are our equals, neither inferior nor superior, and devise ways to use each others' difference to enrich our visions and our joint struggles. The future of our earth may depend upon the ability of all women to identify and develop new definitions of power and new patterns of relating across difference. The old definitions have not served us, nor the earth that supports us. The old patterns, no matter how cleverly rearranged to imitate progress, still condemn us to cosmetically altered repetitions of the same old exchanges, the same old guilt, hatred, recrimination, lamentation, and suspicion.

For we have, built into all of us, old blueprints of expectation and response, old structures of oppression, and these must be altered at the same time as we alter the living conditions which are a result of those structures. For the master's tools will never dismantle the master's house.

As Paulo Freire shows so well in the *Pedagogy of the Oppressed*, the true focus of revolutionary change is never merely the oppressive situations which we seek to escape, but that piece of the oppressor which is planted deep within each of us, and which knows only the oppressors' tactics, the oppressors' relationships.

Change means growth, and growth can be painful. But we sharpen self-definition by exposing the self in work and struggle together with those whom we define as different from ourselves, although sharing the same goals. For Black and white, old and young, lesbian and heterosexual women alike, this can mean new paths to our survival.

We have chosen each other
and the edge of each other's battles
the war is the same
if we lose
someday women's blood will congeal
upon a dead planet
if we win
there is no telling
we seek beyond history
for a new and more possible meeting.

2

From a Native Daughter

HAUNANI-KAY TRASK

E noi'i wale mai nō ka haole, a,
'a'ole e pau nō hana a Hawai'i 'imi loa /
Let the haole freely research us in detail
But the doings of deep delving Hawai'i
will not be exhausted.

—*Kepelino*
19th-century Hawaiian historian

When I was young, the story of my people was told twice: once by my parents, then again by my school teachers. From my *'ohana* (family), I learned about the life of the old ones: how they fished and planted by the moon; shared all the fruits of their labors, especially their children; danced in great numbers for long hours; and honored the unity of their world in intricate genealogical chants. My mother said Hawaiians had sailed over thousands of miles to make their home in these sacred islands. And they had flourished, until the coming of the *haole* (whites).

At school, I learned that the "pagan Hawaiians" did not read or write, were lustful cannibals, traded in slaves, and could not sing. Captain Cook had "discovered" Hawai'i and the ungrateful Hawaiians had killed him. In revenge, the Christian god had cursed the Hawaiians with disease and death.

I learned the first of these stories from speaking with my mother and father. I learned the second from books. By the time I left for college, the books had won out over my parents, especially since I spent four long years in a missionary boarding school for Hawaiian children.

When I went away I understood the world as a place and a feeling divided in two: one *haole* (white), and the other *kānaka* (Native). When I returned ten years later with a Ph.D., the division was sharper, the lack of connection more painful. There was the world that we lived in—my ancestors, my family, and my people—and then there was the world historians described. This world, they had written, was the truth. A primitive group, Hawaiians had been ruled by blood-thirsty priests and despotic kings who owned all the land and kept our people in feudal subjugation. The chiefs were cruel, the people poor.

SOURCE: Trask, Haunani-Kay. 1993. *From a Native Daughter*. Monroe, ME: Common Courage Press. Reprinted by permission.

But this was not the story my mother told me. No one had owned the land before the *haole* came; everyone could fish and plant, except during sacred periods. And the chiefs were good and loved their people.

Was my mother confused? What did our *kūpuna* (elders) say? They replied: Did these historians (all *haole)* know the language? Did they understand the chants? How long had they lived among our people? Whose stories had they heard?

None of the historians had ever learned our mother tongue. They had all been content to read what Europeans and Americans had written. But why did scholars, presumably well-trained and thoughtful, neglect our language? Not merely a passageway to knowledge, language is a form of knowing by itself; a people's way of thinking and feeling is revealed through its music.

I sensed the answer without needing to answer. From years of living in a divided world, I knew the historian's judgment: *There is no value in things Hawaiian; all value comes from things haole.*

Historians, I realized, were very much like missionaries. They were a part of the colonizing horde. One group colonized the spirit; the other, the mind. Frantz Fanon had been right, but not just about Africans. He had been right about the bondage of my own people: "By a kind of perverted logic, [colonialism] turns to the past of the oppressed people, and distorts, disfigures, and destroys it" (1963:210). The first step in the colonizing process, Fanon had written, was the deculturation of a people. What better way to take our culture than to remake our image? A rich historical past became small and ignorant in the hands of Westerners. And we suffered a damaged sense of people and culture because of this distortion.

Burdened by a linear, progressive conception of history and by an assumption that Euro-American culture flourishes at the upper end of that progression, Westerners have told the history of Hawai'i as an inevitable if occasionally bittersweet triumph of Western ways over "primitive" Hawaiian ways. A few authors—the most sympathetic—have recorded with deep-felt sorrow the passing of our people. But in the end, we are repeatedly told, such an eclipse was for the best.

Obviously it was best for Westerners, not for our dying multitudes. This is why the historian's mission has been to justify our passing by celebrating Western dominance. Fanon would have called this missionizing, intellectual colonization. And it is clearest in the historian's insistence that *pre-haole* Hawaiian land tenure was "feudal"—a term that is now applied, without question, in every monograph, in every schoolbook, and in every tour guide description of my people's history.

From the earliest days of Western contact my people told their guests that *no one* owned the land. The land—like the air and the sea—was for all to use and share as their birthright. Our chiefs were *stewards* of the land; they could not own or privately possess the land any more than they could sell it.

But the *haole* insisted on characterizing our chiefs as feudal landlords and our people as serfs. Thus, a European term which described a European practice founded on the European concept of private property—feudalism—was imposed upon the people halfway around the world from Europe and vastly different from her in every conceivable way. More than betraying an ignorance of Hawaiian culture and history, however, this misrepresentation was malevolent in design.

By inventing feudalism in ancient Hawai'i, Western scholars quickly transformed a spiritually based, self-sufficient economic system of land use and occupancy into an oppressive, medieval European practice of divine right owner-ship, with the common people tied like serfs to the land. By claiming that the Pacific people lived under a European system—that the Hawaiians lived under feudalism—Westerners could then degrade a successful system of shared land use with a pejorative and inaccurate Western term. Land tenure changes instituted by Americans and in line with current Western notions of private property were then made to appear beneficial to the Hawaiians. But in practice, such changes benefited the *haole,* who alienated the people from the land, taking it for themselves.

The prelude to this land alienation was the great dying of the people. Barely half a century after contact with the West, our people had declined in number by eighty percent. Disease and death were rampant. The sandalwood forests had been stripped bare for international commerce between England and China. The missionaries had insinuated themselves everywhere. And a debt-ridden Hawaiian king (there had been no king before Western contact) succumbed to enormous pressure from the Americans and followed their schemes for dividing the land.

This is how private property land tenure entered Hawai'i. The common people, driven from their birthright, received less than one percent of the land. They starved while huge haole-owned sugar plantations thrived.

And what had the historians said? They had said that the Americans "liberated" the Hawaiians from an oppressive "feudal" system. By inventing a false feudal past, the historians justify—and become complicitous in—massive American theft.

Is there "evidence"—as historians call it—for traditional Hawaiian concepts of land use? The evidence is in the sayings of my people and in the words they wrote more than a century ago, much of which has been translated. However, historians have chosen to ignore any references here to shared land use. But there is incontrovertible evidence in the very structure of the Hawaiian language. If the historians had bothered to learn our language (as any American historian of France would learn French) they would have discovered that we show posses-sion in two ways: through the use of an "a" possessive, which reveals acquired status, and through the use of an "o" possessive, which denotes inherent status. My body (*ko'u kino*) and my parents (*ko'u mākua*), for example, take the "o" form; most material objects, such as food *(ka'u mea'ai)* take the "a" form. But land, like one's body and one's parents, takes the "o" possessive (*ko'u 'āina*). Thus, in our way of speaking, land is inherent to the people; it is like our bodies and our parents. The people cannot exist without the land, and the land cannot exist without the people.

Every major historian of Hawai'i has been mistaken about Hawaiian land tenure. The chiefs did not own the land: they *could not* own the land. My mother was right and the *haole* historians were wrong. If they had studied our language they would have known that no one owned the land. But was their failing merely ignorance, or simple ethnocentric bias?

No, I did not believe them to be so benign. As I read on, a pattern emerged in their writing. Our ways were inferior to those of the West, to those of the historians' own culture. We were "less developed," or "immature," or "authoritarian." In some tellings we were much worse. Thus, Gavan Daws (1968), the most famed modern historian of Hawai'i, had continued a tradition established earlier by missionaries Hiram Bingham (1848; reprinted, 1981) and Sheldon Dibble (1909), by referring to the old ones as "thieves" and "savages" who regularly practiced infanticide and who, in contrast to "civilized" whites, preferred "lewd dancing" to work. Ralph Kuykendall (1938), long considered the most thorough if also the most boring of historians of Hawai'i, sustained another fiction—that my ancestors owned slaves, the outcast *kauwā*. This opinion, as well as the description of Hawaiian land tenure as feudal, had been supported by respected sociologist Andrew Lind…. Finally, nearly all historians had refused to accept our genealogical dating of A.D. 400 or earlier for our arrival from the South Pacific. They had, instead, claimed that our earliest appearance in Hawai'i could only be traced to A.D. 1100. Thus at least seven hundred years of our history were repudiated by "superior" Western scholarship. Only recently have archaeological data confirmed what Hawaiians had said these many centuries (Tuggle 1979).[1]

Suddenly the entire sweep of our written history was clear to me. I was reading the West's view of itself through the degradation of my own past. When historians wrote that the king owned the land and the common people were bound to it, they were saying that ownership was the only way human beings in their world could relate to the land, and in that relationship, some one person had to control both the land and the interaction between humans.

And when they said that our chiefs were despotic, they were telling of their own society, where hierarchy always results in domination. Thus any authority or elder is automatically suspected of tyranny.

And when they wrote that Hawaiians were lazy, they meant that work must be continuous and ever a burden.

And when they wrote that we were promiscuous, they meant that lovemaking in the Christian West is a sin.

And when they wrote that we were racist because we preferred our own ways to theirs, they meant that their culture needed to dominate other cultures.

And when they wrote that we were superstitious, believing in the *mana* of nature and people, they meant that the West has long since lost a deep spiritual and cultural relationship to the earth.

And when they wrote that Hawaiians were "primitive" in their grief over the passing of loved ones, they meant that the West grieves for the living who do not walk among their ancestors.

For so long, more than half my life, I had misunderstood this written record, thinking it described my own people. But my history was nowhere present. For we had not written. We had chanted and sailed and fished and built and prayed. And we had told stories through the great blood lines of memory: genealogy.

To know my history, I had to put away my books and return to the land. I had to plant *taro* in the earth before I could understand the inseparable bond between people and *'āina*. I had to feel again the spirits of nature and take gifts of plants and fish to the ancient altars. I had to begin to speak my language with our elders and leave long silences for wisdom to grow. But before anything else, I needed to learn the language like a lover so that I could rock within her and lie at night in her dreaming arms.

There was nothing in my schooling that had told me of this, or hinted that somewhere there was a longer, older story of origins, of the flowing of songs out to a great but distant sea. Only my parents' voices, over and over, spoke to me of a Hawaiian world. While the books spoke from a different world, a Western world.

And yet, Hawaiians are not of the West. We are of *Hawai'i Nei*, this world where I live, this place, this culture, this *'āina*.

What can I say, then, to Western historians of my place and people? Let me answer with a story.

A while ago I was asked to appear on a panel on the American overthrow of our government in 1893. The other panelists were all *haole*. But one was a *haole* historian from the American continent who had just published a book on what he called the American anti-imperialists. He and I met briefly in preparation for the panel. I asked him if he knew the language. He said no. I asked him if he knew the record of opposition to our annexation to America. He said there was no real evidence for it, just comments here and there. I told him that he didn't understand and that at the panel I would share the evidence. When we met in public and spoke, I said this:

There is a song much loved by our people. It was written after Hawai'i had been invaded and occupied by American marines. Addressed to our dethroned Queen, it was written in 1893, and tells of Hawaiian feelings for our land and against annexation to the United States. Listen to our lament:

Kaulana nā pua a'o Hawai'i	Famous are the children of Hawai'i
Kupa'a ma hope o ka 'āina	Who cling steadfastly to the land
Hiki mai ka 'elele o ka loko 'ino	Comes the evil-hearted with
Palapala 'ānunu me ka pākaha	A document greedy for plunder
Pane mai Hawai'i moku o Keawe	Hawai'i, island of Keawe, answers
Kokua nā hono a'o Pi'ilani	The bays of Pi'ilani [of Maui, Moloka'i, and Lana'i] help
Kako'o mai Kaua'i Mano	Kaua'i of Mano assists
Pau pu me ke one o Kakuhihewa	Firmly together with the sands of Kakuhihewa
'A'ole a'e kau i ka pūlima	Do not put the signature
Maluna o ka pepa o ka 'enemi	On the paper of the enemy
Ho'ohui 'āina kū'ai hewa	Annexation is wicked sale
I ka pono sivila a'o ke kānaka	Of the civil rights of the Hawaiian people
Mahope mākou o Lili'ūlani	We support Lili'uokalani
A loa'a 'e ka pono o ka 'āina	Who has earned the right to the land
Ha'ina 'ia mai ana ka puana	The story is told
'O ka po'e i aloha i ka 'āina	Of the people who love the land

This song, I said, continues to be sung with great dignity at Hawaiian political gatherings, for our people still share the feelings of anger and protest that it conveys.

But our guest, the *haole* historian, answered that this song, although beautiful, was not evidence of either opposition or of imperialism from the Hawaiian perspective.

Many Hawaiians in the audience were shocked at his remarks, but, in hindsight, I think they were predictable. They are the standard response of the historian who does not know the language and has no respect for its memory.

Finally, I proceeded to relate a personal story, thinking that surely such a tale could not want for authenticity since I myself was relating it. My *tūtū* (grandmother) had told my mother who had told me that at the time of the overthrow a great wailing went up throughout the islands, a wailing of weeks, a wailing of impenetrable grief, a wailing of death. But he remarked again, this too is not evidence.

And so, history goes on, written in long volumes by foreign people. Whole libraries begin to form, book upon book, shelf upon shelf. At the same time, the stories go on, generation to generation, family to family.

Which history do Western historians desire to know? Is it to be a tale of writings by their own countrymen, individuals convinced of their "unique" capacity for analysis, looking at us with Western eyes, thinking about us within Western philosophical contexts, categorizing us by Western indices, judging us by Judeo-Christian morals, exhorting us to capitalist achievements, and finally, leaving us an authoritative-because-Western record of their complete misunderstanding?

All this has been done already. Not merely a few times, but many times. And still, every year, there appear new and eager faces to take up the same telling, as if the West must continue, implacably, with the din of its own disbelief. But there is, as there has been always, another possibility. If it is truly our history Western historians desire to know, they must put down their books, and take up our practices. First, of course, the language. But later, the people, the *āina*, the stories. Above all, in the end, the stories. Historians must listen, they must hear the generational connections, the reservoir of sounds and meanings.

They must come, as American Indians suggested long ago, to understand the land. Not in the Western way, but in the indigenous way, the way of living within and protecting the bond between people and *'āina*. This bond is cultural, and it can be understood only culturally. But because the West has lost any cultural understanding of the bond between people and land, it is not possible to know this connection through Western culture. This means that the history of indigenous people cannot be written from within Western culture. Such a story is merely the West's story of itself.

Our story remains unwritten. It rests within the culture, which is inseparable from the land. To know this is to know our history. To write this is to write of the land and the people who are born from her.

NOTES

1. See also Fornander (1878–85; reprinted, 1981). Lest one think these sources antiquated, it should be noted that there exist only a handful of modern scholarly works on the history of Hawai'i. The most respected are those by Kuykendall (1938) and

Daws (1968), and a social history of the 20th century by Lawrence Fuchs (1961). Of these, only Kuykendall and Daws claim any knowledge of pre-*haole* history, while concentrating on the 19th century. However, countless popular works have relied on these two studies which, in turn, are themselves based on primary sources written in English by extremely biased, anti-Hawaiian Westerners such as explorers, traders, missionaries (e.g., Bingham [1848; reprinted, 1981] and Dibble [1909]), and sugar planters. Indeed, a favorite technique of Daws's—whose *Shoal of Time* was once the most acclaimed and recent general history—is the lengthy quotation without comment of the most racist remarks by missionaries and planters. Thus, at one point, half a page is consumed with a "white man's burden" quotation from an 1886 *Planters Monthly* article ("It is better here that the white man should rule ...," etc., p. 213). Daws's only comment is, "The conclusion was inescapable." To get a sense of such characteristic contempt for Hawaiians, one has but to read the first few pages, where Daws refers several times to the Hawaiians as "savages" and "thieves" and where he approvingly has Captain Cook thinking, "It was a sensible primitive who bowed before a superior civilization" (p. 2). See also—among examples too numerous to cite—his glib description of sacred *hula* as a "frivolous diversion," which, instead of work, the Hawaiians "would practice energetically in the hot sun for days on end ... their bare brown flesh glistening with sweat" (pp. 65–66). Daws, who repeatedly displays an affection for descriptions of Hawaiian skin color, taught Hawaiian history for some years at the University of Hawai'i. He once held the Chair of Pacific History at the Australian National University's Institute of Advanced Studies.

Postscript: Since this article was written, the first scholarly history by a Native Hawaiian was published in English: *Native Land and Foreign Desires* by Lilikalà Kame'eleihiwa (Honolulu: Bishop Museum Press, 1992).

REFERENCES

Bingham, Hiram. 1981. *A Residence of Twenty-one Years in the Sandwich Islands.* Tokyo: Charles E. Tuttle.

Daws, Gavan. 1968. *Shoal of Time: A History of the Hawaiian Islands.* Honolulu: University of Hawai'i Press.

Dibble, Sheldon. 1909. *A History of the Sandwich Islands.* Honolulu: Thrum Publishing.

Fanon, Frantz. 1963. *The Wretched of the Earth.* New York: Grove Press.

Fornander, Abraham. 1981. *An Account of the Polynesian Race, Its Origins, and Migrations and the Ancient History of the Hawaiian People to the Times of Kamehameha I.* Routledge, Vermont: Charles E. Tuttle.

Fuchs, Lawrence H. 1961. *Hawaii Pono: A Social History.* New York: Harcourt Brace & World.

Kuykendall, Ralph. 1938. *The Hawaiian Kingdom, 1778–1854: Foundation and Transformation.* Honolulu: University of Hawai'i Press.

Tuggle, H. David. 1979. "Hawai'i," in *The Prehistory of Polynesia*, ed. Jessie D. Jennings. Cambridge: Harvard University Press.

3

Label Us Angry

JEREMIAH TORRES

It hurts to know that the most painful and shocking event of my life happened in part because of my race—something I can never change. On October 23, 1998, my friend and I experienced what would forever change our perceptions of our hometown and society in general.

We both attended elementary, middle, and high school in the quiet, prosperous, seemingly sophisticated college town of Palo Alto. In the third grade, we happily sang "It's a Small World," holding hands with the children of professors, graduate students, and professionals of the area, oblivious to our diversity in race, culture, or experience. Our small world grew larger as we progressed through the school system, each year learning more about what made us different from each other. But on that October evening, the world grew too large for us to handle.

Carlos and I were ready for a night out with the boys. It was his seventeenth birthday, and we were about to celebrate at the pool hall. I pulled out of the Safeway driveway as a speeding driver delivered a jolting honk. I followed him out, speeding to catch up with him, my immediate anger getting the better of me.

We lined up at the stoplight, and the passenger, a young white man dressed for the evening, rolled down his window; I followed. He looked irritated.

"He wasn't honking at you, you stupid fuck!"

His words slapped me across the face. I opened my stunned mouth, only to deliver an empty breath, so I gave him my middle finger until I could return some angry words. He grimaced and reached under his seat to pull out a bottle of mace, spraying it directly in my face, barely missing Carlos, who witnessed the bizarre scene in shock. It burned.

"Take that you fucking lowlifes! Stupid chinks!"

Carlos instinctively bolted out the door at those words. He started pounding the white guy without a second thought, with a new anger he had never known or felt before. Pssssht! The white guy hit Carlos point blank in the face with the mace. He screamed; tires squealed; "fuck you's" were exchanged.

We spent the next ten minutes half-blind, clutching our eyes in the burning pain, cursing in raging anger that made us forget for moments the intense, throbbing fire on our faces. I crawled out of my car to follow Carlos's screams and curses, opening my eyes to the still, spectating traffic surrounding us. I stumbled

SOURCE: Torres, Jeremiah. 2004. "Label Us Angry." In *Asian American X: An Intersection of 21st Century Asian American Voices*, edited by A. Har and J. Hsu. Ann Arbor, MI: University of Michigan Press. Reprinted by permission.

to the sidewalk, where Carlos pounded the ground and recalled the words of the white guy. We needed water.

I stumbled further to a nearby house that had lights in the living room. I doorbelled frantically, but nobody answered. I appealed to the traffic for help. They just watched, forming a new route around my car to continue about their evening. The mucous membranes in our sinuses cut loose, and we spit every few seconds to sustain our gasping breaths. After nearly five minutes of appeals, a kind woman stopped to call the cops and give us water to quench the burning.

The cops came within minutes with advice for dealing with the mace. We tried to identify the car and the white guy who had sprayed us, and they sent out the obligatory all points bulletin. They questioned us soon after, asking if we were in a gang. I returned a blank stare with a silent "no." Apparently, two Filipino teenagers finding trouble on a Friday evening raised suspicions of a new Filipino gang in Palo Alto—yeah, all five of us.

I often ask myself if it would have been different had I been driving a BMW and dressed in an ironed polo shirt and slacks, like a typical Palo Alto kid. Maybe then the white guy would not have been afraid and called us lowlifes and chinks. I don't think so. He wasn't afraid of us; he initiated the curses and maced us from a safe distance. He reached out to hurt us because he was having a bad day and we looked different.

That night was our first encounter with overt racism that stems from a hatred of difference. We hadn't seen it through the smiles and happy songs of elementary school or the isolated cliques of middle and high school, but now we knew it was there. We hadn't seen it through the clean-cut, sophisticated facade of the Palo Alto white guy, but now we knew it was there. The "lowlife," "chink," and "gangster" labels made us different, marginalizing us from the town we called home.

Those labels made us angry, but we hesitated to project that anger. At first, we didn't tell anyone except our closest friends, afraid our parents would find out and react irrationally by locking us in our rooms to keep us away from trouble. But then we realized that the trouble had found us, and we decided to voice our anger.

We wrote an anonymous article in the school newspaper narrating the incident and the underlying racism that had come to surface. We noted that the incident wasn't purely racial, or a hate crime, but proof that racist tendencies still exist, even in open-minded suburban towns like Palo Alto. Parents, students, and teachers were shocked, maybe because they knew the truth in what we were saying. Many asked if it was Carlos and me who had been maced, but I responded, "Does it matter? What matters is that some people in this town still can't accept diversity. It's sad." We confronted the community with an issue previously reserved for hypothetical classroom discussions and brought it into the open. It was the least we could do to release our anger and expose its roots, hoping for a change in those who chose to label us.

After the article, Carlos and I took different routes. I continued with my studies, complying with my regimen of high school classes and activities as my anger subsided. I tried to lay the incident aside, having exposed it and promoted self-inspection and possible change in others through writing. Carlos remained angry. Why not? He got a face full of mace and racist labels for his seventeenth birthday. He alienated himself from the white majority and returned the mean gestures of the white guy to the yuppie congregation of Palo Alto. He became an outsider. Whenever someone would look at him funny, he would stare back, sometimes too harshly.

On the day after finals, he was making his way through the front parking lot of school when a parent looked at him funny. He stared back. The parent called him a punk. Carlos exploded. He cursed and gestured all he could at the father, and when he sped away in his Suburban, Carlos followed. Carlos couldn't keep up with the Suburban, so he took a quarter from his pocket and threw it at the back window, shattering it to pieces. Carlos ran away when the cops came to school.

Within two days, students had identified Carlos as the perpetrator, and he was suspended from school as the father called his lawyer, indicting Carlos of "assault with the intent to hurt." Weeks passed until a court hearing, and Carlos attended anger management counseling, but he was still angry—angry that he was being tried over throwing a quarter and that once again "the white guys were winning." His mother scraped up the little money she had to spare to afford him a lawyer for the trial, but there was no contesting the father's accusations. Carlos was sentenced to a night in juvenile hall and two hundred hours of community service over some angry words and throwing a quarter. He became a convicted felon.

He had learned once again that he couldn't win against the labels thrown at him, the labels that hurt him more than the mace or the night in juvy, and so he became more of an outsider. In both cases, the labels distanced us from the "normal" Palo Altans: white, clean-cut, wealthy. That division didn't always exist, however; it was created by the generalizations "normal" Palo Altans made through labels. To them, we looked like lowlifes, chinks, gangsters, and punks. In truth, we were two Filipino Americans headed toward Stanford and Berkeley, living in a town that swiftly disowned us with four reckless labels after raising us for ten years. Label us angry.

4

"It Looks Like a Demon"

Black Masculinity and Spirituality in the Age of Ferguson

JAMIE D. HAWLEY AND STAYCIE L. FLINT

It. Demon. These are the words Darren Wilson, a White (former) Ferguson, Missouri, police officer used to describe why he killed Michael Brown, an unarmed Black 18-year-old on August 9, 2014. The same words were espoused by early Englishmen once they arrived in Africa and encountered the natives in the 16th century. One might argue Wilson saw a hyper-masculine brute, an animal, an untamed "it." Whether it be the Buck of the 16th century or the Thug of the 21st century, Black men are the *invisibly visible*—caught in a paradox of being devoid of participation in socioeconomic and political structures but always visible enough to incite public fear. In this commentary, we contribute to broader men's studies scholarship by problematizing and linking Black masculinity and Black male spirituality as it relates to potential tensions between the police and Black men.

Wilson's testimony to the grand jury, in particular his language used to describe Michael Brown, is laced with theological underpinnings revealing warped messages of spirituality in modern U.S. culture that affect how broader society views Black boys and men. There is a dearth of scholarship affirming Black male spirituality that unpacks the nuance and complicated factors that describe our most recent police state. Leveraging the voices of hospital trauma chaplains, we confront dominant pathological and theological understandings of Black men. Our commentary engages research and thought on the Black male experience locating Black masculinity and Black spiritual identity as the potential site of trauma and soul murder.

Let us carefully examine the variation in language used to describe Michael Brown. Leslie McSpadden, Michael Brown's mother described her recently deceased son as "a gentle giant." Meanwhile, Darren Wilson describes brown as an "it" and "demon" and as one having the "most intense aggressive face." How might two individuals have polar opposite experiences of the same individual? The unarmed Black male perceived as a threat has become so familiar to the American psyche that it resembles a standardized rehearsed script. However,

SOURCE: Hawley, Jamie D. and Staycie L. Flint. 2016. "'It Looks Like a Demon': Black Masculinity and Spirituality in the Age of Ferguson." *The Journal of Men's Studies*, 24(2): 208–212. Copyright © 2016. Reprinted by permission of SAGE Publications, Inc.

these descriptors of Black males are not a new phenomenon. Psychohistories of White racism have always called attention to the tension between the construction of Black male body as danger and the underlying eroticization that always then imagines that body as a location for transgressive pleasure (Hooks, 2004). As such, the Black male body must be tamed and/or destroyed.

Overtime, cultural critics, writers, historians, and filmmakers have deemphasized and emasculated Black manhood. For example, some writers have argued that the institution of slavery was needed to keep Black men tame (Higgs, 1977). The critique against Black manhood is further evidenced in D. W. Griffith's 1915 film *The Birth of a Nation*, which depicts Black men as hypersexual, brutal menaces that White society has no choice but to control through the use of violent force, intimidation, and fear. Animalistic imagery continues to be associated with Black people in the 21st century (Goff, Eberhardt, Williams, & Jackson, 2008). Immediately following the Michael Brown shooting, as protesters took to the streets, they were often referred to by the mainstream media as "rioting" as opposed to "revolting." Simultaneously, the hashtag #Chimpout—a term used to describe the bad behavior of Black people, especially when they behave like animals (Urban Dictionary) started trending as a description to describe the Black protesters. Not only has animalistic language been used historically to *other* Black males (and females in many instances), but it is also a part of the fabric of American history (Nott & Gliddon, 1855).

The denigration of Black manhood and the Black male experience does not begin in adulthood. A recent study published in the *Journal of Personality and Social Psychology* suggests that Black boys as young as 10 years old were significantly less likely to be viewed as children than were their White peers (Goff, 2014). The neglect of seeing a child when meeting a black youth is on par with the state sanctioned death of 14-year-old George Stinney Jr. in 1944. The dismissal of childhood is hauntingly echoed by one of Emmett Till's killers who stated the 14-year-old "looked like a man" (Huie, 2009). Likewise, 12-year-old Tamir Rice was gunned down while playing in a public park and described by the officer who shot him as being "in his late teens or early 20s" (Dolan, 2015). In similar fashion, Wilson, who is 6 feet 4, 210 pounds, testified before a grand jury he "felt like a 5 year old holding onto Hulk Hogan." He continued, "That's how big he felt and how small I felt just from grasping his arm" (Bzdeck, 2015). Indeed, Wilson's narrative conjures up the familiar trope of the infamous dehumanized "Magical Negro." The magical negro is defined as a Black (usually male) character in film or literature possessed of supernatural power, folksy wisdom, and a kind of primordial goodness that he or she uses in the service and aid of White people to help them to self-actualization. Ascribing supernatural powers to Black men is not beneficial, and instead, we argue, deprives them of a real presence in the world, of bodies that have spirits that exist *apart from and independent* of Whiteness, the White gaze, and White expectation. The hypersexualized superhuman brute and the "Magical Negro" caricatures reduce Black males to extremes: The evil monster who must be controlled and/or wiped out or as benevolent figures completely at the service of others. Although on its surface this trope is benevolent, its underbelly is quite sinister in that Black people end up using the formidable powers ascribed to them to maintain the status quo of default Whiteness and White supremacy. Furthermore, slavery itself thrived on the image of the happy and content slave. The "Magical Negro" is nothing more than a spiritualized variation of this. Sinister in its underlying assertion that even as a spiritual being, their spiritual power and authority is in the service of Whiteness. Conversely, after the civil war, many White writers argued that the institution of slavery was what kept the savagery of Black men subdued and therefore a necessity for the safety of the public (Findlay, 2000). S. Opotow, a psychological theorist, and other historians of genocide have argued that dehumanization is a necessary component for culturally and/or state-sanctioned violence. Furthermore, Goff and colleagues also make a strong argument that it is the process of dehumanization inflicted by "in-groups" (the majority) on "out-groups" (minorities) that make possible state-sanctioned violence. When Black male bodies are constructed in the psyche as a problem to be solved, they invite violence—violence manifested in the form of overt racism that can be lethal and/or violence expressions that are more nuanced, that is, microaggressions (such as stop and frisk, driving while Black, being followed around a store, etc.). The roots of this particular construction within the cultural imaginary are both evident in and a consequence of the history of Black bodies within the United States. Beginning, of course, with the theologized, monetized, commodified, and finally mundane psychic, spiritual, and physical violation of Black bodies that was slavery; on through the era of Jim Crow; into the contemporary iterations, the routine violation of Black bodies has moved beyond the mundane to the expected in the cultural imaginary and other workings of material life in the United States (Ray, 2014).

The authors, as trauma chaplains, are well acquainted with grief, suffering, sorrow, and death; and have found disproportionate negative treatment of Black men during this process. The first author vividly recalls one early morning on Columbus Day and being the chaplain paged to assist a family experiencing the death of an actively dying loved one. The patient was an older White identified woman who eventually died surrounded by her husband, children, and grandchildren. Simultaneously and unknown to the chaplain was a young African American identified man who lay actively dying and completely alone in the neighboring room. Security had been asked to remove the family from the premises for "crying too loudly." The African American patient and his family's human expression of grief was seen by the predominantly White staff as a "problem." The perception of black grief as problematic illuminates how mourning practices and grief itself has been colonized.

A large body of research in the health care field suggests a girth of cultural disparities between Blacks and Whites. In almost every disease studied and published in the Institute of Medicine (IOM), Blacks received less effective care than Whites. The IOM suggest that clinical uncertainty, biases (implicit or explicit), stereotypes, and prejudice are central to Blacks receiving less effective health care (Smedley, Stith, Nelson, & IOM, 2000). In addition, evidence notes that not only are Blacks far less likely to be treated for pain (Todd, 2007), but there is also a noticeable gap in staff's ability to perceive pain in those with Black and brown bodies. People perceive Blacks as feeling less pain than Whites (Trawalter, Hoffman, & Waytz, 2012).

Those raising the Black men of the future know that the age of Ferguson is an ancient age. The killing of Mike Brown by Darren Wilson simply placed a national spotlight on the warped messages of identity and social location that permeate U.S. culture and affect centuries old societal views that dominate perspectives of Black boys and men. It has long been noted that individualism is the binding that holds social identity together in U.S. culture, and therefore, it becomes difficult to see the centrality of systemic forces at work that academic and activist bell hooks and many others teach us is an empire held together by White supremacist capitalistic heterosexist patriarchal structures of corrective violence. It is not the individual who creates large systems such as "capitalism," "patriarchy," and "racism," but individuals are the cogs that keep such system mechanics in motion. The militarized responses in Ferguson and the U.S. Department of Justice-Civil Rights Division report on the Ferguson policing system reveal populus support of systemic inequities (DOJ, 2015). Indeed, buried in his iconic speech delivered at the 1963 March on Washington that has become known as the "I Have a Dream" speech, The Reverend Martin Luther King Jr. speaking of the police state says, "We can never be satisfied as long as the Negro is the victim of the unspeakable horrors of police brutality."

REFERENCES

Bzdeck, V. (2015). Michael Brown's mother blames governor for violence. *Washington Post.* Retrieved from https://www.washingtonpost.com/blogs/liveblog-live/live-blog/live-updates-ferguson-grand-jury-decision/

Dolan, M. (2015, June 13). Cleveland police officer who shot Tamir Rice said he had "no choice," probe finds. *The Wall Street Journal.* Retrieved from http://www.wsj.com/articles/sherrifs-report-doesnt-say-whether-cleveland-boys-death-warrants-charges-against-police-1434224512

Findlay, J. B. (2000). *Drapetomania—A disease called freedom: An exhibition of 18th-, 19th-, & early 20th-century.* Fort Lauderdale, FL: Bienes Center for the Literary Arts.

Goff, P. A. (2014). The essence of innocence: Consequences of dehumanizing Black children. *Journal of Personality and Social Psychology, 106,* 526–545.

Goff, P. A., Eberhardt, J. L., Williams, M. J., & Jackson, M. C. (2008). Not yet human: Implicit knowledge, historical dehumanization, and contemporary consequences. *Journal of Personality and Social Psychology, 94,* 292–306.

Higgs, R. (1977). *Competition and coercion: Blacks in the American economy 1865–1914.* New York, NY: Cambridge University Press.

Hooks, B. (2004). *We real cool.* New York, NY: Routledge.

Huie, W. B. (2009). *The shocking story of approved killing in Mississippi.* Retrieved from http://www.pbs.org/wgbh/amex/till/sfeature/sf_look_confession.html

Nott, J. C., & Gliddon, G. R. (1855). *Types of mankind: Or, ethnological researches, based upon the ancient monuments, paintings, sculptures, and crania of races, and upon their natural, geographical, philological and biblical history.* Philadelphia, PA: Lippincott, Grambo.

Ray, S. (2014, April 15). *Black & blue: Violence against Blacks.* Retrieved from https://www.ctschicago.edu/about/cts-publications

Smedley, B., Stith, A., Nelson, A., & Institute of Medicine. (2000). *Unequal treatment: Confronting racial and ethnic disparities in health care.* Washington, DC: The National Academy Press.

Todd, K. H. (2007). Pain in the emergency department: Results of the Pain and Emergency Medicine Initiative (PEMI) multicenter study. *The Journal of Pain, 8,* 460–466.

Trawalter, S., Hoffman, K., & Waytz, A. (2012). Racial bias in perceptions of others' pain. *PLoS ONE, 7*(11), e48546. doi:10.1371/journal.pone.0048546

U.S. Department of Justice Civil Rights Division. (2015). *Investigation of the Ferguson Police Department.* Washington, DC: U.S. Department of Justice.

PART II

Systems of Power and Inequality

MARGARET L. ANDERSEN
AND PATRICIA HILL COLLINS

O ne of the most important things to learn about race, class, and gender is that they are *systemic forms of power and inequality*. Although many people tend to think of each in terms of individual identities, they are more than that. Race, class, and gender are built into the very structure of society. It is this social fact that drives the analysis of race, class, and gender as *intersectional systems of inequality*. This does not make them irrelevant as individual identities, but it points you to the importance of *social structure* in understanding how race, class, and gender—along with ethnicity and sexuality—influence people's lives.

People do, of course, have various identities based on their racial and ethnic, class, gender, and sexual identities, along with other individual characteristics. Obviously, your specific identity has an enormous impact on your experience in society. But if you look at identities solely as individual attributes or experiences, you miss how profoundly they are embedded in the structure and power of social institutions. Institutions may be hard to see as "things," but they are present in your everyday experience. Institutional structures become more apparent when they pose obstacles to you, although even then many people tend to blame individual behaviors for what are, much of the time, institutional phenomena.

The intersectional analysis we employ here explains race, class, and gender as *systems of power that are structured into social institutions.* Together, this interlocking system of power advantages and disadvantages groups differently, as evidenced by many social indicators: economic status, health and well-being, and educational attainment, to name some. At the core of the intersectional perspective is the idea that no one of these social facts singularly determines anyone's lived experiences. To illustrate, not all men are equally powerful, nor are all women equally disadvantaged. Some women, indeed, have more power and money than many men. In other words, gender alone is not sufficient to understand men's or women's social status.

Likewise, race is highly significant in determining one's life chances, but it does not operate in a vacuum. Class matters, too, even though your race may be particularly salient at a given point in time—such as when a middle- or upper-class Black man is stopped by the police. Together, race, class, and gender—as explained in Part I—are foundations in people's lives, but they work together along with other social factors in shaping how you live. Understanding this requires an intersectional perspective.

Race, class, and gender operate within a system of simultaneous, interrelated social relationships—what we have earlier called the *matrix of domination* (Collins 2000). This means that they also engage other social realities—ethnicity, sexuality, age, disability, regional location, and so forth. Any one person's or group's experience depends on the particular configuration of all of these factors at different points in time. Each shapes the others. Sexuality, for example, has long served to buttress the beliefs that support racial and gender subordination (Collins 2000, 2004). Likewise, a system of racial subordination has historically been a primary way that class structure was created. That is, some White people accumulated property through the appropriation of slave labor, even while Black slaves were denied basic rights, including the right to vote, the right to marry, the right to own property, and the right to be a citizen.

To put it simply, *race, class, and gender are social structures, not just individual identities or attributes.* Furthermore, these structures are supported by ideological beliefs that make things appear "normal" and "acceptable," thereby clouding our awareness of how the social structure operates. Philosopher Marilyn Frye likens this to a birdcage. If you look myopically at only a single wire, you might not understand why the bird does not just fly away, but when you see the complete structure of the cage, you realize there is a whole structure of containment enclosing the bird (Frye 1983). Today, because formal barriers to racial discrimination have been removed, many believe that racism is largely a thing of the past. But racism persists, even when it takes on new forms. Perhaps one "wire" from the cage has been removed—laws endorsing racial discrimination—but a whole structure of racism, including new belief systems, is in place.

Understanding the intersections among race, class, gender, ethnicity, and sexuality requires knowing how to conceptualize each. We discuss them separately here only to first learn what each means and how each is manifested in different group experiences. As you read the articles in this section, you will notice several common themes.

First, *each is a socially constructed category.* That is, their significance stems not from some "natural" state, but from what they have become as the result of social and historical processes.

Second, *each constructs groups in binary (or opposite; "either/or") terms:* Black/White; rich/poor; man/woman; gay/straight, or citizen/alien. These binary constructions create a notion of "otherness" for those who are subordinated by this constellation of inequalities.

Third, these are categories of individual and group identity, but *their significance stems from the fact that they are located in a system of institutional power.* Social structure, not individual identity, is the basis for the inequalities that we can identify in society. That social structure is fundamentally organized around inequalities of race, class, and gender, which in turn connect to other systems of power, such as ethnicity, sexuality, physical ability, and even age.

Finally, *race, ethnicity, class, gender, and sexuality are social constructions.* This means that they are not fixed categories of social relationships. As fluid social constructions, their form, their meaning, and their interrelationship change over time. This also means that social change is possible.

Here is one way to illustrate an intersectional analysis. Imagine a typical sports event. There are different components of the "game"—the players on the field or court, the cheerleaders moving about on the side, the band playing, fans cheering, boosters watching from the best seats, and perhaps a television crew. Everybody seems to have a place in the game. Each system of power we examine here shapes the totality of the game. You can see that race and ethnicity matter because many players are African American, Latino, and, increasingly, international recruits. Sports provide an opportunity structure, even though only for the rare few. For people whose opportunities are otherwise limited, sports can be a path to social mobility. Yet, the odds of success are poor and, at the college level, graduation rates for Black athletes are often abysmal, especially in the powerhouse teams.

Gender matters, too, given the status of women in the sports world. Women athletes are ignored more than men, rarely reported on in sports pages unless they are superstars. Even then, they may be described in sexual terms, as has regularly happened to Serena Williams. Most often, women are relegated to the sidelines as scantily clad cheerleaders or as occasional sports announcers who rarely report the play of the game, commenting instead on player's injuries. Sexuality intersects with gender in the sports world. Where else are men able to put their arms around one another, slap one another's buttocks, hug one another, or cry in public without having their "masculinity" questioned?

Class matters too. In professional sports, those who benefit the most are team owners and corporate sponsors who bring in millions of dollars based on the labor of others. And who are those workers? Of course, it is the athletes—some, but not all, of whom make fabulous salaries. But, sometimes what you don't usually see can be just as revealing as what you do see. People of color and immigrants do much of the work in the concession stands. And who cleans the restrooms and locker rooms, picks up the trash, and closes up the stadium after the crowd goes home?

This discussion of sports demonstrates how each factor—race, ethnicity, class, gender, and sexuality—provides an important, yet partial, perspective on any given institution or situation. Using an intersectional model, we can see each of them in turn but, more importantly, we also see the connections among them. Understanding any institution or social issue requires thinking about all of them and how they work together.

Race, ethnicity, class, gender, and sexuality affect all levels of our experience—our consciousness and ideas, our interaction with others, and the social institutions we live within. Because they are so interconnected, no one can be subsumed under the other. In this section of the book, we focus on each one to provide conceptual grounding. Keep in mind, though, that race, ethnicity, class, gender, and sexuality are connected and overlapping—in all realms of society, including the realm of ideas, everyday social interactions, and social institutions (Glenn 2002).

You might begin by considering a few facts:

- Women of color, including Latinas, African American women, Native American women, some Asian American women, and immigrant women are concentrated in the bottom rungs of the labor market (Bureau of Labor Statistics 2018).

- The United States is in the midst of a sizable redistribution of wealth, with a greater concentration of wealth and income in the hands of a few than ever before. At the same time, a declining share of income is going to the middle class—a class that finds its position slipping relative to years past. As Thomas E. Shapiro will show in his essay included here ("Toxic Inequality"), the wealth gap is not just about class but also divides people by race and ethnicity.

- Women in the top 5 percent of women earners have seen their wages grow by 67 percent since 1980 while women in the bottom 10 percent have seen their wages drop by 3 percent (Mishel et al. 2012). Class differences among women are hidden, if you think of women as a monolithic group.

- The most common hate crimes (at least as reported to the FBI) occur against Black Americans, followed in number by those against gay men, and then Jewish people. The vast majority of offenders are White. Hate crime against Muslims has also dramatically increased in recent years (Federal Bureau of Investigation 2017; Potok 2017).

None of these facts can be explained through an analysis that focuses on one factor only. Clearly, race matters. Ethnicity matters. Class matters. Gender matters. Sexuality matters. And, they matter together. We turn now to defining these different systems of power and inequality.

RACE AND RACISM

In some ways, the United States has shed many of the racially exclusionary practices of the past. Legally mandated racial segregation has been outlawed. Overt expressions of racial prejudice are less common than in the past. The Black and Latinx middle class have expanded and people of color are more visible in positions of leadership in every social institution. Young people—the millennial generation, in particular—see increased racial and ethnic diversity as good for the country (Pew Research Center 2010). In some ways, we seem to be moving toward a more tolerant, racially just, and integrated society.

Yet, signs of racial injustice are also very apparent, and increasingly so in recent years. Police shootings of Black men and woman are all too common. Calls to "build walls" to keep out immigrants or to ban entry to people from some Muslim-majority nations inflame and divide public opinion. Immigrants have been referred to by our nation's president as "infesting" the nation. Voting rights, so valiantly fought for in the civil rights movement, are being infringed upon through various forms of voter suppression. Incarceration of African American, Latino, and Native American men and, increasingly, women, is shockingly high. Poverty afflicts parts of our nation's cities and, now, suburbs too. Although overt prejudice may be lessened, people of color also experience "microaggressions"—daily and routine encounters with racism.

These and other facts point to the ongoing structure of **racial stratification**— the inequality between and among racial groups. Racial stratification is experienced personally, but its roots lie in social institutions. No doubt, racial stratification changes over time, but it is rooted in the practices of the past and interacts with social forces of the present.

People often think of racism as individual attitudes—that is, **prejudice**. But racism takes different forms and is not always expressed as overt prejudice. Racism is structured into society, not just in people's minds. In fact, a White person can be against racism and yet still benefit from racial advantages. **Institutional racism** is built into the very fabric of the social structure of the United States. This has been true since the founding of the nation. Institutional racism may be less visible than individual expressions of racial prejudice, but it is just as significant—actually, more so—in shaping the experiences of people of color. And, as Joe Feagin shows ("The Persistence of White Nationalism in America"), even though people have been shocked by the rise of white nationalist movements, such as the march in Charlottesville, Virginia in 2017, white nationalism has long been part of the *systemic racism* that permeates U.S. social institutions.

Understanding racism and racial inequality means understanding that *race is a social construction*. That is, the meaning and significance of race stems from specific social, historical, and political contexts. It is the social and historical context that makes race meaningful, not whatever physical differences may exist among groups (Andersen 2017). Although many people think of race as biologically based, the idea of race is more social than biological. Scientists have concluded that there is no "race" gene, but you should not conclude from this that race is not "real." The reality of race stems from its social, not its biological, significance.

In the opening essay of this section, Omi and Winant explain this through the concept of racial formation. **Racial formation** is "the sociohistorical process by which racial categories are created, inhabited, transformed, and destroyed" (1994: 55). Put differently, society constructs ideas, rules, and practices that define some groups in racial terms—differently at different points in time, and for the explicit purpose of group oppression. It is exploitation that defines some groups as a "race." As institutions evolve, prevailing racial definitions can actually change. Some groups are "racialized"; others are not. The concept of race is actually quite recent in human history. It evolved as the global system of slavery was created. White people defined Black Africans as a race to justify their enslavement and exploitation as unpaid labor. The very concept of race evolved only because of slavery, not the other way around.

Thinking about race in this way means asking how racial categories are created, by whom, and for what purposes. Racial classification systems, for example, reflect prevailing views of race, socially constructing groups that are then presumed to be "natural." These constructed racial categories then serve as the basis for allocating resources and become the basis for social and political conflicts (Omi and Winant 1994).

How people have been defined and categorized in the U.S. census provides an example. In 1860, only three "races" were presumed to exist—Whites,

Blacks, and mulattoes. By 1890, however, these original three races had been joined by five others—quadroon, octoroon, Chinese, Japanese, and Indian. Ten short years later, this list shrank to five races—White, Black, Chinese, Japanese, and Indian—a situation reflecting the growth of strict segregation in the South (Rodriguez 2000). The U.S. government now allows people to check multiple boxes to identify themselves as more than one race, reflecting the growing number of multiracial people in the United States. The census categories are not just a matter of accurate statistics; they have significant consequences for the apportionment of societal resources.

Although some might argue that we should not "count" race at all, doing so is important because data on racial groups are used to enforce voting rights, to regulate equal employment opportunities, to determine various governmental supports, and to be able to track the socioeconomic status of various groups, among other things.

The concept of race is changing. For many years, race and racism in the United States have been thought of primarily in terms of Black/White relations. Changes in the racial landscape of the United States now make such a framework obsolete. Some would say that Latinos are in the process of becoming *racialized*—even though many Latino groups lived within the borders of the United States long before their lands were considered part of this nation. Chicanos originally held land in what is now the American Southwest, but it was seized following the Mexican American War because in 1848, Mexico ceded huge parts of what are now California, New Mexico, Nevada, Colorado, Arizona, and Utah to the United States (for $15 million). Mexicans living there were one day Mexicans, the next day, "Americans" but without all the rights of citizens and defined as "other."

To sum up, race and racism are located in the shifting relations of power that mark a social structure of racial inequality. And, as that social structure changes, so do the forms that racism takes. Charles Gallagher ("Color-Blind Privilege: The Social and Political Functions of Erasing the Color Line in Post-Race America") shows this in his discussion of a now common form of racism—*color-blind racism*.

As Gallagher argues, color-blind racism is a form of racism whereby dominant groups assume that race no longer matters—even when society remains highly structured by racial inequality. Many people believe that being nonracist means being color-blind—that is, refusing to recognize or treat a person's racial background and identity as significant. But ignoring the significance of race in a society where racial groups have distinct historical and contemporary experiences denies the reality of the impact of racial inequality. Blindness to the persistent realities of race also leads to an idea that there is nothing we should be doing about it, either individually or collectively. Thus, racism is perpetuated.

Understanding that racism is institutionally based also means people may not see themselves as individually racist but can nonetheless be benefiting from a system that is organized to privilege some at the expense of others, as shown in Peggy McIntosh's now classic article, "Unpacking the Invisible Knapsack." McIntosh likens racism to an "invisible knapsack" of white privilege that typically goes unnoticed by Whites. She argues that White Americans benefit from a system of racism, *even when they do not intend to do so.* Of course, not all Whites benefit equally or in the same way, as an intersectional perspective points out. Yet, white privilege is also a part of the social structure on which racial inequality rests.

ETHNICITY

Ethnicity refers to groups who share a common culture, including language, customs, religion, and so forth. Ethnic groups tend to share a feeling of common identity and a common history. Although ethnicity has traditionally been thought of as distinct from race, you might have immediately noticed that the same characteristics that define ethnicity (a shared identity and common culture) also apply to so-called racial groups. This tells you that the distinction between race and ethnicity is complex and blurry. Some groups become racialized; others do not. Just as *race is a social construction, so is ethnicity.*

The best example of an ethnic group becoming racialized comes from the horrendous example of the Nazi Holocaust. Jewish people are an ethnic group, collectively bound by their religious beliefs, history, and sense of common identity. But under the Nazi regime, Hitler defined Jewish people as a race—a social construction that became the basis for mass murder. As Abby L. Ferber shows ("What White Supremacists Taught a Jewish Scholar about Identity"), the history of anti-Semitism is a classic example of an ethnic group becoming racialized. Ferber also shows the interconnections between racism and *anti-Semitism* (the hatred of Jewish people), reminding us of the interplay between different systems of oppression.

Ethnicity and race thus can be very closely linked at times. Latinos, for example, are considered an ethnic group, even though they have very diverse origins, cultures, and histories. Some may not even speak Spanish. The ethnic identity as "Latino" stems from people's social and political location in American society. Chicanos may even object to the label "Latino" because it is a label imposed on them by outside groups. And, a Puertoriqueno may hold a different identity in New York City than in San Juan.

Like race, ethnicity develops within systems of power. Thus, the meaning and significance of ethnicity can shift over time and in different social and political contexts. For example, groups may develop a sense of heightened ethnicity in the context of specific historical events. Likewise, their feeling of sharing a common group identity can result from being labeled as "outsiders" by dominant groups. Think, for example, of the heightened sense of group identity that has developed for Arab Americans in the aftermath of 9/11.

Although race and ethnicity are usually treated as separate concepts, the line between the two is not always firm because both are nestled within systems of power. People from outside the United States typically even find the U.S. conception of race to be quite strange (Andersen 2017). Ethnicity also has to be understood in the context of social structures. Some ethnic groups, for example, may have their ethnic identity imposed on them by others. An example would be Vietnamese immigrants who came to the United States as the result of the Vietnam War. Although they may perceive themselves as "Vietnamese," within the United States they get lumped together with other Asians and redefined as "Asian American." Ethnicity can be self-defined, but it can also be imposed.

Bhoomi K. Thakore's article ("Must-See TV: South Asian Characterization in American Popular Media") provides a case in point. She notes that stereotypes of Asian Americans are rampant in the popular U.S. media, adding to the "otherness" with which Asian Americans are perceived by dominant groups. Even as Indians and South Asians attempt to "become American," they do so within the U.S. racial hierarchy, becoming racialized through the stereotypes perpetuated in the mass media and popular culture. Key to Thakore's argument is the idea that racialization serves particular social and political interests, certainly not those of South Asian people. Her work shows how the concepts of race and ethnicity are fluid because of the powerful interests they serve.

The meaning of ethnicity is evolving with the changing contours of the U.S. population. Racial and ethnic groups other than Whites comprise an increasing proportion of the population, and their numbers are expected to rise in the years ahead. One-quarter of the U.S. population is now Black, Hispanic, Asian American, Pacific Islander, or Native American. Latinos have recently exceeded African Americans as the largest minority group in the population. Hispanic, Native American, Asian American, and African American populations are also growing more rapidly than White American populations, with the greatest growth among Hispanics and Asian Americans. By 2050, non-Hispanic Whites are predicted to make up only slightly more than half the total population.

In this context, the meaning of both ethnicity and race is quite politically charged. Some Whites perceive their dominance as winnowing with the increased diversity in the population. Although not directly expressed as such, the mobilization of some white groups to "restore America" has a distinct racial ring to it. Some White people feel threatened by the changes taking place. This creates fertile ground for reactionary movements seeking to maintain white supremacy.

In this context, what groups will be defined as races and how will that map onto concepts of citizenship? What will it mean to be American in a nation where racial-ethnic diversity is so much a part of the national fabric? Eduardo Bonilla-Silva explore these questions in his article "We Are All Americans! The Latin Americanization of Racial Stratification in the United States." His suggestion that the United States could develop into a tripartite racial society is based on his analysis of how different groups are perceived and treated within the U.S. racial hierarchy. He shows that racial stratification is not fixed, but evolves while protecting dominant groups interests.

Questions about both race and ethnic identity are now taking on heightened meaning at a time when some White people fear a loss of their historic position. International tensions and terrorist attacks add to this dynamic, making for a potentially explosive mix. International and "ethnic" students are increasingly under surveillance. Calls for more restrictive immigration practices and policies and restrictions on many civil liberties are indicative of the perceived threats to national safety and identity. How will the nation maintain its great ethnic diversity within a framework of racial and ethnic justice?

Some groups will continue to celebrate ethnicity without penalty, as Mary C. Waters points out in "Optional Ethnicities: For Whites Only?" Waters contends that White Americans of European ancestry have *symbolic ethnicity*, meaning that their ethnic identity does not influence their lives unless they want it to. She contrasts this with the socially enforced and imposed racial identity among African Americans. Because race operates as a physical marker in the United States, the intersection of race and ethnicity operates differently for Whites and for people of color. White ethnics can thus have "ethnicity" without cost, but people of color pay the price for their racial and ethnic identity.

CLASS

What is social class? Most think of class as a rank held by an individual, but class is structured into society, shaping our social institutions and our relationships with one another. **Class** is a hierarchical arrangement in society that gives groups different access to the economic, social, political, and cultural resources (Andersen and Taylor 2015). Power and privilege are key components of a class system. Class advantages some, disadvantages others. Where you live, how you dress, where you attend school, where you work, how much leisure you have, and how others perceive you are all shaped by the class system.

In the United States, class is formed based on the economic system of *capitalism*, which is based on the principles of profit and private ownership. Within capitalism, profit comes from the undervaluing of certain groups of people who work for low wages and are socially devalued as well.

As Joan Acker argues in the lead article of this section ("Is Capitalism Gendered and Racialized?"), race and gender inequality are fundamental to the structure of capitalism. Race and gender segregation in the labor force fuel the formation of capitalism, a process that has been in place throughout the development of the U.S. economy. As capitalism developed, the labor force was strictly organized around race and gender, with Black people as slaves and women by and large excluded from paid labor. As Acker shows, although strict race and gender exclusion are no longer with us, they have become deeply embedded in the operation of capitalism. Race and gender form the scaffolding of a capitalist economy. Her essay explains that race and gender are critical components of capitalism.

Herbert J. Gans's article ("Race as Class") shows how race has been used as a marker of the capitalist class hierarchy. There is such a strong correlation between class and race that the effect of each is almost impossible to untangle. Like Acker, Gans shows how much race and class are built into the very structure of society. They are not just the result of individual prejudices or irrational evils. As the title of this section shows, class and race, like gender (as we will see), are systems of power that differentiate people's access to society's material and symbolic resources.

Class is particularly slippery as a concept in U.S. society because of the widely held cultural belief that your class origin should not determine your eventual station in life. The American dream imagines that class mobility, regardless of origins, is possible if you work hard enough and hold the right values. This ideal has a firm hold on American's consciousness, despite the reality that most of the time, people end up in exactly the same class where they started (Economic Mobility Report 2012). Of course, there are exceptions, and those exceptions are extolled as cultural heroes. You might think of Oprah Winfrey, Bill Gates, or Serena and Venus Williams.

This idea of an open class society is reproduced in many ways, but especially through the mass media, as Gregory Mantsios shows in "Media Magic: Making Class Invisible." Class ideology renders some groups simply invisible, especially the working class and the poor. Have you ever noticed that working-class people are most often portrayed in the media as comic figures, rarely in serious drama? As Mantsios shows, the working class is widely stigmatized in popular culture and the media, thus distorting how people imagine class realities. The media make it appear that everyone is middle class, a myth that is exposed through any serious look at the evidence.

Despite the cultural ideal, class is highly significant—along with race and gender—in determining one's chances in life. Clear class differences persist and, indeed, are growing. The most common way of showing and documenting class inequality is through data on income inequality. **Income** is the amount of money brought into a household in one year. Measures of income in the United States are based on annually reported census data drawn from a representative sample of the population.

Median income is the income level above and below which half of the population lies. It is the best measure of group income standing. In 2017, median income, counting all U.S. households, was $61,372. This means that half of all households earned more than this amount and half earned less; this is the "middle." As you can see in Figure 1, however, median income varies significantly by race with Black and Hispanic households lagging behind White, non-Hispanic households. Further, the gap between Black and White Americans is unchanged since 1970.

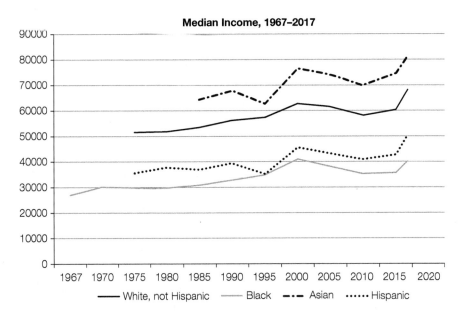

FIGURE 1 Median Income by Race, 1967–2017

Source: Data from: Fontenot, Kayla, Jessica Semega, and Melissa Kollar. 2018. *Income and Poverty in the United States: 2017*. Washington, DC: U.S. Census Bureau. www.census.gov

You have to be careful, however, in interpreting these aggregate data. They give you a broad picture of group differences, but aggregate groupings do not show differences within these racial and ethnic labels. Class experiences even *within* racial groups vary widely. Not all Whites are well-off, just as not all African Americans, Native Americans, or Latinos are poor.

The label Asian American, for example, includes many different groups whose class status varies widely. You should not conclude, as the model minority myth implies, that Asians are all as well-off as the aggregate median income figure suggests. Although Asian American median income is, in the aggregate, higher than that for White Americans, not all Asian American families are better off than White families. In fact, Asian immigrants from Vietnam, Laos, and Cambodia have rates of poverty that match that of African Americans and Native Americans (Guo 2016). In sum, median income data reveal a lot about group standing, but do not show the details of particular group's class standing.

To get a feel for what a median income actually means, you can do a simple exercise. Take the national median income figure of $61,372 and divide by twelve to get a monthly income. Then figure out a household budget that accounts for all things you think are essential—housing, food, transportation, health care, childcare, clothing, and so forth. Don't forget to pay your taxes. Use the best figures you can find for the region where you live. Can you make ends meet for a family of four on a median income? How would you live? Would you drive a new car? Take a vacation? Have a pet? What would you do in an emergency? Of course, your experiment is only a rough estimate of what median income feels like. It does not account for such thing as tax credits, regional variation, savings, and other financial resources. Still, it can tell you a lot about the reality of class in shaping how people live.

As important as income is in shaping one's living situation, it only tells part of the story. Wealth is even more revealing in determining people's class standing. **Wealth** is the sum of all financial assets minus all debt. The result is one's net worth, or wealth.

To understand this, imagine two recent college graduates. They graduate in the same year, from the same college, with the same major and identical grade point averages. Both get jobs with the same salary in the same company. One student's parents paid all college expenses and gave her a car upon graduation. The other student worked while in school and has graduated with substantial debt from student loans. This student's family has no money to help support the new worker. Who is better-off? Same salary, same credentials, but one person has a clear advantage—one that will be played out many times over as the young worker buys a home, finances her own children's education, and perhaps inherits additional assets. This shows you the significance of wealth—not just income—in structuring social class.

The above scenario is not just hypothetical. Switching to the real world, a recent study has found that after four years from college graduation, the gap between Black and White student debt is even greater than it was at the time of graduation (see Figure 2). This suggests that White graduates are better able to pay debts down sooner, while Black graduates get deeper in debt. And, if one does not keep up with debt payments, interest charges accumulate, putting one deeper and deeper into debt, and thus making it harder to get out. If high debt levels happen at a time when young people are best served by saving to start families, buy homes, or plan for retirement, there can be long-term consequences (Scott-Clayton and Li 2016).

Wealth provides, a *cumulative* advantage to those who have it. Wealth produces more wealth because people with such resources can make investments that usually payoff over time (such as purchasing a home that holds its value). Wealth can also be transmitted from one generation to the next, thus passing on advantages that people without wealth do not have. Even modest amounts of wealth can cushion the impact of emergencies, such as unexpected unemployment or sudden health problems. Buying a home, investing, being free of debt, affording the best education, and transferring economic assets to the next generation are all pieces of class advantage that add up over time and enhance one's class standing—above and beyond income.

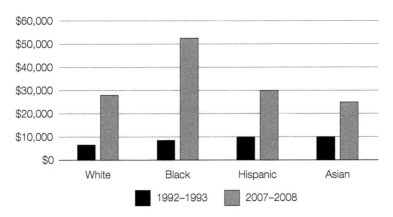

F I G U R E 2 Student Debt, Four Years after Graduation

Source: Scott-Clayton, Judith, and Jing Li. 2016. *Black-White Disparity in Student Loan Debt More Than Triples after Graduation.* Washington, DC: The Brookings Institute. https://www.brookings.edu/research/black-white-disparity-in-student-loan-debt-more-than-triples-after-graduation/

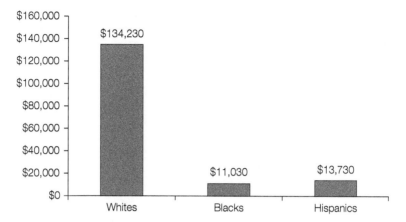

FIGURE 3 The Racial Wealth Gap

Source: McKernan, Signe-Mary, Caroline Ratcliffe, C. Eugene Steuerle, Emma Kalish, and Caleb Quakenbush. 2015. *Nine Charts about Wealth Inequality in America.* Washington, DC: The Urban Institute. http://datatools.urban.org/Features/wealth-inequality-charts/

Research finds that race has a huge impact on the class advantages that wealth provides. Currently, White Americans have *twelve times* the wealth of Black Americans and almost *ten times* as much as Hispanics, as shown in Figure 3 (McKernan et al. 2015). Furthermore, even at the same income level and with the same educational and occupational status, Black and White Americans differ substantially in their financial assets (Oliver and Shapiro 2006). Clearly, income alone does not determine class status.

Income and wealth data reflect the entanglement of both race and class—and, as we will see, also gender. Furthermore, the differences that these data show appear at a time when class inequality has reached unprecedented levels. The increasing division between the "haves" and "have-nots" is leaving some people very well-off, while many are struggling to hold on to middle class status, and others are mired in poverty.

Thomas M. Shapiro ("Toxic Inequality") documents the rising inequality that is occurring, especially around wealth. He particularly shows how wealth inequality is deeply entangled with racial inequality. According to Shapiro, wealth inequality is highly "toxic" because it spreads over generations. And, while current growth in wealth inequality threatens to erode people's standard of living, it also further divides us along lines of race and class.

Growing inequality in the United States cannot be fully understood without an intersectional analysis that accounts for race, class, *and* gender. You can see this especially when looking at poverty. Nineteen percent of Black

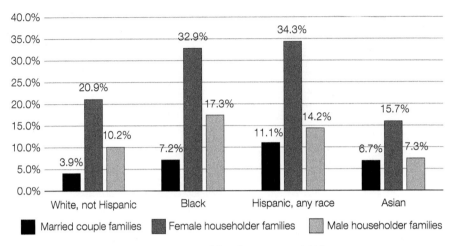

FIGURE 4 Poverty Rate by Race and Family Structure, 2017

Source: Data from: U.S. Census Bureau. 2018. *POV-02-People in Families by Family Structure, Age and Sex, Iterated by Income-to-Poverty Ratio and Race.* Washington, DC: U.S. Census Bureau. www.census.gov

families and 17 percent of Hispanics are poor, as defined by the federal poverty line.[1] This is compared to only 6.3 percent of White, non-Hispanic, and 7.3 percent of Asian American families. Even more telling is the rate of poverty when you consider gender as indicated by family structure. Female-headed households of any race are more likely to be poor than male-headed and married couple families in every racial-ethnic group. And, as you can see in Figure 4, poverty rates for African American and Hispanic female-headed households are staggering—over 30 percent of such households.

Such levels of inequality produce some of our most intransigent social problems. The safety net that is supposed to protect people in need is also shrinking. Inequality in the United States is also greater than in other Western, industrialized nations. Understanding how this is happening needs the intersectional analysis that we provide in this book. We now turn to two other important systems of power: gender and sexuality.

---

[1]The federal poverty line is calculated based on a formula established in the 1930s by the U.S. Department of Agriculture. That calculation took the price of a low-cost food budget and multiplied by three, assuming that food was one-third of a family's budget. That amount has since been adjusted every year for the cost of living. In 2017, the official poverty line was $24,858 for a family of four (with two children). There are different thresholds of poverty based on family size.

GENDER

Gender, like the other systems of power examined here, is a social construction, not a biological imperative. **Gender** is rooted in social institutions and results in patterns within society that structure the relationships between women and men and the advantages and disadvantages that women and men receive. Gender is also an identity, but a learned one. Like race, however, gender cannot be understood solely at the individual level.

Gender is built into social institutions, just as it is built into our identity and our ideas. The concept of a **gendered institution** refers to the total patterns of gender relations that are "present in the processes, practices, images, and ideologies, and distribution of power in the various sectors of social life" (Acker 1992: 567). This means that institutions take on the characteristics of the dominant gender such that the institution itself becomes gendered. When men dominate, the institution becomes culturally "masculine"—hierarchical, rational, and unfeeling, making anyone (regardless of their own gender) who does not fit that dominant cultural norm feel like they do not really belong. A workplace, for example, becomes unfriendly to those with family commitments, including men. Because the workplace is "gendered masculine," workplace institutions recognize only the public world of work, not the emotional and caring characteristics of "family."

The concept of a gendered institution brings a social structural analysis of gender to the forefront. Rather than seeing gender only as a matter of interpersonal relationships and learned identities, the gendered institution framework focuses the analysis of gender on relations of power—just as thinking about institutional racism focuses on power relations and economic and political subordination—not just interpersonal relations. Thus, changing gender relations is not just a matter of changing individuals. As with race and class, change requires transformation of institutional structures.

The institutional structure of gender is most apparent when looking at such things as the persistent pay gap between men and women. As you can see, gender alone does not determine economic status. Gender intersects with race in shaping the economic status of both women and men. Thus, something as simple as claiming that women earn 80 percent of what men earn changes when you include both race and gender. For example, while White women earn 59 percent of White men's earnings, but Black women earn only 54 percent of White men's earnings, and Hispanic women, 44 percent (see Figures 5A and 5B).

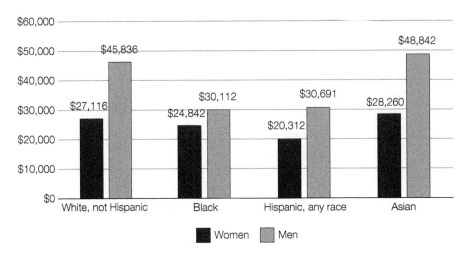

FIGURE 5A   Median Income by Sex and Race, 2017

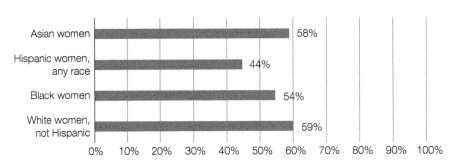

FIGURE 5B   Women's Income as Percent of White Men's Income

Source: Data for both charts from: U.S. Census Bureau. 2017. *PINC-01. Selected Characteristics of People 15 Years and Over, by Total Money Income, Work Experience, Race, Hispanic Origin, and Sex.* Washington, DC: U.S. Census Bureau. www.census.gov

In this one example, you can see that outcomes for women differ depending on both their race and their gender (as well as their ethnicity, age, disability status, and other social factors). Gender is simply not a monolithic category. It is manifested differently depending on the different systems of power we examine here. There is no other way to explain the fact that, on virtually every measure of social and economic well-being, women of color face the harshest impact of race, class, and gender inequities. Although gender may seem particularly salient in a given situation, such as rape or sexual harassment, race and class also shape the gendered experiences that women and men have.

Many people think of gender as an immutable part of our being, but gender is produced through social institutions and social interaction. Cultural variations in gender relations prove this. But using a careful eye, you can also see the socially constructed character of gender within your own culture.

Karen D. Pyke and Denise L. Johnson illustrate this in "Asian American Women and Racialized Femininities: 'Doing Gender' across Cultural Worlds." Pyke and Johnson demonstrate how gender is enacted in experiences of second-generation Korean and Vietnamese women in the United States. As daughters of immigrants, these women straddle two worlds, each world with its own set of gendered expectations. The women must navigate these competing gender and ethnic norms. The analysis that Pyke and Johnson provide shows the connection of gender to other systems of power, such as ethnicity, as well as race and class. As racial-ethnic women, Asian American women have to negotiate gender within particular social contexts, contexts that shift for them from home to school. As Pyke and Johnson show, gender is refracted through the particular social locations that race, class, *and* gender produce.

Pyke and Johnson's article is framed by a theoretical perspective known as "doing gender." *Doing gender* means that gender is not a fixed attribute of persons, but is instead accomplished through routine social interaction (West and Zimmerman 1987; West and Fenstermaker 1995). Put differently, gender is something that we "do," not just something we "are." We dress certain ways; we talk in gendered patterns of speech; we interact with others based on well-documented patterns of gendered behavior; we reward and value people differently based on their gender. "Doing gender" is built into the daily interactions and ideas that we have. This means that gender is a fluid social category—one that emerges through everyday interaction. Thus, gender is not somehow "fixed" in our bodies or our minds and, thus can be changed.

Although it is common for people to think of gender in binary categories—that is, as either woman or man, as new awareness of and new thinking about transpeople shows how fluid the social construction of gender can be. Joelle Ruby Ryan explains this in her personal narrative "From Transgender to Trans." Ryan describes her personal journey of coming out as a transgender person. She writes about the affirmation she felt as she learned that gender identities do not have to fit within binary categories—as if differently gendered people are opposites of each other. She also shows how class and race privilege certain transgender people. In the end, she asks her readers to challenge the gender binary system in order to bring greater gender and sexual freedom to all.

Understanding gender as a social construction, as Ryan argues, debunks the notion of gender as a fixed identity. When you understand this, you will also see that women are not the only gendered subjects. Men are gendered too, and in particular ways, given their class and race status. Thus, not all men benefit equally from gender privilege, just as not all women are disadvantaged in the same way. To see this all you need to do is think about the experiences of men of color, disabled men, gay men, and working-class men.

Aida Hurtado and Mrinal Sinha ("More than Men: Latino Feminist Masculinities and Intersectionality") show that men are not a monolithic group. Hurtado and Sinha's interviews with working-class Latinos reveal the complex ways that Latinos define masculinity and how they identify as feminists. **Hegemonic masculinity** is the dominant form of masculinity that permeates social institutions. But, as Hurtado and Sinha show in their research, some men challenge dominant systems of masculinity by defining manhood in ways that are specific to their race and gender identities. Many Latinos reject hegemonic masculinity because their understandings of manhood reflect their standing in a system structured by racial, class, and heterosexual privilege.

As you come to understand gender as a social and fluid category, remember that its effects can still be punishing. You can witness this in the persistent discrimination that women, and women of color especially, experience in every social institution, including in the high rates of violence against women even on college campuses. Women of color also feel the impact of gender and its interaction with race in their everyday interactions within institutions as Marlese Durr and Adia M. Harvey Wingfield show ("Keep Your 'N' in Check: African American Women and the Interactive Effects of Etiquette and Emotional Labor"). By doing research about professional African American women in the workplace, Durr and Wingfield document the many ways that African American women have to resist the controlling images that others have of them and the microaggressions that are a daily feature of their lives at work.

Altogether, the articles in this section show that gender operates as a system of power that is found in individual lives, but is anchored in social structure. That social structure is formed not by gender alone, but within a hierarchical system of race, class, and gender relationships—as well as sexuality, as we will see in the next section.

SEXUALITY

The linkage between race, class, and gender is also revealed within studies of sexuality, just as sexuality is a dimension of each. For example, constructing images about Black sexuality is central to maintaining institutional racism.

Similarly, beliefs about women's sexuality structure gender oppression. Sexuality operates as a system of power and inequality comparable to and intersecting with the systems of race, class, and gender.

The opening essay here by Patricia Hill Collins ("Prisons for Our Bodies, Closets for Our Minds: Racism, Heterosexism, and Black Sexuality") explores the connection between race, class, gender, and sexuality. Collins shows how racism has historically been buttressed by beliefs about Black sexuality. Sexuality has been used as the vehicle to support racial fears and racial subordination. Racial subordination was built on the exploitation of Black bodies—both as labor and as sexual objects. Strictures against certain interracial, sexual relationships (but not those between White men and African American women) are a way to maintain White racism and patriarchy. Collins also shows how stigmatizing the sexuality of African Americans can in some ways be likened to the experiences of lesbian, gay, bisexual, and transgendered people. This is not to say that racial and sexual oppression are the same, but it says that these systems of oppression operate together, producing social beliefs and social actions that oppress and exploit African American men and women in particular ways, while also oppressing sexual minorities.

Beliefs about sexuality are not free-floating ideas. They emerge within institutional structures, just as gender, race, and class ideas reside within institutions, not just individual minds. Moreover, dominant institutional structures are heterosexist. As Jonathan Ned Katz points out ("The Invention of Heterosexuality"), **heterosexism** is a specific historic construction. Its meaning and presumed significance have evolved and changed at distinct historical times. Heterosexism means that, like gendered institutions, institutions are also organized around beliefs and structures that define and enforce heterosexual behavior as the only natural and permissible form of sexual expression.

Heterosexism also relies on **homophobia**, defined as the fear and hatred of homosexuality. If only heterosexual forms of gender identity are labeled "normal," then gays, lesbians, transpeople, and bisexuals are ostracized, oppressed, and defined as "socially deviant." Such beliefs emerge only within an institutional system that affords power and privilege to presumed heterosexual behavior and identification.

Homophobia affects heterosexuals as well because it is part of the gender ideology used to distinguish "normal" men and women from those deemed deviant. Thus, young boys learn a rigid view of masculinity—one often associated with violence, bullying, and degrading others—to avoid being perceived as a "fag." The fag banter that boys use to insult each other is not necessarily directed just at gay boys or men. Rather, it is directed at men, especially adolescents, who are perceived as not conforming to the dominant norms of masculinity (Pascoe 2011). In this way, *fag* is a mechanism of social control—a controlling image that attempts to regulate the normative gender power structure. The hatred directed toward lesbians, gays, and bisexuals is thus part of the system by which gender is created and maintained. In this regard, sexuality and gender are deeply linked.

Just as "fag" is used to sustain dominant constructions of heterosexuality, Elizabeth A. Armstrong, Laura T. Hamilton, Elizabeth M. Armstrong, and J. Lotus Seely ("'Good Girls': Gender, Social Class, and Slut Discourse on Campus") show how the sexual label of "slut" buttresses a class system that privileges some to the disadvantage of others. Armstrong's research on sexuality on a college campus shows that sexuality does not exist outside the class system. By observing students' usage of the term slut, Armstrong and her colleagues show how the label "others" women who do not have the class privileges of other students. The label "slut," like the label "fag," does not actually correspond to sexual behavior, but it reinforces the class system that is embedded in educational institutions. Their discussion shows that sexuality is not just behavior, but is part of an institutional system that interlocks sexuality with class, race, and gender inequality.

Queer theory destabilizes the thinking that buttresses heterosexism. **Queer theory** is the idea we can subvert fixed gender and sexual identities by challenging them as fixed categories. Sexual identity is often thought of in binary terms—as if one must be either gay or straight, male or female. Queer theory teaches us otherwise. By "queering" the either/or construction of gender and sexuality, queer theory opens up the possibility of multiple, perhaps even conflicting, definitions of sexual and gendered identities.

Donovan Lessard explores this in an interesting way in "Queering the Sexual and Racial Politics of Urban Revitalization." Lessard shows that even at a time when many public attitudes are becoming more accepting of LGBTQ people, not all LGBTQ people are equally tolerated or valued. He makes his case by comparing two urban spaces where LGBTQ youth congregate—one a predominantly White queer space for the arts, the other populated by poor and working-class LGBTQ youth in a blighted neighborhood. He shows that some LGBTQ people become more valued citizens because of their position in the intersectional system of race, class, gender, and sexuality. Working-class and poor LGBTQ youth of color experience more policing and social control. Lessard also debunks the idea that gay people's movement into blighted urban neighborhoods revitalizes that space and increases property values as he disentangles the influence of race and class in the experiences of LGBTQ youth.

Lessard's research reminds us again that sexuality, gender, race, ethnicity, and class are more than individual identities. They are fundamental social structures that frame institutions and dominant belief systems. The cost to different people cannot then be simply noted through one or the other, but must acknowledge the overlapping and intersecting influence of these social factors altogether.

REFERENCES

Acker, Joan. 1992. "Gendered Institutions: From Sex Roles to Gendered Institutions." *Contemporary Sociology* 21 (September): 565–569.

Andersen, Margaret L. 2017. *Race in Society: The Enduring American Dilemma.* Lanham, MD: Rowman and Littlefield.

Andersen, Margaret L., and Howard F. Taylor. 2019. *Sociology: The Essentials,* 9th ed. San Francisco: Cengage.

Bureau of Labor Statistics. 2018. *Employment and Earnings.* Washington, DC: U.S. Department of Labor. www.bls.gov

Collins, Patricia Hill. 2004. *Black Sexual Politics: African Americans, Gender, and the New Racism.* New York: Routledge.

Collins, Patricia Hill. 2000. *Black Feminist Thought: Knowledge, Consciousness, and the Politics of Empowerment.* New York: Routledge.

Economic Mobility Report. 2012. *Pursuing the American Dream: Economic Mobility across Generations.* Washington, DC: Pew Charitable Trust. www.pewtrusts.org

Federal Bureau of Investigation. 2018. *2018 Hate Crime Statistics.* Washington, DC: Federal Bureau of Investigation. www.fbi.gov

Frye, Marilyn. 1983. *The Politics of Reality.* Trumansburg, NY: The Crossing Press.

Glenn, Evelyn Nakano. 2002. *Unequal Freedom: How Race and Gender Shaped American Citizenship and Labor.* Cambridge, MA: Harvard University Press.

Guo, Jeff. 2016. "The Staggering Difference between Rich Asian Americans and Poor Asian Americans." *The Washington Post,* December 20. www.washingtonpost.com

McKernan, Signe-Mary, Caroline Ratcliffe, C. Eugene Steuerle, Emma Kalish, and Caleb Quakenbush. 2015. *Nine Charts about Wealth Inequality in America.* Washington, DC: The Urban Institute.

Mishel, Lawrence, Josh Bivens, Elise Gould, and Heidi Shierholz. 2012. *The State of Working America,* 12th ed. Ithaca, NY: Cornell University Press.

Oliver, Melvin L., and Thomas M. Shapiro. 2006. *Black Wealth/White Wealth: A New Perspective on Racial Inequality,* 2nd ed. New York: Routledge.

Omi, Michael, and Howard Winant. 1994. *Racial Formation in the United States: From the 1960s to the 1990s,* 2nd ed. New York: Routledge.

Pascoe, C. J. 2008. *Dude, You're a Fag: Masculinity and Sexuality in High School.* Berkeley: University of California Press.

Pew Research Center. 2010. "Millennials Judgments about Recent Trends Not So Different." Washington, DC: Pew Research Center. www.pewresearch.org

Potok, Mark. 2017. "The Year in Hate and Extremism." *Intelligence Report* (February), Montgomery, AL: Southern Poverty Law Center. www.splcenter.org

Rodriguez, Clara. 2000. *Changing Race: Latinos, the Census, and the History of Ethnicity.* New York: New York University Press.

Scott-Clayton, Judith, and Jing Li. 2016. *Black-White Disparity in Student Loan Debt More Than Triples after Graduation.* Washington, DC: The Brookings Institute. https://www.brookings.edu/research/black-white-disparity-in-student-loan-debt-more-than-triples-after-graduation/

West, Candace, and Sarah Fenstermaker. 1995. "Doing Difference." *Gender & Society* 9 (February): 8–37.

West, Candace, and Don Zimmerman. 1987. "Doing Gender." *Gender & Society* 1 (June): 125–151.

5

Racial Formation

MICHAEL OMI AND HOWARD WINANT

The concept of racial formation, first developed by Omi and Winant, has become central to the sociological study of race. It refers to the social and historical processes by which groups come to be defined in racial terms, and it specifically locates those processes in state-based institutions, such as the law.

In 1982–83, Susie Guillory Phipps unsuccessfully sued the Louisiana Bureau of Vital Records to change her racial classification from black to white. The descendant of an 18th-century white planter and a black slave, Phipps was designated "black" in her birth certificate in accordance with a 1970 state law which declared anyone with at least 1/32nd "Negro blood" to be black.

The Phipps case raised intriguing questions about the concept of race, its meaning in contemporary society, and its use (and abuse) in public policy. Assistant Attorney General Ron Davis defended the law by pointing out that some type of racial classification was necessary to comply with federal recordkeeping requirements and to facilitate programs for the prevention of genetic diseases. Phipps's attorney, Brian Begue, argued that the assignment of racial categories on birth certificates was unconstitutional and that the 1/32nd designation was inaccurate. He called on a retired Tulane University professor who cited research indicating that most Louisiana whites have at least 1/20th "Negro" ancestry.

In the end, Phipps lost. The court upheld the state's right to classify and quantify racial identity. . . .

Phipps's problematic racial identity, and her effort to resolve it through state action, is in many ways a parable of America's unsolved racial dilemma. It illustrates the difficulties of defining race and assigning individuals or groups to racial categories. It shows how the racial legacies of the past—slavery and bigotry—continue to shape the present. It reveals both the deep involvement of the state in the organization and interpretation of race, and the inadequacy of state institutions to carry out these functions. It demonstrates how deeply Americans both as individuals and as a civilization are shaped, and indeed haunted, by race.

Having lived her whole life thinking that she was white, Phipps suddenly discovers that by legal definition she is not. In U.S. society, such an event is indeed catastrophic. But if she is not white, of what race is she? The state claims that she is black, based on its rules of classification . . . and another state agency, the court, upholds this judgment. But despite these classificatory standards which have imposed an either-or logic on racial identity, Phipps will not in fact "change color." Unlike what would have happened during slavery times if one's claim to whiteness was successfully challenged, we can assume that despite the outcome of her legal challenge, Phipps will remain in most of the social relationships she had occupied before the trial. Her socialization, her familial and friendship networks, her cultural orientation, will not change. She will simply have to wrestle with her newly acquired "hybridized" condition. She will have to confront the "Other" within.

The designation of racial categories and the determination of racial identity is no simple task. For centuries, this question has precipitated intense debates and conflicts, particularly in the United States—disputes over natural and legal rights, over the distribution of resources, and indeed, over who shall live and who shall die.

A crucial dimension of the Phipps case is that it illustrates the inadequacy of claims that race is a mere matter of variations in human physiognomy, that it is simply a matter of skin color. But if race cannot be understood in this manner, how can it be understood? We cannot fully hope to address this topic—no less than the meaning of race, its role in society, and the forces which shape it—in one [article], nor indeed in one book. Our goal in this [article], however, is far from modest: we wish to offer at least the outlines of a theory of race and racism.

WHAT IS RACE?

There is a continuous temptation to think of race as an essence, as something fixed, concrete, and objective. And there is also an opposite temptation: to imagine race as a mere illusion, a purely ideological construct which some ideal non-racist social order would eliminate. It is necessary to challenge both these positions, to disrupt and reframe the rigid and bipolar manner in which they are posed and debated, and to transcend the presumably irreconcilable relationship between them.

The effort must be made to understand race as an unstable and "decentered" complex of social meanings constantly being transformed by political struggle. With this in mind, let us propose a definition: race is a concept which signifies and symbolizes social conflicts and interests by referring to different types of human bodies. Although the concept of race invokes biologically based human characteristics (so-called "phenotypes"), selection of these particular human features for purposes of racial signification is always and necessarily a social and historical process. In contrast to the other major distinction of this type, that of gender, there is no biological basis for distinguishing among human groups along the lines of race.... Indeed, the categories employed to differentiate among human groups along racial lines reveal themselves, upon serious examination, to be at best imprecise, and at worst completely arbitrary.

If the concept of race is so nebulous, can we not dispense with it? Can we not "do without" race, at least in the "enlightened" present? This question has been posed often, and with greater frequency in recent years. . . . An affirmative answer would of course present obvious practical difficulties: it is rather difficult to jettison widely held beliefs, beliefs which moreover are central to everyone's identity and understanding of the social world. So the attempt to banish the concept as an archaism is at best counterintuitive. But a deeper difficulty, we believe, is inherent in the very formulation of this schema, in its way of posing race as a *problem*, a misconception left over from the past, and suitable now only for the dustbin of history.

A more effective starting point is the recognition that, despite its uncertainties and contradictions, the concept of race continues to play a fundamental role in structuring and representing the social world. The task for theory is to explain this situation. It is to avoid both the utopian framework which sees race as an illusion we can somehow "get beyond," and also the essentialist formulation which sees race as something objective and fixed, a biological datum. Thus, we should think of race as an element of social structure rather than as an irregularity within it; we should see race as a dimension of human representation rather than an illusion. These perspectives inform the theoretical approach we call racial formation.

Racial Formation

We define *racial formation* as the sociohistorical process by which racial categories are created, inhabited, transformed, and destroyed. Our attempt to elaborate a theory of racial formation will proceed in two steps. First, we argue that racial formation is a process of historically situated projects in which human bodies and social structures are represented and organized. Next we link racial formation to the evolution of hegemony, the way in which society is organized and ruled. Such an approach, we believe, can facilitate understanding of a whole range of contemporary controversies and dilemmas involving race, including the nature of racism, the relationship of race to other forms of differences, inequalities, and oppression such as sexism and nationalism, and the dilemmas of racial identity today.

From a racial formation perspective, race is a matter of both social structure and cultural representation. Too often, the attempt is made to understand race simply or primarily in terms of only one of these two analytical dimensions. . . . For example, efforts to explain racial inequality as a purely social structural phenomenon are unable to account for the origins, patterning, and transformation of racial difference.

Conversely, many examinations of racial difference—understood as a matter of cultural attributes à la ethnicity theory, or as a society-wide signification system, à la some poststructuralist accounts—cannot comprehend such structural phenomena as racial stratification in the labor market or patterns of residential segregation.

An alternative approach is to think of racial formation processes as occurring through a linkage between structure and representation. *Racial projects do the ideological "work" of making these links. A racial project is simultaneously an interpretation, representation, or explanation of racial dynamics, and an effort to recognize and redistribute resources along particular racial lines.* Racial projects connect what race means in a particular discursive practice and the ways in which both social structures and everyday experiences are racially *organized,* based upon that meaning. Let us consider this proposition, first in terms of large-scale or macro-level social processes, and then in terms of other dimensions of the racial formation process.

Racial Formation as a Macro-Level Social Process

To interpret the meaning of race is to frame it social structurally. Consider, for example, this statement by Charles Murray on welfare reform:

> My proposal for dealing with the racial issue in social welfare is to repeal
> every bit of legislation and reverse every court decision that in any way
> requires, recommends, or awards differential treatment according to
> race, and thereby put us back onto the track that we left in 1965. We
> may argue about the appropriate limits of government intervention in
> trying to enforce the ideal, but at least it should be possible to identify
> the ideal: Race is not a morally admissible reason for treating one person
> differently from another. Period. . . .

Here there is a partial but significant analysis of the meaning of race: it is not a morally valid basis upon which to treat people "differently from one another." We may notice someone's race, but we cannot act upon that awareness. We must act in a "color-blind" fashion. This analysis of the meaning of race is immediately linked to a specific conception of the role of race in the social structure: it can play no part in government action, save in "the enforcement of the ideal." No state policy can legitimately require, recommend, or award different status according to race. This example can be classified as a particular type of racial project in the present-day United States—a "neoconservative" one.

Conversely, *to recognize the racial dimension in social structure is to interpret the meaning of race.* Consider the following statement by the late Supreme Court Justice Thurgood Marshall on minority "set-aside" programs:

> A profound difference separates governmental actions that themselves
> are racist, and governmental actions that seek to remedy the effects of
> prior racism or to prevent neutral government activity from perpetuat-
> ing the effects of such racism. . . .

Here the focus is on the racial dimensions of social structure—in this case of state activity and policy. The argument is that state actions in the past and present have treated people in very different ways according to their race, and thus the government cannot retreat from its policy responsibilities in this area. It cannot suddenly declare itself "color-blind" without in fact perpetuating the same type of differential, racist treatment. . . . Thus, race continues to signify difference and structure inequality. Here, racialized social structure is immediately linked to an interpretation of the meaning of race. This example too can be classified as a particular type of racial project in the present-day United States—a "liberal" one.

To be sure, such political labels as "neoconservative" or "liberal" cannot fully capture the complexity of racial projects, for these are always multiply determined, politically contested, and deeply shaped by their historical context. Thus encapsulated within the neoconservative example cited here are certain egalitarian commitments which derive from a previous historical context in which they played a very different role, and which are rearticulated in neoconservative racial discourse precisely to oppose a more open-ended, more capacious conception of the meaning of equality. Similarly, in the liberal example, Justice Marshall recognizes that the contemporary state, which was formerly the architect of segregation and the chief enforcer of racial difference, has a tendency to reproduce those patterns of inequality in a new guise. Thus, he admonishes it (in dissent, significantly) to fulfill its responsibilities to uphold a robust conception of equality. These particular instances, then, demonstrate how racial projects are always concretely framed, and thus are always contested and unstable. The social structures they uphold or attack, and the representations of race they articulate, are never invented out of the air, but exist in a definite historical context, having descended from previous conflicts. This contestation appears to be permanent in respect to race.

These two examples of contemporary racial projects are drawn from mainstream political debate; they may be characterized as center-right and center-left expressions of contemporary racial politics. . . . . We can, however, expand the discussion of racial formation processes far beyond these familiar examples. In fact, we can identify racial projects in at least three other analytical dimensions: first, the political spectrum can be broadened to include radical projects, on both the left and right, as well as along other political axes. Second, analysis of racial projects can take place not only at the macro-level of racial policy-making, state activity, and collective action, but also at the micro-level of everyday experience. Third, the concept of racial projects can be applied across historical time, to identify racial formation dynamics in the past.

DISCUSSION QUESTIONS

1. What do Omi and Winant mean by *racial formation?* What role does the law play in such a process? How is this shown in the history of the United States?

2. What difference does it make to conceptualize race as a property of social structures versus as a property (or attribute) of individuals?

6

Color-Blind Privilege

The Social and Political Functions of Erasing the Color Line in Post-Race America

CHARLES A. GALLAGHER

The young white male sporting a FUBU (African-American owned apparel company "For Us By Us") shirt and his white friend with the tightly set, perfectly braided cornrows blended seamlessly into the festivities at an all white bar mitzvah celebration. A black model dressed in yachting attire peddles a New England, yuppie boating look in Nautica advertisements. It is quite unremarkable to observe white, Asian or African Americans with dyed purple, blond or red hair. White, black and Asian students decorate their bodies with tattoos of Chinese characters and symbols. In cities and suburbs young adults across the color line wear hip-hop clothing and listen to white rapper Eminem and black rapper 50-cent. It went almost unnoticed when a north Georgia branch of the NAACP installed a white biology professor as its president. Subversive musical talents like Jimi Hendrix, Bob Marley and The Who are now used to sell Apple Computers, designer shoes and SUVs. Du-Rag kits, complete with bandana headscarf and elastic headband, are on sale for $2.95 at hip-hop clothing stores and family centered theme parks like Six Flags. Salsa has replaced ketchup as the best selling condiment in the United States. Companies as diverse as Polo, McDonalds, Tommy Hilfiger, Walt Disney World, Master Card, Skechers sneakers, IBM, Giorgio Armani and Neosporin antibiotic ointment have each crafted advertisements that show an integrated, multiracial cast of characters interacting and consuming their products in [a] post-race, color-blind world.

Americans are constantly bombarded by depictions of race relations in the media, which suggests that discriminatory racial barriers have been dismantled. Social and cultural indicators suggest that America is on the verge, or has already become, a truly color-blind nation. National polling data indicate that a majority of whites now believe discrimination against racial minorities no longer exists. A majority of whites believe that blacks have "as good a chance as whites" in procuring housing and employment or achieving middle class status while a 1995 survey of white adults found that a majority of whites (58%) believed that African Americans were "better off" finding jobs than whites (Gallup, 1997; Shipler, 1998). Much of white America now see[s] a level playing field, while a majority of black

SOURCE: Gallagher, Charles A. 2003. *Race, Gender & Class* 10: 22–37. Reprinted by permission of the author.

Americans sees a field [that] is still quite uneven.... The colorblind or race neutral perspective holds that in an environment where institutional racism and discrimination have been replaced by equal opportunity, one's qualifications, not one's color or ethnicity, should be the mechanism by which upward mobility is achieved. Color as a cultural style may be expressed and consumed through music, dress or vernacular but race as a system which confers privileges and shapes life chances is viewed as an atavistic and inaccurate accounting of U.S. race relations.

Not surprisingly, this view of society blind to color is not equally shared. Whites and blacks differ significantly, however, on their support for affirmative action, the perceived fairness of the criminal justice system, the ability to acquire the "American Dream," and the extent to which whites have benefited from past discrimination (Moore, 1995; Moore & Saad, 1995; Kaiser Foundation, 1995). This article examines the social and political functions colorblindness serves for whites in the United States. Drawing on interviews and focus groups with whites from around the country I argue that color-blind depictions of U.S. race relations serves [sic] to maintain white privilege by negating racial inequality. Embracing a color-blind perspective reinforces whites' belief that being white or black or brown has no bearing on an individual's or a group's relative place in the socioeconomic hierarchy.

DATA AND METHOD

I use data from seventeen focus groups and thirty individual interviews with whites from around the country. Thirteen of the seventeen focus groups were conducted in a college or university setting, five in a liberal arts college in the Rocky Mountains and the remaining eight at a large urban university in the Northeast. Respondents in these focus groups were selected randomly from the student population. Each focus group averaged six respondents ... equally divided between males and females. An overwhelming majority of these respondents were between the ages of eighteen and twenty-two years of age. The remaining four focus groups took place in two rural counties in Georgia and were obtained through contacts from educational and social service providers in each county. One county was almost entirely white (99.54%) and in the other county whites constituted a racial minority. These four focus groups allowed me to tap rural attitudes about race relations in environments where whites had little or consistent contact with racial minorities....

COLORBLINDNESS AS NORMATIVE IDEOLOGY

The perception among a majority of white Americans that the socio-economic playing field is now level, along with whites' belief that they have purged themselves of overt racist attitudes and behaviors, has made colorblindness the dominant lens through which whites understand contemporary race relations.

Colorblindness allows whites to believe that segregation and discrimination are no longer [an] issue because it is now illegal for individuals to be denied access to housing, public accommodations or jobs because of their race. Indeed, lawsuits alleging institutional racism against companies like Texaco, Denny's, Coke, and Cracker Barrel validate what many whites know at a visceral level is true: firms which deviate from the color-blind norms embedded in classic liberalism will be punished. As a political ideology, the commodification and mass marketing of products that signify color but are intended for consumption across the color line further legitimate colorblindness. Almost every household in the United States has a television that, according to the U.S. Census, is on for seven hours every day (Nielsen 1997). Individuals from any racial background can wear hip-hop clothing, listen to rap music (both purchased at Wal-Mart) and root for their favorite, majority black, professional sports team. Within the context of racial symbols that are bought and sold in the market, colorblindness means that one's race has no bearing on who can purchase a Jaguar, live in an exclusive neighborhood, attend private schools or own a Rolex.

The passive interaction whites have with people of color through the media creates the impression that little, if any, socio-economic difference exists between the races....

Highly visible and successful racial minorities like [former] Secretary of State Colin Powell and ... [Secretary of State] Condelleeza Rice are further proof to white America that the state's efforts to enforce and promote racial equality have been accomplished.

The new color-blind ideology does not, however, ignore race; it acknowledges race while disregarding racial hierarchy by taking racially coded styles and products and reducing these symbols to commodities or experiences that whites and racial minorities can purchase and share. It is through such acts of shared consumption that race becomes nothing more than an innocuous cultural signifier. Large corporations have made American culture more homogenous through the ubiquitousness of fast food, television, and shopping malls but this trend has also created the illusion that we are all the same through consumption. Most adults eat at national fast food chains like McDonalds, shop at mall anchor stores like Sears and J.C. Penney's and watch major league sports, situation comedies or television drama. Defining race only as cultural symbols that are for sale allows whites to experience and view race as nothing more than a benign cultural marker that has been stripped of all forms of institutional, discriminatory or coercive power. The post-race, color-blind perspective allows whites to imagine that depictions of racial minorities working in high status jobs and consuming the same products, or at least appearing in commercials for products whites desire or consume, is the same as living in a society where color is no longer used to allocate resources or shape group outcomes. By constructing a picture of society where racial harmony is the norm, the color-blind perspective functions to make white privilege invisible while removing from public discussion the need to maintain any social programs that are race-based.

How then, is colorblindness linked to privilege? Starting with the deeply held belief that America is now a meritocracy, whites are able to imagine that the socio-economic success they enjoy relative to racial minorities is a function of individual hard work, determination, thrift and investments in education. The color-blind perspective removes from personal thought and public discussion any taint or suggestion of white supremacy or white guilt while legitimating the existing social, political and economic arrangements which privilege whites. This perspective insinuates that class and culture, and not institutional racism, are responsible for social inequality. Colorblindness allows whites to define them- selves as politically and racially tolerant as they proclaim their adherence to a belief system that does not see or judge individuals by the "color of their skin." This perspective ignores, as Ruth Frankenberg puts it, how whiteness is a "loca- tion of structural advantage" (2001, p. 1).... Colorblindness hides white privilege behind a mask of assumed meritocracy while rendering invisible the institutional arrangements that perpetuate racial inequality. The veneer of equality implied in colorblindness allows whites to present their place in the racialized social struc- ture as one that was earned.

OPPORTUNITY HAS NO COLOR

Given this norm of colorblindness it was not surprising that respondents in this study believed that using race to promote group interests was a form of (reverse) racism....

Believing and acting as if America is now color-blind allows whites to imag- ine a society where institutional racism no longer exists and racial barriers to upward mobility have been removed. The use of group identity to challenge the existing racial order by making demands for the amelioration of racial inequi- ties is viewed as racist because such claims violate the belief that we are a nation that recognizes the rights of individuals not rights demanded by groups....

The logic inherent in the color-blind approach is circular; since race no longer shapes life chances in a color-blind world there is no need to take race into account when discussing differences in outcomes between racial groups. This approach erases America's racial hierarchy by implying that social, economic and political power and mobility is [*sic*] equally shared among all racial groups. Ignoring the extent or ways in which race shapes life chances validates whites' social location in the existing racial hierarchy while legitimating the political and economic arrangements that perpetuate and reproduce racial inequality and privilege.

REFERENCES

Frankenberg, R. (2001). The mirage of an unmarked whiteness. In B. B. Rasmussen, E. Klineberg, I. J. Nexica & M. Wray (eds.) *The making and unmaking of whiteness.* Durham: Duke University Press.

Gallup Organization. (1997). *Black/white relations in the U.S.* June 10, pp. 1–5.

Kaiser Foundation. (1995). *The four Americas: Government and social policy through the eyes of America's multi-racial and multi-ethnic society.* Menlo Park, CA: Kaiser Family Foundation.

Moore, D. (1995). "Americans' most important sources of information: Local news." *The Gallup Poll Monthly*, September, pp. 2–5.

Moore, D. & Saad, L. (1995). No immediate signs that Simpson trial intensified racial animosity. *The Gallup Poll Monthly*, October, pp. 2–5.

Nielsen, A. C. (1997). *Information please almanac.* Boston: Houghton Mifflin.

Shipler, D. (1998). *A country of strangers: Blacks and whites in America.* New York: Vintage Books.

7

White Privilege

Unpacking the Invisible Knapsack

PEGGY MCINTOSH

Through work to bring materials from Women's Studies into the rest of the curriculum, I have often noticed men's unwillingness to grant that they are overprivileged, even though they may grant that women are disadvantaged. They may say they will work to improve women's status, in the society, the university, or the curriculum, but they can't or won't support the idea of lessening men's. Denials which amount to taboos surround the subject of advantages which men gain from women's disadvantages. These denials protect male privilege from being fully acknowledged, lessened, or ended.

Thinking through unacknowledged male privilege as a phenomenon, I realized that since hierarchies in our society are interlocking, there was most likely a phenomenon of white privilege which was similarly denied and protected. As a white person, I realized I had been taught about racism as something which puts others at a disadvantage, but had been taught not to see one of its corollary aspects, white privilege, which puts me at an advantage.

I think whites are carefully taught not to recognize white privilege, as males are taught not to recognize male privilege. So I have begun in an untutored way to ask what it is like to have white privilege. I have come to see white privilege as an invisible package of unearned assets which I can count on cashing in each day, but about which I was "meant" to remain oblivious. White privilege is like an invisible weightless knapsack of special provisions, maps, passports, codebooks, visas, clothes, tools, and blank checks.

Describing white privilege makes one newly accountable. As we in Women's Studies work to reveal male privilege and ask men to give up some of their power, so one who writes about having white privilege must ask, "Having described it, what will I do to lessen or end it?"

After I realized the extent to which men work from a base of unacknowledged privilege, I understood that much of their oppressiveness was unconscious. Then I remembered the frequent charges from women of color that white women whom they encounter are oppressive. I began to understand why we are justly seen as oppressive, even when we don't see ourselves that way. I began to count

SOURCE: McIntosh, Peggy. 1989. "White Privilege: Unpacking the Invisible Knapsack." *Peace and Freedom Magazine* (July/August): 10–12. Women's International League for Peace and Freedom, Philadelphia. Reprinted by permission of the author.

the ways in which I enjoy unearned skin privilege and have been conditioned into oblivion about its existence.

My schooling gave me no training in seeing myself as an oppressor, as an unfairly advantaged person, or as a participant in a damaged culture. I was taught to see myself as an individual whose moral state depended on her individual moral will. My schooling followed the pattern my colleague Elizabeth Minnich has pointed out: whites are taught to think of their lives as morally neutral, normative, and average, and also ideal, so that when we work to benefit others, this is seen as work which will allow "them" to be more like "us."

I decided to try to work on myself at least by identifying some of the daily effects of white privilege in my life. I have chosen those conditions which I think in my case *attach somewhat more to skin-color privilege* than to class, religion, ethnic status, or geographical location, though of course all these other factors are intricately intertwined. As far as I can see, my African American co-workers, friends and acquaintances with whom I come into daily or frequent contact in this particular time, place, and line of work cannot count on most of these conditions.

1. I can if I wish arrange to be in the company of people of my race most of the time.

2. If I should need to move, I can be pretty sure of renting or purchasing housing in an area which I can afford and in which I would want to live.

3. I can be pretty sure that my neighbors in such a location will be neutral or pleasant to me.

4. I can go shopping alone most of the time, pretty well assured that I will not be followed or harassed.

5. I can turn on the television or open to the front page of the paper and see people of my race widely represented.

6. When I am told about our national heritage or about "civilization," I am shown that people of my color made it what it is.

7. I can be sure that my children will be given curricular materials that testify to the existence of their race.

8. If I want to, I can be pretty sure of finding a publisher for this piece on white privilege.

9. I can go into a music shop and count on finding the music of my race represented, into a supermarket and find the staple foods which fit with my cultural traditions, into a hairdresser's shop and find someone who can cut my hair.

10. Whether I use checks, credit cards, or cash, I can count on my skin color not to work against the appearance of financial reliability.

11. I can arrange to protect my children most of the time from people who might not like them.

12. I can swear, or dress in secondhand clothes, or not answer letters, without having people attribute these choices to the bad morals, the poverty, or the illiteracy of my race.

13. I can speak in public to a powerful male group without putting my race on trial.

14. I can do well in a challenging situation without being called a credit to my race.

15. I am never asked to speak for all the people of my racial group.

16. I can remain oblivious of the language and customs of persons of color who constitute the world's majority without feeling in my culture any penalty for such oblivion.

17. I can criticize our government and talk about how much I fear its policies and behavior without being seen as a cultural outsider.

18. I can be pretty sure that if I ask to talk to "the person in charge," I will be facing a person of my race.

19. If a traffic cop pulls me over or if the IRS audits my tax return, I can be sure I haven't been singled out because of my race.

20. I can easily buy posters, postcards, picture books, greeting cards, dolls, toys, and children's magazines featuring people of my race.

21. I can go home from most meetings of organizations I belong to feeling somewhat tied in, rather than isolated, out-of-place, outnumbered, unheard, held at a distance, or feared.

22. I can take a job with an affirmative action employer without having co-workers on the job suspect that I got it because of my race.

23. I can choose public accommodation without fearing that people of my race cannot get in or will be mistreated in the places I have chosen.

24. I can be sure that if I need legal or medical help, my race will not work against me.

25. If my day, week, or year is going badly, I need not ask of each negative episode or situation whether it has racial overtones.

26. I can choose blemish cover or bandages in "flesh" color and have them more or less match my skin.

I repeatedly forgot each of the realizations on this list until I wrote it down. For me white privilege has turned out to be an elusive and fugitive subject. The pressure to avoid it is great, for in facing it I must give up the myth of meritocracy. If these things are true, this is not such a free country; one's life is not what one makes it; many doors open for certain people through no virtues of their own.

In unpacking this invisible knapsack of white privilege, I have listed conditions of daily experience which I once took for granted. Nor did I think of any of these perquisites as bad for the holder. I now think that we need a more finely differentiated taxonomy of privilege, for some of these varieties are only what one would want for everyone in a just society, and others give license to be ignorant, oblivious, arrogant and destructive.

I see a pattern running through the matrix of white privilege, a pattern of assumptions which were passed on to me as a white person. There was one main piece of cultural turf; it was my own turf, and I was among those who could control the turf. *My skin color was an asset for any move I was educated to want to make.* I could think of myself as belonging in major ways, and of making social systems work for me. I could freely disparage, fear, neglect, or be oblivious to anything outside of the dominant cultural forms. Being of the main culture, I could also criticize it fairly freely.

In proportion as my racial group was being made confident, comfortable, and oblivious, other groups were likely being made inconfident, uncomfortable, and alienated. Whiteness protected me from many kinds of hostility, distress, and violence, which I was being subtly trained to visit in turn upon people of color.

For this reason, the word "privilege" now seems to me misleading. We usually think of privilege as being a favored state, whether earned or conferred by birth or luck. Yet some of the conditions I have described here work to systematically overempower certain groups. Such privilege simply *confers dominance* because of one's race or sex.

I want, then, to distinguish between earned strength and unearned power conferred systemically. Power from unearned privilege can look like strength when it is in fact permission to escape or to dominate. But not all of the privileges on my list are inevitably damaging. Some, like the expectation that neighbors will be decent to you, or that your race will not count against you in court, should be the norm in a just society. Others, like the privilege to ignore less powerful people, distort the humanity of the holders as well as the ignored groups.

We might at least start by distinguishing between positive advantages which we can work to spread, and negative types of advantages which unless rejected will always reinforce our present hierarchies. For example, the feeling that one belongs within the human circle, as Native Americans say, should not be seen as privilege for a few. Ideally it is an *unearned entitlement*. At present, since only a few have it, it is an *unearned advantage* for them. This paper results from a process of coming to see that some of the power which I originally saw as attendant on being a human being in the U.S. consisted in *unearned advantage* and *conferred dominance.*

I have met very few men who are truly distressed about systemic, unearned male advantage and conferred dominance. And so one question for me and others like me is whether we will be like them, or whether we will get truly distressed, even outraged, about unearned race advantage and conferred dominance and if so, what we will do to lessen them. In any case, we need to do more work in identifying how they actually affect our daily lives. Many, perhaps most, of our white students in the U.S. think that racism doesn't affect them because they are not people of color; they do not see "whiteness" as a racial identity. In addition, since race and sex are not the only advantaging systems at work, we need similarly to examine the daily experience of having age advantage, or ethnic advantage, or physical ability, or advantage related to nationality, religion, or sexual orientation.

Difficulties and dangers surrounding the task of finding parallels are many. Since racism, sexism, and heterosexism are not the same, the advantaging associated with them should not be seen as the same. In addition, it is hard to disentangle aspects of unearned advantage which rest more on social class, economic class, race, religion, sex and ethnic identity than on other factors. Still, all of the oppressions are interlocking, as the Combahee River Collective Statement of 1977 continues to remind us eloquently.

One factor seems clear about all of the interlocking oppressions. They take both active forms which we can see and embedded forms which as a member of the dominant group one is taught not to see. In my class and place, I did not see myself as a racist because I was taught to recognize racism only in individual acts of meanness by members of my group, never in invisible systems conferring unsought racial dominance on my group from birth.

Disapproving of the systems won't be enough to change them. I was taught to think that racism could end if white individuals changed their attitudes. [But] a "white" skin in the United States opens many doors for whites whether or not we approve of the way dominance has been conferred on us. Individual acts can palliate, but cannot end, these problems.

To redesign social systems we need first to acknowledge their colossal unseen dimensions. The silences and denials surrounding privilege are the key political tool here. They keep the thinking about equality or equity incomplete, protecting unearned advantage and conferred dominance by making these taboo subjects. Most talk by whites about equal opportunity seems to me now to be about equal opportunity to try to get into a position of dominance while denying that *systems* of dominance exist.

It seems to me that obliviousness about white advantage, like obliviousness about male advantage, is kept strongly enculturated in the United States so as to maintain the myth of meritocracy, the myth that democratic choice is equally available to all. Keeping most people unaware that freedom of confident action is there for just a small number of people props up those in power, and serves to keep power in the hands of the same groups that have most of it already.

Though systemic change takes many decades, there are pressing questions for me and I imagine for some others like me if we raise our daily consciousness on the perquisites of being light-skinned. What will we do with such knowledge? As we know from watching men, it is an open question whether we will choose to use unearned advantage to weaken hidden systems of advantage, and whether we will use any of our arbitrarily-awarded power to try to reconstruct power systems on a broader base.

8

The Persistence of White Nationalism in America

JOE FEAGIN

Understanding recent White nationalist views, demonstrations, and riots requires historical and sociological perspectives. Although these White nationalists are often presented as deviant from "American values," they in fact often exhibit old values and perspectives that are common among many Whites across the country.

Consider our centuries-old White racist history and foundation. The United States was founded by leading slaveholders such as George Washington, James Madison, and Thomas Jefferson who operated from a White supremacist and White nationalist perspective in crafting founding documents such as the U.S. Constitution and in developing major political, economic, and legal institutions. Only three human lifetimes have passed since the ostensibly egalitarian Declaration of Independence, yet a document principally authored by the outspoken White supremacist slaveholder Thomas Jefferson. Two human lifetimes have passed since the 13th Amendment (1865) to the Constitution ended centuries of human slavery (about 60% of our history). Only one human lifetime has passed since White mobs enforced legal segregation with thousands of lynchings of Americans of color (another 23% of our history).

In that Jim Crow era millions of Whites supported the world's oldest terrorist group, the violent White nationalist KKK. Over early 20th century decades many White politicians belonged; Senate and House members were often elected with Klan help. Warren Harding was closely linked to the Klan before he was president. Across the country hundreds of White supremacist marches and violent actions—cross burnings, assaults, lynchings—targeted African, Jewish, and Catholic Americans. Only since the 1960s Civil Rights Act has the United States been officially "free" of legally imposed racial discrimination.

Understanding that the extreme oppression of slavery and Jim Crow constitutes all but 17% of our history is essential to comprehending why White racism today remains *extensive, foundational,* and *systemic.* By no means has enough time passed or enough effort been taken to eradicate the deep continuing impacts of centuries of systemic racism.

SOURCE: Feagin, Joe. 2017. "The Persistence of White Nationalism in America." *Contexts*, 17:(Summer). Republished with permission of SAGE Publications Inc.

Links between this racist past and the racist present have been obvious in recent decades, those filled with the growth and violent actions of organized White supremacists. The Southern Poverty Law Center has recently counted more than 500 U.S. "hate groups" with specifically White supremacist links, including Klan, neo-Nazi, skinhead, anti-immigrant, and anti-Semitic organizations. An estimated 300,000 Whites are active or passive supporters. Such groups have grown because of supremacists' obsession with the immigration of people of color and so-called "browning of America," as well as the election of a Black president in 2008.

Members of White supremacist groups are often implicated in White-racist-framed crimes. One recent report of the Southern Poverty Law Center (https://www.splcenter.org/) noted that the modest number of hate crime incidents reported to policing agencies in a given year is likely a fraction of the 195,000 such crimes estimated by experts to be perpetrated annually. In these crimes, African and Jewish Americans are often major targets, joined recently by immigrants of color.

Contemporary White nationalist groups have made substantial use of internet websites and social media to recruit supporters to White supremacist versions of the dominant U.S. White racial framing. These websites reveal direct ties to our old White-supremacist foundation. The "Stormfront" nationalist website—probably the largest with hundreds of thousands of users and millions of racist posts—has had a portrait labeled "Thomas Jefferson, White nationalist," with a Jefferson quote that if slavery ends, "the two races, equally free, cannot live in the same government." The Council of Europe has estimated there are about 4,000 hate-based websites worldwide, with 2,500 in the United States. Many spread a White-supremacist ideology and themes of White racial violence.

Additionally, many Whites, especially males, are attracted to White supremacist views by playing racist video games online and offline. Many popular video games reproduce for their players images of White dominance in U.S. society's racial framing and hierarchy. In addition, openly racist commentary utilized by White players of popular video games increases an acceptance of a White racist framing of society similar to that of White nationalist groups.

Another key to understanding the power and persistence of these centuries-old White nationalist groups is that some major aspects of their racist framing of society is not significantly different from that of Whites unsupportive of extreme nationalism. Most ordinary Whites, surveys indicate, hold numerous White-racist views of U.S. racial matters, including their views on African Americans. One Pew survey found that more than 80% of Black respondents reported widespread racial discrimination in at least one societal area. In contrast, a majority of Whites in that poll denied there was significant anti-Black discrimination. A CNN/ORC poll found 61% of Black respondents felt that anti-Black discrimination was serious in their local area, compared to 25% of Whites. Leslie Picca and I have reported on brief semester diaries that 626 White students at numerous colleges and universities kept on racial events they saw around them. Most of these well-educated young Whites recorded racist events, discussions, and performances they encountered—altogether about 7,500 instances of clearly racist events. African Americans appear in a substantial majority of often blatantly racist commentaries, jokes, and performances reported by White students across the country. Moreover, in online research involving interviews with mostly college-educated White men in several regions, sociologist Brittany Slatton found that most openly expressed a negative and racist framing of Black women, especially of their bodies and lifestyles.

Clearly, much historical and contemporary research evidence demonstrates that the White-racist views expressed by White nationalists in current publications and demonstrations are held by many other Whites. Today's White nationalists are in many ways *not* radically deviant from the deep reality of White supremacy on which this country was founded, and which is still evident in many areas of U.S. society beyond extremist White nationalism.

9

What White Supremacists Taught a Jewish Scholar about Identity

ABBY L. FERBER

A few years ago, my work on white supremacy led me to the neo-Nazi tract *The New Order,* which proclaims: "The single serious enemy facing the white man is the Jew." I must have read that statement a dozen times. Until then, I hadn't thought of myself as the enemy.

When I began my research for a book on race, gender, and white supremacy, I could not understand why white supremacists so feared and hated Jews. But after being immersed in newsletters and periodicals for months, I learned that white supremacists imagine Jews as the masterminds behind a great plot to mix races and, thereby, to wipe the white race out of existence.

The identity of white supremacists, and the white racial purity they espouse, requires the maintenance of secure boundaries. For that reason, the literature I read described interracial sex as "the ultimate abomination." White supremacists see Jews as threats to racial purity, the villains responsible for desegregation, integration, the civil-rights movement, the women's movement, and affirmative action—each depicted as eventually leading white women into the beds of black men. Jews are believed to be in control everywhere, staging a multi-pronged attack against the white race. For *WAR,* the newsletter of White Aryan Resistance, the Jew "promotes a thousand social ills ... [f]or which you'll have to foot the bills."

Reading white-supremacist literature is a profoundly disturbing experience, and even more difficult if you are one of those targeted for elimination. Yet, as a Jewish woman, I found my research to be unsettling in unexpected ways. I had not imagined that it would involve so much self-reflection. I knew white supremacists were vehemently anti-Semitic, but I was ambivalent about my Jewish identity and did not see it as essential to who I was. Having grown up in a large Jewish community, and then having attended a college with a large Jewish enrollment, my Jewishness was invisible to me—something I mostly ignored. As I soon learned, to white supremacists, that is irrelevant.

Contemporary white supremacists define Jews as non-white: "not a religion, they are an Asiatic *race,* locked in a mortal conflict with Aryan man," according to *The New Order.* In fact, throughout white-supremacist tracts, Jews are

SOURCE: Ferber, Abby L. From *The Chronicle of Higher Education,* May 7, 1999, pp. B6–B7. Reprinted with permission of the author.

described not merely as a separate race, but as an impure race, the product of mongrelization. Jews, who pose the ultimate threat to racial boundaries, are themselves imagined as the product of mixed-race unions.

Although self-examination was not my goal when I began, my research pushed me to explore the contradictions in my own racial identity. Intellectually, I knew that the meaning of race was not rooted in biology or genetics, but it was only through researching the white-supremacist movement that I gained a more personal understanding of the social construction of race. Reading white-supremacist literature, I moved between two worlds: one where I was white, another where I was the non-white seed of Satan; one where I was privileged, another where I was despised; one where I was safe and secure, the other where I was feared and thus marked for death.

According to white-supremacist ideology, I am so dangerous that I must be eliminated. Yet, when I put down the racist, anti-Semitic newsletters, leave my office, and walk outdoors, I am white.

Growing up white has meant growing up privileged. Sure, I learned about the historical persecutions of Jews, overheard the hushed references to distant relatives lost in the Holocaust. I knew of my grandmother's experiences with anti-Semitism as a child of the only Jewish family in a Catholic neighborhood. But those were just stories to me. Reading white supremacists finally made the history real.

While conducting my research, I was reminded of the first time I felt like an "other." Arriving in the late 1980s for the first day of graduate school in the Pacific Northwest, I was greeted by a senior graduate student with the welcome: "Oh, you're the Jewish one." It was a jarring remark, for it immediately set me apart. This must have been how my mother felt, I thought, when, a generation earlier, a college classmate had asked to see her horns. Having lived in predominantly Jewish communities, I had never experienced my Jewishness as "otherness." In fact, I did not even *feel* Jewish. Since moving out of my parents' home, I had not celebrated a Jewish holiday or set foot in a synagogue. So it felt particularly odd to be identified by this stranger as a Jew. At the time, I did not feel that the designation described who I was in any meaningful sense.

But whether or not I define myself as Jewish, I am constantly defined by others that way. Jewishness is not simply a religious designation that one may choose, as I once naïvely assumed. Whether or not I see myself as Jewish does not matter to white supremacists.

I've come to realize that my own experience with race reflects the larger historical picture for Jews. As whites, Jews today are certainly a privileged group in the United States. Yet the history of the Jewish experience demonstrates precisely what scholars mean when they say that race is a social construction.

At certain points in time, Jews have been defined as a non-white minority. Around the turn of the last century, they were considered a separate, inferior race, with a distinguishable biological identity justifying discrimination and even genocide. Today, Jews are generally considered white, and Jewishness is largely considered merely a religious or ethnic designation. Jews, along with other European ethnic groups, were welcomed into the category of "white" as beneficiaries of one of the largest affirmative-action programs in history—the 1944 GI Bill of Rights. Yet, when I read white-supremacist discourse, I am reminded that my ancestors were expelled from the dominant race, persecuted, and even killed.

Since conducting my research, having heard dozens of descriptions of the murders and mutilations of "race traitors" by white supremacists, I now carry with me the knowledge that there are many people out there who would still wish to see me dead. For a brief moment, I think that I can imagine what it must feel like to be a person of color in our society … but then I realize that, as a white person, I cannot begin to imagine that.

Jewishness has become both clearer and more ambiguous for me. And the questions I have encountered in thinking about Jewish identity highlight the central issues involved in studying race today. I teach a class on race and ethnicity, and usually, about midway through the course, students complain of confusion. They enter my course seeking answers to the most troubling and divisive questions of our time, and are disappointed when they discover only more questions. If race is not biological or genetic, what is it? Why, in some states, does it take just one black ancestor out of 32 to make a person legally black, yet those 31 white ancestors are not enough to make that person white? And, always, are Jews a race?

I have no simple answers. As Jewish history demonstrates, what is and is not a racial designation, and who is included within it, is unstable and changes over time—and that designation is always tied to power. We do not have to look far to find other examples: The Irish were also once considered non-white in the United States, and U.S. racial categories change with almost every census.

My prolonged encounter with the white-supremacist movement forced me to question not only my assumptions about Jewish identity, but also my assumptions about whiteness. Growing up "white," I felt raceless. As it is for most white people, my race was invisible to me. Reflecting the assumption of most research on race at the time, I saw race as something that shaped the lives of people of color—the victims of racism. We are not used to thinking about whiteness when we think about race. Consequently, white people like myself have failed to recognize the ways in which our own lives are shaped by race. It was not until others began identifying me as the Jew, the "other," that I began to explore race in my own life.

Ironically, that is the same phenomenon shaping the consciousness of white supremacists: They embrace their racial identity at the precise moment when they feel their privilege and power under attack. Whiteness historically has equaled power, and when that equation is threatened, their own whiteness becomes visible to many white people for the first time. Hence, white supremacists seek to make racial identity, racial hierarchies, and white power part of the natural order again. The notion that race is a social construct threatens that order. While it has become an academic commonplace to assert that race is socially constructed, the revelation is profoundly unsettling to many, especially those who benefit most from the constructs.

My research on hate groups not only opened the way for me to explore my own racial identity, but also provided insight into the question with which I began this essay: Why do white supremacists express such hatred and fear of Jews? This ambiguity in Jewish racial identity is precisely what white supremacists find so threatening. Jewish history reveals race as a social designation, rather than a God-given or genetic endowment. Jews blur the boundaries between whites and people of color, failing to fall securely on either side of the divide. And it is ambiguity that white supremacists fear most of all.

I find it especially ironic that, today, some strict Orthodox Jewish leaders also find that ambiguity threatening. Speaking out against the high rates of intermarriage among Jews and non-Jews, they issue dire warnings. Like white supremacists, they fear assaults on the integrity of the community and fight to secure its racial boundaries, defining Jewishness as biological and restricting it only to those with Jewish mothers. For both white supremacists and such Orthodox Jews, intermarriage is tantamount to genocide.

For me, the task is no longer to resolve the ambiguity, but to embrace it. My exploration of white-supremacist ideology has revealed just how subversive doing so can be: Reading white-supremacist discourse through the lens of Jewish experience has helped me toward new interpretations. White supremacy is not a movement just about hatred, but even more about fear: fear of the vulnerability and instability of white identity and privilege. For white supremacists, the central goal is to naturalize racial identity and hierarchy, to establish boundaries.

Both my own experience and Jewish history reveal that to be an impossible task. Embracing Jewish identity and history, with all their contradictions, has given me an empowering alternative to white-supremacist conceptions of race. I have found that eliminating ambivalence does not require eliminating ambiguity.

10

Must-See TV

South Asian Characterizations in American Popular Media

BHOOMI K. THAKORE

INTRODUCTION

Why are there so many Indians on TV all of a sudden?
–Nina Shen Rastogi (2010)

In June 2010, Rastogi pondered the above question in her article, "Beyond Apu: Why are there suddenly so many Indians on Television?" for the online magazine, *Slate*. The increasing number of Indians and South Asians... in American popular television has been hard not to notice.... For example, among *TV Guide's* top 15 television shows of 2010 (the year Rastogi wrote this article); four had a South Asian character or actor—*Community*, *Glee*, *The Good Wife*, and *Parks and Recreation*. Since then, more characters have entered the fold—including those in such shows as *The Big Bang Theory*, *Royal Pains*, *Outsourced*, and *The Mindy Project*.

As Rastogi (2010) noted, the popularity of the 2008 film *Slumdog Millionaire* helped propel Indians to become a noteworthy ethnic group in popular media. These days, South Asian characters tend to be presented as the minority alongside majority-white characters. South Asians are also a good stand-in for Arab and Muslim characters in this post-9/11 reality of fear....

Contemporary South Asian media characters tend to reflect the characteristics of one of the two South Asian demographic groups. In the 1960s and 1970s, highly educated South Asians were allowed to immigrate to the United States after the passage of the 1965 Immigration and Nationality Act.... Soon after that, less educated family members of these immigrants arrived to the United States and worked in service positions, including behind the counters of convenience stores, franchises, and motels.... To date, there has been relatively little scholarship addressing the reasons behind this increasing trend of South Asian characters in the media and, more specifically, the ways in which these characters

SOURCE: Thakore, Bhoomi K. 2014. "Must-See TV: South Asian Characterizations in American Popular Media." *Sociology Compass*, 8(2): 149–156.

have been created, written, and produced. In this review, I discuss these representations by bridging the gaps between discussions in the fields of immigration studies, race/ethnicity studies, and critical media studies. In my discussion, I identify the concept of "(ethnic) characterization" and illustrate how studying the media representations of this ethnic group are in line with the concerns of sociology....

SOUTH ASIAN IMMIGRATION AND ASSIMILATION

The immigration and assimilation experiences of South Asians are relevant when understanding their influence on the representation of South Asians in American media. Those ethnic characteristics that media producers (and most Americans) know as "South Asian characteristics" will be used in the media characterization of these characters. This is evident in those media examples of South Asian characters that rely on overt stereotypes of this group, such as convenience store clerks or cab drivers. Thus, it is important to understand how South Asians, as a new and growing demographic, have assimilated in their new society and negotiated their own hyphenated-American identity.

Like all other immigrant and ethnic groups before them, Indians and South Asians have experienced an uphill battle in conceptualizing their identity in the United States and claiming their place in the American racial hierarchy. After President Johnson signed the Immigration and Nationality Act in 1965 that allowed technical professionals from Asia to immigrate to the United States, most South Asians immigrated in order to achieve financial and professional success. Many were able to experience upward social mobility as a result of the educational capital they brought with them. Additionally, their success influenced the success of their children and proceeding generations.

Experiences of assimilation are particularly salient for the second-generation American born children of first-generation South Asian immigrants, many of whom are portrayed in American media, and also happen to be the actors of these media characters. As Alba and Nee (2003) argued, while immigrants of the late 20th century overall have equal chances for social success as compared to their non-immigrant counterparts, the experiences of these immigrants collectively are not always the same. Their experiences are influenced by the various forms of capital a particular first or second-generation immigrant possesses, and the extent to which that capital can be useful within economic and labor markets. These experiences of segmented assimilation explain not only differences in the social mobility of immigrant groups, but also the differences by individuals within an immigrant group.

Segmented assimilation, as it relates to the social mobility between first-generation parents and second-generation children is noteworthy for all post-1965 South Asian immigrant families in the United States today (e.g. Alba and Nee 2003; Haller et al. 2011; Zhou and Xiong 2007). Additionally, it is important to note that the characterizations of South Asians in American media do not occur in a vacuum, but are informed in large part by the extent to which they assimilate into American society. However, traditional theories of assimilation fail to take into account the everyday experiences of racism that occur in the labor market and throughout society. These experiences inform the level at which ethnic groups can integrate into their new society, which in turn inform their acceptance by mainstream (white) Americans. Both dynamics are influenced by ideologies inherent in the pre-existing U.S. racial hierarchy.

THE RACIALIZATION OF SOUTH ASIANS

In the history of the United States, race relations have been fluid in order to serve particular political or social interests. The extent to which immigrants are able to assimilate into mainstream American culture will influence their place within the American racial hierarchy. Their place within the American racial hierarchy will also determine their social success in American society. As I argue, racial perceptions play a significant role in the characterization of South Asians in the media.

Some contemporary race scholars have argued that the U.S. racial hierarchy is developing into a three-tiered system consisting of Whites at the top, Blacks at the bottom, and (South) Asians as honorary Whites in the middle (Bonilla-Silva 2004; Feagin 2001; Kim 1999). As Kim (1999) argued, Asian Americans are triangulated between Blacks and Whites in terms of perceived superiority but outside of both groups for their perceived foreignness. Tuan (1999) identified these physical differences as the "forever foreigner" syndrome that Asians in the United States are subjected to regardless of their immigrant or citizenship status. However, as Bonilla-Silva (2004) argued, light skin tone and high class can help "whiten" the position and experiences of South Asians in the United States. These dynamics are particularly influential when considering the "types" of South Asian characters found in American popular media.

Historically, immigrants and racial minorities (including South Asians) have been subjected to a fluid racial hierarchy in the United States. As Omi and Winant (1994) suggested, these racial formations are the result of social, political, and economic forces that determine the social status of racial and ethnic minorities. These statuses have formed over time and are dependent on various social and historical circumstances. Examples of such racial formations include everything from Jim Crow slavery to changes in immigration policy.

Racial formations are further influenced by the level of racialization that immigrants and minorities experience. Racialization is a process by which individuals are categorized into racial groups based on their physical appearance. Additionally, these racial categorizations are then used as units of analysis to explain social relations.... Bonilla-Silva (1997) develops upon the idea of racialization in his racialized social systems theory, which supposes that political, economic, and social structures are dependent on a racialized society and the racialization of individuals into it.

... The racialization of Indian and South Asian Americans tends to be a negative process, specifically through the perpetuation of those negative stereo-types and assumptions associated with this group.... While slavery and Jim Crow shaped the racialization of African-Americans in the United States, immigration and legislative policy have shaped the racialization of South Asians and determined the extent to which they compare to Whites in America.... Contemporary ideologies, including the post-9/11 rhetoric in America and the global West, have contributed to a racialization of South Asians that portray them in popular media as foreigners and "others." These dynamics are further intersected by overt racialized perceptions that use obvious differences in skin color, religion, and ethnicity as markers of difference....

... Whites "commit" racialization of South Asians through the negative, stereotypical, and secondary ways in which they perceive them. Not only are they seen as "others" in American society, but they are also, seen as "less than" in the American racial hierarchy.... Racialization exists separate from the segmented assimilation based on skin color and class to which South Asians are subjected. Racialization is a process imposed upon all Indians and South Asians as another way to maintain the perceptions of this group as outside of American norms. As Kibria (1998) suggested, the development of such hyphenated umbrella identities as "South Asian American" is a result of racialization and these racialized experiences.

South Asian media characterizations are informed by the degree at which South Asians are racialized in society. This is evident in the examples of South Asian characters that are characterized solely around overt stereotypes. As I argue in the next section, these stereotypes not only serve the purpose of maintaining the perception of South Asians as foreigners in society, but are used consciously by media producers in their characterizations and ultimately reflect how South Asians are already perceived in the United States.

SOUTH ASIAN CHARACTERIZATIONS IN AMERICAN POPULAR MEDIA AND SOCIETY

Racial and ethnic minorities have historically been stereotyped in American popular media.... While these representations have generally improved in recent years, insofar as there are significantly fewer examples of overt stereotypes, reflections of covert and subtle stereotypes remain. As I argue, the characterization of South Asians and other minorities in the media is informed by an intentional characterization that is dependent in large part on the racial ideologies that are reproduced in the representations. This is evident in the historical trajectory of South Asian characterizations in American popular media.

South Asian media characters began appearing sporadically in films throughout the 20th century.... These early examples generally consisted of savage Indians in India who were defeated by the White star and savior (e.g. Vera and Gordon 2003). What was unique about these early representations was the location of the story itself—most were represented as Indians in India. During the 1980s, Indian and South Asian characters began appearing sporadically as tertiary or non-speaking characters cast in an American, usually urban, environment. Examples of representations included the generic and stereotypical cab driver, convenient store owner, and high-achieving student.

On the one hand, this stereotype of South Asians as a low-level service employee runs counter to the historical realities of South Asian demographics in the United States. Immigrants who arrived from South Asia in the 1960s and 1970s were highly educated.... [Eighty-three percent] of Indian immigrants between 1966 and 1977 had backgrounds in the STEM fields, including approximately 20,000 science PhDs, 40,000 engineers, and 25,000 physicians. However, after these early migrants and their immediate families were settled, many chose to invest in franchises and small businesses, including fast food stores, convenience store, and motels.... Once these businesses became established, Indian American owners took advantage of family reunification immigration policies of the 1970s and 1980s to bring over relatives to work for them. As more and more extended family members were able to settle in the United States, these individuals with few skills took other blue collar jobs working in factories or driving taxi cabs.

Media producers in positions to create media characterizations do so based on their personal experiences. While they may have been less likely to run into South Asians who were scientists, professors, or even doctors, they were more likely to run into South Asians who were behind the counter of a local convenience store or driving their cab in an urban city. Additionally, such representations proved useful to them in the context of the stories they were producing—well-to-do White American characters encountering bumbling, foreign, South Asian immigrants working in jobs most identifiable to viewers, often with ensuing hilarity. These stereotypes proved useful for the story and for the characterization of South Asian characters. As critical media studies scholars argue, White media executives have total control over the major media outlets and consciously reproduce upper class ideologies, which in turn subjugate racial minorities....

In the early 21st century, there were more noteworthy examples of Indian and South Asian media characters that were cast in such roles as highly skilled scientists or medical professionals. It is difficult to identify what led to this change, but it is likely due to the increased awareness of the high economic capital possessed by South Asian Americans, thus identifying them as a group to be coveted by advertisers (the financiers of network television). These new representations were more in line with the "model minority" stereotype, which was originally used in the 1960s to characterize East Asians in the United States.... While it is assumed that the model minority stereotype is a positive one, many scholars have identified its problematic nature. The assumption that Asian Americans are the model minority presupposes that they experience no discrimination in the United States. In fact, South Asians are subjected to the same discriminatory experiences of not being White as are other ethnic minorities in the United States.

One key example of such discrimination is through skin tone, particularly for women. All women of color deal with hegemonic skin tone ideologies in their racial/ethnic communities, with lighter skin tone and Caucasian facial features considered more appealing and attractive…. These same beauty ideals are also reproduced in the media…. This is evident in the examples of South Asian women in the media, particularly those created and cast by White, American producers…. As media producers favor casting women who are attractive, so too do the same media producers favor casting women of color who are attractive in terms of their proximity to White physical characteristics. Not only is this another reproduction of hegemonic ideology that favors one particular type of physical appearance over others, but it creates a social assumption around what an attractive South Asian can "look like."

CONCLUSION

… It is important to acknowledge a few points. First, ethnic media characterizations are intentional decisions made by media producers. These characterizations reflect hegemonic ideologies and also reproduce commonly understood stereotypes. These stereotypes are the by-products of the U.S. racial hierarchy. The racial hierarchy in turn is developed alongside immigration and assimilation trends, which further determine the qualities that an individual needs to become "American." All of these dynamics inform and influence 21st century representations in American popular media….

REFERENCES

Alba, Richard and Victor Nee. 2003. *Remaking the American Mainstream.* Cambridge, MA: Harvard University Press.

Bonilla-Silva, Eduardo. 1997. 'Rethinking Racism: Toward a Structural Interpretation.' *American Sociological Review* 62(3): 465–80.

Bonilla-Silva, Eduardo. 2004. 'From Bi-Racial to Tri-Racial: Towards a New System of Racial Stratification in the USA.' *Ethnic and Racial Studies* 27(6): 931–50.

Feagin, Joe. 2001. *Racist America: Roots, Current Realities, and Future Reparation.* New York: Routledge.

Haller, William, Alejandro Portes, and Scott M. Lynch. 2011. 'Dreams Fulfilled, Dreams Shattered: Determinants of Segmented Assimilation in the Second Generation.' *Social Forces* 89(3): 733–62.

Kibria, Nazli. 1998. "The Contested Meaning of 'Asian American': Racial Dilemmas in the Contemporary U.S.' *Ethnic and Racial Studies* 21(5): 939–58.

Kim, Claire Jean. 1999. 'The Racial Triangulation of Asian Americans.' *Politics and Society* 27(1): 105–38.

Omi, Michael and Howard Winant. 1994. *Racial Formation in the United States: From the 1960s to the 1990s,* 2nd ed. New York: Routledge.

Rastogi, Nina. 2010. Beyond Apu: Why are There Suddenly So Many Indians on Television? *Slate* June 9. Accessed July 7, 2013 (http://www.slate.com/id/2255937/).

Tuan, Mia. 1999. *Forever Foreigners or Honorary Whites? The Asian Ethnic Experience Today.* New Brunswick: Rutgers University Press.

Vera, Hernán and Andrew Gordon. 2003. *Screen Saviors: Hollywood Fictions of Whiteness.* Lanham, MD: Rowman and Littlefield.

Zhou, Min and Yang Sao Xiong. 2007. 'The Multifaceted American Experiences of the Children of Asian Immigrants: Lessons for Segmented Assimilation.' *Ethnic and Racial Studies* 28(6): 1119–52.

11

We are All Americans!

The Latin Americanization of Racial Stratification in the USA

EDUARDO BONILLA-SILVA

Racial stratification in the United States has operated along biracial lines (White–non-White) for centuries. For demographic (the relative large size of the Black population) and historical reasons (the centrality of Blacks to the national economic development from the 17th to the middle part of the 20th century), the biracial order has been anchored on the Black–White experience (Feagin, 2000). Historically, those on the non-White side of the divide have shared similar experiences of colonialism, oppression, exploitation, and racialization (Amott & Matthaei, 1991). Hence, being non-White has meant having a restricted access to the multiple benefits or "wages of whiteness" (Roediger, 1991) such as good housing, decent jobs, and a good education.

Nevertheless, the post-civil rights era has brought changes in how racial stratification seems to operate. For example, significant gaps in status have emerged between groups that previously shared a common denizen position in the racial order.... Yet another instance of the changes in contemporary America is that few Whites endorse in surveys segregationist views. This has been heralded by some as reality as "the end of racism" or pointing to "the declining significance of race" (Wilson, 1978). Lastly, Blacks have been surpassed by Latinos as the largest minority group (by 2001, the Census noted that "Hispanics" were 13 percent of the population and Blacks 12 percent).

I propose that all this reshuffling denotes that the biracial order typical of the United States, which was the exception in the world-racial system, is evolving into a complex triracial stratification system similar to that of many Latin American and Caribbean nations. Specifically, I suggest that the emerging triracial system will be comprised of "Whites" at the top, and intermediary group of "honorary Whites"—similar to the coloreds in South Africa during formal apartheid and a non-White group or the "collective Black" at the bottom. In Figure 1, I sketch how these three groups may look like. I hypothesize that the White group will include "traditional" Whites, new "White" immigrants and, in the near future, assimilated Latinos, some multiracials, and other

SOURCE: Eduardo, Bonilla-Silva. 2002. "We Are All Americans!: The Latin Americanization of Racial Stratification in the USA." *Race and Society*, 5(1): 3–16. Reprinted with permission from Elsevier.

"Whites"

Whites
New Whites (Russians, Albanians, etc.)
Assimilated white Latinos
Some multiracials
Assimilated (urban) Native Americans
A few Asian-origin people

"Honorary Whites"

Light-skinned Latinos
Japanese Americans
Korean Americans
Asian Indians
Chinese Americans
Middle Eastern Americans
Most multiracials

"Collective Black"

Filipinos
Vietnamese
Hmong
Laotians
Dark-skinned Latinos
Blacks
New West Indian and African immigrants
Reservation-bound Native Americans

F I G U R E 1 Preliminary Map of Triracial System in the United States

sub-groups; the intermediate racial group or honorary Whites will comprise most light-skinned Latinos (e.g., most Cubans and segments of the Mexican and Puerto Rican communities), Japanese Americans, Korean Americans, Asian Indians, Chinese Americans, and most Middle Eastern Americans Americans; and, finally, that the collective Black will include Blacks, dark-skinned Latinos, Vietnamese, Cambodians, Laotians, and maybe Filipinos.

As a triracial system (or Latin- or Caribbean-like racial order), race conflict will be buffered by the intermediate group, much like class conflict is when the class structure includes a large middle class. Furthermore, color gradations, which have always been important matters of within-group differentiation, will become more salient factors of stratification. Lastly, Americans, like people in complex racial stratification orders, will begin making nationalists appeals ("We are all Americans"), decry their racial past, and claim they are "beyond race."...

WHY WOULD A TRIRACIAL SYSTEM
BE EMERGING IN THE USA NOW?

Why would race relations in the United States be moving toward a triracial regime at this point in history? The reasons are multiple. First, the demography of the nation is changing. Racial minorities are up to 30 percent of the population and, as population projections suggest, may become a numeric majority in the year 2050....

The rapid darkening of America is creating a situation similar to that of many Latin American and Caribbean nations where the White elites realized their countries were becoming "Black" or "Indian" and devised a number of strategies to whiten their population and maintaining White power. Although whitening the population through immigration or by classifying many newcomers as White is a possible solution to the new American demography, a more plausible accommodation to the new racial reality, and one that would still help maintain "White supremacy" is to (1) create an intermediate racial group to buffer racial conflict, (2) allow some newcomers into the White racial strata, and (3) incorporate most immigrants into the collective Black strata.

Second, as part of the tremendous reorganization that transpired in America in the post-civil rights era, a new kinder and gentler White supremacy emerged which Bonilla-Silva has labeled elsewhere as the "new racism" (Bonilla-Silva, 2001). In post-civil rights America the maintenance of systemic White privilege is accomplished socially, economically, and politically through institutional, covert, and apparently non-racial practices. Whether in banks or Universities, in stores or housing markets, "smiling discrimination" tends to be the order of the day. This kinder and gentler form of White supremacy has produced an accompanying ideology: the ideology of color-blind racism. This ideology denies the salience of race, scorns those who talk about race, and increasingly proclaims that "We are all Americans" (for a detailed analysis of color-blind racism).

Third, race relations have become globalized. The once almost all-White Western nations have now "interiorized the other" (Miles, 1993). The new world-systemic need for capital accumulation has led to the incorporation of "dark" foreigners as "guest workers" and even as permanent workers. Thus, today European nations have racial minorities in their midst who are progressively becoming an underclass have developed an internal "racial structure" (Bonilla-Silva, 1997) to maintain White power, and have a curious racial ideology that combines ethnonationalism with a race-blind ideology similar to the color-blind racism of the United States today.

This new global racial reality will reinforce the trend toward triracialism in the United States as versions of color-blind racism will become prevalent in most Western nations. Furthermore, as many formerly almost-all White Western countries (e.g., Germany, France, England) become more and more racially diverse, triracial divisions may surface in these societies, too.

Fourth, the convergence of the political and ideological actions of the Republican Party, conservative commentators and activists, and the so-called "multi-racial" movement have created the space for the radical transformation of the way we gather racial data in America. One possible outcome of the Census Bureau categorical back-and-forth on racial and ethnic classifications is either the dilution of racial data or the elimination of race as an official category.

Lastly, the attack on affirmative action, which is part of what Steinberg (1995) has labeled as the "racial retreat," is the clarion call signaling the end of race-based social policy in the United States.... Again, this trend reinforces my thesis because the elimination of race-based social policy is, among other things, predicated on the notion that race no longer affects minorities' status....

A LOOK AT THE DATA

Objective Indicators of Standing of the Three Racial Strata

If the racial order in the United States is becoming triracial significant gaps in socioeconomic status between Whites, honorary Whites, and the collective Black should be developing. The available data suggests this is the case. In terms of income, Latino groups that are mostly White (Argentines, Chileans, Costa Ricans, and Cubans) have per capita incomes that are 40–100 percent higher that those of Latino groups that are predominantly comprised of dark-skinned people (Mexicans, Puerto Ricans, Dominicans). The exceptions (Bolivians and Panamanians) are examples of self-selected immigrants. For example, 4 of the largest 10 concentrations of Bolivians are in the state of Virginia, a state with just 7.2 percent Latinos (Census Bureau, 2000). There is a similar pattern for Asians: a severe income gap is emerging among honorary White Asians (Japanese, Koreans, Filipinos, and Chinese) and those I classify as belonging to the collective Black (Vietnamese, Cambodian, Hmong, and Laotians). (Data on educational standing and poverty rates exhibit the same pattern.)

Substantial group differences are also evident in the occupational status of the groups. White Latinos, although far from Whites, are between 50 and 100 percent more likely to be represented in the "Managerial and Professional" and "Technical" categories than dark-skinned Latinos (for example, whereas 32 percent of Costa Ricans are in such categories, only 17 percent of Mexicans are). Along the same lines, elite Asians are even more likely to be well-represented in the higher prestige occupational categories than underclass Asians for example, 45 percent of Asian Indians are in "Professional" and "Technical" jobs, but only 5 percent of Hmong, 9 percent of Laotians, 10 percent of Cambodians, and 23 percent of Vietnamese.

Subjective Indicators of "Consciousness" of Three Racial Strata

Social psychologists have amply demonstrated that it takes very little for groups to form, develop a common view, and adjudicate status positions to nominal characteristics. Thus, it should not be surprising if objective gaps in income, occupational status, and education between these various groups is contributing to group formation. That is, honorary Whites may be classifying themselves as "White" or believing they are better than the "collective Black." If this is happening, this group should also be in the process of developing White-like racial attitudes befitting of their new social position and differentiating (distancing) themselves from the "collective Black." In line with my thesis, I also expect Whites to be making distinctions between honorary Whites and the collective Black, specifically, exhibiting a more positive outlook toward the former than toward the latter. Finally, if a triracial order is emerging, I speculate the "collective Black" will begin to exhibit a diffused and contradictory racial consciousness as Blacks and Indians tend to do throughout Latin America and the Caribbean.

Social Interaction Among Members of the Three Racial Strata

If a triracial system is emerging, one would expect more social (e.g., friendship, associations as neighbors) and intimate (e.g., marriage) contact between Whites and honorary Whites than between Whites and members of the collective Black. A cursory analysis of the interracial marriage and segregation data suggests this seems to be the case.

CONCLUDING REMARKS: TRIRACIAL ORDER, RACIAL POLITICS, AND THE FUTURE OF WHITE SUPREMACY IN AMERICA

... Almost all the objective, subjective, and social interaction indicators I reviewed tend to point in the direction one would expect if a triracial system is emerging. For example, objective indicators on income and education show substantive gaps between the groups I labeled "White," "honorary White," and the "collective Black." Not surprisingly, a variety of subjective indicators signal the emergence of *internal* stratification among racial minorities. This has led some minority groups to develop racial attitudes similar to those of Whites, and others to develop attitudes closer to those of Blacks. Finally, the findings on the objective and subjective indicators have an interactional correlate. Data on interracial marriage and residential segregation show that Whites are significantly more likely to live nearby honorary Whites and intermarry with them than live and intermarry with members of the collective Black.

If my prediction is right, what may the consequences for race relations in the United States? First, racial politics will change dramatically. The "us" versus "them" racial dynamic will lessen as "honorary Whites" grow in size and social importance. This group is likely to buffer racial conflict—or derail it—as intermediate groups do in many Latin American countries....

Second, the ideology of color-blind racism will become even more salient among Whites and honorary Whites and will also impact members of the collective Black. This ideology will help glue the new social system and further buffer racial conflict.

Third, if the state decides to stop gathering racial statistics, the struggle to document the impact of race in a variety of social venues will become monumental. More significantly, because state actions always impact civil society, if the state decides to erase race from above, the *social* recognition of "races" in the polity may become harder....

Fourth, the deep history of Black–White divisions in the United States has been such that the centrality of the Black identity will not dissipate. Research on the "Black elite," for I instance, shows they exhibit racial attitudes in line with their racial rather than class group. That identity may be taken up by dark-skinned Latinos as it is being rapidly taken up by most West Indians and some Latinos....

... Furthermore, the external pressure of "multiracials" in White contexts and the internal pressure of "ethnic" Blacks may change the notion of "Blackness" and even the position of some "Blacks" in the system....

Fifth, the new order will force a reshuffling of *all* racial identities. Certain "racial" claims may dissipate (or, in some cases, decline in significance) as mobility will increasingly be seen as based on (1) whiteness or near-whiteness and (2) intermarriage with Whites (this seems to be the case among many Japanese Americans, particularly those who have intermarried). This dissipation of ethnicity will not be limited to "honorary Whites" as members of the "collective Black" strata strive to position themselves higher in the new racial totem pole based on degrees of proximity or closeness to whiteness. Will Vietnamese, Filipinos, and other members of the Asian underclass coalesce with Blacks and dark-skinned Latinos or will they try to distance themselves from them and struggle to emphasize their "Americanness?"

Lastly, the new racial stratification system will be more effective in maintaining "White supremacy." Whites will still be at the top of the social structure but will face fewer race-based challenges and racial inequality will remain and may even widen as is the case throughout Latin America and the Caribbean. And, to avoid confusion about my claim regarding "honorary Whites," let me clarify that their standing and status will be dependent upon Whites' wishes and practices. "Honorary" means they will remain secondary, will still face discrimination, and will not receive equal treatment in society. For example, although Arab Americans will be regarded as "honorary Whites," their treatment in the post-September 11 era suggests their status as "White" and "American" is tenuous at best. Likewise, albeit substantial segments of the Asian American community may become "honorary White," they will also continue suffering from discrimination and be regarded in many quarters as "perpetual foreigners."

Therein lies the weaknesses of the emerging triracial order and the possibilities for challenging it. Members of the "collective Black" must be the backbone of the movement challenging the new order as they are the ones who will remain literally "at the bottom of the well." However, if they want to be successful, they must wage, in coalition with progressive Asian and Latino organizations, a concerted effort to politicize the segments I label "honorary Whites" and make them aware of the *honorary* character of their status. This is the way out of the impending new racial quandary. We need to short-circuit the belief in near-whiteness as the solution to status differences and create a coalition of all "people of color" and their White allies. If the triracial Latin American- or Caribbean-like model of race prevails and "pigmentocracy" crystallizes, most Americans will scramble for the meager wages that near-whiteness will provide to those willing to play the "We are all American" game.

REFERENCES

Amott, T., & Matthaei, L. (1991). *Race, gender, and work: A multicultural history of women in the United States.* Boston, MA: South End Press.

Feagin, J. R. (2000). *Racist America: Roots, current realities, and future reparations.* London and New York: Routledge.

Miles, R. (1993). *Racism after race relations.* London and New York: Routledge.

Roediger, D. (1991). *The wages of whiteness: Race and the making of the American working class.* New York: Verso.

Wilson, W. J. (1978). *The declining significance of race.* Chicago, IL: The University of Chicago Press.

12

Optional Ethnicities

For Whites Only?

MARY C. WATERS

What does it mean to talk about ethnicity as an option for an individual? To argue that an individual has some degree of choice in their ethnic identity flies in the face of the commonsense notion of ethnicity many of us believe in—that one's ethnic identity is a fixed characteristic, reflective of blood ties and given at birth. However, social scientists who study ethnicity have long concluded that while ethnicity is based on a *belief* in a common ancestry, ethnicity is primarily a *social* phenomenon, not a biological one (Alba 1985, 1990; Barth 1969; Weber [1921] 1968, p. 389). The belief that members of an ethnic group have that they share a common ancestry may not be a fact. There is a great deal of change in ethnic identities across generations through intermarriage, changing allegiances, and changing social categories. There is also a much larger amount of change in the identities of individuals over their lives than is commonly believed. While most people are aware of the phenomenon known as "passing"—people raised as one race who change at some point and claim a different race as their identity—there are similar life course changes in ethnicity that happen all the time and are not given the same degree of attention as racial "passing."

White Americans of European ancestry can be described as having a great deal of choice in terms of their ethnic identities. The two major types of options White Americans can exercise are (1) the option of whether to claim any specific ancestry, or to just be "White" or American, [Lieberson (1985) called these people "unhyphenated Whites"] and (2) the choice of which of their European ancestries to choose to include in their description of their own identities. In both cases, the option of choosing how to present yourself on surveys and in everyday social interactions exists for Whites because of social changes and societal conditions that have created a great deal of social mobility, immigrant assimilation, and political and economic power for Whites in the United States. Specifically, the option of being able to not claim any ethnic identity exists for Whites of European background in the United States because they are the majority group—in terms of holding political and social power, as well as being a numerical majority. The option of choosing among different ethnicities in their

SOURCE: Waters, Mary C. 1996. "Optional Ethnicities: For Whites Only." Pp. 444–454 in *Origins and Destinies: Immigration, Race and Ethnicity in America*, edited by S. Pedraza and R.G. Rumbaut. Belmont, CA: Cengage Learning. © 1996 Cengage Learning.

family backgrounds exists because the degree of discrimination and social distance attached to specific European backgrounds has diminished over time....

SYMBOLIC ETHNICITIES FOR WHITE AMERICANS

What do these ethnic identities mean to people and why do they cling to them rather than just abandoning the tie and calling themselves American? My own field research with suburban Whites in California and Pennsylvania found that later-generation descendants of European origin maintain what are called "symbolic ethnicities." Symbolic ethnicity is a term coined by Herbert Gans (1979) to refer to ethnicity that is individualistic in nature and without real social cost for the individual. These symbolic identifications are essentially leisure-time activities, rooted in nuclear family traditions and reinforced by the voluntary enjoyable aspects of being ethnic (Waters 1990). Richard Alba (1990) also found later-generation Whites in Albany, New York, who chose to keep a tie with an ethnic identity because of the enjoyable and voluntary aspects to those identities, along with the feelings of specialness they entailed. An example of symbolic ethnicity is individuals who identify as Irish, for example, on occasions such as Saint Patrick's Day, on family holidays, or for vacations. They do not usually belong to Irish American organizations, live in Irish neighborhoods, work in Irish jobs, or marry other Irish people. The symbolic meaning of being Irish American can be constructed by individuals from mass media images, family traditions, or other intermittent social activities. In other words, for later-generation White ethnics, ethnicity is not something that influences their lives unless they want it to. In the world of work and school and neighborhood, individuals do not have to admit to being ethnic unless they choose. And for an increasing number of European-origin individuals whose parents and grandparents have intermarried, the ethnicity they claim is largely a matter of personal choice as they sort through all of the possible combinations of groups in their genealogies....

RACE RELATIONS AND SYMBOLIC ETHNICITY

However much symbolic ethnicity is without cost for the individual, there is a cost associated with symbolic ethnicity for the society. That is because symbolic ethnicities of the type described here are confined to White Americans of European origin. Black Americans, Hispanic Americans, Asian Americans, and American Indians do not have the option of a symbolic ethnicity at present in the United States. For all of the ways in which ethnicity does not matter for White Americans, it does matter for non-Whites. Who your ancestors are does affect your choice of spouse, where you live, what job you have, who your friends are, and what your chances are for success in American society, if those ancestors happen not to be from Europe. The reality is that White ethnics have a lot more choice and room for maneuver than they themselves think they do.

The situation is very different for members of racial minorities, whose lives are strongly influenced by their race or national origin regardless of how much they may choose not to identify themselves in terms of their ancestries.

When White Americans learn the stories of how their grandparents and great-grandparents triumphed in the United States over adversity, they are usually told in terms of their individual efforts and triumphs. The important role of labor unions and other organized political and economic actors in their social and economic successes are left out of the story in favor of a generational story of individual Americans rising up against communitarian, Old World intolerance, and New World resistance. As a result, the "individualized" voluntary, cultural view of ethnicity for Whites is what is remembered.

One important implication of these identities is that they tend to be very individualistic. There is a tendency to view valuing diversity in a pluralist environment as equating all groups. The symbolic ethnic tends to think that all groups are equal; everyone has a background that is their right to celebrate and pass on to their children. This leads to the conclusion that all identities are equal and all identities in some sense are interchangeable—"I'm Italian American, you're Polish American. I'm Irish American, you're African American." The important thing is to treat people as individuals and all equally. However, this assumption ignores the very big difference between an individualistic symbolic ethnic identity and a socially enforced and imposed racial identity.

My favorite example of how this type of thinking can lead to some severe misunderstandings between people of different backgrounds is from the *Dear Abby* advice column. A few years back a person wrote in who had asked an acquaintance of Asian background where his family was from. His acquaintance answered that this was a rude question and he would not reply. The bewildered White asked Abby why it was rude, since he thought it was a sign of respect to wonder where people were from, and he certainly would not mind anyone asking HIM about where his family was from. Abby asked her readers to write in to say whether it was rude to ask about a person's ethnic background. She reported that she got a large response, that most non-Whites thought it was a sign of disrespect, and Whites thought it was flattering:

> Dear Abby,
> I am 100 percent American and because I am of Asian ancestry I am often asked "What are you?" It's not the personal nature of this question that bothers me, it's the question itself. This query seems to question my very humanity. "What am I? Why I am a person like everyone else!"
> *Signed, A REAL AMERICAN*

> Dear Abby,
> Why do people resent being asked what they are? The Irish are so proud of being Irish, they tell you before you even ask. Tip O'Neill has never tried to hide his Irish ancestry.
> *Signed, JIMMY.*

(SOURCE: Published by Universal Press Syndicate)

In this exchange Jimmy cannot understand why Asians are not as happy to be asked about their ethnicity as he is, because he understands his ethnicity and theirs to be separate but equal. Everyone has to come from somewhere—his family from Ireland, another's family from Asia—each has a history and each should be proud of it. But the reason he cannot understand the perspective of the Asian American is that all ethnicities are not equal; all are not symbolic, costless, and voluntary. When White Americans equate their own symbolic ethnicities with the socially enforced identities of non-White Americans, they obscure the fact that the experiences of Whites and non-Whites have been qualitatively different in the United States and that the current identities of individuals partly reflect that unequal history.

In the next section I describe how relations between Black and White students on college campuses reflect some of these asymmetries in the understanding of what a racial or ethnic identity means. While I focus on Black and White students in the following discussion, you should be aware that the myriad other groups in the United States—Mexican Americans, American Indians, Japanese Americans—all have some degree of social and individual influences on their identities, which reflect the group's social and economic history and present circumstance.

RELATIONS ON COLLEGE CAMPUSES

Both Black and White students face the task of developing their race and ethnic identities. Sociologists and psychologists note that at the time people leave home and begin to live independently from their parents, often ages eighteen to twenty-two, they report a heightened sense of racial and ethnic identity as they sort through how much of their beliefs and behaviors are idiosyncratic to their families and how much are shared with other people. It is not until one comes in close contact with many people who are different from oneself that individuals realize the ways in which their backgrounds may influence their individual personality. This involves coming into contact with people who are different in terms of their ethnicity, class, religion, region, and race. For White students, the ethnicity they claim is more often than not a symbolic one—with all of the voluntary, enjoyable, and intermittent characteristics I have described above.

Black students at the university are also developing identities through interactions with others who are different from them. Their identity development is more complicated than that of Whites because of the added element of racial discrimination and racism, along with the "ethnic" developments of finding others who share their background. Thus Black students have the positive attraction of being around other Black students who share some cultural elements, as well as the need to band together with other students in a reactive and oppositional way in the face of racist incidents on campus.

Colleges and universities across the country have been increasing diversity among their student bodies in the last few decades. This has led in many cases to strained relations among students from different racial and ethnic backgrounds. The 1980s and 1990s produced a great number of racial incidents and high racial tensions on campuses. While there were a number of racial incidents that were due to bigotry, unlawful behavior, and violent or vicious attacks, much of what happens among students on campuses involves a low level of tension and awkwardness in social interactions.

Many Black students experience racism personally for the first time on campus. The upper-middle-class students from White suburbs were often isolated enough that their presence was not threatening to racists in their high schools. Also, their class background was known by their residence and this may have prevented attacks being directed at them. Often Black students at the university who begin talking with other students and recognizing racial slights will remember incidents that happened to them earlier that they might not have thought were related to race.

Black college students across the country experience a sizeable number of incidents that are clearly the result of racism. Many of the most blatant ones that occur between students are the result of drinking. Sometimes late at night, drunken groups of White students coming home from parties will yell slurs at single Black students on the street. The other types of incidents that happen include being singled out for special treatment by employees, such as being followed when shopping at the campus bookstore, or going to the art museum with your class and the guard stops you and asks for your I.D. Others involve impersonal encounters on the street—being called a nigger by a truck driver while crossing the street, or seeing old ladies clutch their pocketbooks and shake in terror as you pass them on the street. For the most part these incidents are not specific to the university environment, they are the types of incidents middle-class Blacks face every day throughout American society, and they have been documented by sociologists (Feagin 1991).

In such a climate, however, with students experiencing these types of incidents and talking with each other about them, Black students do experience a tension and a feeling of being singled out. It is unfair that this is part of their college experience and not that of White students. Dealing with incidents like this, or the ever-present threat of such incidents, is an ongoing developmental task for Black students that takes energy, attention, and strength of character. It should be clearly understood that this is an asymmetry in the "college experience" for Black and White students. It is one of the unfair aspects of life that results from living in a society with ongoing racial prejudice and discrimination. It is also very understandable that it makes some students angry at the unfairness of it all, even if there is no one to blame specifically. It is also very troubling because, while most Whites do not create these incidents, some do, and it is never clear until you know someone well whether they are the type of person who could do something like this. So one of the reactions of Black students to these incidents is to band together.

In some sense then, as Blauner (1992) has argued, you can see Black students coming together on campus as both an "ethnic" pull of wanting to be together to share common experiences and community, and a "racial" push of banding together defensively because of perceived rejection and tension from Whites. In this way the ethnic identities of Black students are in some sense similar to, say, Korean students wanting to be together to share experiences. And it is an ethnicity that is generally much stronger than, say, Italian Americans. But for Koreans who come together there is generally a definition of themselves as "different from" Whites. For Blacks reacting to exclusion, there is a tendency for the coming together to involve both being "different from" but also "opposed to" Whites.

The anthropologist John Ogbu (1990) has documented the tendency of minorities in a variety of societies around the world, who have experienced severe blocked mobility for long periods of time, to develop such oppositional identities. An important component of having such an identity is to describe others of your group who do not join in the group solidarity as devaluing and denying their very core identity. This is why it is not common for successful Asians to be accused by others of acting "White" in the United States, but it is quite common for such a term to be used by Blacks and Latinos. The oppositional component of a Black identity also explains how Black people can question whether others are acting "Black enough." On campus, it explains some of the intense pressures felt by Black students who do not make their racial identity central and who choose to hang out primarily with non-Blacks. This pressure from the group, which is partly defining itself by not being White, is exacerbated by the fact that race is a physical marker in American society. No one immediately notices the Jewish students sitting together in the dining hall, or the one Jewish student sitting surrounded by non-Jews, or the Texan sitting with the Californians, but everyone notices the Black student who is or is not at the "Black table" in the cafeteria.

An example of the kinds of misunderstandings that can arise because of different understandings of the meanings and implications of symbolic versus oppositional identities concerns questions students ask one another in the dorms about personal appearances and customs. A very common type of interaction in the dorm concerns questions Whites ask Blacks about their hair. Because Whites tend to know little about Blacks, and Blacks know a lot about Whites, there is a general asymmetry in the level of curiosity people have about one another. Whites, as the numerical majority, have had little contact with Black culture; Blacks, especially those who are in college, have had to develop bicultural skills—knowledge about the social worlds of both Whites and Blacks. Miscommunication and hurt feelings about White students' questions about Black students' hair illustrate this point. One of the things that happens freshman year is that White students are around Black students as they fix their hair. White students are generally quite curious about Black students' hair—they have basic questions such as how often Blacks wash their hair, how they get it straightened or curled, what products they use on their hair, how they comb it, etc. Whites often wonder to themselves whether they should ask these questions. One thought experiment Whites perform is to ask themselves whether a particular question would upset them. Adopting the "do unto others" rule, they ask themselves, "If a Black person was curious about my hair would I get upset?" The answer usually is "No, I would be happy to tell them." Another example is an Italian American student wondering to herself, "Would I be upset if someone asked me about calamari?" The answer is no, so she asks her Black roommate about collard greens, and the roommate explodes with an angry response such as, "Do you think all Black people eat watermelon too?" Note that if this Italian American knew her friend was Trinidadian American and asked about peas and rice the situation would be more similar and would not necessarily ignite underlying tensions.

Like the debate in *Dear Abby,* these innocent questions are likely to lead to resentment. The issue of stereotypes about Black Americans and the assumption that all Blacks are alike and have the same stereotypical cultural traits has more power to hurt or offend a Black person than vice versa. The innocent questions about Black hair also bring up a number of asymmetries between the Black and White experience. Because Blacks tend to have more knowledge about Whites than vice versa, there is not an even exchange going on, the Black freshman is likely to have fewer basic questions about his White roommate than his White roommate has about him. Because of the differences historically in the group experiences of Blacks and Whites there are some connotations to Black hair that don't exist about White hair. (For instance, is straightening your hair a form of assimilation, do some people distinguish between women having "good hair" and "bad hair" in terms of beauty and how is that related to looking "White"?) Finally, even a Black freshman who cheerfully disregards or is unaware that there are these asymmetries will soon slam into another asymmetry if she willingly answers every innocent question asked of her. In a situation where Blacks make up only 10 percent of the student body, if every non-Black needs to be educated about hair, she will have to explain it to nine other students. As one Black student explained to me, after you've been asked a couple of times about something so personal you begin to feel like you are an attraction in a zoo, that you are at the university for the education of the White students.

INSTITUTIONAL RESPONSES

Our society asks a lot of young people. We ask young people to do something that no one else does as successfully on such a wide scale—that is to live together with people from very different backgrounds, to respect one another, to appreciate one another, and to enjoy and learn from one another. The successes that occur every day in this endeavor are many, and they are too often overlooked. However, the problems and tensions are also real, and they will not vanish on their own. We tend to see pluralism working in the United States in much the same way some people expect capitalism to work. If you put together people with various interests and abilities and resources, the "invisible hand" of capitalism is supposed to make all the parts work together in an economy for the common good.

There is much to be said for such a model—the invisible hand of the market can solve complicated problems of production and distribution better than any "visible hand" of a state plan. However, we have learned that unequal power relations among the actors in the capitalist marketplace, as well as "externalities" that the market cannot account for, such as long-term pollution, or collusion between corporations, or the exploitation of child labor, means that state regulation is often needed. Pluralism and the relations between groups are very similar. There is a lot to be said for the idea that bringing people who belong to different ethnic or racial groups together in institutions with no interference will have good consequences. Students from different backgrounds will make friends if they share a dorm room or corridor, and there is no need for the institution to do any more than provide the locale. But like capitalism, the invisible hand of pluralism does not do well when power relations and externalities are ignored. When you bring together individuals from groups that are differentially valued in the wider society and provide no guidance, there will be problems. In these cases the "invisible hand" of pluralist relations does not work, and tensions and disagreements can arise without any particular individual or group of individuals being "to blame." On college campuses in the 1990s some of the tensions between students are of this sort. They arise from honest misunderstandings, lack of a common background, and very different experiences of what race and ethnicity mean to the individual.

The implications of symbolic ethnicities for thinking about race relations are subtle but consequential. If your understanding of your own ethnicity and its relationship to society and politics is one of individual choice, it becomes harder to understand the need for programs like affirmative action, which recognize the ongoing need for group struggle and group recognition, in order to bring about social change. It also is hard for a White college student to understand the need that minority students feel to band together against discrimination. It also is easy, on the individual level, to expect everyone else to be able to turn their ethnicity on and off at will, the way you are able to, without understanding that ongoing discrimination and societal attention to minority status makes that impossible for individuals from minority groups to do. The paradox of symbolic ethnicity is that it depends upon the ultimate goal of a pluralist society, and at the same time makes it more difficult to achieve that ultimate goal. It is dependent upon the concept that all ethnicities mean the same thing, that enjoying the traditions of one's heritage is an option available to a group or an individual, but that such a heritage should not have any social costs associated with it.

As the Asian Americans who wrote to *Dear Abby* make clear, there are many societal issues and involuntary ascriptions associated with non-White identities. The developments necessary for this to change are not individual but societal in nature. Social mobility and declining racial and ethnic sensitivity are closely associated. The legacy and the present reality of discrimination on the basis of race or ethnicity must be overcome before the ideal of a pluralist society, where all heritages are treated equally and are equally available for individuals to choose or discard at will, is realized.

REFERENCES

Alba, Richard D. 1985. *Italian Americans: Into the of Twilight Ethnicity.* Englewood Cliffs, NJ: Prentice-Hall.

Alba, Richard D. 1990. *Ethnic Identity: The Transformation of White America.* New Haven: Yale University Press.

Barth, Frederick. 1969. *Ethnic Groups and Boundaries.* Boston: Little, Brown.

Blauner, Robert. 1992. "Talking Past Each Other: Black and White Languages of Race." *American Prospect* (Summer): 55–64.

Feagin, Joe R. 1991. "The Continuing Significance of Race: Anti-Black Discrimination in Public Places." *American Sociological Review* 56: 101–17.

Gans, Herbert. 1979. "Symbolic Ethnicity: The Future of Ethnic Groups and Cultures in America." *Ethnic and Racial Studies* 2: 1–20.

Lieberson, Stanley. 1985. *Making It Count: The Improvement of Social Research and Theory.* Berkeley: University of California Press.

Ogbu, John. 1990. "Minority Status and Literacy in Comparative Perspective." *Daedalus* 119: 141–69.

Waters, Mary C. 1990. *Ethnic Options: Choosing Identities in America.* Berkeley: University of California Press.

Weber, Max. [1921]/1968. *Economy and Society: An Outline of Interpretive Sociology.* Eds. Guenther Roth and Claus Wittich, trans. Ephraim Fischoff. New York: Bedminister Press.

13

Is Capitalism Gendered and Racialized?

JOAN ACKER

Capitalism is racialized and gendered in two intersecting historical processes. First, industrial capitalism emerged in the United States dominated by white males, with a gender- and race-segregated labor force, laced with wage inequalities, and a society-wide gender division of caring labor. The processes of reproducing segregation and wage inequality changed over time, but segregation and inequality were not eliminated. A small group of white males still dominate the capitalist economy and its politics. The society-wide gendered division of caring labor still exists. Ideologies of white masculinity and related forms of consciousness help to justify capitalist practices. In short, conceptual and material practices that construct capitalist production and markets, as well as beliefs supporting those practices, are deeply shaped through gender and race divisions of labor and power and through constructions of white masculinity.

Second, these gendered and racialized practices are embedded in and replicated through the gendered substructures of capitalism. These gendered substructures exist in ongoing incompatible organizing of paid production activities and unpaid domestic and caring activities. Domestic and caring activities are devalued and seen as outside the "main business" (Smith 1999) of capitalism. The commodification of labor, the capitalist wage form, is an integral part of this process, as family provisioning and caring become dependent upon wage labor. The abstract language of bureaucratic organizing obscures the ongoing impact on families and daily life. At the same time, paid work is organized on the assumption that reproduction is of no concern. The separations between paid production and unpaid life-sustaining activities are maintained by corporate claims that they have no responsibility for anything but returns to shareholders. Such claims are more successful in the United States, in particular, than in countries with stronger labor movements and welfare states. These often successful claims contribute to the corporate processes of establishing their interests as more important than those of ordinary people.

SOURCE: Acker, Joan. 2006. *Class Questions, Feminist Answers*. Lanham, MD: Rowman and Littlefield, pp. 113–117. (Reproduced with permission of Rowman and Littlefield in the format to republish in a book via Copyright Clearance Center.)

THE GENDERED AND RACIALIZED DEVELOPMENT OF U.S. CAPITALISM

Segregations and Wage Inequalities

Industrial capitalism is historically, and in the main continues to be, a white male project, in the sense that white men were and are the innovators, owners, and holders of power. Capitalism developed in Britain and then in Europe and the United States in societies that were already dominated by white men and already contained a gender-based division of labor. The emerging waged labor force was sharply divided by gender, as well as by race and ethnicity with many variations by nation and regions within nations. At the same time, the gendered division of labor in domestic tasks was reconfigured and incorporated in a gendered division between paid market labor and unpaid domestic labor. In the United States, certain white men, unburdened by caring for children and households and already the major wielders of gendered power, buttressed at least indirectly by the profits from slavery and the exploitation of other minorities, were, in the nineteenth century, those who built the U.S. factories and railroads, and owned and managed the developing capitalist enterprise. As far as we know, they were also heterosexual and mostly of Northern European heritage. Their wives and daughters benefited from the wealth they amassed and contributed in symbolic and social ways to the perpetuation of their class, but they were not the architects of the new economy.

Recruitment of the labor force for the colonies and then the United States had always been transnational and often coercive. Slavery existed prior to the development of industrialism in the United States: Capitalism was built partly on profits from that source. Michael Omi and Howard Winant (1994, 265) contend that the United States was a racial dictatorship for 258 years, from 1607 to 1865. After the abolition of slavery in 1865, severe exploitation, exclusion, and domination of blacks by whites perpetuated racial divisions cutting across gender and some class divisions, consigning blacks to the most menial, low-paying work in agriculture, mining, and domestic service. Early industrial workers were immigrants. For example, except for the brief tenure (twenty-five years) of young, native-born white women workers in the Lowell, Massachusetts, mills, immigrant women and children were the workers in the first mass production industry in the United States, the textile mills of Massachusetts and Philadelphia, Pennsylvania (Perrow 2002). This was a gender and racial/ethnic division of labor that still exists, but now on a global basis. Waves of European immigrants continued to come to the United States to work in factories and on farms. Many of these European immigrants, such as impoverished Irish, Poles, and eastern European Jews were seen as non-white or not-quite-white by white Americans and were used in capitalist production as low-wage workers, although some of them were actually skilled workers (Brodkin 1998). The experiences of racial oppression built into industrial capitalism varied by gender within these racial/ethnic groups.

Capitalist expansion across the American continent created additional groups of Americans who were segregated by race and gender into racial and ethnic enclaves and into low-paid and highly exploited work. This expansion included the extermination and expropriation of native peoples, the subordination of Mexicans in areas taken in the war with Mexico in 1845, and the recruitment of Chinese and other Asians as low-wage workers, mostly on the west coast (Amott and Matthaei 1996; Glenn 2002).

Women from different racial and ethnic groups were incorporated differently than men and differently than each other into developing capitalism in the late nineteenth and early twentieth centuries. White Euro-American men moved from farms into factories or commercial, business, and administrative jobs. Women aspired to be housewives as the male breadwinner family became the ideal. Married white women, working class and middle class, were housewives unless unemployment, low wages, or death of their husbands made their paid work necessary (Goldin 1990, 133). Young white women with some secondary education moved into the expanding clerical jobs and into elementary school teaching when white men with sufficient education were unavailable (Cohn 1985). African Americans, both women and men, continued to be confined to menial work, although some were becoming factory workers, and even teachers and professionals as black schools and colleges were formed (Collins 2000). Young women from first- and second-generation European immigrant families worked in factories and offices. This is a very sketchy outline of a complex process (Kessler-Harris 1982), but the overall point is that the capitalist labor force in the United States emerged as deeply segregated horizontally by occupation and stratified vertically by positions of power and control on the basis of both gender and race.

Unequal pay patterns went along with sex and race segregation, stratification, and exclusion. Differences in the earnings and wealth (Keister 2000) of women and men existed before the development of the capitalist wage (Padavic and Reskin 2002). Slaves, of course, had no wages and earned little after abolition. These patterns continued as capitalist wage labor became the dominant form and wages became the primary avenue of distribution to ordinary people. Unequal wages were justified by beliefs about virtue and entitlement. A living wage or a just wage for white men was higher than a living wage or a just wage for white women or for women and men from minority racial and ethnic groups (Figart, Mutari, and Power 2002). African-American women were at the bottom of the wage hierarchy.

The earnings advantage that white men have had throughout the history of modern capitalism was created partly by their organization to increase their wages and improve their working conditions. They also sought to protect their wages against the competition of others, women and men from subordinate groups (for example, Cockburn 1983, 1991). This advantage also suggests a white male coalition across class lines (Connell 2000; Hartmann 1976), based at least partly in beliefs about gender and race differences and beliefs about the superior skills of white men. White masculine identity and self-respect were complexly involved in these divisions of labor and wages. This is another way in which capitalism is a gendered and racialized accumulation process (Connell 2000). Wage differences between white men and all other groups, as well as divisions of labor between these groups, contributed to profit and flexibility, by helping to maintain growing occupational areas, such as clerical work, as segregated and low paid. Where women worked in manufacturing or food processing, gender divisions of labor kept the often larger female work force in low-wage routine jobs, while males worked in other more highly paid, less routine, positions (Acker and Van Houten 1974). While white men might be paid more, capitalist organizations could benefit from this "gender/racial dividend." Thus, by maintaining divisions, employers could pay less for certain levels of skill, responsibility, and experience when the worker was not a white male.

This is not to say that getting a living wage was easy for white men, or that most white men achieved it. Labor-management battles, employers' violent tactics to prevent unionization, [and] massive unemployment during frequent economic depressions characterized the situation of white industrial workers as wage labor spread in the nineteenth and early twentieth centuries. During the same period, new white-collar jobs were created to manage, plan, and control the expanding industrial economy. This rapidly increasing middle class was also stratified by gender and race. The better-paid, more respected jobs went to white men; white women were secretaries and clerical workers; people of color were absent. Conditions and issues varied across industries and regions of the country. But, wherever you look, those variations contained underlying gendered and racialized divisions. Patterns of stratification and segregation were written into employment contracts in work content, positions in work hierarchies, and wage differences, as well as other forms of distribution.

These patterns persisted, although with many alterations, through extraordinary changes in production and social life. After World War II, white women, except for a brief period immediately after the war, went to work for pay in the expanding service sector, professional, and managerial fields. African Americans moved to the North in large numbers, entering industrial and service sector jobs. These processes accelerated after the 1960s, with the civil rights and women's movements, new civil rights laws, and affirmative action. Hispanics and Asian Americans, as well as other racial/ethnic groups, became larger proportions of the population, on the whole finding work in low-paid, segregated jobs. Employers continued, and still continue, to select and promote workers based on gender and racial identifications, although the processes are more subtle, and possibly less visible, than in the past (for example, Brown et al. 2003; Royster 2003). These processes continually recreate gender and racial inequities, not as cultural or ideological survivals from earlier times, but as essential elements in present capitalisms (Connell 1987, 103–106).

Segregating practices are a part of the history of white, masculine-dominated capitalism that establishes class as gendered and racialized. Images of masculinity support these practices, as they produce a taken-for-granted world in which certain men legitimately make employment and other economic decisions that affect the lives of most other people. Even though some white women and people from other-than-white groups now hold leadership positions, their actions are shaped within networks of practices sustained by images of masculinity (Wacjman 1998).

Masculinities and Capitalism

Masculinities are essential components of the ongoing male project, capitalism. While white men were and are the main publicly recognized actors in the history of capitalism, these are not just any white men. They have been, for example, aggressive entrepreneurs or strong leaders of industry and finance (Collinson and Hearn 1996). Some have been oppositional actors, such as self-respecting and tough workers earning a family wage, and militant labor leaders. They have been particular men whose locations within gendered and racialized social relations and practices can be partially captured by the concept of masculinity. "Masculinity" is a contested term. As Connell (1995, 2000), Hearn (1996), and others have pointed out, it should be pluralized as "masculinities," because in any society at any one time there are several ways of being a man. "Being a man" involves cultural images and practices. It always implies a contrast to an unidentified femininity.

Hegemonic masculinity can be defined as the taken-for-granted, generally accepted form, attributed to leaders and other influential figures at particular historical times. Hegemonic masculinity legitimates the power of those who embody it. More than one type of hegemonic masculinity may exist simultaneously, although they may share characteristics, as do the business leader and the sports star at the present time. Adjectives describing hegemonic masculinities closely follow those describing characteristics of successful business organizations, as Rosabeth Moss Kanter (1977) pointed out in the 1970s. The successful CEO and the successful organization are aggressive, decisive, competitive, focused on winning and defeating the enemy, taking territory from others. The ideology of capitalist markets is imbued with a masculine ethos. As R. W. Connell (2000, 35) observes, "The market is often seen as the antithesis of gender (marked by achieved versus ascribed status, etc.). But the market operates through forms of rationality that are historically masculine and involve a sharp split between instrumental reason on the one hand, emotion and human responsibility on the other" (Seidler 1989). Masculinities embedded in collective practices are part of the context within which certain men made and still make the decisions that drive and shape the ongoing development of capitalism. We can speculate that how these men see themselves, what actions and choices they feel compelled to make and they think are legitimate, how they and the world around them define desirable masculinity, enter into that decision making (Reed 1996). Decisions made at the very top reaches of (masculine) corporate power have consequences that are experienced as inevitable economic forces or disembodied social trends. At the same time, these decisions symbolize and enact varying hegemonic masculinities (Connell 1995). However, the embeddedness of masculinity within the ideologies of business and the market may become invisible, seen as just part of the way business is done. The relatively few women who reach the highest positions probably think and act within these strictures.

Hegemonic masculinities and violence are deeply connected within capitalist history: The violent acts of those who carried out the slave trade or organized colonial conquests are obvious examples. Of course, violence has been an essential component of power in many other socioeconomic systems, but it continues into the rational organization of capitalist economic activities. Violence is frequently a legitimate, if implicit, component of power exercised by bureaucrats as well as "robber barons." Metaphors of violence, frequently military violence, are often linked to notions of the masculinity of corporate leaders, as "defeating the enemy" suggests. In contemporary capitalism, violence and its links to masculinity are often masked by the seeming impersonality of objective conditions. For example, the masculinity of top managers, the ability to be tough, is involved in the implicit violence of many corporate decisions, such as those cutting jobs in order to raise profits and, as a result, producing unemployment. Armies and other organizations, such as the police, are specifically organized around violence. Some observers of recent history suggest that organized violence, such as the use of the military, is still mobilized at least partly to reach capitalist goals, such as controlling access to oil supplies. The masculinities of those making decisions to deploy violence in such a way are hegemonic, in the sense of powerful and exemplary. Nevertheless, the connections between masculinity, capitalism, and violence are complex and contradictory, as Jeff Hearn and Wendy Parkin (2001) make clear. Violence is always a possibility in mechanisms of control and domination, but it is not always evident, nor is it always used.

As corporate capitalism developed, Connell (1995) and others (for example, Burris 1996) argue that a hegemonic masculinity based on claims to expertise developed alongside masculinities organized around domination and control. Hegemonic masculinity relying on claims to expertise does not necessarily lead to economic organizations free of domination and violence, however (Hearn and Parkin 2001). Hearn and Parkin (2001) argue that controls relying on both explicit and implicit violence exist in a wide variety of organizations, including those devoted to developing new technology.

Different hegemonic masculinities in different countries may reflect different national histories, cultures, and change processes. For example, in Sweden in the mid-1980s, corporations were changing the ways in which they did business toward a greater participation in the international economy, fewer controls on currency and trade, and greater emphasis on competition. Existing images of dominant masculinity were changing, reflecting new business practices. This seemed to be happening in the banking sector, where I was doing research on women and their jobs (Acker 1994a). The old paternalistic leadership, in which primarily men entered as young clerks expecting to rise to managerial levels, was being replaced by young, aggressive men hired as experts and managers from outside the banks. These young, often technically trained, ambitious men pushed the idea that the staff was there to sell bank products to customers, not, in the first instance, to take care of the needs of clients. Productivity goals were put in place; nonprofitable customers, such as elderly pensioners, were to be encouraged not to come into the bank and occupy the staff's attention. The female clerks we interviewed were disturbed by these changes, seeing them as evidence that the men at the top were changing from paternal guardians of the people's interests to manipulators who only wanted riches for themselves. The confirmation of this came in a scandal in which the CEO of the largest bank had to step down because he had illegally taken money from the bank to pay for his housing. The amount of money was small; the disillusion among employees was huge. He had been seen as a benign father; now he was no better than the callous young men on the way up who were dominating the daily work in the banks. The hegemonic masculinity in Swedish banks was changing as the economy and society were changing.

Hegemonic masculinities are defined in contrast to subordinate masculinities. White working class masculinity, although clearly subordinate, mirrors in some of its more heroic forms the images of strength and responsibility of certain successful business leaders. The construction of working class masculinity around the obligations to work hard, earn a family wage, and be a good provider can be seen as providing an identity that both served as a social control and secured male advantage in the home. That is, the good provider had to have a wife and probably children for whom to provide. Glenn (2002) describes in some detail how this image of the white male worker also defined him as superior to and different from black workers.

Masculinities are not stable images and ideals, but [shift] with other societal changes. With the turn to neoliberal business thinking and globalization, there seem to be new forms. Connell (2000) identifies "global business masculinity," while Lourdes Beneria (1999) discusses the "Davos man," the global leader from business, politics, or academia who meets his peers once a year in the Swiss town of Davos to assess and plan the direction of globalization. Seeing masculinities as implicated in the ongoing production of global capitalism opens the possibility of seeing sexualities, bodies, pleasures, and identities as also implicated in economic relations.

In sum, gender and race are built into capitalism and its class processes through the long history of racial and gender segregation of paid labor and through the images and actions of white men who dominate and lead central capitalist endeavors. Underlying these processes is the subordination to production and the market of nurturing and caring for human beings, and the assignment of these responsibilities to women as unpaid work. Gender segregation that differentially affects women in all racial groups rests at least partially on the ideology and actuality of women as carers. Images of dominant masculinity enshrine particular male bodies and ways of being as different from the female and distanced from caring.... I argue that industrial capitalism, including its present neoliberal form, is organized in ways that are, at the same time, antithetical and necessary to the organization of caring or reproduction and that the resulting tensions contribute to the perpetuation of gendered and racialized class inequalities. Large corporations are particularly important in this process as they increasingly control the resources for provisioning but deny responsibility for such social goals.

REFERENCES

Acker, Joan. 1994a. The Gender Regime of Swedish Banks. *Scandinavian Journal of Management* 10, no. 2: 117–30.

Acker, Joan, and Donald Van Houten. 1974. Differential Recruitment and Control: The Sex Structuring of Organizations. *Administrative Science Quarterly* 19 (June, 1974): 152–63.

Amott, Teresa, and Julie Matthaei. 1996. *Race, Gender, and Work: A Multi-cultural Economic History of Women in the United States.* Revised edition. Boston: South End Press.

Beneria, Lourdes. 1999. Globalization, Gender and the Davos Man. *Feminist Economics* 5, no. 3: 61–83.

Brodkin, Karen. 1998. Race, Class, and Gender: The Metaorganization of American Capitalism. *Transforming Anthropology* 7, no. 2: 46–57.

Brown, Michael K., Martin Carnoy, Elliott Currie, Troy Duster, David B. Oppenheimer, Marjorie M. Shultz, and David Wellman. 2003. *White-Washing Race The Myth of a Color-Blind Society.* Berkeley: University of California Press.

Burris, Beverly H. 1996. Technocracy, Patriarchy and Management. In *Men as Managers, Managers as Men,* ed. David L. Collinson and Jeff Hearn. London: Sage.

Cockburn, Cynthia. 1983. *Brothers.* London: Pluto Press.

_____. 1991. *In the Way of Women: Men's Resistance to Sex Equality in Organization.* Ithaca, N.Y.: ILR Press.

Cohn, Samuel. 1985. *The Process of Occupational Sex-Typing: The Femininization of Clerical Labor in Great Britain.* Philadelphia: Temple University Press.

Collins, Patricia Hill. 2000. *Black Feminist Thought,* second edition. New York and London: Routledge.

Collinson, David L., and Jeff Hearn. 1996. Breaking the Silence: On Men, Masculinities and Managements. In *Men as Managers, Managers as Men,* ed. David L. Collinson and Jeff Hearns. London: Sage.

Connell, R. W. 1987. *Gender & Power.* Stanford, Calif.: Stanford University Press.

_____. 1995. *Masculinities.* Berkeley: University of California Press.

_____. 2000. *The Men and the Boys.* Berkeley: University of California Press.

Figart, Deborah M., Ellen Mutari, and Marilyn Power. 2002. *Living Wages, Equal Wages.* London and New York: Routledge.

Glenn, Evelyn Nakano. 2002. *Unequal Freedom: How Race and Gender Shaped American Citizenship and Labor.* Cambridge: Harvard University Press.

Goldin, Claudia. 1990. *Understanding the Gender Gap: An Economic History of American Women.* New York and Oxford: Oxford University Press.

Hartmann, Heidi. 1976. "Capitalism, Patriarchy, and Job Segregation by Sex," *Signs: Journal of Women in Culture and Society* 1(3), part 2, spring: 137–167.

Hearn, Jeff. 1996. Is Masculinity Dead? A Critique of the Concept of Masculinity/ Masculinities. In *Understanding Masculinities: Social Relations and Cultural Arenas,* ed. M. Mac an Ghaill. Buckingham: Oxford University Press.

_____. 2004. From Hegemonic Masculinity to the Hegemony of Men. *Feminist Theory* 5, no. 1: 49–72.

Hearn, Jeff, and Wendy Parkin. 2001. *Gender, Sexuality and Violence in Organizations.* London: Sage.

Kanter, Rosabeth Moss. 1977. *Men and Women of the Corporation.* New York: Basic Books.

Keister, Lisa. 2000. *Wealth in America: Trends in Wealth Inequality.* Cambridge: Cambridge University Press.

Kessler-Harris, Alice. 1982. *Out to Work: A History of Wage-Earning Women in the United States.* New York: Oxford University Press.

Omi, Michael, and Howard Winant. 1994. *Racial Formation in the United States*. New York: Routledge.

Padavic, Irene, and Barbara Reskin. 2002. *Women and Men at Work*, second edition. Thousand Oaks, Calif.: Pine Forge Press.

Perrow, Charles. 2002. *Organizing America*. Princeton and Oxford: Princeton University Press.

Reed, Rosslyn. 1996. Entrepreneurialism and Paternalism in Australian Management: A Gender Critique of the "Self-Made" Man. In *Men as Managers, Managers as Men*, ed. David L. Collinson and Jeff Hearn. London: Sage.

Royster, Deirdre A. 2003. *Race and the Invisible Hand: How White Networks Exclude Black Men from Blue-Collar Jobs*. Berkeley: University of California Press.

Seidler, Victor J. 1989. *Rediscovering Masculinity: Reason, Language, and Sexuality*. London and New York: Routledge.

Smith, Dorothy. 1999. *Writing the Social: Critique, Theory, and Investigation*. Toronto: University of Toronto Press.

Wacjman, Judy. 1998. *Managing Like a Man*. Cambridge: Polity Press.

14

Race as Class

HERBERT J. GANS

Humans of all colors and shapes can make babies with each other. Consequently most biologists, who define races as subspecies that cannot inter-
breed, argue that scientifically there can be no human races. Nonetheless,
laypeople still see and distinguish between races. Thus, it is worth asking again
why the lay notion of race continues to exist and to exert so much influence in
human affairs.

Laypersons are not biologists, nor are they sociologists who argue these days
that race is a social construction arbitrary enough to be eliminated if "society"
chose to do so. The laity operates with a very different definition of race. They
see that humans vary, notably in skin color, the shape of the head, nose, and lips,
and quality of hair, and they choose to define the variations as individual races.

More important, the lay public uses this definition of race to decide whether
strangers (the so-called "other") are to be treated as superior, inferior, or equal.
Race is even more useful for deciding quickly whether strangers might be threatening and thus should be excluded. Whites often consider dark-skinned strangers
threatening until they prove otherwise, and none more than African Americans.

Scholars believe the color differences in human skins can be traced to cli-
matic adaptation. They argue that the high levels of melanin in dark skin originally protected people living outside in hot, sunny climates, notably in Africa and
South Asia, from skin cancer. Conversely, in cold climates, the low amount of
melanin in light skins enabled the early humans to soak up vitamin D from a sun
often hidden behind clouds. These color differences were reinforced by millen-
nia of inbreeding when humans lived in small groups that were geographically
and socially isolated. This inbreeding also produced variations in head and nose
shapes and other facial features so that Northern Europeans look different from
people from the Mediterranean area, such as Italians and, long ago, Jews. Likewise, East African faces differ from West African ones, and Chinese faces from
Japanese ones. (Presumably the inbreeding and isolation also produced the
DNA patterns that geneticists refer to in the latest scientific revival and redefinition of race.)

Geographic and social isolation ended long ago, however, and human population movements, intermarriage, and other occasions for mixing are eroding
physical differences in bodily features. Skin color stopped being adaptive too

after people found ways to protect themselves from the sun and could get their vitamin D from the grocery or vitamin store. Even so, enough color variety persists to justify America's perception of white, yellow, red, brown, and black races.

Never mind for the moment that the skin of "whites," as well as many East Asians and Latinos is actually pink; that Native Americans are not red; that most African Americans come in various shades of brown; and that really black skin is rare. Never mind either that color differences within each of these populations are as great as the differences between them, and that, as DNA testing makes quite clear, most people are of racially mixed origins, even if they do not know it. But remember that this color palette was invented by whites. Nonwhite people would probably divide the range of skin colors quite differently.

Advocates of racial equality use these contradictions to fight against racism. However, the general public also has other priorities. As long as people can roughly agree about who looks "white," "yellow," or "black" and find that their notion of race works for their purposes, they ignore its inaccuracies, inconsistencies, and other deficiencies.

Note, however, that only some facial and bodily features are selected for the lay definition of race. Some, like the color of women's nipples or the shape of toes (and male navels), cannot serve because they are kept covered. Most other visible ones, like height, weight, hairlines, ear lobes, finger or hand sizes—and even skin texture—vary too randomly and frequently to be useful for categorizing and ranking people or judging strangers. After all, your own child is apt to have the same stubby fingers as a child of another skin color or, what is equally important, a child from a very different income level.

RACE, CLASS, AND STATUS

In fact, the skin colors and facial features commonly used to define race are selected precisely because, when arranged hierarchically, they resemble the country's class-and-status hierarchy. Thus, whites are on top of the socioeconomic pecking order as they are on top of the racial one, while variously shaded nonwhites are below them in socioeconomic position (class) and prestige (status).

The darkest people are for the most part at the bottom of the class-status hierarchy. This is no accident, and Americans have therefore always used race as a marker or indicator of both class and status. Sometimes they also use it to enforce class position, to keep some people "in their place." Indeed, these uses are a major reason for its persistence.

Of course, race functions as more than a class marker, and the correlation between race and the socioeconomic pecking order is far from statistically perfect: All races can be found at every level of that order. Still, the race-class correlation is strong enough to utilize race for the general ranking of others. It also becomes more useful for ranking dark-skinned people as white poverty declines so much that whiteness becomes equivalent to being middle or upper class.

The relation between race and class is unmistakable. For example, the 1998–2000 median household income of non-Hispanic whites was $45,500; of Hispanics (currently seen by many as a race) as well as Native Americans, $32,000; and of African Americans, $29,000. The poverty rates for these same groups were 7.8 percent among whites, 23.1 among Hispanics, 23.9 among blacks, and 25.9 among Native Americans. (Asians' median income was $52,600—which does much to explain why we see them as a model minority.)

True, race is not the only indicator used as a clue to socioeconomic status. Others exist and are useful because they can also be applied to ranking co-racials. They include language (itself a rough indicator of education), dress, and various kinds of taste, from given names to cultural preferences, among others.

American English has no widely known working-class dialect like the English Cockney, although "Brooklynese" is a rough equivalent, as is "black vernacular." Most blue-collar people dress differently at work from white-collar, professional, and managerial workers. Although contemporary American leisure-time dress no longer signifies the wearer's class, middle-income Americans do not usually wear Armani suits or French haute couture, and the people who do can spot the knockoffs bought by the less affluent.

Actually, the cultural differences in language, dress, and so forth that were socially most noticeable are declining. Consequently, race could become yet more useful as a status marker, since it is so easily noticed and so hard to hide or change. And in a society that likes to see itself as classless, race comes in very handy as a substitute.

THE HISTORICAL BACKGROUND

Race became a marker of class and status almost with the first settling of the United States. The country's initial holders of cultural and political power were mostly WASPs (with a smattering of Dutch and Spanish in some parts of what later became the United States). They thus automatically assumed that their kind of whiteness marked the top of the class hierarchy. The bottom was assigned to the most powerless, who at first were Native Americans and slaves. However, even before the former had been virtually eradicated or pushed to the country's edges, the skin color and related facial features of the majority of colonial America's slaves had become the markers for the lowest class in the colonies.

Although dislike and fear of the dark are as old as the hills and found all over the world, the distinction between black and white skin became important in America only with slavery and was actually established only some decades after the first importation of black slaves. Originally, slave owners justified their enslavement of black Africans by their being heathens, not by their skin color.

In fact, early Southern plantation owners could have relied on white indentured servants to pick tobacco and cotton or purchased the white slaves that were available then, including the Slavs from whom the term slave is derived. They also had access to enslaved Native Americans. Blacks, however, were cheaper, more plentiful, more easily controlled, and physically more able to survive the intense heat and brutal working conditions of Southern plantations.

After slavery ended, blacks became farm laborers and sharecroppers, de facto indentured servants, really, and thus they remained at the bottom of the class hierarchy. When the pace of industrialization quickened, the country needed new sources of cheap labor. Northern industrialists, unable and unwilling to recruit southern African Americans, brought in very poor European immigrants, mostly peasants. Because these people were near the bottom of the class hierarchy, they were considered nonwhite and classified into races. Irish and Italian newcomers were sometimes even described as black (Italians as "guineas"), and the eastern and southern European immigrants were deemed "swarthy."

However, because skin color is socially constructed, it can also be reconstructed. Thus, when the descendants of the European immigrants began to move up economically and socially, their skins apparently began to look lighter to the whites who had come to America before them. When enough of these descendants became visibly middle class, their skin was seen as fully white. The biological skin color of the second and third generations had not changed, but it was socially blanched or whitened. The process probably began in earnest just before the Great Depression and resumed after World War II. As the cultural and other differences of the original European immigrants disappeared, their descendants became known as white ethnics.

This pattern is now repeating itself among the peoples of the post-1965 immigration. Many of the new immigrants came with money and higher education, and descriptions of their skin color have been shaped by their class position. Unlike the poor Chinese who were imported in the 19th century to build the West and who were hated and feared by whites as a "yellow horde," today's affluent Asian newcomers do not seem to look yellow. In fact, they are already sometimes thought of as honorary whites, and later in the 21st century they may well turn into a new set of white ethnics. Poor East and Southeast Asians may not be so privileged, however, although they are too few to be called a "yellow horde."

Hispanics are today's equivalent of a "swarthy" race. However, the children and grandchildren of immigrants among them will probably undergo "whitening" as they become middle class. Poor Mexicans, particularly in the Southwest, are less likely to be whitened, however. (Recently a WASP Harvard professor came close to describing these Mexican immigrants as a brown horde.)

Meanwhile, black Hispanics from Puerto Rico, the Dominican Republic, and other Caribbean countries may continue to be perceived, treated, and mistreated as if they were African American. One result of that mistreatment is their low median household income of $35,000, which was just $1,000 more than that of non-Hispanic blacks but $4,000 below that of so-called white Hispanics.

Perhaps South Asians provide the best example of how race correlates with class and how it is affected by class position. Although the highly educated Indians and Sri Lankans who started coming to America after 1965 were often darker than African Americans, whites only noticed their economic success.

They have rarely been seen as nonwhites, and are also often praised as a model minority.

Of course, even favorable color perceptions have not ended racial discrimination against newcomers, including model minorities and other affluent ones. When they become competitors for valued resources such as highly paid jobs, top schools, housing, and the like, they also become a threat to whites. California's Japanese-Americans still suffer from discrimination and prejudice four generations after their ancestors arrived here.

AFRICAN-AMERICAN EXCEPTIONALISM

The only population whose racial features are not automatically perceived differently with upward mobility are African Americans: Those who are affluent and well educated remain as visibly black to whites as before. Although a significant number of African Americans have become middle class since the civil rights legislation of the 1960s, they still suffer from far harsher and more pervasive discrimination and segregation than nonwhite immigrants of equivalent class position. This not only keeps whites and blacks apart but prevents blacks from moving toward equality with whites. In their case, race is used both as a marker of class and, by keeping blacks "in their place," an enforcer of class position and a brake on upward mobility.

In the white South of the past, African Americans were lynched for being "uppity." Today, the enforcement of class position is less deadly but, for example, the glass ceiling for professional and managerial African Americans is set lower than for Asian Americans, and on-the-job harassment remains routine.

Why African-American upward economic mobility is either blocked or, if allowed, not followed by public blanching of skin color remains a mystery. Many explanations have been proposed for the white exceptionalism with which African Americans are treated. The most common is "racism," an almost innate prejudice against people of different skin color that takes both personal and institutional forms. But this does not tell us why such prejudice toward African Americans remains stronger than that toward other nonwhites.

A second explanation is the previously mentioned white antipathy to blackness, with an allegedly primeval fear of darkness extrapolated into a primordial fear of dark-skinned people. But according to this explanation, dark-skinned immigrants such as South Asians should be treated much like African Americans.

A better explanation might focus on "Negroid" features. African as well as Caribbean immigrants with such features—for example, West Indians and Haitians—seem to be treated somewhat better than African Americans. But this remains true only for new immigrants; their children are generally treated like African Americans.

Two additional explanations are class-related. For generations, a majority or plurality of all African Americans were poor, and about a quarter still remain so. In addition, African Americans continue to commit a proportionally greater share of the street crime, especially street drug sales—often because legitimate job opportunities are scarce. African Americans are apparently also more often arrested without cause. As one result, poor African Americans are more often considered undeserving than are other poor people, although in some parts of America, poor Hispanics, especially those who are black, are similarly stigmatized.

The second class-based explanation proposes that white exceptionalist treatment of African Americans is a continuing effect of slavery: They are still perceived as ex-slaves. Many hateful stereotypes with which today's African Americans are demonized have changed little from those used to dehumanize the slaves. (Black Hispanics seem to be equally demonized, but then they were also slaves, if not on the North American continent.) Although slavery ended officially in 1864, ever since the end of Reconstruction subtle efforts to discourage African-American upward mobility have not abated, although these efforts are today much less pervasive or effective than earlier.

Some African Americans are now millionaires, but the gap in wealth between average African Americans and whites is much greater than the gap between incomes. The African-American middle class continues to grow, but many of its members barely have a toehold in it, and some are only a few paychecks away from a return to poverty. And the African-American poor still face the most formidable obstacles to upward mobility. Close to a majority of working-age African-American men are jobless or out of the labor force. Many women, including single mothers, now work in the low-wage economy, but they must do without most of the support systems that help middle-class working mothers. Both federal and state governments have been punitive, even in recent Democratic administrations, and the Republicans have cut back nearly every antipoverty program they cannot abolish.

Daily life in a white-dominated society reminds many African Americans that they are perceived as inferiors, and these reminders are louder and more relentless for the poor, especially young men. Regularly suspected of being criminals, they must constantly prove that they are worthy of equal access to the American Dream. For generations, African Americans have watched immigrants pass them in the class hierarchy, and those who are poor must continue to compete with current immigrants for the lowest-paying jobs. If unskilled African Americans reject such jobs or fail to act as deferentially as immigrants, they justify the white belief that they are less deserving than immigrants. Blacks' resentment of such treatment gives whites additional evidence of their unworthiness, thereby justifying another cycle of efforts to keep them from moving up in class and status.

Such practices raise the suspicion that the white political economy and white Americans may, with the help of nonwhites who are not black, use African Americans to anchor the American class structure with a permanently lower-class population. In effect, America, or those making decisions in its name, could be seeking, not necessarily consciously, to establish an undercaste that cannot move out and up. Such undercastes exist in other societies: the gypsies of Eastern Europe, India's untouchables, "indigenous people," and "aborigines" in yet other places. But these are far poorer countries than the United States.

SOME IMPLICATIONS

The conventional wisdom and its accompanying morality treat racial prejudice, discrimination, and segregation as irrational social and individual evils that public policy can reduce but only changes in white behavior and values can eliminate. In fact, over the years, white prejudice as measured by attitude surveys has dramatically declined, far more dramatically than behavioral and institutional discrimination.

But what if discrimination and segregation are more than just a social evil? If they are used to keep African Americans down, then they also serve to eliminate or restrain competitors for valued or scarce resources, material and symbolic. Keeping African Americans from decent jobs and incomes as well as quality schools and housing makes more of these available to all the rest of the population. In that case, discrimination and segregation may decline significantly only if the rules of the competition change or if scarce resources, such as decent jobs, become plentiful enough to relax the competition, so that the African-American population can become as predominantly middle class as the white population. Then the stigmas, the stereotypes inherited from slavery, and the social and other arrangements that maintain segregation and discrimination could begin to lose their credibility. Perhaps "black" skin would eventually become as invisible as "yellow" skin is becoming.

THE MULTIRACIAL FUTURE

One trend that encourages upward mobility is the rapid increase in interracial marriage that began about a quarter century ago. As the children born to parents of different races also intermarry, more and more Americans will be multiracial, so that at some point far in the future the current quintet of skin colors will be irrelevant. About 40 percent of young Hispanics and two-thirds of young Asians now "marry out," but only about 10 percent of blacks now marry nonblacks—yet another instance of the exceptionalism that differentiates blacks.

Moreover, if race remains a class marker, new variations in skin color and in other visible bodily features will be taken to indicate class position. Thus, multiracials with "Negroid" characteristics could still find themselves disproportionately at the bottom of the class hierarchy. But what if at some point in the future everyone's skin color varied by only a few shades of brown? At that point, the dominant American classes might have to invent some new class markers.

If in some utopian future the class hierarchy disappears, people will probably stop judging differences in skin color and other features. Then lay Americans would probably agree with biologists that race does not exist. They might even insist that race does not need to exist.

15

Media Magic
Making Class Invisible

GREGORY MANTSIOS

Of the various social and cultural forces in our society, the mass media is arguably the most influential in molding public consciousness. Americans spend an average twenty-eight hours per week watching television. They also spend an undetermined number of hours reading periodicals, listening to the radio, and going to the movies. Unlike other cultural and socializing institutions, ownership and control of the mass media is highly concentrated. Twenty-three corporations own more than one-half of all the daily newspapers, magazines, movie studios, and radio and television outlets in the United States.[1] The number of media companies is shrinking and their control of the industry is expanding. And a relatively small number of media outlets is producing and packaging the majority of news and entertainment programs. For the most part, our media is national in nature and single-minded (profit-oriented) in purpose. This media plays a key role in defining our cultural tastes, helping us locate ourselves in history, establishing our national identity, and ascertaining the range of national and social possibilities. In this essay, we will examine the way the mass media shapes how people think about each other and about the nature of our society.

The United States is the most highly stratified society in the industrialized world. Class distinctions operate in virtually every aspect of our lives, determining the nature of our work, the quality of our schooling, and the health and safety of our loved ones. Yet remarkably, we, as a nation, retain illusions about living in an egalitarian society. We maintain these illusions, in large part, because the media hides gross inequities from public view. In those instances when inequities are revealed, we are provided with messages that obscure the nature of class realities and blame the victims of class-dominated society for their own plight. Let's briefly examine what the news media, in particular, tells us about class.

ABOUT THE POOR

The news media provides meager coverage of poor people and poverty. The coverage it does provide is often distorted and misleading.

SOURCE: Mantsios, Gregory. 1998. "Media Magic: Making Class Invisible." In *Race, Class, and Gender in the United States: An Integrated Study*, 4th ed., edited by Paula Rothenberg. New York: St. Martin's Press. Reprinted with permission of the author.

The Poor Do Not Exist

For the most part, the news media ignores the poor. Unnoticed are forty million poor people in the nation—a number that equals the entire population of Maine, Vermont, New Hampshire, Connecticut, Rhode Island, New Jersey, and New York combined. Perhaps even more alarming is that the rate of poverty is increasing twice as fast as the population growth in the United States. Ordinarily, even a calamity of much smaller proportion (e.g., flooding in the Midwest) would garner a great deal of coverage and hype from a media usually eager to declare a crisis, yet less than one in five hundred articles in the *New York Times* and one in one thousand articles listed in the *Readers Guide to Periodic Literature* are on poverty. With remarkably little attention to them, the poor and their problems are hidden from most Americans.

When the media does turn its attention to the poor, it offers a series of contradictory messages and portrayals.

The Poor Are Faceless

Each year the Census Bureau releases a new report on poverty in our society and its results are duly reported in the media. At best, however, this coverage emphasizes annual fluctuations (showing how the numbers differ from previous years) and ongoing debates over the validity of the numbers (some argue the number should be lower, most that the number should be higher). Coverage like this desensitizes us to the poor by reducing poverty to a number. It ignores the human tragedy of poverty—the suffering, indignities, and misery endured by millions of children and adults. Instead, the poor become statistics rather than people.

The Poor Are Undeserving

When the media does put a face on the poor, it is not likely to be a pretty one. The media will provide us with sensational stories about welfare cheats, drug addicts, and greedy panhandlers (almost always urban and Black). Compare these images and the emotions evoked by them with the media's treatment of middle-class (usually white) "tax evaders," celebrities who have a "chemical dependency," or wealthy businesspeople who use unscrupulous means to "make a profit." While the behavior of the more affluent offenders is considered an "impropriety" and a deviation from the norm, the behavior of the poor is considered repugnant, indicative of the poor in general, and worthy of our indignation and resentment.

The Poor Are an Eyesore

When the media does cover the poor, they are often presented through the eyes of the middle class. For example, sometimes the media includes a story about community resistance to a homeless shelter or storekeeper annoyance with panhandlers. Rather than focusing on the plight of the poor, these stories are about middle-class opposition to the poor. Such stories tell us that the poor are an inconvenience and an irritation.

The Poor Have Only Themselves to Blame

In another example of media coverage, we are told that the poor live in a personal and cultural cycle of poverty that hopelessly imprisons them. They routinely center on the Black urban population and focus on perceived personality or cultural traits that doom the poor. While the women in these stories typically exhibit an "attitude" that leads to trouble or a promiscuity that leads to single motherhood, the men possess a need for immediate gratification that leads to drug abuse or an unquenchable greed that leads to the pursuit of fast money. The images that are seared into our mind are sexist, racist, and classist. Census figures reveal that most of the poor are white not Black or Hispanic, that they live in rural or suburban areas not urban centers, and hold jobs at least part of the year.[2] Yet, in a fashion that is often framed in an understanding and sympathetic tone, we are told that the poor have inflicted poverty on themselves.

The Poor Are Down on Their Luck

During the Christmas season, the news media sometimes provides us with accounts of poor individuals or families (usually white) who are down on their luck. These stories are often linked to stories about soup kitchens or other charitable activities and sometimes call for charitable contributions. These "Yule time" stories are as much about the affluent as they are about the poor: they tell us that the affluent in our society are a kind, understanding, giving people—which we are not. The series of unfortunate circumstances that have led to impoverishment are presumed to be a temporary condition that will improve with time and a change in luck.

Despite appearances, the messages provided by the media are not entirely disparate. With each variation, the media informs us what poverty is not (i.e., systemic and indicative of American society) by informing us what it is. The media tells us that poverty is either an aberration of the American way of life (it doesn't exist, it's just another number, it's unfortunate but temporary) or an end product of the poor themselves (they are a nuisance, do not deserve better, and have brought their predicament upon themselves).

By suggesting that the poor have brought poverty upon themselves, the media is engaging in what William Ryan has called "blaming the victim."[3] The media identifies in what ways the poor are different as a consequence of deprivation, then defines those differences as the cause of poverty itself. Whether blatantly hostile or cloaked in sympathy, the message is that there is something fundamentally wrong with the victims—their hormones, psychological make up, family environment, community, race, or some combination of these—that accounts for their plight and their failure to lift themselves out of poverty.

But poverty in the United States is systemic. It is a direct result of economic and political policies that deprive people of jobs, adequate wages, or legitimate support. It is neither natural nor inevitable: there is enough wealth in our nation to eliminate poverty if we chose to redistribute existing wealth or income. The plight of the poor is reason enough to make the elimination of poverty the nation's first priority. But poverty also impacts dramatically on the non-poor. It has a dampening effect on wages in general (by maintaining a reserve army of unemployed and underemployed anxious for any job at any wage) and breeds crime and violence (by maintaining conditions that invite private gain by illegal means and rebellion-like behavior, not entirely unlike the urban riots of the 1960s). Given the extent of poverty in the nation and the impact it has on us all, the media must spin considerable magic to keep the poor and the issue of poverty and its root causes out of the public consciousness.

ABOUT EVERYONE ELSE

Both the broadcast and the print news media strive to develop a strong sense of "we-ness" in their audience. They seek to speak to and for an audience that is both affluent and like-minded. The media's solidarity with affluence, that is, with the middle and upper class, varies little from one medium to another. Benjamin DeMott points out, for example, that the *New York Times* understands affluence to be intelligence, taste, public spirit, responsibility, and a readiness to rule and "conceives itself as spokesperson for a readership awash in these qualities."[4] Of course, the flip side to creating a sense of "we," or "us," is establishing a perception of the "other." The other relates back to the faceless, amoral, undeserving, and inferior "underclass." Thus, the world according to the news media is divided between the "underclass" and everyone else. Again the messages are often contradictory.

The Wealthy Are Us

Much of the information provided to us by the news media focuses attention on the concerns of a very wealthy and privileged class of people. Although the concerns of a small fraction of the populace, they are presented as though they were the concerns of everyone. For example, while relatively few people actually own stock, the news media devotes an inordinate amount of broadcast time and print space to business news and stock market quotations. Not only do business reports cater to a particular narrow clientele, so do the fashion pages (with $2,000 dresses), wedding announcements, and the obituaries. Even weather and sports news often have a class bias. An all news radio station in New York City, for example, provides regular national ski reports. International news, trade agreements, and domestic policies issues are also reported in terms of their impact on business climate and the business community. Besides being of practical value to the wealthy, such coverage has considerable ideological value. Its message: the concerns of the wealthy are the concerns of us all.

The Wealthy (as a Class) Do Not Exist

While preoccupied with the concerns of the wealthy, the media fails to notice the way in which the rich as a class of people create and shape domestic and foreign policy. Presented as an aggregate of individuals, the wealthy appear without special interests, interconnections, or unity in purpose. Out of public view are the class interests of the wealthy, the interlocking business links, the concerted actions to preserve their class privileges and business interests (by running for public office, supporting political candidates, lobbying, etc.). Corporate lobbying is ignored, taken for granted, or assumed to be in the public interest. (Compare this with the media's portrayal of the "strong arm of labor" in attempting to defeat trade legislation that is harmful to the interests of working people.) It is estimated that two-thirds of the U.S. Senate is composed of millionaires.[5] Having such a preponderance of millionaires in the Senate, however, is perceived to be neither unusual nor antidemocratic; these millionaire senators are assumed to be serving "our" collective interests in governing.

The Wealthy Are Fascinating and Benevolent

The broadcast and print media regularly provide hype for individuals who have achieved "super" success. These stories are usually about celebrities and superstars from the sports and entertainment world. Society pages and gossip columns serve to keep the social elite informed of each others' doings, allow the rest of us to gawk at their excesses, and help to keep the American dream alive. The print media is also fond of feature stories on corporate empire builders. These stories provide an occasional "insider's" view of the private and corporate life of industrialists by suggesting a rags to riches account of corporate success. These stories tell us that corporate success is a series of smart moves, shrewd acquisitions, timely mergers, and well thought out executive suite shuffles. By painting the upper class in a positive light, innocent of any wrongdoing (labor leaders and union organizations usually get the opposite treatment), the media assures us that wealth and power are benevolent. One person's capital accumulation is presumed to be good for all. The elite, then, are portrayed as investment wizards, people of special talent and skill, who even their victims (workers and consumers) can admire.

The Wealthy Include a Few Bad Apples

On rare occasions, the media will mock selected individuals for their personality flaws. Real estate investor Donald Trump and New York Yankees owner George Steinbrenner, for example, are admonished by the media for deliberately seeking publicity (a very un-upper class thing to do); hotel owner Leona Helmsley was caricatured for her personal cruelties; and junk bond broker Michael Milkin was condemned because he had the audacity to rob the rich. Michael Parenti points out that by treating business wrongdoings as isolated deviations from the socially beneficial system of "responsible capitalism," the media overlooks the features of the system that produce such abuses and the regularity with which they occur. Rather than portraying them as predictable and frequent outcomes of corporate power and the business system, the media treats abuses as if they were isolated and atypical. Presented as an occasional aberration, these incidents serve not to challenge, but to legitimate, the system.[6]

The Middle Class Is Us

By ignoring the poor and blurring the lines between the working people and the upper class, the news media creates a universal middle class. From this perspective, the size of one's income becomes largely irrelevant: what matters is that most of "us" share an intellectual and moral superiority over the disadvantaged. As *Time* magazine once concluded, "Middle America is a state of mind."[7] "We are all middle class," we are told, "and we all share the same concerns": job security, inflation, tax burdens, world peace, the cost of food and housing, health care, clean air and water, and the safety of our streets. While the concerns of the wealthy are quite distinct from those of the middle class (e.g., the wealthy worry about investments, not jobs), the media convinces us that "we [the affluent] are all in this together."

The Middle Class Is a Victim

For the media, "we" the affluent not only stand apart from the "other"—the poor, the working class, the minorities, and their problems—"we" are also victimized by the poor (who drive up the costs of maintaining the welfare roles), minorities (who commit crimes against us), and by workers (who are greedy and drive companies out and prices up). Ignored are the subsidies to the rich, the crimes of corporate America, and the policies that wreak havoc on the economic well-being of middle America. Media magic convinces us to fear, more than anything else, being victimized by those less affluent than ourselves.

The Middle Class Is Not a Working Class

The news media clearly distinguishes the middle class (employees) from the working class (i.e., blue collar workers) who are portrayed, at best, as irrelevant, outmoded, and a dying breed. Furthermore, the media will tell us that the hardships faced by blue collar workers are inevitable (due to progress), a result of bad luck (chance circumstances in a particular industry), or a product of their own doing (they priced themselves out of a job). Given the media's presentation of reality, it is hard to believe that manual, supervised, unskilled, and semiskilled workers actually represent more than 50 percent of the adult working population.[8] The working class, instead, is relegated by the media to "the other."

In short, the news media either lionizes the wealthy or treats their interests and those of the middle class as one in the same. But the upper class and the middle class do not share the same interests or worries. Members of the upper class worry about stock dividends (not employment), they profit from inflation and global militarism, their children attend exclusive private schools, they eat and live in a royal fashion, they call on (or are called upon by) personal physicians, they have few consumer problems, they can escape whenever they want from environmental pollution, and they live on street and travel to other areas under the protection of private police forces.[*][9]

The wealthy are not only a class with distinct life-styles and interests, they are a ruling class. They receive a disproportionate share of the country's yearly income, own a disproportionate amount of the country's wealth, and contribute a disproportionate number of their members to governmental bodies and decision-making groups—all traits that William Domhoff, in his classic work *Who Rules America,* defined as characteristic of a governing class.[10]

This governing class maintains and manages our political and economic structures in such a way that these structures continue to yield an amazing proportion of our wealth to a minuscule upper class. While the media is not above referring to ruling classes in other countries (we hear, for example, references to Japan's ruling elite),[11] its treatment of the news proceeds as though there were no such ruling class in the United States.

Furthermore, the news media inverts reality so that those who are working class and middle class learn to fear, resent, and blame those below, rather than those above them in the class structure. We learn to resent welfare, which accounts for only two cents out of every dollar in the federal budget (approximately $10 billion) and provides financial relief for the needy,[**] but learn little about the $11 billion the federal government spends on individuals with incomes in excess of $100,000 (not needy),[12] or the $17 billion in farm subsidies, or the $214 billion (twenty times the cost of welfare) in interest payments to financial institutions.

Middle-class whites learn to fear African Americans and Latinos, but most violent crime occurs within poor and minority communities and is neither inter-racial[†] nor interclass. As horrid as such crime is, it should not mask the destruction and violence perpetrated by corporate America. In spite of the fact that 14,000 innocent people are killed on the job each year, 100,000 die prematurely, 400,000 become seriously ill, and 6 million are injured from work-related accidents and diseases, most Americans fear government regulation more than they do unsafe working conditions.

Through the media, middle-class—and even working-class—Americans learn to blame blue collar workers and their unions for declining purchasing power and economic security. But while workers who managed to keep their jobs and their unions struggled to keep up with inflation, the top 1 percent of American families saw their average incomes soar 80 percent in the last decade.[13]

---

[*]The number of private security guards in the United States now exceeds the number of public police officers. (Robert Reich, "Secession of the Successful," *New York Times Magazine,* February 17, 1991, p. 42.)

[**]A total of $20 billion is spent on welfare when you include all state funding. But the average state funding also comes to only two cents per state dollar.

[†]In 92 percent of the murders nationwide the assailant and the victim are of the same race (46 percent are white/white, 46 percent are black/black), 5.6 percent are black on white, and 2.4 percent are white on black. (FBI and Bureau of Justice Statistics, 1985–1986, quoted in Raymond S. Franklin, *Shadows of Race and Class,* University of Minnesota Press, Minneapolis, 1991, p. 108.)

Much of the wealth at the top was accumulated as stockholders and corporate executives moved their companies abroad to employ cheaper labor (56 cents per hour in El Salvador) and avoid paying taxes in the United States. Corporate America is a world made up of ruthless bosses, massive layoffs, favoritism and nepotism, health and safety violations, pension plan losses, union busting, tax evasions, unfair competition, and price gouging, as well as fast buck deals, financial speculation, and corporate wheeling and dealing that serve the interests of the corporate elite, but are generally wasteful and destructive to workers and the economy in general.

It is no wonder Americans cannot think straight about class. The mass media is neither objective, balanced, independent, nor neutral. Those who own and direct the mass media are themselves part of the upper class, and neither they nor the ruling class in general have to conspire to manipulate public opinion. Their interest is in preserving the status quo, and their view of society as fair and equitable comes naturally to them. But their ideology dominates our society and justifies what is in reality a perverse social order—one that perpetuates unprecedented elite privilege and power on the one hand and widespread deprivation on the other. A mass media that did not have its own class interests in preserving the status quo would acknowledge that inordinate wealth and power undermines democracy and that a "free market" economy can ravage a people and their communities.

NOTES

1. Martin Lee and Norman Solomon, *Unreliable Sources*, Lyle Stuart (New York, 1990), p. 71. See also Ben Bagdikian, *The Media Monopoly*, Beacon Press (Boston, 1990).

2. Department of Commerce, Bureau of the Census, "Poverty in the United States: 92," *Current Population Reports, Consumer Income*, Series *P60–185*, pp. xi, xv, 1.

3. William Ryan, *Blaming the Victim*, Vintage (New York, 1971).

4. Benjamin Demott, *The Imperial Middle*, William Morrow (New York, 1990), p. 123.

5. Fred Barnes, "The Zillionaires Club," *The New Republic*, January 29, 1990, p. 24.

6. Michael Parenti, *Inventing Reality*, St. Martin's Press (New York, 1986), p. 109.

7. *Time*, January 5, 1979, p. 10.

8. Vincent Navarro, "The Middle Class—A Useful Myth," *The Nation*, March 23, 1992, p. 1.

9. Charles Anderson, *The Political Economy of Social Class*, Prentice Hall (Englewood Cliffs, N.J., 1974), p. 137.

10. William Domhoff, *Who Rules America*, Prentice Hall (Englewood Cliffs, N.J., 1967), p. 5.

11. Lee and Solomon, *Unreliable Sources*, p. 179.

12. *Newsweek*, August 10, 1992, p. 57.

13. *Business Week*, June 8, 1992, p. 86.

16

Toxic Inequality

How America's Wealth Gap Destroys Mobility, Deepens the Racial Divide, and Threatens Our Future

THOMAS M. SHAPIRO

In recent years, as living standards for many families have declined and productivity, income, and wealth gains have flowed to the very top, a new conversation about inequality has emerged in the United States. The Occupy Wall Street movement, which began in the fall of 2011, splashed inequality across the front pages and provided space for discussions about historically high income and wealth disparities and their causes. The movement pitted the wealthiest and most powerful 1 percent against 99 percent of Americans. Thomas Piketty's best-selling 2014 book, *Capital in the Twenty-First Century,* brought attention to a different kind of inequality with a focus on capital. Yet many popular and academic accounts of inequality, spurred by media coverage and the emerging national discourse, continued to focus on income disparities, economic class, and the mega-rich. A pre-occupation with income led to an insufficient understanding of the new inequality that left wealth out of the picture. President Barack Obama provided perhaps the crowning moment in this new public attention to economic inequality when he proclaimed in a December 2014 speech that inequality "is the defining challenge of our time."[1] But the president's speech referenced income inequality eleven times and wealth inequality once. Leaving wealth out of the conversation is a crucial mistake, giving fodder to those who would make personal poverty the result of personal failings.

Wealth inequality in the United States is uncommonly high. The wealthiest 1 percent owned 42 percent of all wealth in 2012 and took in 18 percent of all income. Each year the Allianz Group, the world's largest financial service company, calculates each country's Gini coefficient—a measure of inequality in which zero indicates perfect equality and one hundred perfect inequality, or one person owning all the wealth. In 2015, the United States had the highest wealth inequality among industrialized nations, with a score of 80.56.[2] Allianz dubbed the USA the "Unequal States of America."

Wealth concentration has followed a U-shaped pattern over the last hundred years. It was high in the beginning of the twentieth century, with wealth

SOURCE: Thomas M. Shapiro. 2017. *Toxic Inequality: How America's Wealth Gap Destroys Mobility, Deepens the Racial Divide, and Threatens Our Future.* New York: Basis Books, pp. 12–21, 230–231. Reprinted by permission of Basic Books, an imprint of Hachette Book Group, Inc.

inequality reaching its previous peak during the Depression, in 1929. It fell from 1929 to 1978 and has continuously increased since then. By 2012, the share of wealth owned by the top 0.1 percent was three times higher than in the late 1970s, growing from 7 percent in 1979 to 22 percent in 2012. The bottom 90 percent's wealth share has steadily declined since the mid-1980s.[3]

The rise of wealth inequality is almost entirely due to the increase in the top 0.1 percent's wealth share. The steady decline in the bottom 90 percent's wealth share has struck middle-class families in particular. Half the population has less than $500 in savings. In our interviews, we heard the concerns of those who had more month than paycheck.[4]

Wealth is not just a matter of money. As our interviews revealed, wealth is also about power, status, opportunity, identity, and self-image. Wealth confers transformative advantages, while lack of it brings tremendous disadvantages. A family's income reflects educational and occupational achievements, but wealth is needed to solidify these achievements to build a solid foundation of economic security. Wealth is a fundamental pillar of economic security, and without it, as many of the families we interviewed experienced firsthand, hard-won gains are easily lost.

The explanations for economic inequality are many. One prominent line holds that individual values and characteristics either promote or hinder achievement and prosperity. Inequality, in this view, results from poor people's laziness and lack of work ethic, the decline of traditional marriage, an influx of unskilled, uneducated immigrants, and dependence on welfare. Our interviews contradict such arguments—the people we spoke with, rich and poor, had broadly similar values and aspirations—and reveal instead the importance of policy and institutional factors. Other theories focus on such factors as market forces in a globalizing economy, technological change, policies, and politics.[5]

This article takes a different tack, arguing that we must understand wealth and income inequality together with racial inequality. Despite recent attention to racial disparities in policing, mass deportation, persistent residential segregation, attacks on voting rights, and other manifestations of racial injustice, the conversation about widening economic inequality largely leaves out race, as if that gap's causes, its harshest consequences, and its potential solutions are race neutral. Whether they focus on the widening gulf between the very top and various segments further down the distribution ladder, on the fortunes of the bottom 40 percent, on the dwindling of the middle class, or simply on the growing share garnered by the best-off, traditional accounts emphasize class and economics as the central (and sometimes only) explanation. As a result, much of our national discourse about inequality sees disparities as universals that impact all groups in the same ways, and many of the policy ideas proposed to address it fail to recognize the racially disparate distributional impact of universal-sounding solutions.[6] Recent movements such as the Color of Change, the Dreamers, and Black Lives Matter are vigorously trying to recenter the inequality conversation to include race, ethnicity, and immigration. I have been inspired and heartened by the new public conversation about inequality. At the same time, I am frustrated that once again it looks like attention to class is trumping a reckoning with race.

For it is crucial to understand that the trends toward greater income and wealth inequality are converging with a widening racial wealth gap. The typical African American family today has less than a dime of wealth for every dollar of wealth owned by a typical white family. The civil rights movement and the landmark legislation of the 1960s helped to open educational and professional opportunities and to produce an African American middle class. But despite these hard-won advances, as a study following the same set of families for twenty-nine years shows, the gap between white and black family wealth has widened at an alarming pace, increasing nearly threefold over the past generation (see Figure 1). Looking at a representative sample of Americans in 2013, the median net wealth of white families was $142,000, compared to $11,000 for African American families and $13,700 for Hispanic families. This racial wealth gap means that even black families with incomes comparable to those of white families have much less wealth to use to cushion unemployment or a personal crisis, to apply as a down payment on a home, to secure a place for their families in a strong, resource-rich neighborhood, to send their children to private schools, to start a business, or to plan for retirement.

In short, the basic pillars of economic security—wealth and income—are today distributed vastly inequitably along racial and ethnic lines. African Americans' historical disadvantage has become baked into the American economy. African Americans are effectively stymied from generating and retaining wealth of their own not simply by continuing racial discrimination but also by senseless policies that protect existing wealth—wealth that often originated at times of even more intense racial discrimination, if not specifically from racial plunder. Race and wealth have intertwined throughout our nation's history. Too often missing in today's dialogue about inequality is this binding race and wealth linkage. Failure to tackle the nexus of race and wealth will lead, at best, to only small ameliorations at the worst edges of inequality.

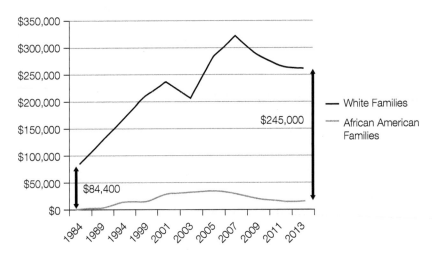

F I G U R E 1 Median Net Wealth by Race, 1984–2013

Source: IASP calculations from PSID; 2013 dollars

Major demographic shifts that are increasingly diversifying America and threatening to sharpen racial and ethnic fault lines further exacerbate the dangers of historically high wealth and income inequality and a widening racial wealth gap. America is becoming a majority-minority nation. Newborns of color outnumbered white newborns for the first time in 2013. America's population growth stems from higher birth rates among families of color and from immigration, especially among Asians and Latinos, while its white population is aging. In 2014, only 21 percent of seniors, but 47 percent of youth, were nonwhite. Demographics are not destiny; yet our institutions, from schools to the workforce to communities to government, are just beginning to confront challenges of racial diversity they were not designed to face, are ill prepared to meet, and often resist. Our institutions grew out of an assumed everlasting, politically dominant white majority. The nation has not yet imagined who we are together.

The phrase "toxic inequality" describes a powerful and unprecedented convergence: historic and rising levels of wealth and income inequality in an era of stalled mobility, intersecting with a widening racial wealth gap, all against the backdrop of changing racial and ethnic demographics.

I call this kind of inequality toxic because, over time and generations, it builds upon itself. Wealth and race map together to consolidate historic injustices, which now weave through neighborhoods and housing markets, educational institutions, and labor markets, creating an increasingly divided opportunity structure. So long as we have entrenched wealth inequality intertwined with racial inequality, we cannot even begin to bend the arc toward equity.

Toxic inequality is also noxious in that it makes these challenges harder to tackle. High levels of material inequality are inherently destabilizing, heightening social tensions. Janet Yellen, chair of the board of governors of the Federal Reserve System, has warned that economic inequality "can shape [and] determine the ability of different groups to participate equally in a democracy and have grave effects on social stability over time."[7] Thomas Piketty argues that extremely high levels of wealth inequality are "incompatible with the meritocratic values and principles of social justice fundamental to modern democratic societies" and warns that a drift toward oligarchy is a real danger.[8] The new inequality is especially politically poisonous because most people of all races feel stuck in place, finding it harder to believe that hard work, sacrifice, and innovation are going to pay off and lead to a better life. People are apt to look for someone to blame, and America's changing demographics encourage racial division, resentment of other groups, and prejudice. These forces have complicated economic policymaking throughout our history, but they are especially dangerous today, given the urgent need to address the particular economic disadvantages facing people of color.

We are just beginning to understand one further dimension of toxic inequality: a phenomenon we might call "toxic inequality syndrome." Are there emotional and even physiological consequences for families and individuals exposed to repeated, persistent economic trauma, frustrated ambitions, and cumulative downward spirals? We know that there is a strong relationship between adversity and social outcomes throughout the life course, with greater frequency of adverse events leading to worse outcomes.[9] One adverse event increases the likelihood of a cascade of other stressful and traumatic events. Research has documented the negative impact of a wide variety of stress-inducing events, including community violence, accidents, life-threatening illnesses, loss of economic status, and incidences of racism. We also know that financial resources shield families from economic and social trauma, lessen the impact of some trauma, enable more rapid recovery, and reduce the risk of subsequent adverse events.[10] Yet many of the families we spoke to experienced multiple forms of adversity—foreclosure, violence, unsafe neighborhoods, incarceration, disability, sudden or chronic family illness, family breakup, unemployment or loss of wages, declining living standards—without adequate wealth resources and without the sorts of family, institutional, community, or policy support that can also foster family resiliency. In the stories in subsequent chapters, we will encounter amazing resiliency, and we also will meet families who became overwhelmed by the stress and trauma associated with toxic inequality.

America's response to toxic inequality will set our future course for generations. The current magnitude of inequality robs the nation of human potential and promise, sapping aspirations and distorting futures. Earned achievements have become uncoupled from financial rewards and personal well-being. Frustrated ambitions and stalled social mobility foment racial anxieties. Without bold changes, we will keep heading toward greater inequality and become even more polarized along class and racial lines. The tiny segments of the population that are doing well will continue to do so, and the vast majority will try even harder just to stay in place. The rich and powerful will continue to write rules that protect and expand their vast advantages at the expense of those struggling to keep pace, especially younger adults and families and communities of color. As differences magnify, those groups facing the brunt of inequality, stalled mobility, and lost status will more critically interrogate the legitimacy of governmental and economic systems. Such an interrogation of deep structures is necessary and productive as long as it uncovers drivers of inequality. However, an explanation that does nothing more than pander to racial, ethnic, and class fears will short-circuit solutions. To avoid this bleak future and bend current trends in the direction of shared prosperity, we must transform the deep structures that foster inequality. Policy solutions must be bold, transformative, and at a scale sufficient to reach the families and communities most affected by toxic inequality.

NOTES

1. Signe-Mary McKernan and Caroline Ratcliffe, "Asset Building for Today's Stability and Tomorrow's Security," Urban Institute, 2009, http://www.urban.org/sites/default/files/alfresco/publication-pdfs/1001374-Asset-Building-for-Today-s-Stability-and-Tomorrow-s-Security.PDF.

2. Shapiro et al., "The Asset Security and Opportunity Index." Costs associated with each of these three different types of mobility investments amount to about $14,000.

3. Emmanuel Saez and Gabriel Zucman, "Wealth Inequality in the United States Since 1913: Evidence from Capitalized Income Tax Data," National Bureau of Economic Research, Working Paper No. 20625, 2014, http://www.nber.org/papers/w20625.pdf.

4. Ibid.

5. IASP analysis of "2013 Survey of Consumer Finances," Board of Governors of the Federal Reserve System, http://www.federalreserve.gov/econresdata/scf/scfindex.htm.

6. Thomas M. Shapiro et al., "The Roots of the Widening Racial Wealth Gap: Explaining the Black-White Economic Divide," Institute on Assets and Social Policy (IASP), 2013, http://iasp.brandeis.edu/pdfs/Author/shapiro-thomas-m/racial-wealthgapbrief.pdf; Signe-Mary McKernan et al., "Less Than Equal: Racial Disparities in Wealth Accumulation," Urban Institute, 2013, http://www.urban.org/research/publication/less-equal-racial-disparities-wealth-accumulation/view/full_report; Paul Taylor et al., "Wealth Gaps Rise to Record Highs Between Whites, Blacks and Hispanics," Pew Research Center, 2011, http://www.pewsocialtrends.org/2011/07/26/wealth-gaps-rise-to-record-highs-between-whites-blacks-hispanics; Dalton Conley, *Being Black, Living in the Red: Race, Wealth, and Social Policy in America* (Berkeley: University of California Press, 1999).

7. Building Strong Families Project, "The Long-Term Effects of Building Strong Families: A Relationship Skills Education Program for Unmarried Parents," MDRC, November 2012, http://www.mdrc.org/sites/default/files/bsf_36_mo_impact_exec_summ_0.pdf.

8. Marco Rubio, "Reclaiming the Land of Opportunity: Conservative Reforms for Combatting Poverty," Marco Rubio US Senator for Florida, http://www.rubio.senate.gov/public/index.cfm/press-releases?ID=958d06fe-16a3-4e8e-bl78-664fcl0745bf.

9. Elise Gould, Alyssa Davis, and Will Kimball, "Broad Based Wage Growth Is a Key Tool in the Fight Against Poverty," Economic Policy Institute, 2015, http://www.epi.org/publication/broad-based-wage-growth-is-a-key-tool-in-the-fight-against-poverty.

10. Rakesh Kocchar and Richard Fry, "Wealth Inequality Has Widened Along Racial, Ethnic Lines Since End of Great Recession," Pew Research Center, 2014, http://www.pewresearch.org/fact-tank/2014/12/12/racial-wealth-gaps-great-recession.

17

Asian American Women and Racialized Femininities

"Doing" Gender across Cultural Worlds

KAREN D. PYKE AND DENISE L. JOHNSON

The study of gender in recent years has been largely guided by two orienting approaches: (1) a social constructionist emphasis on the day-to-day production or doing of gender (Coltrane 1989; West and Zimmerman 1987), and (2) attention to the interlocking systems of race, class, and gender (Espiritu 1997; Hill Collins 2000). Despite the prominence of these approaches, little empirical work has been done that integrates the doing of gender with the study of race. A contributing factor is the more expansive incorporation of social constructionism in the study of gender than in race scholarship where biological markers are still given importance despite widespread acknowledgment that racial oppression is rooted in social arrangements and not biology (Glenn 1999). In addition, attempts to theoretically integrate the doing of gender, race, and class around the concept of "doing difference" (West and Fenstermaker 1995) tended to downplay historical macro-structures of power and domination and to privilege gender over race and class (Hill Collins et al. 1995). Work is still needed that integrates systems of oppression in a social constructionist framework without granting primacy to any one form of inequality or ignoring larger structures of domination.

The integration of gender and race within a social constructionist approach directs attention to issues that have been overlooked. Little research has examined how racially and ethnically subordinated women, especially Asian American women, mediate cross-pressures in the production of femininity as they move between mainstream and ethnic arenas, such as family, work, and school, and whether distinct and even contradictory gender displays and strategies are enacted across different arenas. Many, if not most, individuals move in social worlds that do not require dramatic inversions of their gender performances, thereby enabling them to maintain stable and seemingly unified gender strategies. However, members of communities that are racially and ethnically marginalized and who regularly traverse interactional arenas with conflicting gender expectations might engage different gender performances depending on the local context in which

SOURCE: Pyke, Karen D. and Denise L. Johnson. 2003. "Asian American Women and Racialized Femininities: 'Doing' Gender across Cultural Worlds." *Gender & Society*, 17(1): 33–53. Copyright © 2003 by Sociologists for Women in Society. Reprinted by permission of SAGE Publications, Inc.

they are interacting. Examining the ways that such individuals mediate conflicting expectations would address several unanswered questions. Do marginalized women shift their gender performances across mainstream and subcultural settings in response to different gender norms? If so, how do they experience and negotiate such transitions? What meaning do they assign to the different forms of femininities that they engage across settings? Do racially subordinated women experience their production of femininity as inferior to those forms engaged by privileged white women and glorified in the dominant culture?

We address these issues by examining how second-generation Asian American women experience and think about the shifting dynamics involved in the doing of femininity in Asian ethnic and mainstream cultural worlds. We look specifically at their assumptions about gender dynamics in the Euro-centric mainstream and Asian ethnic social settings, the way they think about their gendered selves, and their strategies in doing gender....

CONSTRUCTING FEMININITIES

Current theorizing emphasizes gender as a socially constructed phenomenon rather than an innate and stable attribute (Lorber 1994; Lucal 1999; West and Zimmerman 1987). Gender is regarded as something people do in social interaction. Gender is manufactured out of the fabric of culture and social structure and has little, if any, causal relationship to biology (Kessler and McKenna 1978; Lorber 1994). Gender displays are "culturally established sets of behaviors, appearances, mannerisms, and other cues that we have learned to associate with members of a particular gender" (Lucal 1999, 784). These displays "cast particular pursuits as expressions of masculine and feminine 'natures'" (West and Zimmerman 1987, 126). The doing of gender involves its display as a seemingly innate component of an individual....

GENDER IN ETHNIC AND MAINSTREAM
CULTURAL WORLDS

We study Korean and Vietnamese Americans, who form two of the largest Asian ethnic groups in southern California, the site of this research. We focus on the daughters of immigrants as they are more involved in both ethnic and mainstream cultures than are members of the first generation....

Respondents dichotomized the interactional settings they occupy as ethnic, involving their immigrant family and other coethnics, and mainstream, involving non–Asian Americans in peer groups and at work and school. They grew up juggling different cultural expectations as they moved from home to school and often felt a pressure to behave differently when among Asian Americans and non–Asian Americans. Although there is no set of monolithic, stable norms in either setting, there are certain pressures, expectations, and structural arrangements that can affect different gender displays (Lee 1996). Definitions of gender and the constraints that patriarchy imposes on women's gender production can vary from culture to culture. The Confucian moral code, which accords male superiority, authority, and power over women in family and social relations, has influenced the patriarchal systems of Korea and Vietnam (Kibria 1993; Min 1998). Women are granted little decision-making power and are not accorded an individual identity apart from their family role, which emphasizes their service to male members. A woman who violates her role brings shame to herself and her family. Despite Western observers' tendency to regard Asian families as uniformly and rigidly patriarchal, variations exist (Ishii-Kuntz 2000). Women's resistance strategies, like the exchange of information in informal social groups, provide pockets of power (Kibria 1990). Women's growing educational and economic opportunities and the rise of women's rights groups in Korea and Vietnam challenge gender inequality (Palley 1994). Thus, actual gender dynamics are not in strict compliance with the prescribed moral code.

As they immigrate to the United States, Koreans and Vietnamese experience a shift in gender arrangements centering on men's loss of economic power and increased dependency on their wives' wages (Kibria 1993; Lim 1997; Min 1998). Immigrant women find their labor in demand by employers who regard them as a cheap labor source. With their employment, immigrant women experience more decision-making power, autonomy, and assistance with domestic chores from their husbands. However, such shifts are not total, and male dominance remains a common feature of family life (Kibria 1993; Min 1998). Furthermore, immigrant women tend to stay committed to the ethnic patriarchal structure as it provides resources for maintaining their parental authority and resisting the economic insecurities, racism, and cultural impositions of the new society (Kibria 1990, 1993; Lim 1997). The gender hierarchy is evident in parenting practices. Daughters are typically required to be home and performing household chores when not in school, while sons are given greater freedom.

Native-born American women, on the other hand, are perceived as having more equality, power, and independence than women in Asian societies, reflecting differences in gender attitudes.... Indeed, the belief that gender equality is the norm in U.S. society obscures the day-to-day materiality of American patriarchy. Despite cultural differences in the ideological justification of patriarchy, gender inequality is the reality in both Asian and mainstream cultural worlds.

METHOD

Our sample ($N = 100$) consists of 48 daughters of Korean immigrants and 52 daughters of Vietnamese immigrants. Respondents ranged in age from 18 to 34 and averaged 22 years of age. Respondents either were U.S. born ($n = 25$) or immigrated prior to the age of 16 ($n = 74$), with 1 respondent having arrived at 18. Both parents of respondents were born in Korea or Vietnam. The data consist of 81 individual interviews and seven group interviews with 26 women—7 of whom were also individually interviewed. Data were collected in California between 1996 and

1999 using a convenience sample located through interviewers' networks and announcements posted at a university campus. We tried to diversify the sample by recruiting community college students and those who had terminated their education prior to receiving a college degree. College graduates or currently enrolled university and community college students compose 81 percent of the sample, and 19 percent are college dropouts or women who never attended college....

Gender loomed large in the accounts of female respondents, who commonly complained about parents' gender attitudes, especially the stricter rules for girls than for boys. We noted that women tended to denigrate Asian ethnic realms and glorify mainstream arenas. They did so in ways both subtle and overt and typically focused on gender behavior, although not always. Some respondents described different behavior and treatment in settings with coethnics compared to those dominated by whites and other non–Asian Americans. We began asking about gender in ethnic and mainstream settings in later interviews. In addition to earlier questions about family dynamics and ethnic identity, we asked if respondents ever alter their behavior around people of different ethnicities, whether people of different ethnicities treat them differently, and if being American and Vietnamese or Korean were ever in opposition. When necessary to prompt a discussion of gender, we also asked respondents to describe any time someone had certain stereotypical expectations of them, although their responses often focused on gender-neutral racial stereotypes of Asians as good at math, bad drivers, or unable to speak English....

GENDER ACROSS CULTURAL TERRAINS: "I'M LIKE A CHAMELEON. I CHANGE MY PERSONALITY"

The 44 respondents who were aware of modifying their gender displays or being treated differently across cultural settings framed their accounts in terms of an oppressive ethnic world and an egalitarian mainstream. They reaffirmed the ideological constructions of the white-dominated society by casting ethnic and mainstream worlds as monolithic opposites, with internal variations largely ignored. Controlling images that denigrate Asian femininity and glorify white femininity were reiterated in many of the narratives. Women's behavior in ethnic realms was described as submissive and controlled, and that in white-dominated settings as freer and more self-expressive.

Some respondents suggested they made complete personality reversals as they moved across realms. They used the behavior of the mainstream as the standard by which they judged their behavior in ethnic settings. As Elizabeth (19, VA) said,

> I feel like when I'm amongst other Asians... I'm much more reserved and I hold back what I think.... But when I'm among other people like at school, I'm much more out-spoken. I'll say whatever's on my mind. It's like a diametric character altogether.... I feel like when I'm with other Asians that I'm the *typical* passive [Asian] person and I feel like

that's what's expected of me and if I do say something and if I'm the
normal person that I am, I'd stick out like a sore thumb. So I just blend
in with the situation. (emphasis added)

Elizabeth juxtaposes the "typical passive [Asian] person" and the "normal,"
outspoken person of the mainstream culture, whom she claims to be. In so
doing, she reaffirms the stereotypical image of Asians as passive while glorifying
Americanized behavior, such as verbal expressiveness, as "normal." This implies
that Asian ethnic behavior is aberrant and inferior compared to white behavior,
which is rendered normal. This juxtaposition was a recurring theme in these data
(Pyke 2000). It contributed to respondents' attempts to distance themselves from
racialized notions of the typical Asian woman who is hyperfeminine and submissive by claiming to possess those traits associated with white femininity, such as
assertiveness, self-possession, confidence, and independence. Respondents often
described a pressure to blend in and conform with the form of gender that
they felt was expected in ethnic settings and that conflicted with the white standard of femininity. Thus, they often described such behavior with disgust and
self-loathing. For example, Min-Jung (24, KA) said she feels "like an idiot"
when talking with Korean adults:

With Korean adults, I act more shy and more timid. I don't talk until
spoken to and just act shy. I kind of speak in a higher tone of voice than
I usually do. But then when I'm with white people and white adults,
I joke around, I laugh, I talk, and I communicate about how I feel. And
then my voice gets stronger. But then when I'm with Korean adults,
my voice gets really high. ...I just sound like an idiot and sometimes
when I catch myself I'm like, "Why can't you just make conversation
like you normally do?"

Many respondents distanced themselves from the compliant femininity associated with their Asianness by casting their behavior in ethnic realms as a mere
act not reflective of their true nature. Repeatedly, they said they cannot be who
they really are in ethnic settings and the enactment of an authentic self takes
place only in mainstream settings. Teresa (23, KA) provides an example. She said,

I feel like I can be myself when I'm around white people or mixed
people. The Korean role is forced on me; it doesn't feel natural. I always
feel like I have to put on this act so that I can be accepted by Korean
people. I think whites are more accepting of other people. Maybe that's
why I feel more comfortable with them.

... The glorification of white femininity and controlling images of Asian
women can lead Asian American women to believe that freedom and equity
can be acquired only in the white-dominated world. For not only is white
behavior glorified as superior and more authentic, but gender relations among
whites are constructed as more egalitarian. Katie (21, KA) explained,

Like when I'm with my family and stuff, I'm treated like my ideas or
feelings of things really don't make a difference. I have to be more

submissive and quiet. I really can't say how I feel about things with guys if it goes against them in public because that is like disrespectful. With Caucasians, I don't quite feel that way. I feel my opinion counts more, like I have some pull. I think society as a whole—America—still treats me like I'm inferior as a girl but I definitely feel more powerful with other races than I do with my own culture because I think at least with Americans it's like [politically correct] to be equal between men and women.

Controlling images of Asian men as hypermasculine further feed presumptions that whites are more egalitarian. Asian males were often cast as uniformly domineering in these accounts. Racialized images and the construction of hegemonic (white) and subordinated (Asian) forms of gender set up a situation where Asian American women feel they must choose between white worlds of gender equity and Asian worlds of gender oppression. Such images encourage them to reject their ethnic culture and Asian men and embrace the white world and white men so as to enhance their power (Espiritu 1997). This was the basis on which some respondents expressed a preference for interacting with whites. As Ha (19, VA) remarked,

Asians would expect me to be more quiet, shy.... But with white friends, I can act like who I am.... With Asians, I don't like it at all because they don't take me for who I am. They treat me differently just because I'm a girl. And whites ...I like the way they treat me because it doesn't matter what you do.

In these accounts, we can see the construction of ethnic and mainstream cultural worlds—and Asians and whites—as diametrically opposed. The perception that whites are more egalitarian than Asian-origin individuals and thus preferred partners in social interaction further reinforces anti-Asian racism and white superiority. The cultural dominance of whiteness is reaffirmed through the co-construction of race and gender in these narratives. The perception that the production of gender in the mainstream is more authentic and superior to that in Asian ethnic arenas further reinforces the racialized categories of gender that define white forms of femininity as ascendant. In the next section, we describe variations in gender performances within ethnic and mainstream settings that respondents typically overlooked or discounted as atypical.

GENDER VARIATIONS WITHIN CULTURAL WORLDS

Several respondents described variations in gender dynamics within mainstream and ethnic settings that challenge notions of Asian and American worlds as monolithic opposites. Some talked of mothers who make all the decisions or fathers who do the cooking. These accounts were framed as exceptions to Asian male dominance. For example, after Vietnamese women were described in a group interview as confined to domesticity, Ngâ (22, VA), who immigrated at 14 and spoke in Vietnamese-accented English, defined her family as gender egalitarian. She related,

> I guess I grow up in a *different* family. All my sisters doesn't have to cook, her husbands cooking all the time. Even my oldest sister. Even my mom—my dad is cooking. . . . My sisters and brothers are all very strong. (emphasis added)

Ngâ does not try to challenge stereotypical notions of Vietnamese families but rather reinforces such notions by suggesting that her family is different. Similarly, Heidi (21, KA) said, "Our family was kind of *different* because... my dad cooks and cleans and does dishes. He cleans house" (emphasis added). Respondents often framed accounts of gender egalitarianism in their families by stating they do not belong to the typical Asian family, with "typical" understood to mean male dominated. This variation in gender dynamics within the ethnic community was largely unconsidered in these accounts.

Other respondents described how they enacted widely disparate forms of gender across sites within ethnic realms, suggesting that gender behavior is more variable than generally framed. Take, for example, the case of Gin (29, KA), a law student married to a Korean American from a more traditional family than her own. When she is with her husband's kin, Gin assumes the traditional obligations of a daughter-in-law and does all the cooking, cleaning, and serving. The role exhausts her and she resents having to perform it. When Gin and her husband return home, the gender hierarchy is reversed. She said,

> When I come home, I take it all out on him. "Your parents are so traditional, look what they are putting me through ... ?" That's when I say, "You vacuum. [Laughing] You deserve it." And sometimes when I'm really mean, "Take me out to dinner. I don't want to cook for a while and clean for a while." So he tries to accommodate that.... Just to be mean I will say I want this, he will buy me something, but I will return it. I want him to do what I want, like I want to be served because I serve when I'm with them.... [It's] kind of like pay back time. It's [a] strategy, it works.

Gin trades on the subservience and labor she performs among her in-laws to boost her marital power. She trades on her subservience to her in-laws to acquire more power in her marriage than she might otherwise have. Similar dynamics were described by Andrea (23, VA). She remarked,

> When I'm with my boyfriend and we're over at his family's house or at a church function, I tend to find myself being a little submissive, kind of like yielding or letting him make the decisions. But we know that at home it ain't gonna happen. ...I tend to be a strong individual. I don't like to conform myself to certain rules even though I know sometimes in public I have to conform... like being feminine and being submissive.

But I know that when I get home, he and I have that understanding that I'm not a submissive person. I speak my own mind and he likes the fact that I'm strong.

Controlling images of Asian men as hyperdomineering in their relations with women obscures how they can be called on to compensate for the subservience exacted from their female partners in some settings. Although respondents typically offered such stories as evidence of the patriarchy of ethnic arenas, these examples reveal that ethnic worlds are far more variable than generally described. Viewing Asian ethnic worlds through a lens of racialized gender stereotypes renders such variation invisible or, when acknowledged, atypical.

Gender expectations in the white-dominated mainstream also varied, with respondents sometimes expected to assume a subservient stance as Asian women. These examples reveal that the mainstream is not a site of unwavering gender equality as often depicted in these accounts and made less so for Asian American women by racial images that construct them as compliant. Many respondents described encounters with non-Asians, usually whites, who expected them to be passive, quiet, and yielding. Several described non-Asian (mostly white) men who brought such expectations to their dating relationships. Indeed, the servile Lotus Blossom image bolsters white men's preference for Asian women (Espiritu 1997). As Thanh (22, VA) recounted,

Like the white guy that I dated, he expected me to be the submissive one—the one that was dependent on the guy. Kind of like the "Asian persuasion," that's what he'd call it when he was dating me. And when he found out that I had a spirit, kind of a wild side to me, he didn't like it at all. Period. And when I spoke up—my opinions—he got kind of scared.

So racialized images can cause Asian American women to believe they will find greater gender equality with white men and can cause white men to believe they will find greater subservience with Asian women. This dynamic promotes Asian American women's availability to white men and makes them particularly vulnerable to mistreatment.

There were other sites in the mainstream, besides dating relationships, where Asian American women encountered racialized gender expectations. Several described white employers and coworkers who expected them to be more passive and deferential than other employees and were surprised when they spoke up and resisted unfair treatment. Some described similar assumptions among non-Asian teachers and professors. Diane (26, KA) related,

At first one of my teachers told me it was okay if I didn't want to talk in front of the class. I think she thought I was quiet or shy because I'm Asian. . . . [Laughing.] I am very outspoken, but that semester I just kept my mouth shut. I figured she won't make me talk anyway, so why try. I kind of went along with her.

Diane's example illustrates how racialized expectations can exert a pressure to display stereotyped behavior in mainstream interactions. Such expectations can subtly coerce behavioral displays that confirm the stereotypes, suggesting a kind of self-fulfilling prophecy. Furthermore, as submissiveness and passivity are denigrated traits in the mainstream, and often judged to be indicators of incompetence, compliance with such expectations can deny Asian American women personal opportunities and success. Not only is passivity unrewarded in the mainstream; it is also subordinated. The association of extreme passivity with Asian women serves to emphasize their otherness. Some respondents resist this subordination by enacting a more assertive femininity associated with whiteness. Lisa (18, KA) described being quiet with her relatives out of respect, but in mainstream scenes, she consciously resists the stereotype of Asian women as passive by adjusting her behavior. She explained,

> I feel like I have to prove myself to everybody and maybe that's why I'm always vocal. I'm quite aware of that stereotype of Asian women all being taught to be submissive. Maybe I'm always trying to affirm that I'm not like that. Yeah, I'm trying to say that if anything, I don't fit into that image and I don't want that to be labeled on me.

... To act Asian by being reserved and quiet would be to "stand out in a negative way" and to be regarded as "not cool." It means one will be denigrated and cast aside. Katie consciously engages loud and gregarious behavior to prove she is not the typical Asian and to be welcomed by white friends. While many respondents describe their behavior in mainstream settings as an authentic reflection of their personality, these examples suggest otherwise. Racial expectations exert pressure on these women's gender performances among whites. Some go to great lengths to defy racial assumptions and be accepted into white-dominated social groups by engaging a white standard of femininity. As they are forced to work against racial stereotypes, they must exert extra effort at being outspoken and socially gregarious. Contrary to the claim of respondents, gender production in the mainstream is also coerced and contrived. The failure of some respondents to recognize variations in gender behavior within mainstream and ethnic settings probably has much to do with the essentialization of gender and race. That is, as we discuss next, the racialization of gender renders variations in behavior within racial groups invisible.

THE RACIALIZATION OF GENDER: BELIEVING IS SEEING

... Among our 100 respondents, there was a tendency to rely on binary categories of American (code for white) and Asian femininity in describing a wide range of topics, including gender identities, personality traits, and orientations toward domesticity or career. Racialized gender categories were deployed as an interpretive template in giving meaning to experiences and organizing a worldview. Internal variation was again ignored, downplayed, or regarded as exceptional. White femininity, which was glorified in accounts of gender behavior across cultural settings, was also accorded superiority in the more general discussions of gender.

Respondents' narratives were structured by assumptions about Asian women as submissive, quiet, and diffident and of American women as independent, self-assured, outspoken, and powerful. That is, specific behaviors and traits were racialized. As Ha (19, VA) explained, "sometimes I'm quiet and passive and shy. That's a Vietnamese part of me." Similarly, domesticity was linked with Asian femininity and domestic incompetence or disinterest, along with success in the work world, with American femininity. Several women framed their internal struggles between career and domesticity in racialized terms. Min-Jung said,

> I kind of think my Korean side wants to stay home and do the cooking and cleaning and take care of the kids whereas my American side would want to go out and make a difference and become a strong woman and become head of companies and stuff like that.

This racialized dichotomy was central to respondents' self-identities. Amy (21, VA) said, "I'm not Vietnamese in the way I act. I'm American because I'm not a good cook and I'm not totally ladylike." In fact, one's ethnic identity could be challenged if one did not comply with notions of racialized gender. In a group interview, Kimberly (21, VA) described "joking around" with coethnic dates who asked if she cooked by responding that she did not. She explained,

> They're like, "You're Vietnamese and you're a girl and you don't know how to cook?" I'm like, "No, why? What's wrong with that?" [Another respondent is laughing.] And they go, "Oh, you're not a Vietnamese girl."

… American (white) women and Asian American women are constructed as diametric opposites. Although many respondents were aware that they contradicted racialized notions of gender in their day-to-day lives, they nonetheless view gender as an essential component of race. Variation is ignored or recategorized so that an Asian American woman who does not comply is no longer Asian. This was also evident among respondents who regard themselves as egalitarian or engage the behavioral traits associated with white femininity. There was the presumption that one cannot be Asian and have gender-egalitarian attitudes. Asian American women can engage those traits associated with ascendant femininity to enhance their status in the mainstream, but this requires a rejection of their racial/ethnic identity. This is evident by the use of words such as "American," "whitewashed," or "white"—but not Asian—to describe such women. Star (22, KA) explained, "I look Korean but I don't act Korean. I'm whitewashed. [Interviewer asks, 'How do you mean you don't act Korean?'] I'm loud. I'm not quiet and reserved."

As a result, struggles about gender identity and women's work/family trajectories become superimposed over racial/ethnic identity. The question is not simply whether Asian American women like Min-Jung want to be outspoken and career oriented or quiet and family oriented but whether they want to be American (white-washed) or Asian. Those who do not conform to racialized expectations risk challenges to their racial identity and charges that they are not really Asian, as occurs with Lisa when she interacts with her non-Asian peers. She said,

> They think I'm really different from other Asian girls because I'm so outgoing. They feel that Asian girls have to be the shy type who is very passive and sometimes I'm not like that so they think, "Lisa, are you Asian?"

These data illustrate how the line drawn in the struggle for gender equality is superimposed over the cultural and racial boundaries dividing whites and Asians. At play is the presumption that the only path to gender equality and assertive womanhood is via assimilation to the white mainstream.... This construction obscures gender inequality in mainstream U.S. society and constructs that sphere as the only place where Asian American women can be free. Hence, the diversity of gender arrangements practiced among those of Asian origin, as well as the potential for social change within Asian cultures, is ignored. Indeed, there were no references in these accounts to the rise in recent years of women's movements in Korea and Vietnam. Rather, Asian ethnic worlds are regarded as unchanging sites of male dominance and female submissiveness.

DISCUSSION AND SUMMARY

Our analysis reveals dynamics of internalized oppression and the reproduction of inequality that revolve around the relational construction of hegemonic and subordinated femininities. Respondents' descriptions of gender performances in ethnic settings were marked by self-disgust and referred to as a mere act not reflective of one's true gendered nature. In mainstream settings, on the other hand, respondents often felt a pressure to comply with caricatured notions of Asian femininity or, conversely, to distance one's self from derogatory images of Asian femininity to be accepted. In both cases, the subordination of Asian femininity is reproduced.

... Our findings illustrate the powerful interplay of controlling images and hegemonic femininity in promoting internalized oppression. Respondents draw on racial images and assumptions in their narrative construction of Asian cultures as innately oppressive of women and fully resistant to change against which the white-dominated mainstream is framed as a paradigm of gender equality. This serves a proassimilation function by suggesting that Asian American women will find gender equality in exchange for rejecting their ethnicity and adopting white standards of gender. The construction of a hegemonic femininity not only (re)creates a hierarchy that privileges white women over Asian American women but also makes Asian American women available for white men.

... These findings underscore the crosscutting ways that gender and racial oppression operates such that strategies and ideologies focused on the resistance of one form of domination can reproduce another form.

REFERENCES

Coltrane, Scott. 1989. Household labor and the routine production of gender. *Social Problems* 36:473–90.

Glenn, Evelyn Nakano. 1999. The social construction and institutionalization of gender and race. In *Revisioning gender*, edited by Myra Marx Ferree, Judith Lorber, and Beth B. Hess. Thousand Oaks, CA: Sage.

Hill Collins, Patricia. 2000. *Black feminist thought*. New York: Routledge.

Hill Collins, Patricia, Lionel A. Maldonado, Dana Y. Takagi, Barrie Thorne, Lynn Weber, and Howard Winant. 1995. Symposium: On West and Fenstermaker's "Doing difference." *Gender & Society* 9:491–513.

Kessler, Suzanne, and Wendy McKenna. 1978. *Gender: An ethnomethodological approach.* Chicago, IL: University of Chicago Press.

Kibria, Nazli. 1990. Power, patriarchy, and gender conflict in the Vietnamese immigrant community. *Gender & Society* 4:9–24.

———. 1993. *Family tightrope: The changing lives of Vietnamese Americans.* Princeton, NJ: Princeton University Press.

Lim, In-Sook. 1997. Korean immigrant women's challenge to gender inequality at home: The interplay of economic resources, gender, and family. *Gender & Society* 11:31–51.

Lorber, Judith. 1994. *Paradoxes of gender.* New Haven, CT: Yale University Press.

Lucal, Betsy. 1999. What it means to be gendered me: Life on the boundaries of a dichotomous gender system. *Gender & Society* 13:781–97.

Min, Pyong Gap. 1998. *Changes and conflicts.* Boston, MA: Allyn & Bacon.

Palley, Marian Lief. 1994. Feminism in a Confucian society: The women's movement in Korea. In *Women of Japan and Korea*, edited by Joyce Gelb and Marian Lieff. Philadelphia, PA: Temple University Press.

West, Candace, and Sarah Fenstermaker. 1995. Doing difference. *Gender & Society* 9:8–37.

West, Candace, and Don H. Zimmerman. 1987. Doing gender. *Gender & Society* 1:125–51.

18

From Transgender to Trans*

The Ongoing Struggle for the Inclusion, Acceptance and Celebration of Identities Beyond the Binary

JOELLE RUBY RYAN

In April 2015, Caitlyn Jenner (born Bruce Jenner) sat down with Diane Sawyer on ABC News for an in-depth, two-hour interview about her transition from male to female. The ABC interview was watched by over 17 million people and catapulted not only Caitlyn into the national consciousness, but the issue of transgender identity, gender transition, and the overall acceptance of gender diversity in our families and communities. The Diane Sawyer interview was followed up by a cover story in *Vanity Fair* magazine featuring Caitlyn, for the first time seen dressed as a woman, and shot by well-known celebrity photographer Annie Liebovitz. The cover quickly went viral and garnered widespread support, condemnation, curiosity, and fascination.

The response to Jenner within the transgender community was decidedly mixed. On the one hand, many were excited to have such a famous public figure come out so visibly and give the community a larger platform. With the exception of several other recent high-profile transgender personalities such as Laverne Cox, Chaz Bono, and Janet Mock, the media attention given to Jenner has been unprecedented in the modern epoch, and it has generated many conversations and dialogues around the globe about what it means to be transgender. In fact, local news affiliates across the United States seized on Caitlyn's coming out as an opportunity to publish and televise sympathetic portraits of local transgender communities, further helping to lift up a marginalized population from the shadows of invisibility into the light. However, multiple critics within trans communities were quick to address the case of Caitlyn Jenner with caution due to her overall social location and political orientation. Issues of fame, privilege, wealth, race and whiteness, appearance, medical versus nonmedical transitioning, femininity, and mass media were among the many topics that surfaced in transgender communities in the blogosphere and on social media.

---

*Editors' note: Some use the asterisk (*) with the term trans as an inclusive acknowledgment of the diversity among trans people.

These debates made me think about the topic of diversity within the transgender community, both in terms of reflecting upon my own identities and my many years of transgender activism, as well as to analyzing Caitlyn Jenner and other high profile transgender people in relation to the binary gender system. In this chapter, I would like to, in the spirit of the second wave feminist mantra, "combine the personal and the political" in order to analyze the ways in which contemporary iterations of transgender identity have generally failed to expand options beyond the binary of "man" or "woman" in the mainstream culture. Despite lip service to the contrary, I assert that the "umbrella" formulations of "transgender", "trans" and most recently "trans★" have often served to relegate those of us with nonbinary identities to the margins due to political expediency, internalized oppression within the trans community, and the desire for media validation and approval from mainstream culture. Greater acceptance for binary, medically transitioned transsexuals in the mainstream culture has caused GLBT advocacy organizations to center these experiences in their political work. The lack of nonbinary visibility has often caused trans★ folks who don't or can't "pass" to internalize dominant messages about their group. The impetus is strong to conform to mainstream narratives of transgender identity if one seeks any kind of sociocultural acceptance, validation, or encouragement.

To begin, I would like to reflect on my own history of coming out as transgender, and how my identities have affected my place within the political and activist spheres. Then, I would like to discuss how the media privileges famous, binary trans people and other privileged social identities in the new, supposedly more accepting cultural climate of gender and sexuality. Finally, I would like to proffer ideas for political, social and cultural interventions that move the conversation beyond the stale man/woman divide, to one that deploys an intersectional, queer, feminist approach in the fight for gender and sexual justice and liberation.

TRANSGENDER IDENTITIES IN THE 1990S: PERSONAL AND POLITICAL AWAKENINGS

I creaked open the closet door at age 17 when I told my therapist that I thought I was a transsexual. Back in the early 1990s, there was not yet a "transgender" community, but a "TV/TS" (transvestite/transsexual) community. This meant that you were either a heterosexual male ("TV") who cross-dressed on occasion for personal fulfillment, or a transsexual who completely transitioned to the "opposite" gender through hormones and surgery and totally disavowed their assigned gender role. I knew that my gender identity was in permanent incongruence with my assigned gender and body, and that cross-dressing on occasion would not suffice. Therefore, I came to identify as a transsexual, as that was the only other option that was known to me. At the time, a transsexual was widely understood as a person who was "trapped in the wrong body," i.e., a person who needed to make gender identity and body congruent through hormones, surgery, and a seamless transition. As soon as possible, I wanted to jump up onto the nearest operating table to be able to get sex reassignment surgery to make mind and body line up in the expected way. This was more difficult in those early days because there were not as many young people who were transitioning, nor was the use of hormone blockers even a possibility (blockers were not a possibility until 2007 in the United States). When I went to my first transgender support groups and social events in my early 20s, I found that I was the youngest person in attendance and that many of the other transsexual women (there were seldom transsexual men in attendance) that were present were in their 40s, 50s, and 60s. I believe that this is because transsexualism was so stigmatized and despised at the time that many people could not truly make the decision to come out as transsexual or to transition until middle-age or older.

In 1992, I first stepped foot onto the campus of my undergraduate institution, and I finally began to deal more intensely with my gender identity issues. I started to attend the GLBT group (which had just become trans-inclusive through my own efforts) and to talk seriously with my new therapist about my identity as a transsexual. However, my journey ended up being quite different than what I had previously thought or anticipated it would be.

In my sophomore year, I declared a Women's Studies major, and it truly transformed my life. It coincided with my discovery of the new emerging literature about gender diversity. Three of the first books that I discovered in the early to mid-1990s were *Gender Outlaw* by Kate Bornstein, *Stone Butch Blues* by Leslie Feinberg, and *The Apartheid of Sex* by Martine Rothblatt. All three of these brilliant authors introduced the idea of living beyond the gender binary to me by bravely defending gender complexity, multiplicity, and fluidity. Through fiction, memoir and theory, they discussed their desire to claim an identity that was neither man nor woman but something else altogether. Their texts revolutionized my own thinking as a member of the gender-variant community. Quite quickly, the "TV/TS" community was transforming into the *transgender* community. And what I found was the idea that there were more than two options (male OR female, cross-dressing OR complete medical transition), which was a startling and joyful revelation that prompted me to view my own identity in a very different fashion.

Through reading these and other authors and theorists, I discovered that it was possible to *not* identify as a man or as a woman, that there was a continuum and spectrum of infinite possibilities beyond our limited cultural concepts of gender and sexuality. I could take hormones, obtain surgeries or do neither. I could change my mind, and my identity could be fluid, shifting and transgressive. As soon as I read the work of Feinberg, Bornstein, and Rothblatt, as well as other authors writing cutting-edge gender theory like Judith Butler and Riki Wilchins, the idea of genders beyond the binary spoke to me in such an incredibly powerful way. This radical notion resonated with the very core of my being, and I knew that I was, in fact, a nonbinary transgender person who melded aspects of masculinity and femininity in one being. Despite my sexed assignment, I knew quite clearly that I was not a man. In this culture, that meant I was then a woman by default. But maybe I was not a woman either but something completely different, a horse of a different color. Understanding nonbinary gender identities enabled me to realize that I could express my gender however I wished, and I could customize my own transition in ways that resonated with my own complexity rather than conform to a cookie-cutter formulation of transsexualism created by non-transgender "experts."

However, what I have learned over time is that there is tremendous resistance to the notion of third, fourth, fifth, or any number of genders beyond one and two. I think a lot of this has to do with the fact that people cannot "wrap their heads around" identities beyond the gender binary because as genderqueer people we continue to be so culturally illegible. While there may be a small amount of space now allotted to trans people who go clearly from one gender to the other, those who are both or neither gender continue to be beyond the scope of many people's understanding. And sadly this misunderstanding and stigma happens in both the heteronormative society and within GLB communities. Nonbinary, genderqueer, gender-fluid, and neutrois identities, among others, represent a threat to the presumed tidiness of hegemonic categorizations. This "category crisis" may cause too much dissonance for some onlookers, the profound unintelligibility of our identities ultimately resulting in stigma, discrimination, and even violence in the most extreme cases.

In the gay and lesbian community, which often glorifies masculinity, trans women and trans-feminine people in general are particularly stigmatized. Part of what was so powerful for me as a young trans woman coming into feminism and Women's Studies was that it allowed me to connect the dots between my own marginalized identity and the discrimination I had faced my entire life. I was able to see that my identity as a trans-feminine subject was linked to the oppression that cisgender women and queer people face under a patriarchal system of gender and sexuality. Early trans-feminist voices like Susan Stryker, Sandy Stone, and Beth Elliott wrote rich and ground-breaking texts that weaved the personal, political, and theoretical strands of trans women's vibrant experiences into the feminist canon.

Becoming a feminist also helped me to begin to link gender oppression to other forms of oppression like racism, classism, and ableism. From my early 20s, intersectional feminism, particularly the work of women of color writers and theorists like Audre Lorde, Gloria Anzaldúa, bell hooks, and Patricia Hill Collins, gave me critical tools to analyze power differentials in society and envision ways of mounting efforts for equality and social justice that were inclusive of multiple identities and cognizant of multiple axes of oppression. I feel lucky that from the very beginning of my feminist education, these authors were front and center. This helped me to see the problem of white, middle-class bias in feminism and Women's Studies and quickly deduce that there was a similar problem with cisgender privilege and bias. The work of intersectional feminists thus illuminated these ideological and political blind spots, but also demonstrated the need to fight on for inclusion of our marginalized groups to change the face of feminism and the future of feminism.

By the mid-1990s, the notion of a distinct "transgender community" started to solidify in a much more visible and systemic way. I think this was important because under the previous medical model of transsexualism, the entire idea was that people were programmed to essentially disappear into their target genders as seamlessly as possible. So, for example, if somebody was assigned male at birth the goal was to completely transition by jumping through the required bureaucratic hoops, obtain hormones and surgery, complete the "real life test," pass well as women, and eventually be able to start an entirely new life for themselves, preferably in a new town, with a new job, and even abandoning previous social relationships. This was often referred to as "going stealth" or "woodworking." The older medical model of transsexualism essentially wanted people's transness to be effectively erased. This required going from one closet to another, hardly a progressive political approach to managing gender diversity. The emergence of transgender liberation helped to drive home the idea that the previous model perpetuated shame and self-hatred, and that to be open, visible and "out" as transgender people helped to promote self-acceptance and pride, an idea that completely resonated with who I was as a person, as a feminist and as an emerging activist.

TEARS IN THE UMBRELLA: LIMITS TO TRANS DIVERSITY AND INCLUSION

At that hopeful but naïve point in time of the nascent transgender movement, I believed that transgender would emerge victoriously as a truly inclusive movement with a "big tent" approach that would encompass all people who were in some way "gender outlaws." When I did education programs, for instance, I always talked about transgender as an umbrella term that included diverse gender constituencies like drag queens, drag kings, cross-dressers, transgenderists, transsexuals, bigender people, genderqueer folks and many more. In the beginning, transgender was very inclusive, at least rhetorically, of all of the various oppressed gender identities found in our society. However, even from the beginning, there were people who simply substituted "transgender" for "transsexual." Now, over 20 years into this social movement, what I find is that the transgender umbrella is seemingly not as inclusive as I had originally hoped, with more and more rips and tears appearing over time. As we have gotten more attention, and our issues have increasingly been taken up by popular media, the realities of people who live beyond the gender binary have been minimized in a race for trans acceptance and assimilation into mainstream society.

Just recently, my rural state of New Hampshire had an educational event on transgender issues with members of the state assembly. Within my state, I have been very involved with transgender community building, activism, and education since 1993. We currently do not have a law that prohibits discrimination in employment, housing, credit, and public accommodations on the basis of gender identity and gender expression in our state. We are the only state in New England that still does not have such protections. In 2009, such a bill was introduced in the state legislature, but right-wing conservative and Christian fundamentalist activists took what was a simple nondiscrimination bill, redubbed it as a "bathroom bill" and made the spurious claim that the bill would enable transgender women to go into women's restrooms to assault girls and women with impunity. The atmosphere of fear mongering and misinformation was ultimately successful and promptly killed the bill.

In the intervening years, there has been a push to build trans community in the state to promote our visibility and to conduct education efforts at the local and state level to foster acceptance of trans people in the Granite State. When I heard about this educational program, I was excited that the groundwork was being laid once again to go forward with introducing a new nondiscrimination bill. I expressed my interest in being on the panel with the organizer, who was a cisgender gay man. As someone who has been involved in transgender politics for such a long time in New Hampshire, I wanted to provide context for the history of our community and our efforts to end discrimination. After not hearing about whether I was participating, I was finally told that as a nonbinary person, my perspective would be too confusing for the people at the session, who were still struggling to understand transgender issues in the most basic sense. This was truly heartbreaking and infuriating to me. Approaches based in political expediency are very invested in particular outcomes and have a hyper-practical tactical strategy. Within political expediency, the end justifies the means. Often, the means are not ethical, principled or fair, with the idea being that politics requires constant compromise to reach desired objectives, thus legitimizing these problematic strategies. Needless to say, I find both the tactics themselves as well as their justification abhorrent and completely antithetical to the way in which I wish to engage in community and in social justice.

And it also felt like déjà vu. Back in my undergraduate days, many of us GLBT students would sit on educational panels for classes, in residence halls and in various departments across campus. There was a concerted attempt in those panels to put forward people who were gender conforming or "straight-acting." This meant that as a transgender person, I was frequently not invited. Likewise, my friends who were masculine lesbians, feminine gay men and other queers (including bisexuals) who did not conform to hegemonic gender norms and sexual binaries were also excluded. Sadly, this desire to hide or conceal the gender and sexual variance amongst us has a long history in the gay and lesbian community. My idealized notion of a utopic GLBT community was quickly dashed as I experienced the prevalence of biphobia and transphobia amongst many gay men and lesbians. "It is hard enough for people to understand being gay; bisexual and transgender people are just too confusing" was a frequent refrain that made many of my friends and I feel like outsiders even within the queer community.

Assimilationist politics are not new to the GLBT community, going back to the early days of the homophile movement in the United States. Assimilationism refers to the political strategy of targeted social groups trying to conform as much as possible to the culture of the dominant group. The idea is that adherence to heteronormative culture and values will make GLBT more acceptable to the dominant society and encourage their integration into the mainstream. Assimilationism is also closely related to respectability politics, upholding the idea that if people from marginalized groups act as much as possible like their oppressor, they will be better treated and more easily accepted. Like political expediency, assimilationism is a deeply flawed tactic that will always leave people behind and implicitly and explicitly endorse the notion that the culture of the dominant group is superior, preferable, and universal. The fact that now some parts of the transgender community are engaging in the same sort of assimilationist tactics of manufacturing a limited and conservatizing image of the community has been exasperating to me in so many ways. Clearly, we have not learned the lessons of our history that we can never throw enough people overboard, never jettison enough of the "different" people, to win approval from our enemies. For example, way back in 1973, radical trans woman of color and Stonewall combatant Sylvia Rivera was nearly booed off the stage of New York City Pride because of the assimilationist and anti-trans politics of gay and lesbian organizers. As I was completing this essay in June 2015, this tension was again perfectly exemplified by a GLBT Pride event held at the White House. While Obama was giving a speech to a bunch of well-heeled GLBT advocates, a transgender woman of color immigrant interrupted him to talk about the plight of immigrant transgender women and their horrific treatment in U.S. detention centers designated for men, by other detainees but also by Immigration and Customs Enforcement (ICE) officials. The woman, Jennicet Gutiérrez, was shushed, booed, and ultimately ejected from the White House. Obama gave her a thorough dressing down, to the applause of the many white, cisgender gays and lesbians present at the event. The media described the woman as a "heckler" rather than a transgender activist taking a principled stand against xenophobic, racist, and transmisogynistic state-sponsored violence. Mainstream media, including some GLBT media, are complicit with this silencing and desire to erase and hide women like Gutiérrez from the public consciousness.

DOMINANT CULTURE, MAINSTREAM MEDIA, AND THE POLITICS OF ASSIMILATION

As the above examples of the marginalization of Sylvia Rivera and Jennicet Gutiérrez show, there has to be a unified community, a big tent approach in which no one is left behind. Despite working extremely hard for transgender human rights for over two decades, in many ways I find myself on the outside looking in of GLBT politics. As the trans community inches ever closer toward acceptance and civil rights, I find that this acceptance is predicated on certain basic assumptions, expectations, and requirements. And we need only look once again to Caitlyn Jenner, to see how this so-called "acceptance" is actually quite limited and highly conditional.

If we're going to talk about public acceptance of trans people, I believe that Caitlyn Jenner is a particularly problematic person to focus on. Caitlyn Jenner, other than being a transgender woman, won the "privilege and power sweepstakes" and thus is hardly representative of the vast majority of transgender people. Caitlyn Jenner is white, wealthy, a former Olympic athlete, famous, camera-ready, thin, able-bodied, and a Republican. The vast majority of trans people in the United States are markedly different from Caitlyn Jenner along multiple vectors of difference. Many trans people in the United States are people of color, poor or working-class, liberals, progressives and radicals, people who are unable to access proper medical care, including life-saving surgeries, people who are nonbinary or unable or unwilling to "pass" and people whose marginality places them at risk of violence and severe discrimination at every turn.

While I saw many cisgender people seemingly acting like Caitlyn Jenner somehow invented transgender identity, it struck me that the desire to extend acceptance to Jenner was also because she was clearly locatable within the binary gender system. In fact, Jenner seemed to want to magically go from being a clearly masculine male to a clearly feminine female as is evidenced by the Diane Sawyer interview versus the *Vanity Fair* cover and 22-page article. While Jenner's transition has been many years in the making (and in fact the media and the paparazzi *did* notice that her body and appearance were changing), Jenner seemed to want a very unambiguous arrival to womanhood rather than easing into transition from male to female, thus concealing and making invisible any hint of gender ambiguity. This is, of course, her choice, and I do not mean to valorize one mode of transition over another. Rather, I am pointing out that it is vital to place Jenner's binary transition in the context of popular media culture, which has always privileged hyper-masculine men becoming hyper-feminine women, and vice versa, complete with the obligatory before-and-after photos. When I went to purchase the *Vanity Fair* magazine at my local drugstore, several of the employees could not wait to tell me how great they thought Caitlyn looked. While I appreciate that they indeed had something so positive to say about a transgender woman, I find it troubling that so much emphasis and attention has been placed upon Caitlyn Jenner's appearance, including her "passability" as a woman and ability to manifest adequate femininity.

I want to be clear that I unambiguously support every person's rights to their own unique gender expression. The most masculine man to the most feminine woman, and every unique and fabulous permutation in between and outside of hegemonic gender, should be fully embraced and celebrated. The problem has never been with people's gender expression, per se, including trans women who are high femme and trans men who are very masculine.

The problem is the framing of certain trans identities within the media, within the medical establishment, and within the dominant, popular imagination. I also reject the old radical feminist canard advocated by ideologues like Janice Raymond that transgender folks conserve or reinforce conservative gender roles, as if such a disenfranchised and relatively small minority has the power to do that rather than the cisgender and heterosexual men and women who make up the bulk of normative gender subjectivities. The problem is not that some people fit into traditional binary gender roles but the assumption that everyone *can or should* fit into such roles. In addition, the notions that trans women must "pass" as cisgender women to be fully accepted as women and conform to dominant notions of feminine beauty illustrate a conditional acceptance that is hardly transformative.

We need to affirm and accept people's gender identities, including their names and pronouns, whether or not they are legible to us as members of the gender they identify with. Let me return to my own personal experience to underscore this point. I am a 6'6", 400 pound, asexual, genderqueer, gender-fluid, disabled, visibly trans woman. Due to these intersecting identities, people have perceived me as male, female, and something else throughout my entire life. In addition, I have struggled with self-esteem and with finding role models that affirm my beauty as a fat, non-passing and nonbinary trans woman who is not hyper-feminine or high femme. I have already seen some of my trans women friends on social media who are remarking how they feel badly about themselves because they have not been able to achieve the femininity or the passability of Caitlyn Jenner. They also experience great sadness that hormone blockers or even early transition were not an option for them, further making it much harder to be perceived and accepted as women. The emphasis on passing, beauty, and femininity can have a very damaging effect on transgender women, and it also leaves out people who are unable to transition or who do not want to medically transition.

So many trans folks are economically marginalized, under- or unemployed, and struggle for the most basic dignity and respect in their day-to-day life. In Diane Sawyer's interview of Jenner, the plight of transgender women of color was very briefly mentioned but not at all adequately fleshed out or elaborated. My hope was that the interview would be used as a kickoff to further the conversation about the experiences of the most marginalized trans people in the United States and the violence and oppression they contend with each and every day. Unfortunately, as with what often happens in the media, Caitlyn's personal story took complete precedence over the realities of how our society is constructed in terms of not only gender identity, but race and class as well. It became about personal individual acceptance and acceptance from her family, but failed to look at the realities of systemic, institutional, macro-level discrimination and oppression like the fact that many transgender women must work in the underground economies to survive (sex work, dealing drugs, etc.) or face the violent horrors wrought by being put into men's units in gender-segregated facilities such as prisons, immigration detention centers, homeless shelters, etc.

I believe that this media interview resulted not only from the fact that Caitlyn Jenner is so incredibly privileged in multiple ways, but also because media likes to perpetuate the notion that we live in a society of equality and opportunity rather than one that is rife with bigotry, prejudice and systematic oppression. Emphasis is placed on whether one achieves self-acceptance and approval from one's family, as well as a "convincing" gendered appearance, rather than on the many societal barriers that lower trans people's life chances in multiple institutions from cradle to grave.

This new media hype includes trans men as well. Trans man Aydian Dowling recently made a splash by vying to become the first transsexual man on the cover of *Men's Health* magazine. This was preceded by a photo circulating of Dowling that features him in a pose that recreates nude photographs of musician and TV personality Adam Levine in *Cosmopolitan UK* magazine. In the original photo, Adam Levine is completely naked, save two entwined hands sporting bright red nail polish covering his genitals. In the new photo, we see a white man in the exact same pose, with similarly manicured and presumably female hands also covering his genital region. Many people might look at the photo and may not even know that Dowling is a trans man. The only possible giveaway might be the faint scars on his chest from chest surgery.

However, once again, as with Caitlyn Jenner, we see Dowling as having privilege in practically every other way except for his trans-ness. He is white, very masculine, medically transitioned through hormones and surgery, passable and he possesses the body and appearance which society considers "hot" (as with Levine) and therefore the confluence of these social identities make him of acceptable value in our patriarchal system. It is also easily understood as a part of assimilationist politics. "Look, he is just like a 'real man' except for his genitals!" The photograph also ironically recenters the genitals and the body more generally as the determinant of gender, a notion that the trans community has been fighting against for decades. The media's love affair with Dowling's photo homage to Levine and his desire to be on the cover of *Men's Health* was capped off by a well-publicized appearance on *The Ellen DeGeneres Show*.

At a workshop I gave on weightism, lookism and appearance in the trans community, we talked about how images such as those of Caitlyn Jenner, Aydian Dowling, and other high profile trans people like Chaz Bono, Laverne Cox, and Janet Mock affect the lives and emotions of everyday trans people. This included extended discussion about how these images negatively affected self-esteem, self-worth, and even hope for the future. Many trans people live in such dire economic jeopardy that they are not even able to access the health care that they need, including hormones and surgeries to alter their bodies in ways that can be life-sustaining in the face of persistent gender dysphoria. In addition, people of size and people whose appearance is not considered beautiful, handsome, "hot," sexy, etc., suffer from these images in that it reinforces the dominant cultural ideal of what constitutes attractive people and even who is worthy of dating, love, and sex. Weightism, lookism, and transphobia can combine as a lethal force in trans people's lives, furthering vulnerability for depression, suicide, exploitation, and even violence. This is not to demonize Caitlyn Jenner or Aydian Dowling or other public transgender people in any way as it is not really about them as individuals. Rather, it is about a social system that values certain bodies and certain identities while systematically devaluing, erasing, and marginalizing other bodies and identities.

From transgender or trans we have now arrived at trans★. Trans★ was specifically coined to call attention to nonbinary identities under the trans umbrella. The desire is to reclaim trans★ as a way that is not reducible to transsexuals or medically transitioning people firmly locatable within the binary gender system. Under the wide spectrum of gender-diverse, gender-expansive and gender-variant people, it was calling attention to and emphasizing those of us who are genderqueer, neutrois, gender nonconforming, multi-gendered, agender, gender fluid, etc. While I often now use trans★ in writing and very much like the impetus behind its creation and usage, I once again fear that whatever moniker we use, transgender, trans, or trans★, that we will fail as a movement and as a society to truly be inclusive of nonbinary gender identities. Moreover, we come up short of establishing gender multiplicity and fluidity as a foundational world view of the culture. Because of the confluence of corporate media culture and "reality" TV, the (often conservative) medical establishment, and increasingly assimilationist transgender politics, including within the GLBT nonprofit world, I am skeptical of genuine inclusion of nonbinary genders anytime soon. Those of us who stand outside of male and female or even trans man and trans woman continue to be relegated to the margins. My goal in writing this chapter is to articulate an increasing frustration that I have had as a transgender woman with feeling that even in my own LGBT community, a community which I have fought for and with for two decades, I and many others do not represent those who deserve liberation and civil and human rights because of entrenched political, social and ideological attitudes that devalue the most marginalized queer and trans★ community members among us.

CONCLUSION: TRANSFORMING TRANS★ POLITICS AND CELEBRATING QUEER DIFFERENCES

To conclude, I would like to sketch out a few preliminary ideas for interventions that push back against these trans- and homonormative tendencies and call for a renewed insurgence of queer and trans★ militancy and radicalism. While there are no easy answers to these complex problems, I would like to suggest that we question the state of contemporary GLBT politics, and its emphasis on assimilation, normalization, and respectability politics (see Warner 1999, for instance). Across the GLBT spectrum, there has been tremendous emphasis on marriage equality, military service, adoption of children, and other issues that strived to assimilate GLBT people into mainstream heteronormative institutions and dominant cultural life. While I understand the need for these equal legal protections and institutional access, we must continue to stress that often this work benefits the sectors of the GLBT community that already hold the most power and privilege. The "top down" approach to equal rights prioritizes long and protracted political battles that benefit the few at the expense of the many. They privilege moneyed, white, cisgender gay men, and lesbians while marginalizing people of color, poor and working-class queers and trans★ folks. Rather than a top down approach, we need a bottom up model that works to empower and protect those who are most materially disenfranchised, including nonbinary folks. Our visible queerness and gender nonconformity place us at the most risk for discrimination, economic jeopardy, and violence.

Part of this is moving away from the single-issue, liberal approach that GLBT politics has often clung to. Intersectional, queer transfeminism can advocate for political strategies that are attentive to the heterogeneity of our communities and that proffer sharp and incisive *analyses* of power differentials. We need to embrace dissidents, nonconformists, and outlaws and lift up their voices rather than rush to showcase the most normative among us to win respectability from our oppressors. From Compton's, to Stonewall, to Queer Nation, Act-Up and the Lesbian Avengers, we have a rich history of radical queer and trans★ resistance to oppression to inform our current efforts. We need to acknowledge that while the "incrementalism" of the mainstream GLBT movement has benefit for some, many queers are struggling mightily from class oppression, homelessness, the criminalization of HIV/AIDS, the lack of culturally competent medical care, police profiling and violence, the prison system, immigration issues such as detention and deportation, hate crimes, and much more.

We need to push back against a media that only allows people like Caitlyn Jenner and Aydian Dowling to narrowly represent the transgender community. In addition to demanding coverage of nonbinary identities, we need to challenge the stale media obsession with medical transition and individualized acceptance from cisgender family members and society and demand that we look at macro-level societal discrimination and oppression and what we need to do to transform our society to one based in liberation and justice. In media and beyond, we need to push for genderqueer and nonbinary inclusion and visibility, but we also need to demand that *everyone* look at the binary gender system and challenge it in their day-to-day lives. Issues of gender and sexuality are not only the province of LGBT people; they are issues for everyone and that matter to everyone. Beginning in kindergarten, we need to teach children that there is more to humanity than boy and girl, and that whatever their own identity, they can be and do whatever they wish in life. While the current transgender movement has made incredible strides toward creating greater gender and sexual freedom in our society, there is still tremendous work to liberate us all from the constraints of hegemonic gender scripts, sexual norms, and traditional models of family, relationships, and community.

While being an outsider and a misfit can be difficult and even painful, I also see it as the site of radical possibility that will continue to push the envelope of who we can be as human beings and the kind of world we can create. Too much of society is straitjacketed by gray, unimaginative and crushing demands for conformity, muting the potential of humanity. As queer people, we have the rainbow as our international symbol of strength and pride, but we still have tremendous work to ensure that people represented by *all* of the beautiful and dazzling colors of the spectrum have the ability to self-actualize and live a life unencumbered by systemic hate, violence, and oppression. While I struggle to engender such a world for myself and my comrades in struggle, I take strength from the resolve and perseverance of my trans★ and queer warrior ancestors like Sylvia Rivera and Leslie Feinberg. It is up to us to continue their rich and revolutionary legacy. Are you ready to join me?

BIBLIOGRAPHY

Bergado, Gabe. 2015. "Meet Aydian Dowling, the Trans Hunk Aiming for a 'Men's Health' Cover." *The Daily Beast.* 14 April 2015. Web. Accessed 10 July 2015.

Bissinger, Buzz. 2015. "He Says Goodbye, She Says Hello." *Vanity Fair.* July 2015. 50–69 and 105–106.

Bornstein, Kate. 1995. *Gender Outlaw: On Men, Women and the Rest of Us.* New York: Vintage Books.

Bornstein, Kate and S. Bear Bergman, eds. 2010. *Gender Outlaws: The Next Generation.* Berkeley, CA: Seal Press.

Bornstein, Kate (performer) and Susan Marenco (director). 1993. *Adventures in the Gender Trade.* Filmmakers Library, VHS.

"Bruce Jenner: The Interview." 2015. *20/20 with Diane Sawyer.* ABC News. 24 April 2015.

Butler, Judith. 1990. *Gender Trouble: Feminism and the Subversion of Identity.* New York: Routledge.

Collins, Patricia Hill. 1990. *Black Feminist Thought: Knowledge, Consciousness and the Politics of Empowerment.* Boston, MA: Unwin Hyman.

Elliott, Beth. 1996. *Mirrors: Portrait of a Lesbian Transsexual.* New York: Masquerade Books.

Feinberg, Leslie. 1993. *Stone Butch Blues: A Novel.* Ithaca, NY: Firebrand Books.

Gendernauts: A Journey Through Shifting Identities. 1999. Dir. Monika Treut. Perf. Sandy Stone, Susan Stryker, Texas Tomboy, Annie Sprinkle. First Run Features.

Nestle, Joan, Clare Howell and Riki Anne Wilchins, eds. 2002. *GenderQueer: Voices From Beyond the Sexual Binary.* Los Angeles, CA: Alyson Books.

Outlaw. 1994. Dir. Alisa Lebow. Perf. Leslie Feinberg. Women Make Movies.

Raymond, Janice. 1979. *The Transsexual Empire: The Making of the She-Male.* Boston: Beacon Press.

Rothblatt, Martine Aliana. 1995. *The Apartheid of Sex: A Manifesto on the Freedom of Gender.* New York: Crown Publishers.

Stone, Sandy. 1991. "The Empire Strikes Back: A Post-Transsexual Manifesto." Julia Epstein and Kristina Straub, eds. *Body Guards: The Cultural Politics of Gender Ambiguity*. New York: Routledge.

Stryker, Susan. 1994. "My Words to Victor Frankenstein above the Village of Chamounix: Performing Transgender Rage." *GLQ: A Journal of Lesbian and Gay Studies* 1.3: 237–254.

Warner, Michael. 1999. *The Trouble with Normal: Sex, Politics and the Ethics of Queer Life*. New York: Free Press.

Wilchins, Riki Anne. 1997. *Read My Lips: Sexual Subversion and the End of Gender*. Ithaca, NY: Firebrands Books.

19

More than Men

Latino Feminist Masculinities and Intersectionality

AÍDA HURTADO AND MRINAL SINHA

INTRODUCTION

During this historical moment, hegemonic masculinity is embodied at the specific intersections of race, class, and sexuality. It is currently defined as white, rich, and heterosexual. Because these social identities are privileged ones, they interact in ways that exclude specific groups of men from systems of privilege on the basis of their devalued group memberships. In other words, being a man of Color, gay, and working class or poor creates various obstacles to accessing the full range of male privilege. Hegemonic masculinity not only excludes certain groups of men from accessing aspects of male privilege in this way, it is an impossible ideal that many men are socialized to strive to attain, but cannot. According to Connell (1995), hegemonic masculinity "can be defined as the configuration of gender practices which embodies the currently accepted answer to the problem of the legitimacy of patriarchy which guarantees (or is taken to guarantee) the dominant position of men and the subordination of women" (p. 77). Hegemonic masculinity not only oppresses women, it also restricts men from engaging in certain behaviors, particularly those that would undermine the dominant position of men as a group.

Adhering to hegemonic conceptions of masculinity is associated with negative social and psychological consequences at the same time that it provides material privileges (Hooks 1992; Hurtado and Sinha 2005; Messner 1997).... Given the negative consequences associated with adherence to hegemonic masculinity, feminist engagement on the part of men can be a more constructive (and social justice oriented) response to the oppressive restrictiveness of masculinity as a social construct.

The present study explores the definitions of manhood provided by a sample of feminist identified, working class Latino men.... One of the insights that can be gleaned from these previous examinations is the idea that, while there are certain commonalities across different groups of men, masculinity is defined

SOURCE: Hurtado, Aída and Mrinal Sinha. 2008. "More than Men: Latino Feminist Masculinities and Intersectionality." *Sex Roles*, 59, 337–349. Reprinted by permission from Springer Science and Business Media.

in racially, culturally, and class specific ways. In other words, the fact that participants belong to different social groups (as organized around their common membership to the category of "man") has implications for how they define manhood, and fundamentally produces commonalities and differences in the way that this social construct is perceived. For example, young boys raised in immigrant families in the United States (or those who themselves immigrated) many times must contend with competing (and in some cases contradicting) discourses as they are socialized into masculinity (España-Maram 2006). The content of these competing discourses are dependent on the cultural background of individuals, which, along with the value attached to the specific culture in the host society, can function to create differences in the way that manhood is viewed. This socialization process is further complicated by the racial and class-based oppression experienced by many young men of Color, particularly when considered in conjunction with the male privilege bestowed on them within families (Hurtado and Sinha 2005). For young men experiencing privilege in the home, the harsh reality of material deprivation, growing up in dangerous neighborhoods, and witnessing violence against women (especially female family members) can create commonalities in experience (Collins 2000), which can potentially facilitate similarities in the way that manhood is perceived. In this way, race, class, and ethnicity can interact to produce complicated, group specific experiences of the social world. Such experiences can contribute to complex perceptions of what it means to be a man in contemporary U.S. society. These perceptions become even more complex when considered in light of feminist consciousness and highlights the importance of intersectionality for the present project—it allows for an exploration of participants' views toward manhood at the intersections of feminist consciousness, race, class, and ethnicity. More specifically, the present study uses an intersectional framework in examining the role of Latino culture and its affects on the complex perceptions of manhood with an educated sample of working class Latino men who identify as feminist.

The few qualitative studies with feminist identified men have focused predominantly on the experiences of working class and middle class white men. Findings from these projects have indicated four major dimensions associated with feminist masculinities—these include emphases on being an ethical human being, having emotionally healthy relationships with others (both women and men), being involved in activism and social justice oriented activities, and rejecting aspects of hegemonic masculinity. The aspects of hegemonic masculinity that participants rejected included male bonding around the objectification of women, physical and sexual domination of women, and homophobia. For many participants, the rejection of hegemonic masculinity was linked to their own experiences of class- and sexuality-based oppression. We attempt to document the effects of racial and class based oppression, in conjunction with feminist consciousness, on views toward masculinity....

The present study seeks to provide insights as to the way that *feminist* Latino men define masculinity. How do feminist, working class Latino men view manhood as a social construct? Do these views differ from their

African American and white counterparts? What discourses do they draw from in defining manhood? Further, we attempt to explicate the role that participants' multiple and intersecting social identities have in shaping definitions of masculinity. Intersectionality allows for such an analysis with men of Color in that it provides the opportunity to explore contradictory experiences of power (and disadvantage)....

METHOD

Participants

The data for this project come from a larger study of interviews conducted with 105 Latino men (Hurtado and Sinha 2006).... The topics addressed in both studies included issues related to early adolescence and dating, sexuality, gender, relationships with parents, political participation, and educational achievement. In this study, we examine only the gender issues portion of the interview. The interviews were semi-structured, utilizing open and closed ended questions, however, participants were encouraged to share experiences that they deemed relevant to the topics addressed.

Latino was defined as participants who had at least one parent of Latino ancestry. The majority of the participants (72%) were of Mexican descent, 4% were of Puerto Rican descent, 2% were of Central American descent, 7% were of South American descent, and 15% were of mixed ancestry. The participants were between the ages of 19 and 33, the average age was 24, and all were attending or had attended an institution of higher education. They were interviewed in five southwestern states—California, Colorado, New Mexico, Texas, and Arizona—as well as in Illinois, Massachusetts, Michigan, New York, and Washington, DC.

Of the 105 participants, 36 considered themselves feminist and identified their class background as poor or working class.... Participants were allowed to identify as feminist according to their subjective understanding of the term. Further, participants were asked to identify the economic background of their family while they were growing up....

Participants were contacted using the social networks of counselors, professors, student organizations, and personnel of student affairs offices on various college campuses. Specifically, an electronic message was sent describing the study and outlined the age and ethnic requirements for potential participants....

All interviews were conducted in English by the second author of this article (a 30-year old man of East-Indian descent)....

The transcripts from the interviews conducted with the 36 participants were coded and data analyses were conducted on the participants' responses to the question "What does the word 'manhood' mean to you?"...

RESULTS

Participants' Social Identifications—Embodying Intersectionality

… Participants provided identifications with their various significant social groups to varying degrees. All 36 participants identified as men and Latino as a prerequisite to be a part of the study. Further, all of the 36 participants considered in this study also identified the economic background of their family as working class or poor. In order to further explore the ways in which participants embodied intersectionality via identifications with their multiple social groups, we conducted a content analysis of the entire gender issues section of the interviews, providing a count of each time they referenced their race, ethnicity, class, and sexuality. The gender issues portion of the interview contained a total of 11 open and closed ended questions. The questions asked about participants' views on feminism and manhood, whether they felt anyone in their family was feminist, if there were strong women and men in their family, if they could provide examples of people they admired as men (as well as the reasons for admiring them), their views toward male privilege, and whether they considered themselves to be "men of Color." The frequencies presented below are in relationship to the interview questions outlined above.

… All 36 participants referenced their race in their narratives.… They talked extensively about being racialized by their families, communities, and society in general based on their phenotype—that is, whether they were light- or dark-skinned and whether they looked "indigenous" or "European" to others. If participants were fair skinned, they were aware how they were often confused as not being Latino both by other Latinos and non-Latinos. Further, a few participants discussed the ways in which their race had affected their experiences of higher education. This was particularly salient for participants who had attended private institutions, as they were many times one of the few working class students of Color in their classes and at graduation ceremonies. In some cases, participants wove references to race throughout their discussions of male privilege. These responses illustrated the way that their race interacted with their gender to complicate their experience of male privilege. Some felt as though their race kept them from being able to access patriarchal privilege in the same way that white men could. Albert Dominguez III, who was 27-years-old and working as a program coordinator at George Washington University, provides an illustrative example:

> Let's not forget we're Latinos. I am not a white male…if I was a white man I could say "Hey, I have certain privileges" in terms of societies I could get into or a certain door you can open a little bit easier…I am a Latino male…let me give you a better example of what I am trying to say. It's as if you are Black, you're Jewish and you're gay…that's the ultimate minority right there, right? So I feel like to a certain extent I am a male but I am a Latino male so if I was just a male… being of a different ethnicity or a different nationality there may be a little bit of extra perks.

Albert's narrative illustrates an awareness of the stigma attached to his racial identity and an understanding of how this stigma intersects with and limits his access to male privilege. He repeatedly brings up the fact that he is "a Latino male" in his discussion of male privilege. In other words, in talking about his gender identity, he also mentions his racial identity, and does so repeatedly in the same passage. Albert also demonstrates an understanding of the way that various disparaged social identities (e.g., being Black, Jewish, and gay) can act in combination to limit people's opportunities. His narrative is illustrative of the way many participants felt about their race as an important (stigmatized) social identity influencing their views of gender. Overall, participants were aware of their subordinate status in society based on their racial categorization, and viewed this as influencing their experience of being men.

Thirty-two participants mentioned their ethnicity in their narratives.... Participants discussing their ethnicity talked about speaking Spanish in the home, cultural practices, and their parents' immigration experiences. Jorge Morales, a 27-year-old doctoral student in comparative literature at the University of California, Berkeley emphasized the role of culture in the way that masculinity was constructed by saying:

> I guess it depends on what kind of manhood you're talking about,
> whether it's manhood as constructed in American culture or as it's con-
> structed in Mexican culture. I think they're very different constructions.

Twenty-two participants referenced their class in their narratives.... These participants discussed their parents' education levels, occupations and the economic hardships their families experienced while growing up. Jose "Nike" Martinez, who was 22-years-old and had graduated from California State University, Monterey Bay with a degree in Computer Science, was unemployed and looking for work at the time of the interview. He described the struggles his father went through while raising their family:

> He went to like the second or third grade and then he had to drop out
> of school to support his family. Once he gained his own family, which
> is us, he immigrated to the United States in search of work...back in
> Mexico, we had it really hard...it's just hard to make a living over
> there...he's pretty much worked all his life, he's worked in the fields,
> like lettuce and strawberries... and I've seen him get up everyday at like
> three or four in the morning and come back at like five or six in the
> evening and everyday doing this backbreaking job all sunburned...that's
> all he's done all his life, is work in the fields... he's done that for us.

Jose's narrative describes the reasons that he admired his father *as a man*, alluding to the way that ethnicity and class have interacted in his life to influence his views of what it means to be a good father (and ultimately what it means to be a man). It is interesting to note the way that class and ethnicity, via discussions of the physical hardships his father endured for the sake of the family, are interwoven throughout his response. As is seen in the above quotation, Jose references his class (via his father's educational level, occupation, and long work hours) and ethnicity multiple times in talking about why he admires his father as a man. In other words, Jose's narrative, like many of the other participants in the sample, illustrates the way that his Mexican immigrant background (his ethnicity) is inextricably tied to his class background to shape his views toward gender.

Finally, only eight participants referenced their sexuality in their narratives.... Two participants discussed the fact that they were gay, some talked about their heterosexuality in unproblematic ways (e.g., their relationships with girlfriends or partners), while others discussed how their heterosexuality complicated their notions of masculinity. This was particularly the case in terms of how heterosexuality bestowed them with unearned structural privilege and was stigmatized in racially specific ways. Issaac, a 25-year old elementary school teacher who obtained a master's degree from Colombia University, in discussing his sexuality, said that he had to "fight more stereotypes because" he was not a "macho male of color." He thought that "in people's minds" the prototypical "Latino male is Ricky Martin" or "Antonio Banderas," both of whom were represented in the media as that "suave model." Issaac felt as though this way of thinking was a "paradigm" that "still exists" and one that his version of heterosexuality did not "fit into." Issaac's response illustrates intersectionality in that his discussion of heterosexuality is intimately tied to his membership in a disparaged racial category—his experience of being heterosexual (a dominant social identity) cannot be separated from his experience of being a racialized man of Color (a stigmatized social identity).

These results suggest that participants were aware that they belonged to various social categories, and further, that some of these categories were disparaged and problematic. The way that they identified with these categories was indicative that their social identities were linked to their views toward gender. Participants illustrated this link by including discussions of their devalued social identities in their narratives addressing gender. This was particularly the case when considered in light of the way that ethnicity, race, social class, and to lesser degree, sexuality, interacted in their views of the way that masculinity was constructed in culturally specific ways, experiences of male privilege, and reasons for admiring men *as men*. An important finding here was that the social identities that participants discussed most (e.g., race, ethnicity, and class) were ones that were disparaged in certain contexts....

Definitions of Manhood: Relational

Eighteen participants mentioned relational definitions of manhood in their responses.... The participants who emphasized this theme gave long elaborate explanations of manhood being a developmental process that unfolded as individuals matured. The end point of the process was when an individual got married and raised children. As Andrés Elenes, a 26-year-old senior at MIT majoring in managerial science indicated:

> The word manhood, it's when... our mind matures enough that you start thinking as a grown adult... It's a person who from now on instead of thinking about himself is someone who starts thinking about repercussions about his actions for his family and for [his] community.

Andrés emphasis on manhood being a developmental process that culminates in a commitment to the family and community with which one lives is consistent with the way that manhood has been defined with middle and working class African American men (Hammond and Mattis 2005).

A second component of this theme was the notion that manhood can only be understood in relationship to cultural and community practices within families and groups of individuals. As Alberto Barragan, a 27-year-old medical student at the University of Michigan stated:

> Manhood to me is a culture... the ring of men at our family functions— manhood is being able to stand in that ring. And when you stand in that ring that means that you have a job... adolescents are able to stand in the ring even though they are low ranking members. You're a full member of that ring when you're married and have children.... Like I said, the men [in his family] tend to be quiet and passive. You don't brag about things. Manhood means being able to stand in that ring and talk and be respected, have an opinion... my father having all of his children in college is an incredible booster in the manhood ring.

Alberto's response demonstrates the way that ethnicity interacts with gender to influence his definition of masculinity. In particular, the fact that he emphasizes manhood as being constructed in the context of "a culture" and among "the ring of men" at family functions highlights the relational and culturally specific nature of such definitions, and echoes the words of Jorge Morales quoted above (that constructions of manhood vary from one cultural context to another). The emphasis on manhood being constructed in the context of relationships with family, culture, and community is also consistent with the findings of studies conducted with African American men (Hammond and Mattis 2005; Hunter and Davis 1992).

Definitions of Manhood: Positive Ethical Positionings

Twenty-two participants mentioned definitions of manhood based in positive ethical positions in their responses.... This theme was mentioned the most times across all three themes addressing participants' definitions of manhood. These participants felt that manhood entailed being ethical and standing behind one's word and not cheating or being untruthful, being a good human being, and respecting others. For example, Hugo Hernandez, a 21-year-old junior at the University of Arizona, said "manhood would be to work hard to respect people." Furthermore manhood was a commitment to viewing everyone as equals and honoring people as people and emphasizing their "humanhood." Issaac, the 25-year-old elementary school teacher quoted above discussing his sexuality, eloquently stated his views on the definition of "manhood":

> It's coming into one's own about being open to change and to new ideas but also staying strong to principles or values that you have set out for yourself... like Gandhi says, "being the peace that you wish to see";

it's like being the man that you wish to see in others... walking through the world in a way that is open but strong. In that sense, it's not only men; all people should be [that way]; kind of like a peoplehood, where we all learn to be strong but also collaborative and open to help and conversation, being open to dialogue about those things but also holding strong to whatever it is you bring to the table in whatever conversations you engage in; knowing who you are...To me that's coming into one's own about being a man or womanhood or peoplehood or personhood, I guess that's how I define it.

Issaac's views toward manhood highlight the influence of his feminist orientation insofar as the ethical characteristics he highlights directly contradict aspects of hegemonic masculinity. Specifically, he emphasizes "being open to change and new ideas" and equates this openness not only as a positive quality that men should strive for, but that all people, regardless of gender, should try and attain.... He emphasizes a version of masculinity predicated on a version of selfhood that stresses the importance of connections (as opposed to isolation), particularly in terms of being collaborative and receiving help from others.

Participants also felt that part of being a good and ethical person entailed being comfortable with one's self, being independent, and approaching things more confidently. Jesse Obas, who was 30-years-old and working for the Educational Partnership Center at the University of California, Santa Cruz, said:

Manhood is when you are comfortable with you're identity. I'm not saying complacent or that's all you want to achieve, I'm not saying it's the pinnacle of your manliness, but...for the longest time I was uncomfortable with who I was as a man and who I was as a person...I feel like right now I'm probably the closest I've ever been to the man, the person, the Chicano, the Filipino, that I've ever been...encompassing all those identities...I think that's what manhood is.

Jesse's response directly connects his views of manhood with his other, disparaged social identities. In particular, he defines manhood as something that is achieved when one is comfortable with their memberships in various social categories—in his case, this amounts to becoming comfortable with his (bi)racial identity as a Chicano, a Filipino, and ultimately as a person of Color....

Definitions of Manhood: Rejection of Hegemonic Masculinity

Fourteen participants provided definitions of manhood who rejected aspects of hegemonic masculinity in their responses.... These participants felt that definitions of manhood are in flux because of the intense questioning of gender and sexual roles and, as a result, definitions need to go beyond biology and the objectification of women as the basis for manhood. In particular, participants were concerned that hegemonic definitions excluded others from the rubric of manhood if they did not meet the "physical" requisites. Some were especially worried about excluding women who had the responsibilities usually assigned to men, such as being the main breadwinners in their families, and gay men because of their sexuality. Jesse Obas, the participant quoted above, also indicated that "manhood to me doesn't mean heterosexual, educated man...it could also mean gay, white or whatever." Other participants rejected particular behaviors associated with hegemonic masculinity. Ryan Ramírez, a 20-year-old sophomore majoring in philosophy at the University of Colorado, Boulder, provides an illustrative example:

> There's always that whole thing when you are younger...the whole virginity thing, you know, if you don't sleep with someone by the time you're this age, then you're not a man and I'm like, well whatever...I consider myself a man because I've done things.

Ryan explicitly rejects the notion that in order to achieve manhood, one has had to have engaged in sexual intercourse with a woman—an indication of heterosexuality. Instead, he considers himself a man because he has "done things." In so doing, he refers specifically to the fact that he has overcome economic obstacles in order to attain an education. Ryan traveled from Denver to Boulder, Colorado by himself, enrolled in courses, and was working his way through school independent of any financial support from his mother because she was the single head of household and had to take care of his sisters. His narrative is illustrative of the ways that participants refused to objectify women in defining masculinity, instead celebrating their educational accomplishments in light of their working class social identities. Ryan's rejection of the gender-specific, developmental ritual of sex with a woman fundamentally runs counter to one of the core behaviors associated with hegemonic masculinity (Collins 2004).

Furthermore, participants were concerned that hegemonic and normative definitions of manhood reinforced the negative aspects of masculinity. In particular they were concerned about personal negative characteristics of manhood that entailed harshness and domination of others. Among the dispositional characteristics enumerated were such things as being rude, aggressive, insulting others' beliefs, and not listening. Instead of including the negative characteristics as part of their definitions of manhood, participants enumerated the desirable ones that men should ascribe to, for example, being supportive, being less selfish and expressing emotions.

The last theme mentioned by participants was the rejection of manhood because it is a social construction that has no value—a construction that participants were openly rebelling against by deconstructing its meaning. As Jonathan Rosa, a 23-year-old doctoral student in anthropology at the University of Chicago stated:

> [Manhood] means a constructed image of masculinity... it means an idea that I am trying to fight against; something that I am trying to unsettle personally, and in the world individually, and among the social networks where I occupy different positions.... manhood is bullshit, basically.

From Jonathan's perspective, as well as from the perspectives of other participants, manhood was equated with patriarchy and undeserved male privileges. Patriarchy hurts everyone and therefore manhood is a false ideology that should be questioned and eventually obliterated and replaced with more equitable arrangements between people. Jonathan's views toward manhood, in addition to the other participants endorsing this theme, echo the words of other feminist men (Vicario 2003; White 2008) in that he is not only advocating for a fundamental restructuring and transformation of social relationships, he provides an outright rejection of masculinity as a social construct. This point is made most poignantly in his summation that "manhood is bullshit, basically."

DISCUSSION

... In defining what manhood meant to them, participants applied their feminist and class consciousness in complex ways. They wove in and out of definitions that were relational, ethical, and that rejected aspects of hegemonic masculinity. For example, some participants would discuss relational and ethical definitions of masculinity, while rejecting aspects of hegemonic masculinity *in the same narrative*. In doing so, they provided definitions of manhood that ran counter to mainstream ideas of what it means to be a man in the United States, while simultaneously drawing on positive aspects of hegemonic masculinity. This was particularly the case in terms of the relational and ethical definitions of manhood provided by some participants. Aspects of each of these themes (e.g., sacrificing and being committed to the welfare of the family, being respectful and standing up for one's word), have been conceptualized as positive components of hegemonic masculinity worth rescuing in the reconstruction of masculinity.... Further, participants rejecting aspects of hegemonic masculinity demonstrated the impact of ascribing to feminist ideology by positioning themselves in direct opposition to dominant conceptions of manhood....

Participants provided definitions of manhood that were also influenced by their membership in various social groups. This was especially salient in relationship to their race, ethnicity, social class, and sexuality (although sexuality played a lesser role in the context of this sample). They defined manhood in ways that integrated their cultural background, racial categorizations, social class, and in some cases, the questioning of their heterosexual, male privilege....

Participants in this sample defined manhood in ways that emphasized emotional connections with others, being open to change and help from others, being collaborative, and being comfortable with one's multiple (and in some cases, derogated) social identities. In other words, participants redefined masculinity in ways that allow men to experience the full range of the human experience (e.g., emotional expression, meaningful relationships with family, community, and others) unencumbered by the restrictions imposed by traditional masculine gender roles (Pleck 1981). They defined manhood in ways that let men be *more than men*.

REFERENCES

Collins, P. H. (2004). *Black sexual politics: African-Americans, gender, and the new racism.* New York: Routledge.

Connell, R. W. (1995). *Masculinities.* Oxford: Polity Press.

Hammond, W. P., & Mattis, J. S. (2005). Being a man about it: Manhood meaning among African-American men. *Psychology of Men & Masculinity,* 6, 114–126.

Hooks, B. (1992). *Black looks: Race and representation.* Boston, MA, USA: South End Press.

Hunter, A. G., & Davis, J. E. (1992). Constructing gender: An exploration of Afro-American men's conceptualization of manhood. *Gender & Society,* 6, 464–479.

Hurtado, A., & Sinha, M. (2005). Restriction and freedom in the construction of sexuality: Young Chicanas and Chicanos speak out. *Feminism & Psychology,* 15, 33–38.

Messner, M. A. (1997). *Politics of masculinity: men in movements.* Thousand Oaks, CA, USA: Sage Publications.

Pleck, J. (1981). *The myth of masculinity.* Cambridge, MA, USA: MIT Press.

Vicario, B. A. (2003). *A qualitative study of profeminist men.* Doctoral Dissertation. Auburn University.

White, A. (2008). *Ain't I a feminist: African-American men speak out on fatherhood, friendship, forgiveness, and freedom.* Albany, NY: SUNY Press.

20

Keep Your "N" in Check

African American Women and the Interactive Effects of Etiquette and Emotional Labor

MARLESE DURR AND ADIA M. HARVEY WINGFIELD

INTRODUCTION

Perhaps more than any other First Lady since Hillary Clinton, Michelle Obama faces considerable pressure to transform, change, and adapt her persona to become more palatable to a broad spectrum of voters. She has been alternately cast as unpatriotic and as an angry, dangerous black woman (Blitt, 2008). Even while she is admired for her fashion sense, she is simultaneously carefully scrutinized and often perceived as too aggressive and pushy.

While many of the same criticisms were applied to Hillary Clinton in 1992, intersections of gender and race create a particular and somewhat unique situation for Michelle Obama. Hillary Clinton was cast as an overbearing, emasculating feminist, but Michelle Obama faces gendered racial stereotypes of the "angry black woman," an image grounded in the Sapphire stereotype of black women as domineering, vociferous, and curt. Like the image of the angry black woman, Sapphire serves to reinforce ideas of black women's inherent lack of femininity and worth (Collins, 2004). Thus, Michelle Obama has struggled to distance herself from these stereotypical images and behaviors in hopes of altering the way she is perceived.

Her challenge is familiar to many professional black women, who like her must transform or alter themselves to be welcomed and accepted in their workplaces. The nature of social relationships in the office is dictated by historical customs, which have been a traditionally white male citadel. Not until after Executive Order 11246 mandating affirmative action was implemented... did employers actively seek ways to include women and racial minorities within organizations' and agencies' hierarchies. Men, the carriers of organizational culture and authority (Acker, 1990; Kanter, 1977) created this new bureaucratic arm

SOURCE: Durr, Marlese, and Adia M. Harvey Wingfield. 2011. "Keep Your 'N' in Check." *Critical Sociology* 37: 557–571. Copyright © 2011. Reprinted by Permission of SAGE Publications.

of professional and managerial expansion, making decisions regarding acceptable behavior, communication, skin color, style and dress, (e.g. dark suits, conservative dresses, white shirts, low heels, and no flashy jewelry, hair, or make-up) concentrating on who their clients are and their cultural tastes. This has had implications for women of all races who become employed within this new-found bureaucratic configuration of organizational norms, but has had particular consequences for women of color, who do not fit either the gendered or racialized norms of these environments.

By establishing an implicitly gendered and racialized culture, obstacles remain for women. Acker (1990) and Kanter (1977) argue that organizations are not the gender-neutral bureaucracies they purport to be. Rather, they are gendered in ways that often locate women in dead-end jobs, with exposure to the organizational hierarchy as tokens. Yet this hierarchy is two-tiered. For women of color, especially those in professional posts, these disadvantages are further complicated by race. These women often explain that upon entering the labor market as professionals, they alter their behavior by changing their look, conversation content, and style to fit in, but also to be promoted (Jones and Shorter-Gooden, 2003). They often speak of performance weariness when describing their spoken and unspoken communicative interaction exchanges with white colleagues. Many simply state that they feel they are in a "parade" where they are being judged for appearance, personal decorum, communication skills, and emotion management in addition to work productivity.

To address this, senior professional women like Kenya counsel: "That's not professional. Remember they got the s[hit] that'll get you bit! Keep your Negro in check! Don't let it jump up and show anger, disapproval, or difference of opinion. They have to like you and think that you are as close to them as possible in thought, ideas, dress, and behavior." Her advice discloses the appropriate etiquette, behavior, and emotion management, but also instructs other black women to blend manners, behavior, and reaction to fashion satisfactory workplace deportment. The counseling given to these women is directly linked to handling stress and alienation while balancing a need for survival and safety in the workplace or remaining employed without a row.

Many of these women have stated, "The work is too much. I get tired of being 'on' for [white] colleagues who scrutinize every behavior. So every now and then, I lose it." Others have said, "They [white supervisors] make deals about the next position or talk about my future being bright. They say I have time, so the next promotion available is mine. But it never happens. For some reason an organizational change erases the promised promotion. It just never happens, unless there is pressure to promote black." Even then, according to many of these women, African American "institutional gatekeepers" are consulted, who are often black women and men that possess institutional acceptance, but are not advisors that navigate new employees through their frosty work environs. Consequently, the challenges these women face involve doing the necessary work to fit in, and managing their feelings in an often inhospitable workplace in hopes of capturing some degree of professional success in the form of promotion.

Working in predominantly white agencies, organizations, and institutions, while living and working as "black," may cause part of these women's apprehension and estrangement. The ever present reminder of their master status—skin color—makes the work week a bit more difficult and requires a bit more strength. Many believe they continue to carry stereotypical and media-based depictions of them as domineering, unaccomplished breeders, whores, welfare queens, as well as confrontational. Bebe reports an administrator at her school saying "I am not afraid of aggressive women." So, "performance" becomes their safety mechanism.

These workplace contexts exact an additional toll as black women also must manage the demands of emotional labor in these work environments. Coined by Hochschild (1983), emotional labor describes women's work experiences in service economies as producing emotions in themselves or others. Emotions, which are typically gendered, become commodified and sold for a wage. Doing emotional labor, women are often expected to recreate gender appropriate feelings, e.g. paralegals are expected to make men attorneys feel cared for (Pierce, 1999), while male bill collectors induce fear and intimidation (Hochschild, 1983). However, for African American women, performance related to whites' gendered and racialized expectations is rarely accounted for or described.

Research that describes the down-to-earth contexts and contests for African American women in organizations, institutions, and agencies continues to be limited and comparative when describing mobility outcomes for black and white women. We seek to situate black women's voices and experiences at the center of discussions and research on occupational mobility by examining the interactive effects of manner, behavior, and reaction or etiquette and emotional labor. We argue that this aperture in the literature lacks a definition of informal and/or formal race-based boundary maintenance in the workplace for black women. Etiquette and emotional labor for African American women is defined as performance to describe two levels of personal deportment:

1. a generalized bureaucratic passive aggressive level; and

2. a race-based set of expectations grounded in survival strategies to cope with challenges they face in environments that are unwelcoming and possess concrete ceilings across organizations and occupations....

DATA AND METHODS

Data for this article was collected through two sources. One involved using direct participant observation from 2005 to 2007 while in conversation with 20 African American women over occupational mobility issues. Because conversations addressed issues in the workplace and their feelings about executives, managers, supervisors, and co-workers, one author listened attentively to concerns presented, speaking only when directly pulled into the discussions, as a sociologist who examined race/gender and would write about these tete-a-tetes. These women were employed as lower-level executives, middle managers, and administrators in public and private organizations, state and city government, and universities.

The use of a snowball or probabilistic sampling procedure was not possible, because many women did not wish to complete surveys and stated their desire to remain anonymous, and in many cases used coded responses or vernacular terms which shielded bureaucratic personalities they spoke of (e.g., "boyfriend," "girl-friend"), while ever so briefly and scantily describing issues in their workplaces. Participant observation allowed one author to capture etiquette and emotional labor responses in their work-social relationships. Conversations took place at sorority and fraternity dinner dances, church and community luncheons, occasional meetings at work or while shopping at local grocery or department stores. Pseudonyms will be used for confidentiality.

All respondents were college graduates, with master's and doctoral degrees, ranging in age from 30 to 55 years of age, five were married, five were divorced, and 10 had never married. Respondents were from southern New England, mid-Atlantic, and Midwestern states and employed in professional or managerial positions that contained less than 10 black employees throughout the levels of administration or management, and required working effectively in workgroups as an integral factor for employment success and /or future promotability.

Additional data came from a larger sample of 25 in-depth, semi-structured interviews with African American professionals. These respondents were located through a snowball sample. Interviews focused on the use of emotional labor in predominantly white workplaces. These respondents were employed in various professional posts in work environments where they estimated African Americans constituted 10 percent or fewer of professional employees. Our questions centered on whether and when respondents had to control emotions at work, in response to what occurrences, the frequency of this practice, and the nature of the emotions being controlled. Respondents also discussed producing emotions in others and the context in which this occurred....

FINDINGS

For professional black women, the performances that they feel compelled to give are shaped by the ways intersections of race and gender isolate them and place them under greater scrutiny. As they take stock of their work environments and perceive colleagues' stereotypes, beliefs, and preconceptions, these women learn that, like Michelle Obama, they must repackage themselves in ways that are more palatable to their white co-workers. As these colleagues' goodwill and collegiality is necessary for advancement and occupational stability, black women professionals find themselves doing both surface acting and emotional labor in order to successfully integrate their work spaces (Hochschild, 1983)....

Many of these women work in administrative roles. To an extent, this shapes the ways in which they must engage in careful self-presentation. Their positions may not originate within the organization, or link to its mission and objectives, but the series of administrative posts they have held or hold, regardless of their nature, provides an avenue and motivation for advancement. Tandy states, "I got my job as a manager because they had no black managers, but guess what I manage? I manage compensatory education staff and programs." Elizabeth says, "I do the same, but it's the only way to become a manager at ... This place does not care about their students, just the federal aid and visibility of their darkies. That's when they get them and if they stay. You are on for them all the time. But you take what you can get."

When, like Michelle Obama, these women advance, their self-presentation and communication style are scrutinized. Barbara reports, "Being direct and speaking your mind is never encouraged. In fact if you do, you encounter a world of silence and avoidance, which is one of the most severe penalties. You are placed outside of the loop, and you may stay there for a long time. Quite possibly—permanently. So despite the fact you may have a contribution, it is not welcomed." So, as they learn the verbal and body language of bureaucracy, they must negate values and styles of communication developed as a survival skill in their community. Most say they feel defenseless.

Nevertheless, for many, negotiation within this environment in most cases occurs at a cost. Sharon reports, "You learn to remain quiet and speak when spoken to and never verbalize your thoughts on an issue or policy." Others, such as Rhonda, suggest that "You go with the flow, since you realize this may not be a battle you can win, despite the fact you may be correct in assumptions and remedies you have in mind." They apply a survival-safety analysis and render their verbal participation to a lower-level of priority. They smile on cue, remaining expressionless, unmoved by the content of conversation even if the subject is distressing or controversial, and carefully couching responses in the language of the workplace. The emotion that is concealed is evident in their voices, body language, and style of conversation. Keena states, "But you have to endure this if you want to get ahead, regardless of where you might be promoted to." Mary says, "Promotions aren't everything, surviving them is. I got kids in college, a mortgage, and my hair to keep up." Laughter follows Mary's comments, but all understand what she means and silently accept her pronouncement.

Black women also engage in etiquette and emotional labor to cope with feelings of alienation and loneliness that stem from being the only, or one of few. Harlow (2003) has documented the existence of emotion management among black college and university professors. Feelings of anger, frustration, and aggravation are often stifled to conform to colleague expectations. Janice, an assistant Dean at a small liberal arts college, states, "I've dealt with people who were so dismissive, and you knew race was at the core of it. But I would have to grin and bear it, because I needed to work ... very rarely do you see me expressing my true feelings, and when I do, my reaction tells me it scares the hell out of [my white colleagues]." Irene offers a very astute assessment of the importance of concealing emotions: "If you don't play by their rules in terms of your behavior ... modulating your emotions [and] if you don't do certain things the way they want you to do them, it has a direct impact on your career and your economic stability." Similarly, Gina states, "I have to be very congenial. Sometimes I don't want to deal with [racism]. But I just have to hide my real emotions."

For many of these women, "performance" becomes their safety mechanism. In some cases, these performances are conscious and intentional—what Hochschild (1983) describes as "surface acting," where the individual is well aware that they are putting on a show. In an example of this, Barbara states, "I always prepare because I want to make sure my temper is in check. You know, I'm mellow. I have unhooked from personal feelings, previous conversations to deal with what's ahead. I do this because so few [black] women are in managerial and executive level posts." Barbara's performance is purposely crafted, specifically designed for the racial parameters of her work setting.

More often than not, workplace dynamics place these women in positions where routine interaction with co-workers heightens these performances. For instance, Deidre, a faculty member at a southern New England university, says "There is a "double standard" in departmental decisions regarding promotion and tenure." She reports, despite her service on the departmental committee to determine promotion procedures, the procedures have been used arbitrarily by those in authority. She notes, "African Americans are denied promotion while whites with fewer publications and lower scholarly profiles were promoted without question." Sheila reports slights and loud verbal assaults in the hallway by department colleagues and secretarial staff. She felt isolated and alienated by colleagues who made a concerted effort to have her removed from the teaching staff. Like Sheila, Janice, a bi-racial woman found herself in a verbal confrontation with her white female department chair in the doorway of her office. This confrontation was the culmination of several months of what she perceived as constant harassment regarding petty issues. During the verbal confrontation, she was poked in the chest by the chair as she loudly chastised her publicly.

When incidents like these occur, emotional labor becomes a key part of the resulting performance. Charlotte, the only black attorney at a mid-sized firm, states, "I think the one that stands out the most is that I am a black professional woman in [names town]. And as much as I'd like to be upset about it, I can't ever do that in my job because it would come off the wrong way. So I have to be happy-go-lucky, everything's great, and the fact that I don't have a life is wonderful when it really sucks!" Similarly, Giselle states, "At one point I was the only person of color who was not cooking, cleaning, or maintaining the grounds. It got so bad that the [black] men, they had their social support group and they invited me to join because they knew how that felt." This respondent's comments led us to revisit Rosenbaum's (1979) tournament mobility thesis through conversation. Giselle stated, "So you mean we need to be seen and supported as one of the stars." We say, "Yes." She says, "Right, that will take until I retire." Shana remarked that she feels she will never be promoted. For black women, internal promotions are a primary means of career mobility since most organizational vacancies are filled from within organizations, often as a matter of official and unofficial policy, making promotions a high reward for white collar workers (Markham et al., 1985)....

CONCLUSION

The number of African American women who now try to penetrate the "concrete ceiling" recognize promotions as their primary avenue of career mobility, given that a great many organizational vacancies are filled within and/or across organizations, often as a matter of official organizational policy. A great majority of the time these considerations take place without regard for their race and gender. They work even harder to climb the ladder to receive each position's resources and rewards.

Generally, emotional labor enables black women to present the appropriate emotional veneer that allows them to fit in and enhance their compatibility with organizational norms. This is particularly useful when confronted with racial issues. Their desire to move up the ladder in many ways represents a voluntary job shift with the same employer and acts as a proxy for vertical advance in the firm or with different employers. They are aware that occupational mobility also depends upon personal, social, and cultural characteristics. More important, they understand that the process of occupational mobility differs depending upon whether it occurs within a particular organization or involves a change of employers (Femlee, 1982). It may be argued that this is true for all women, but nuanced when it comes to African American women.

Or is it that many African Americans know what Bell and Hartmann (2007) suggest: that race is ever present in workplace diversity discussion, but missing in action? The discussions about race do not necessarily translate into decisive action intended to minimize racial inequality. African American women's need to present the appropriate etiquette and emotional labor is likely shaped by tenuous commitment to racial diversity in professional workplaces.

The levels of behavioral expectation and exceptions are both boundary maintenance mechanisms for whites and social-psychological issues of safety for African American women. While calls for inclusiveness and diversity weave seamlessly at work and in the larger society, rejection of difference remains as well. Our analysis reveals that social expectations shape individuals' fit at work and in society, while prescribing the conditions and consequences for integration within employment and social communities and while strengthening ethnic group solidarity.

They believe their social, cultural, and occupational location remains beleaguered by stereotypes beached in psychological needs. Nowhere is this seen more clearly than in a professional community where interaction is close and constant, varied but integrated, but laced with a strong sense of propagandized social acceptance. In such communities, individuals learn more about themselves based on exchange relationships when a sense of "community" and "belonging" is initiated and achieved for most of its members through working to build the community. However, in some instances, becoming part of such a community, especially for persons of color, is a journey into remembering, as well as understanding, that who we are, what we are, where we fit, and how we are received marks our continued journey.

Moreover, these results suggest that the challenges of the professional workplace are shaped in important ways by race and gender. For black women workers, attempting to perform the appropriate emotional labor while simultaneously conforming to etiquette norms creates specific issues that may not be present for other race/gender groups. Future research should consider whether these social and professional expectations pose the same challenges for Latinas, black men, Asian American women, and others. The expected norms, sanctions, and rules of the professional workplace are not neutral, but raced and gendered in ways that may have a different impact on various groups. Future research should consider how this plays out for other populations....

REFERENCES

Acker J (1990) Hierarchies, jobs, bodies: a theory of gendered organizations. *Gender and Society* 4(2): 139–158.

Bell JM and Hartmann D (2007) Diversity in everyday discourse: the cultural ambiguities and consequences of 'happy talk'. *American Sociological Review* 72(6): 895–912.

Blitt B (2008) The polities of fear. *New Yorker Magazine* 84(21): cover.

Collins PH (2004) *Black Sexual Politics.* New York, NY: Routledge.

Femlee DH (1984) The dynamics of women's job mobility. *Work and Occupations* 11(3): 259–281.

Harlow R (2003) Race doesn't matter but...: the effect of race on professors' experiences and emotion management in the undergraduate college classroom. *Social Psychology Quarterly* 66(4): 348–363.

Hochschild AR (1983) Emotional work, feeling rules, and social structure. *American Journal of Sociology* 85(3): 551–575.

Jones C and Shorter-Gooden K (2003) *Shifting: The Double Lives of African American Women in America.* Darby, PA: Diane Publishing Company.

Kanter RM (1977) *Men and Women of the Corporation.* New York, NY: Basic Books.

Markham WT, South SJ, Bonjean CM, et al. (1985) Gender and opportunity in the Federal bureaucracy. *American Journal of Sociology* 9(1): 129–150.

Pierce JL (1999) Emotional labor among paralegals. *Annals of the American Academy of Political and Social Sciences* 561(1): 127–142.

Rosenbaum JE (1979) Tournament mobility: career patterns in a corporation. *Administrative Science Quarterly* 24(2): 220–241.

21

Prisons for Our Bodies, Closets for Our Minds

Racism, Heterosexism, and Black Sexuality

PATRICIA HILL COLLINS

White fear of black sexuality is a basic ingredient of white racism.

Cornel West

For African Americans, exploring how sexuality has been manipulated in defense of racism is not new. Scholars have long examined the ways in which "white fear of black sexuality" has been a basic ingredient of racism. For example, colonial regimes routinely manipulated ideas about sexuality in order to maintain unjust power relations. Tracing the history of contact between English explorers and colonists and West African societies, historian Winthrop Jordan contends that English perceptions of sexual practices among African people reflected preexisting English beliefs about Blackness, religion, and animals. American historians point to the significance of sexuality to chattel slavery. In the United States, for example, slaveowners relied upon an ideology of Black sexual deviance to regulate and exploit enslaved Africans. Because Black feminist analyses pay more attention to women's sexuality, they too identify how the sexual exploitation of women has been a basic ingredient of racism. For example, studies of African American slave women routinely point to sexual victimization as a defining feature of American slavery. Despite the important contributions of this extensive literature on race and sexuality, because much of the literature assumes that sexuality means heterosexuality, it ignores how racism and heterosexism influence one another.

In the United States, the assumption that racism and heterosexism constitute two separate systems of oppression masks how each relies upon the other for meaning. Because neither system of oppression makes sense without the other, racism and heterosexism might be better viewed as sharing one history with similar yet disparate effects on all Americans differentiated by race, gender, sexuality, class, and nationality. People who are positioned at the margins of both systems

SOURCE: Hill Collins, Patricia. 2004. *Black Sexual Politics: African Americans and the New Racism*. Pp. 87–88, 95–105, 114–116. New York: Routledge.
(Republished with permission of Taylor & Francis Group. Permission conveyed through Copyright Clearance Center, Inc.)

and who are harmed by both typically raise questions about the intersections of racism and heterosexism much earlier and/or more forcefully than those people who are in positions of privilege. In the case of intersections of racism and heterosexism, Black lesbian, gay, bisexual, and transgendered (LGBT) people were among the first to question how racism and heterosexism are interconnected. As African American LGBT people point out, assuming that all Black people are heterosexual and that all LGBT people are White distorts the experiences of LGBT Black people. Moreover, such comparisons misread the significance of ideas about sexuality to racism and race to heterosexism.

Until recently, questions of sexuality in general, and homosexuality in particular, have been treated as crosscutting, divisive issues within antiracist African American politics. The consensus issue of ensuring racial unity subordinated the allegedly crosscutting issue of analyzing sexuality, both straight and gay alike. This suppression has been challenged from two directions. Black women, both heterosexual and lesbian, have criticized the sexual politics of African American communities that leave women vulnerable to single motherhood and sexual assault. Black feminist and womanist projects have challenged Black community norms of a sexual double standard that punishes women for behaviors in which men are equally culpable. Black gays and lesbians have also criticized these same sexual politics that deny their right to be fully accepted within churches, families, and other Black community organizations. Both groups of critics argue that ignoring the heterosexism that underpins Black patriarchy hinders the development of a progressive Black sexual politics....

Developing a progressive Black sexual politics requires examining how racism and heterosexism mutually construct one another.

MAPPING RACISM AND HETEROSEXISM:
THE PRISON AND THE CLOSET

... Racism and heterosexism, the prison and the closet, appear to be separate systems, but LGBT African Americans point out that *both* systems affect their everyday lives. If racism and heterosexism affect Black LGBT people, then these systems affect *all* people, including heterosexual African Americans. Racism and heterosexism certainly converge on certain key points. For one, both use similar state-sanctioned institutional mechanisms to maintain racial and sexual hierarchies. For example, in the United States, racism and heterosexism both rely on segregating people as a mechanism of social control. For racism, segregation operates by using race as a visible marker of group membership that enables the state to relegate Black people to inferior schools, housing, and jobs. Racial segregation relies on enforced membership in a visible community in which racial discrimination is tolerated. For heterosexism, segregation is enforced by pressuring LGBT individuals to remain closeted and thus segregated from one another. Before social movements for gay and lesbian liberation, sexual segregation meant that refusing to claim homosexual identities virtually eliminated any group-based political action to resist heterosexism. For another, the state has played a very important role in sanctioning both forms of oppression. In support of racism, the state sanctioned laws that regulated where Black people could live, work, and attend school. In support of heterosexism, the state maintained laws that refused to punish hate crimes against LGBT people, that failed to offer protection when LGBT people were stripped of jobs and children, and that generally sent a message that LGBT people who came out of the closet did so at their own risk.

Racism and heterosexism also share a common set of practices that are designed to discipline the population into accepting the status quo. These disciplinary practices can best be seen in the enormous amount of attention paid both by the state and organized religion to the institution of marriage. If marriage were in fact a natural and normal occurrence between heterosexual couples and if it occurred naturally within racial categories, there would be no need to regulate it. People would naturally choose partners of the opposite sex and the same race. Instead, a series of laws have been passed, all designed to regulate marriage. For example, for many years, the tax system has rewarded married couples with tax breaks that have been denied to single taxpayers or unmarried couples. The message is clear—it makes good financial sense to get married. Similarly, to encourage people to marry within their assigned race, numerous states passed laws banning interracial marriage. These restrictions lasted until the landmark Supreme Court decision in 1967 that overturned state laws. The state has also passed laws designed to keep LGBT people from marrying. In 1996, the U.S. Congress passed the Federal Defense of Marriage Act that defined marriage as a "legal union between one man and one woman." In all of these cases, the state perceives that it has a compelling interest in disciplining the population to marry and to marry the correct partners.

Racism and heterosexism also manufacture ideologies that defend the status quo. When ideologies that defend racism and heterosexism become taken-for-granted and appear to be natural and inevitable, they become hegemonic. Few question them and the social hierarchies they defend. Racism and heterosexism both share a common cognitive framework that uses binary thinking to produce hegemonic ideologies. Such thinking relies on oppositional categories. It views race through two oppositional categories of Whites and Blacks, gender through two categories of men and women, and sexuality through two oppositional categories of heterosexuals and homosexuals. A master binary of normal and deviant overlays and bundles together these and other lesser binaries. In this context, ideas about "normal" race (Whiteness, which ironically, masquerades as racelessness), "normal" gender (using male experiences as the norm), and "normal" sexuality (heterosexuality, which operates in a similar hegemonic fashion) are tightly bundled together. In essence, to be completely "normal," one must be White, masculine, and heterosexual, the core hegemonic White masculinity. This mythical norm is hard to see because it is so taken-for-granted. Its antithesis, its Other, would be Black, female, and lesbian, a fact that Black lesbian feminist Audre Lorde pointed out some time ago.

Within this oppositional logic, the core binary of normal/deviant becomes ground zero for justifying racism and heterosexism. The deviancy assigned to race and that assigned to sexuality becomes an important point of contact between the two systems. Racism and heterosexism both require a concept of sexual deviancy for meaning, yet the form that deviance takes within each system differs. For racism, the point of deviance is created by a *normalized White hetero-sexuality* that depends on a *deviant Black heterosexuality* to give it meaning. For heterosexism, the point of deviance is created by this very same *normalized White heterosexuality* that now depends on a *deviant White homosexuality*. Just as racial normality requires the stigmatization of the sexual practices of Black people, heterosexual normality relies upon the stigmatization of the sexual prac-tices of homosexuals. In both cases, installing White heterosexuality as normal, natural, and ideal requires stigmatizing alternate sexualities as abnormal, unnatu-ral, and sinful.

The purpose of stigmatizing the sexual practices of Black people and those of LGBT people may be similar, but the content of the sexual deviance assigned to each differs. Black people carry the stigma of *promiscuity* or excessive or unre-strained heterosexual desire. This is the sexual deviancy that has both been assigned to Black people and been used to construct racism. In contrast, LGBT people carry the stigma of *rejecting* heterosexuality by engaging in unrestrained homosexual desire. Whereas the deviancy associated with promiscuity (and, by implication, with Black people as a race) is thought to lie in an *excess* of hetero-sexual desire, the pathology of homosexuality (the invisible, closeted sexuality that becomes impossible within heterosexual space) seemingly resides in the *absence* of it.

While analytically distinct, in practice, these two sites of constructed devi-ancy work together and both help create the "sexually repressive culture" in America.... Despite their significance for American society overall, here I confine my argument to the challenges that confront Black people. Both sets of ideas frame a hegemonic discourse of *Black sexuality* that has at its core ideas about an assumed promiscuity among heterosexual African American men and women and the impossibility of homosexuality among Black gays and lesbians. How have African Americans been affected by and reacted to this racialized system of heterosexism (or this sexualized system of racism)?

AFRICAN AMERICANS AND THE RACIALIZATION
OF PROMISCUITY

Ideas about Black promiscuity that produce contemporary sexualized spectacles such as Jennifer Lopez, Destiny's Child, Ja Rule, and the many young Black men on the U.S. talk show circuit have a long history. Historically, Western science, medicine, law, and popular culture reduced an African-derived aesthetic concerning the use of the body, sensuality, expressiveness, and spirituality to an ideology about *Black sexuality*. The distinguishing feature of this ideology was its reliance on the idea of Black promiscuity. The possibility of distinctive and worthwhile African-influenced worldviews on anything, including sexuality, as well as the heterogeneity of African societies expressing such views, was collapsed into an imagined, pathologized Western discourse of what was thought to be essentially African. To varying degrees, observers from England, France, Germany, Belgium, and other colonial powers perceived African sensuality, eroticism, spirituality, and/or sexuality as deviant, out of control, sinful, and as an essential feature of racial difference....

With all living creatures classified in this way, Western scientists perceived African people as being more natural and less civilized, primarily because African people were deemed to be closer to animals and nature, especially the apes and monkeys whose appearance most closely resembled humans. Like African people, animals also served as objects of study for Western science because understanding the animal kingdom might reveal important insights about civilization, culture, and what distinguished the human "race" from its animal counterparts as well as the human "races" from one another....

Those most proximate to animals, those most lacking civilization, also were those humans who came closest to having the sexual lives of animals. Lacking the benefits of Western civilization, people of African descent were perceived as having a biological nature that was inherently more sexual than that of Europeans. The primitivist discourse thus created the category of "beast" and the sexuality of such beasts as "wild." The legal classification of enslaved African people as chattel (animal-like) under American slavery that produced controlling images of bucks, jezebels, and breeder women drew meaning from this broader interpretive framework.

Historically, this ideology of Black sexuality that pivoted on a Black heterosexual promiscuity not only upheld racism but it did so in gender-specific ways. In the context of U.S. society, beliefs in Black male promiscuity took diverse forms during distinctive historical periods. For example, defenders of chattel slavery believed that slavery safely domesticated allegedly dangerous Black men because it regulated their promiscuity by placing it in the service of slave owners. Strategies of control were harsh and enslaved African men who were born in Africa or who had access to their African past were deemed to be the most dangerous. In contrast, the controlling image of the rapist appeared after emancipation because Southern Whites feared that the unfettered promiscuity of Black freedmen constituted a threat to the Southern way of life....

The events themselves may be over, but their effects persist under the new racism. This belief in an inherent Black promiscuity reappears today. For example, depicting poor and working-class African American inner-city neighborhoods as dangerous urban jungles where SUV-driving White suburbanites come to score drugs or locate prostitutes also invokes a history of racial and sexual conquest. Here sexuality is linked with danger, and understandings of both draw upon historical imagery of Africa as a continent replete with danger and peril to the White explorers and hunters who penetrated it. Just as contemporary safari tours in Africa create an imagined Africa as the "White man's playground" and mask its economic exploitation, jungle language masks social relations of hyper-segregation that leave working-class Black communities iso-lated, impoverished, and dependent on a punitive welfare state and an illegal international drug trade. Under this logic, just as wild animals (and the proximate African natives) belong in nature preserves (for their own protection), unassimi-lated, undomesticated poor and working-class African Americans belong in racially segregated neighborhoods....

African American women also live with ideas about Black women's promis-cuity and lack of sexual restraint. Reminiscent of concerns with Black women's fertility under slavery and in the rural South, contemporary social welfare policies also remain preoccupied with Black women's fertility. In prior eras, Black women were encouraged to have many children. Under slavery, having many children enhanced slave owners' wealth and a good "breeder woman" was less likely to be sold. In rural agriculture after emancipation, having many children ensured a sufficient supply of workers. But in the global economy of today, large families are expensive because children must be educated. Now Black women are seen as producing too many children who contribute less to society than they take. Because Black women on welfare have long been seen as undeserving, long-standing ideas about Black women's promiscuity become recycled and redefined as a problem for the state....

RACISM AND HETEROSEXISM REVISITED

On May 11, 2003, a stranger killed fifteen-year-old Sakia Gunn who, with four friends, was on her way home from New York's Greenwich Village. Sakia and her friends were waiting for the bus in Newark, New Jersey, when two men got out of a car, made sexual advances, and physically attacked them. The women fought back, and when Gunn told the men that she was a lesbian, one of them stabbed her in the chest.

Sakia Gunn's murder illustrates the connections among class, race, gender, sexuality, and age. Sakia lacked the protection of social class privilege. She and her friends were waiting for the bus in the first place because none had access to private automobiles that offer protection for those who are more affluent. In Gunn's case, because her family initially did not have the money for her funeral, she was scheduled to be buried in a potter's grave. Community activists took up a collection to pay for her funeral. She lacked the gendered protection provided by masculinity. Women who are perceived to be in the wrong place at the wrong time are routinely approached by men who feel entitled to harass and proposition them. Thus, Sakia and her friends share with all women the vulner-abilities that accrue to women who negotiate public space. She lacked the protection of age—had Sakia and her friends been middle-aged, they may not have been seen as sexually available. Like African American girls and women, regardless of sexual orientation, they were seen as approachable. Race was a factor, but not in a framework of interracial race relations. Sakia and her friends were African American, as were their attackers. In a context where Black men are encouraged to express a hyper-heterosexuality as the badge of Black masculinity, women like Sakia and her friends can become important players in supporting patriarchy. They challenged Black male authority, and they paid for the transgression of refusing to participate in scripts of Black promiscuity. But the immediate precipitating catalyst for the violence that took Sakia's life was her openness about her lesbianism. Here, homophobic violence was the prime factor. Her death illustrates how deeply entrenched homophobia can be among many African American men and women, in this case, beliefs that resulted in an attack on a teenaged girl.

How do we separate out and weigh the various influences of class, gender, age, race, and sexuality in this particular incident? Sadly, violence against Black girls is an everyday event. What made this one so special? Which, if any, of the dimensions of her identity got Sakia Gunn killed? There is no easy answer to this question, because *all* of them did. More important, how can any Black political agenda that does not take *all* of these systems into account, including sexuality, ever hope adequately to address the needs of Black people as a collectivity? One expects racism in the press to shape the reports of this incident. In contrast to the 1998 murder of Matthew Shepard, a young, White, gay man in Wyoming, no massive protests, nationwide vigils, and renewed calls for federal hate crimes legislation followed Sakia's death. But what about the response of elected and appointed officials? The African American mayor of Newark decried the crime, but he could not find the time to meet with community activists who wanted programmatic changes to retard crimes like Sakia's murder. The principal of her high school became part of the problem. As one activist described it, "students at Sakia's high school weren't allowed to hold a vigil. And the kids wearing the rainbow flag were being punished like they had on gang colors."

Other Black leaders and national organizations spoke volumes through their silence. The same leaders and organizations that spoke out against the police beating of Rodney King by Los Angeles area police, the rape of immigrant Abner Louima by New York City police, and the murder of Timothy Thomas by Cincinnati police said nothing about Sakia Gunn's death. Apparently, she was just another unimportant little Black girl to them. But to others, her death revealed the need for a new politics that takes the intersections of racism and heterosexism as well as class exploitation, age discrimination, and sexism into account. Sakia was buried on May 16 and a crowd of approximately 2,500 people attended her funeral. The turnout was unprecedented: predominantly Black, largely high school students, and mostly lesbians. Their presence says that as long as African American lesbians like high school student Sakia Gunn are vulnerable, then every African American woman is in danger; and if all Black women are at risk, then there is no way that any Black person will ever be truly safe or free.

22

The Invention of Heterosexuality

JONATHAN NED KATZ

Heterosexuality is old as procreation, ancient as the lust of Eve and Adam. That first lady and gentleman, we assume, perceived themselves, behaved, and felt just like today's heterosexuals. We suppose that heterosexuality is unchanging, universal, essential: ahistorical.

Contrary to that common sense conjecture, the concept of heterosexuality is only one particular historical way of perceiving, categorizing, and imagining the social relations of the sexes. Not ancient at all, the idea of heterosexuality is a modern invention, dating to the late nineteenth century. The heterosexual belief, with its metaphysical claim to eternity, has a particular, pivotal place in the social universe of the late nineteenth and twentieth centuries that it did not inhabit earlier. This essay traces the historical process by which the heterosexual idea was created as a historical and taken-for-granted....

By not studying the heterosexual idea in history, analysts of sex, gay and straight, have continued to privilege the "normal" and "natural" at the expense of the "abnormal" and "unnatural." Such privileging of the norm accedes to its domination, protecting it from questions. By making the normal the object of a thoroughgoing historical study, we simultaneously pursue a pure truth and a sex-radical and subversive goal: we upset basic preconceptions. We discover that the heterosexual, the normal, and the natural have a history of changing definitions. Studying the history of the term challenges its power.

Contrary to our usual assumption, past Americans and other peoples named, perceived, and socially organized the bodies, lusts, and intercourse of the sexes in ways radically different from the way we do. If we care to understand this vast past sexual diversity, we need to stop promiscuously projecting our own hetero and homo arrangement. Though lip service is often paid to the distorting, ethnocentric effect of such conceptual imperialism, the category heterosexuality continues to be applied uncritically as a universal analytical tool. Recognizing the time-bound and culturally specific character of the heterosexual category can help us begin to work toward a thoroughly historical view of sex....

SOURCE: Katz, Jonathan Ned. 1990. "The Invention of Heterosexuality." *Socialist Review* 20 (January–March): 7–34. Copyright © 2009. Reprinted by permission of the author.

BEFORE HETEROSEXUALITY: EARLY VICTORIAN
TRUE LOVE, 1820–1860

In the early nineteenth-century United States, from about 1820 to 1860, the heterosexual did not exist. Middle-class white Americans idealized a True Womanhood, True Manhood, and True Love, all characterized by "purity"— the freedom from sensuality.[1] Presented mainly in literary and religious texts, this True Love was a fine romance with no lascivious kisses. This ideal contrasts strikingly with late nineteenth- and twentieth-century American incitements to a hetero sex.[2]

Early Victorian True Love was only realized within the mode of proper procreation, marriage, the legal organization for producing a new set of correctly gendered women and men. Proper womanhood, manhood, and progeny—not a normal male-female eros—was the main product of this mode of engendering and of human reproduction.

The actors in this sexual economy were identified as manly men and womanly women and as procreators, not specifically as erotic beings or heterosexuals. Eros did not constitute the core of a heterosexual identity that inhered, democratically, in both men and women. True Women were defined by their distance from lust. True Men, though thought to live closer to carnality, and in less control of it, aspired to the same freedom from concupiscence.

Legitimate natural desire was for procreation and a proper manhood or womanhood; no heteroerotic desire was thought to be directed exclusively and naturally toward the other sex; lust in men was roving. The human body was thought of as a means towards procreation and production; penis and vagina were instruments of reproduction, not of pleasure. Human energy, thought of as a closed and severely limited system, was to be used in producing children and in work, not wasted in libidinous pleasures.

The location of all this engendering and procreative labor was the sacred sanctum of early Victorian True Love, the home of the True Woman and True Man—a temple of purity threatened from within by the monster masturbator, an archetypal early Victorian cult figure of illicit lust. The home of True Love was a castle far removed from the erotic exotic ghetto inhabited most notoriously then by the prostitute, another archetypal Victorian erotic monster....

LATE VICTORIAN SEX-LOVE: 1860–1892

"Heterosexuality" and "homosexuality" did not appear out of the blue in the 1890s. These two eroticisms were in the making from the 1860s on. In late Victorian America and in Germany, from about 1860 to 1892, our modern idea of an eroticized universe began to develop, and the experience of a hetero-lust began to be widely documented and named....

In the late nineteenth-century United States, several social factors converged to cause the eroticizing of consciousness, behavior, emotion, and identity that became typical of the twentieth-century Western middle class. The transformation of the family from producer to consumer unit resulted in a change in family members' relation to their own bodies; from being an instrument primarily of work, the human body was integrated into a new economy, and began more commonly to be perceived as a means of consumption and pleasure. Historical work has recently begun on how the biological human body is differently integrated into changing modes of production, procreation, engendering, and pleasure so as to alter radically the identity, activity, and experience of that body.[3]

The growth of a consumer economy also fostered a new pleasure ethic. This imperative challenged the early Victorian work ethic, finally helping to usher in a major transformation of values. While the early Victorian work ethic had touted the value of economic production, that era's procreation ethic had extolled the virtues of human reproduction. In contrast, the late Victorian economic ethic hawked the pleasures of consuming, while its sex ethic praised an erotic pleasure principle for men and even for women.

In the late nineteenth century, the erotic became the raw material for a new consumer culture. Newspapers, books, plays, and films touching on sex, "normal" and "abnormal," became available for a price. Restaurants, bars, and baths opened, catering to sexual consumers with cash. Late Victorian entrepreneurs of desire incited the proliferation of a new eroticism, a commoditized culture of pleasure.

In these same years, the rise in power and prestige of medical doctors allowed these upwardly mobile professionals to prescribe a healthy new sexuality. Medical men, in the name of science, defined a new ideal of male-female relationships that included, in women as well as men, an essential, necessary, normal eroticism. Doctors, who had earlier named and judged the sex-enjoying woman a "nymphomaniac," now began to label women's *lack* of sexual pleasure a mental disturbance, speaking critically, for example, of female "frigidity" and "anesthesia."[4]

By the 1880s, the rise of doctors as a professional group fostered the rise of a new medical model of Normal Love, replete with sexuality. The new Normal Woman and Man were endowed with a healthy libido. The new theory of Normal Love was the modern medical alternative to the old Cult of True Love. The doctors prescribed a new sexual ethic as if it were a morally neutral, medical description of health. The creation of the new Normal Sexual had its counterpart in the invention of the late Victorian Sexual Pervert. The attention paid the sexual abnormal created a need to name the sexual normal, the better to distinguish the average him and her from the deviant it.

HETEROSEXUALITY: THE FIRST YEARS,
1892–1900

In the periodization of heterosexual American history suggested here, the years 1892 to 1900 represent "The First Years" of the heterosexual epoch, eight key years in which the idea of the heterosexual and homosexual were initially and tentatively formulated by U.S. doctors. The earliest-known American use of the word "heterosexual" occurs in a medical journal article by Dr. James G. Kiernan of Chicago, read before the city's medical society on March 7, 1892, and published that May—portentous dates in sexual history.[5] But Dr. Kiernan's heterosexuals were definitely not exemplars of normality. Heterosexuals, said Kiernan, were defined by a mental condition, "psychical hermaphroditism." Its symptoms were "inclinations to both sexes." These heterodox sexuals also betrayed inclinations "to abnormal methods of gratification," that is, techniques to insure pleasure without procreation. Dr. Kiernan's heterogeneous sexuals did demonstrate "traces of the normal sexual appetite" (a touch of procreative desire). Kiernan's normal sexuals were implicitly defined by a monolithic other-sex inclination and procreative aim. Significantly, they still lacked a name.

Dr. Kiernan's article of 1892 also included one of the earliest-known uses of the word "homosexual" in American English. Kiernan defined "Pure homosexuals" as persons whose "general mental state is that of the opposite sex." Kiernan thus defined homosexuals by their deviance from a gender norm. His heterosexuals displayed a double deviance from both gender and procreative norms.

Though Kiernan used the new words heterosexual and homosexual, an old procreative standard and a new gender norm coexisted uneasily in his thought. His word heterosexual defined a mixed person and compound urge, abnormal because they wantonly included procreative and non-procreative objectives, as well as same-sex and different-sex attractions.

That same year, 1892, Dr. Krafft-Ebing's influential *Psychopathia Sexualis* was first translated and published in the United States.[6] But Kiernan and Krafft-Ebing by no means agreed on the definition of the heterosexual. In Krafft-Ebing's book, "hetero-sexual" was used unambiguously in the modern sense to refer to an erotic feeling for a different sex. "Homosexual" referred unambiguously to an erotic feeling for a "same sex." In Krafft-Ebing's volume, unlike Kiernan's article, heterosexual and homosexual were clearly distinguished from a third category, a "psycho-sexual hermaphroditism," defined by impulses toward both sexes.

Krafft-Ebing hypothesized an inborn "sexual instinct" for relations with the "opposite sex," the inherent "purpose" of which was to foster procreation. Krafft-Ebing's erotic drive was still a reproductive instinct. But the doctor's clear focus on a different-sex versus same-sex sexuality constituted a historic, epochal move from an absolute procreative standard of normality toward a new norm. His definition of heterosexuality as other-sex attraction provided the basis for a revolutionary, modern break with a centuries-old procreative standard.

It is difficult to overstress the importance of that new way of categorizing. The German's mode of labeling was radical in referring to the biological sex, masculinity or femininity, and the pleasure of actors (along with the procreant purpose of acts). Krafft-Ebing's heterosexual offered the modern world a new norm that came to dominate our idea of the sexual universe, helping to change it from a mode of human reproduction and engendering to a mode of pleasure. The heterosexual category provided the basis for a move from a production-oriented, procreative imperative to a consumerist pleasure principle—an institutionalized pursuit of happiness....

Only gradually did doctors agree that heterosexual referred to a normal, "other-sex" eros. This new standard-model heterosex provided the pivotal term for the modern regularization of eros that paralleled similar attempts to standardize masculinity and femininity, intelligence, and manufacturing.[7] The idea of heterosexuality as the master sex from which all others deviated was (like the idea of the master race) deeply authoritarian. The doctors' normalization of a sex that was hetero proclaimed a new heterosexual separatism—an erotic apartheid that forcefully segregated the sex normals from the sex perverts. The new, strict boundaries made the emerging erotic world less polymorphous—safer for sex normals. However, the idea of such creatures as heterosexuals and homosexuals emerged from the narrow world of medicine to become a commonly accepted notion only in the early twentieth century. In 1901, in the comprehensive *Oxford English Dictionary*, "heterosexual" and "homosexual" had not yet made it.

THE DISTRIBUTION OF THE HETEROSEXUAL MYSTIQUE: 1900–1930

In the early years of this heterosexual century the tentative hetero hypothesis was stabilized, fixed, and widely distributed as the ruling sexual orthodoxy: The Heterosexual Mystique. Starting among pleasure-affirming urban working-class youths, southern blacks, and Greenwich Village bohemians as defensive subculture, heterosex soon triumphed as dominant culture.[8]

In its earliest version, the twentieth-century heterosexual imperative usually continued to associate heterosexuality with a supposed human "need," "drive," or "instinct" for propagation, a procreant urge linked inexorably with carnal lust as it had not been earlier. In the early twentieth century, the falling birth rate, rising divorce rate, and "war of the sexes" of the middle class were matters of increasing public concern. Giving vent to heteroerotic emotions was thus praised as enhancing baby-making capacity, marital intimacy, and family stability. (Only many years later, in the mid-1960s, would heteroeroticism be distinguished completely, in practice and theory, from procreativity and male-female pleasure sex justified in its own name.)

The first part of the new sex norm—hetero—referred to a basic gender divergence. The "oppositeness" of the sexes was alleged to be the basis for a universal, normal, erotic attraction between males and females. The stress on the sexes' "oppositeness," which harked back to the early nineteenth century, by no means simply registered biological differences of females and males. The early twentieth-century focus on physiological and gender dimorphism reflected the deep anxieties of men about the shifting work, social roles, and power of men over women, and about the ideals of womanhood and manhood. That gender anxiety is documented, for example, in 1897, in *The New York Times'* publication of the Reverend Charles Parkhurst's diatribe against female "andro-maniacs," the preacher's derogatory, scientific-sounding name for women who tried to "minimize distinctions by which manhood and womanhood are differentiated."[9] The stress on gender difference was a conservative response to the changing social-sexual division of activity and feeling which gave rise to the independent "New Woman" of the 1880s and eroticized "Flapper" of the 1920s.

The second part of the new hetero norm referred positively to sexuality. That novel upbeat focus on the hedonistic possibilities of male-female conjunctions also reflected a social transformation—a revaluing of pleasure and procreation, consumption and work in commercial, capitalist society. The democratic attribution of a normal lust to human females (as well as males) served to authorize women's enjoyment of their own bodies and began to undermine the early Victorian idea of the pure True Woman—a sex-affirmative action still part of women's struggle. The twentieth-century Erotic Woman also undercut the nineteenth-century feminist assertion of women's moral superiority, cast suspicions of lust on women's passionate romantic friendships with women, and asserted the presence of a menacing female monster, "the lesbian."[10]...

In the perspective of heterosexual history, this early twentieth-century struggle for the more explicit depiction of an "opposite-sex" eros appears in a curious new light. Ironically, we find sex-conservatives, the social purity advocates of censorship and repression, fighting against the depiction not just of sexual perversity but also of the new normal hetero-sexuality. That a more open depiction of normal sex had to be defended against forces of propriety confirms the claim that heterosexuality's predecessor, Victorian True Love, had included no legitimate eros....

THE HETEROSEXUAL STEPS OUT: 1930–1945

In 1930, in *The New York Times*, heterosexuality first became a love that dared to speak its name. On April 20th of that year, the word "heterosexual" is first known to have appeared in *The New York Times Book Review*. There, a critic described the subject of André Gide's *The Immoralist* proceeding "from a hetero-sexual liaison to a homosexual one." The ability to slip between sexual categories was referred to casually as a rather unremarkable aspect of human possibility. This is also the first known reference by *The Times* to the new hetero/homo duo.[11]

In September the second reference to the hetero/homo dyad appeared in *The New York Times Book Review*, in a comment on Floyd Dell's *Love in the Machine Age*. This work revealed a prominent antipuritan of the 1930s using the dire threat of homosexuality as his rationale for greater heterosexual freedom. *The Times* quoted Dell's warning that current abnormal social conditions kept the young dependent on their parents, causing "infantilism, prostitution and homosexuality." Also quoted was Dell's attack on the inculcation of purity" that "breeds distrust of the opposite sex." Young people, Dell said, should be "permitted to develop normally to heterosexual adulthood." "But," *The Times* reviewer emphasized, "such a state already exists, here and now." And so it did. Heterosexuality, a new gender-sex category, had been distributed from the narrow, rarified realm of a few doctors to become a nationally, even internationally, cited aspect of middle-class life.[12]...

HETEROSEXUAL HEGEMONY: 1945–1965

The "cult of domesticity" following World War II—the reassociation of women with the home, motherhood, and child-care; men with fatherhood and wage work outside the home—was a period in which the predominance of the hetero norm went almost unchallenged, an era of heterosexual hegemony. This was an age in which conservative mental-health professionals reasserted the old link between heterosexuality and procreation. In contrast, sex-liberals of the day strove, ultimately with success, to expand the heterosexual ideal to include within the boundaries of normality a wider-than-ever range of nonprocreative, premarital, and extramarital behaviors. But sex-liberal reform actually helped to extend and secure the dominance of the heterosexual idea, as we shall see when we get to Kinsey.

The postwar sex-conservative tendency was illustrated in 1947, in Ferdinand Lundberg and Dr. Marynia Farnham's books, *Modern Woman: The Lost Sex.* Improper masculinity and femininity was exemplified, the authors decreed, by "engagement in heterosexual relations ... with the complete intent to see to it that they do not eventuate in reproduction."[13] Their procreatively defined heterosex was one expression of a postwar ideology of fecundity that, internalized and enacted dutifully by a large part of the population, gave rise to the postwar baby boom.

The idea of the feminine female and masculine male as prolific breeders was also reflected in the stress, specific to the late 1940s, on the homosexual as sad symbol of "sterility"—that particular loaded term appears incessantly in comments on homosex dating to the fecund forties.

In 1948, in *The New York Times Book Review*, sex liberalism was in ascendancy. Dr. Howard A. Rusk declared that Alfred Kinsey's just published report on *Sexual Behavior in the Human Male* had found "wide variations in sex concepts and behavior." This raised the question: "What is 'normal' and 'abnormal'?" In particular, the report had found that "homosexual experience is much more common than previously thought," and "there is often a mixture of both homo and hetero experience."[14]

Kinsey's counting of orgasms indeed stressed the wide range of behaviors and feelings that fell within the boundaries of a quantitative, statistically accounted heterosexuality. Kinsey's liberal reform of the hetero/homo dualism widened the narrow, old hetero category to accord better with the varieties of social experience. He thereby contradicted the older idea of a monolithic, qualitatively defined, natural procreative act, experience, and person.[15]

Though Kinsey explicitly questioned "whether the terms 'normal' and 'abnormal' belong in a scientific vocabulary," his counting of climaxes was generally understood to define normal sex as majority sex. This quantified norm constituted a final, society-wide break with the old qualitatively defined reproductive standard. Though conceived of as purely scientific, the statistical definition of the normal as the-sex-most-people-are-having substituted a new, quantitative moral standard for the old, qualitative sex ethic—another triumph for the spirit of capitalism.

Kinsey also explicitly contested the idea of an absolute, either/or antithesis between hetero and homo persons. He denied that human beings "represent two discrete populations, heterosexual and homosexual." The world, he ordered, "is not to be divided into sheep and goats." The hetero/homo division was not nature's doing: "Only the human mind invents categories and tries to force facts into separated pigeon-holes. The living world is a continuum."[16]

With a wave of the taxonomist's hand, Kinsey dismissed the social and historical division of people into heteros and homos. His denial of heterosexual and homosexual personhood rejected the social reality and profound subjective force of a historically constructed tradition which, since 1892 in the United States, had cut the sexual populaton in two and helped to establish the social reality of a heterosexual and homosexual identity.

On the one hand, the social construction of homosexual persons has led to the development of a powerful gay liberation identity politics based on an ethnic group model. This has freed generations of women and men from a deep, painful, socially induced sense of shame, and helped to bring about a society-wide liberalization of attitudes and responses to homosexuals.[17] On the other hand, contesting the notion of homosexual and heterosexual persons was one early, partial resistance to the limits of the hetero/homo construction. Gore Vidal, rebel son of Kinsey, has for years been joyfully proclaiming:

> ...there is no such thing as a homosexual or a heterosexual person.
> There are only homo- or heterosexual acts. Most people are a mixture
> of impulses if not practices, and what anyone does with a willing partner
> is of no social or cosmic significance.
>
> So why all the fuss? In order for a ruling class to rule, there must be
> arbitrary prohibitions. Of all prohibitions, sexual taboo is the most useful
> because sex involves everyone.... We have allowed our governors to
> divide the population into two teams. One team is good, godly, straight;
> the other is evil, sick, vicious.[18]

HETEROSEXUALITY QUESTIONED: 1965–1982

By the late 1960s, anti-establishment counter culturalists, fledgling feminists, and homosexual-rights activists had begun to produce an unprecedented critique of sexual repression in general, of women's sexual repression in particular, of marriage and the family—and of some forms of heterosexuality....

Heterosexual History: Out of the Shadows

Our brief survey of the heterosexual idea suggests a new hypothesis. Rather than naming a conjunction old as Eve and Adam, heterosexual designates a word and concept, a norm and role, an individual and group identity, a behavior and feeling, and a peculiar sexual-political institution particular to the late nineteenth and twentieth centuries.

Because much stress has been placed here on heterosexuality as word and concept, it seems important to affirm that heterosexuality (and homosexuality) came into existence before it was named and thought about. The formulation of the heterosexual idea did not create a heterosexual experience or behavior; to suggest otherwise would be to ascribe determining power to labels and concepts. But the titling and envisioning of heterosexuality did play an important role in consolidating the construction of the heterosexual's social existence. Before the wide use of the word "heterosexual," I suggest, women and men did not mutually lust with the same profound, sure sense of normalcy that followed the distribution of "heterosexual" as universal sanctifier.

According to this proposal, women and men make their own sexual histories. But they do not produce their sex lives just as they please. They make their sexualities within a particular mode of organization given by the past and altered by their changing desire, their present power and activity, and their vision of a better world. That hypothesis suggests a number of good reasons for the immediate inauguration of research on a historically specific heterosexuality.

The study of the history of the heterosexual experience will forward a great intellectual struggle still in its early stages. This is the fight to pull heterosexuality, homosexuality, and all the sexualities out of the realm of nature and biology [and] into the realm of the social and historical. Feminists have explained to us that anatomy does not determine our gender destinies (our masculinities and femininities). But we've only recently begun to consider that *biology does not settle our erotic fates.* The common notion that biology determines the object of sexual desire, or that physiology and society together cause sexual orientation, are determinisms that deny the break existing between our bodies and situations and our desiring. Just as the biology of our hearing organs will never tell us why we take pleasure in Bach or delight in Dixieland, our female or male anatomies, hormones, and genes will never tell us why we yearn for women, men, both, other, or none. That is because desiring is a self-generated project of individuals within particular historical cultures. Heterosexual history can help us see the place of values and judgments in the construction of our own and others' pleasures, and to see how our erotic tastes—our aesthetics of the flesh—are socially institutionalized through the struggle of individuals and classes.

The study of heterosexuality in time will also help us to recognize the *vast historical diversity of sexual emotions and behaviors*—a variety that challenges the monolithic heterosexual hypothesis. John D'Emilio and Estelle Freedman's *Intimate Matters: A History of Sexuality in America* refers in passing to numerous substantial changes in sexual activity and feeling: for example, the widespread use of contraceptives in the nineteenth century, the twentieth-century incitement of the female orgasm, and the recent sexual conduct changes by gay men in response to the AIDS epidemic. It's now a commonplace of family history that people in particular classes feel and behave in substantially different ways under different, historical conditions. Only when we stop assuming an invariable essence of heterosexuality will we begin the research to reveal the full variety of sexual emotions and behaviors.

The historical study of the heterosexual experience can help us *understand the erotic relationships of women and men in terms of their changing modes of social organization.* Such model analysis actually characterizes a sex history well underway. This suggests that the eros-gender-procreation system (the social ordering of lust, femininity and masculinity, and baby-making) has been linked closely to a society's particular organization of power and production. To understand the subtle history of heterosexuality we need to look carefully at correlations between (1) society's organization of eros and pleasure; (2) its mode of engendering persons as feminine or masculine (its making of women and men); (3) its ordering of human reproduction; and (4) its dominant political economy. This General Theory of Sexual Relativity proposes that substantial historical changes in the social organization of eros, gender, and procreation have basically altered the activity and experience of human beings within those modes.

A historical view locates heterosexuality and homosexuality in time, helping us distance ourselves from them. This distancing can help us formulate new questions that clarify our long-range sexual-political goals: What has been and is the social function of sexual categorizing? Whose interests have been served by the division of the world into heterosexual and homosexual? Do we dare not draw a line between those two erotic species? Is some sexual naming socially necessary? Would human freedom be enhanced if the sex-biology of our partners in lust was of no particular concern, and had no name? In what kind of society could we all more freely explore our desire and our flesh?

As we move [into the present], a new sense of the historical making of the heterosexual and homosexual suggests that these are ways of feeling, acting, and being with each other that we can together unmake and radically remake according to our present desire, power, and our vision of a future political-economy of pleasure.

NOTES

1. Barbara Welter, "The Cult of True Womanhood: 1820–1860," *American Quarterly*, vol. 18 (Summer 1966); Welter's analysis is extended here to include True Men and True Love.

2. Some historians have recently told us to revise our idea of sexless Victorians: their experience and even their ideology, it is said, were more erotic than we previously thought. Despite the revisionists, I argue that "purity" was indeed the dominant, early Victorian, white middle-class standard. For the debate on Victorian sexuality see John D'Emilio and Estelle Freedman, *Intimate Matters: A History of Sexuality in America* (New York: Harper & Row, 1988), p. xii.

3. See, for example, Catherine Gallagher and Thomas Laqueur, eds., "The Making of the Modern Body: Sexuality and Society in the Nineteenth Century," *Representations*, no. 14 (Spring 1986) (republished, Berkeley: University of California Press, 1987).

4. This reference to females reminds us that the invention of heterosexuality had vastly different impacts on the histories of women and men. It also differed in its impact on lesbians and heterosexual women, homosexual and heterosexual men, the middle class and working class, and on different religious, racial, national, and geographic groups.

5. Dr. James G. Kieman, "Responsibility in Sexual Perversion," *Chicago Medical Recorder*, vol. 3 (May 1892), pp. 185–210.

6. R. von Krafft-Ebing, *Psychopathia Sexualis, with Especial Reference to Contrary Sexual Instinct: A Medico-Legal Study*, trans. Charles Gilbert Chaddock (Philadelphia: F. A. Davis, 1892), from the 7th and revised German ed. Preface, November 1892.

7. For the standardization of gender see Lewis Terman and C. C. Miles, *Sex and Personality, Studies in Femininity and Masculinity* (New York: McGraw Hill, 1936). For the standardization of intelligence see Lewis Terman, *Stanford-Binet Intelligence Scale* (Boston: Houghton Mifflin, 1916). For the standardization of work, see "scientific management" and "Taylorism" in Harry Braverman, *Labor and Monopoly Capital: The Degradation of Work in the Twentieth Century* (New York: Monthly Review Press, 1974).

8. See D'Emilio and Freedman, *Intimate Matters*, pp. 194–201, 231, 241, 295–96; Ellen Kay Trimberger, "Feminism, Men, and Modern Love: Greenwich Village, 1900–1925," in *Powers of Desire: The Politics of Sexuality*, ed. Ann Snitow, Christine Stansell, and Sharon Thompson (New York: Monthly Review Press, 1983), pp. 131–52; Kathy Peiss, " 'Charity Girls' and City Pleasures: Historical Notes on Working Class Sexuality, 1880–1920," in *Powers of Desire*, pp. 74–87; and Mary P. Ryan, "The Sexy Saleslady: Psychology, Heterosexuality, and Consumption in the Twentieth Century," in her *Womanhood in America*, 2nd ed. (New York: Franklin Watts, 1979), pp. 151–82.

9. [Rev. Charles Parkhurst], "Woman. Calls Them Andromaniacs. Dr. Parkhurst So Characterizes Certain Women Who Passionately Ape Everything That Is Mannish. Woman Divinely Preferred. Her Supremacy Lies in Her Womanliness, and She Should Make the Most of It—Her Sphere of Best Usefulness the Home," *The New York Times*, May 23, 1897, p. 16:1.

10. See Lisa Duggan, "The Social Enforcement of Heterosexuality and Lesbian Resistance in the 1920s," in *Class, Race, and Sex: The Dynamics of Control*, ed. Amy Swerdlow and Hanah Lessinger (Boston: G. K. Hall, 1983), pp. 75–92; Rayna Rapp and Ellen Ross, "The Twenties Backlash: Compulsory Heterosexuality, the Consumer Family, and the Waning of Feminism," in *Class, Race, and Sex*; Christina Simmons, "Companionate Marriage and the Lesbian Threat," *Frontiers*, vol. 4, no. 3 (Fall 1979), pp. 54–59; and Lillian Faderman, *Surpassing the Love of Men* (New York: William Morrow, 1981).

11. Louis Kronenberger, review of André Gide, *The Immoralist*, *New York Times Book Review*, April 20, 1930, p. 9.

12. Henry James Forman, review of Floyd Dell, *Love in the Machine Age* (New York: Farrar & Rinehart), *New York Times Book Review*, September 14, 1930, p. 9.

13. Ferdinand Lundberg and Dr. Marynia F. Farnham, *Modern Woman: The Lost Sex* (New York: Harper, 1947).

14. Dr. Howard A. Rusk, *New York Times Book Review*, January 4, 1948, p. 3.

15. Alfred Kinsey, Wardell B. Pomeroy, and Clyde E. Martin, *Sexual Behavior in the Human Male* (Philadelphia: W. B. Saunders, 1948), pp. 199–200.

16. Kinsey, *Sexual Behavior*, pp. 637, 639.

17. See Steven Epstein, "Gay Politics, Ethnic Identity: The Limits of Social Constructionism," *Socialist Review* 93/94 (1987), pp. 9–54.

18. Gore Vidal, "Someone to Laugh at the Squares With" [Tennessee Williams], *New York Review of Books*, June 13, 1985; reprinted in his *At Home: Essays, 1982–1988* (New York: Random House, 1988), p. 48.

23

"Good Girls"

Gender, Social Class, and Slut Discourse on Campus*

ELIZABETH A. ARMSTRONG, LAURA T. HAMILTON, ELIZABETH M. ARMSTRONG, AND J. LOTUS SEELEY

Slut shaming, the practice of maligning women for presumed sexual activity, is common among young Americans. For example, Urban Dictionary—a website documenting youth slang—refers those interested in the term *slut* to *whore, bitch, skank, ho, cunt, prostitute, tramp, hooker, easy,* or *slug.*[1] Boys and men are not alone in using these terms (Wolf 1997; Tanenbaum 1999; White 2002). In our ethnographic and longitudinal study of college women at a large, moderately selective university in the Midwest, women labeled other women and marked their distance from "sluttiness."

Women's participation in slut shaming is often viewed as evidence of internalized oppression (Ringrose and Renold 2012). This argument proceeds as follows: slut shaming is based on sexual double standards established and upheld by men, to women's disadvantage. Although young men are expected to desire and pursue sex regardless of relational and emotional context, young women are permitted sexual activity only when in committed relationships and "in love." Women are vulnerable to slut stigma when they violate this sexual standard and consequently experience status loss and discrimination. Slut shaming is thus about sexual inequality and reinforces male dominance and female subordination. Women's participation works at cross-purposes with progress toward gender equality.

... We argue that women's participation in this practice is only indirectly related to judgments about sexual activity. Instead it is about drawing class-based moral boundaries that simultaneously organize sexual behavior and gender presentation. Women's definitions of sluttiness revolve around status on campus, which is largely dictated by class background. High-status women employ slut discourse to assert class advantage, defining their styles of femininity and approaches to sexuality as classy rather than trashy. Low-status women express class resentment—deriding rich, bitchy sluts for their wealth, exclusivity, and

*This work was supported by a National Academy of Education/Spencer Foundation Postdoctoral Fellowship, a Radcliffe Institute Fellowship at Harvard University, and a Spencer Foundation Small Grant awarded to Elizabeth A. Armstrong.

SOURCE: Elizabeth A. Armstrong, Laura T. Hamilton and J. Lotus Seeley. 2014. "'Good Girls': Gender, Social Class, and Slut Discourse on Campus." *Social Psychology Quarterly* 77(2): 100–122. Copyright © 2014 by the American Sociological Association. Reprinted by permission of SAGE Publications, Inc.

participation in casual sexual activity. For high-status women—whose definitions prevail in the dominant social scene—slut discourse enables, rather than constrains, sexual experimentation. In contrast, low-status women are vulnerable to public shaming.

INTERPRETING SLUT DISCOURSE AMONG WOMEN

We outline three explanations of women's participation in slut shaming. These approaches are not mutually exclusive, in part because the concept of status is central to all three. We treat status as the relative positioning of individuals in a hierarchy based on esteem and respect.... Those with high status experience esteem and approval; those with low status are more likely to experience disregard and stigma. While status systems among adults often focus on occupation, among youth they develop in peer cultures. Research on American peer cultures has found that youth status is informed by good looks, social skills, popularity with the other gender, and athleticism—traits that are loosely linked to social class (Adler and Adler 1998). In this case, status is produced and accrued in the dominant social world on campus—the largely Greek-controlled party scene.

Gender Performance and the Circulation of Stigma

The "doing gender" tradition suggests that slut stigma regulates the gender presentations of all girls and women (Eder et al. 1995; Tanenbaum 1999). The emphasis is on how women are sanctioned for failing to perform femininity acceptably (West and Zimmerman 1987). This suggests that slut stigma is more about regulating public gender performance than regulating private sexual practices.

... Pascoe (2007) shows that the ubiquitous threat of being labeled regulates performances by all boys, ensuring conformity with hegemonic masculinity. Boys jockey for rank in peer hierarchies by lobbing the fag label at each other in a game of "hot potato." Fag is not, as Pascoe (2007:54) notes, "a static identity attached to a particular (homosexual) boy" but rather "a discourse with which boys discipline themselves and each other."

Pascoe's discursive model, when extended to our case, suggests that slut discourse serves as a vehicle by which girls discipline themselves and others. It does not require the existence of "real" sluts. Just as any boy can temporarily be a fag, so can any girl provisionally fill the slut position. Slut discourse may even circulate more privately than fag discourse: girls do not need to know they have been labeled for the discourse to work. The fag label does not hinge on sexual identity or practices; similarly, the slut label may have little or nothing to do with the amount or kinds of sex women have. In the same way that the "fluidity of the fag identity" makes it a "powerful disciplinary mechanism" (Pascoe 2007:54), so may the ubiquity of the slut label.

Just as masculinities are hierarchically organized, femininities are also differentially valued. Labeling women as "good" or "bad" is about status—the negotiation of rank among women. Men play a critical role in establishing this rank by rewarding particular femininities. Women confront a double standard that penalizes them for (even the suggestion of) sexual behavior normalized for men (Crawford and Popp 2003; Hamilton and Armstrong 2009). We emphasize, however, that women also sexually evaluate and rank each other. Women's competition is oriented toward both attention from men and esteem among women. We challenge literature in which femininities are seen as wholly derivative of masculinities, where women passively accept criteria established by men.

Status competition among women is in part about femininity. Yet other dimensions of inequality—particularly class and race—intersect with gender to inform sexual evaluation. For example, Patricia Hill Collins argues that black women are often stereotyped as "jezebel, whore, or 'hoochie'" (2004:89). Class and race have no necessary connection with sexual behavior yet are taken as its signifiers. Performances of femininity are shaped by class and race and ranked in ways that benefit women in advantaged categories (McCall 1992). Respectable femininity becomes synonymous with the polite, accommodating, demure style often performed by the white middle class (Bettie 2003; Jones 2010; Garcia 2012).

This suggests that high-status women have an interest in applying sexual stigma to others, thus solidifying their erotic rank. Such an explanation is partial as it does not account for why other women engage in slut shaming. We need a framework that accommodates the interests of all actors, no matter how subordinate, in deflecting existing negative classifications.

Intersectionality, Moral Boundaries, and the Centrality of Class

Another approach highlights the symbolic boundaries people draw to affirm the identities and reputations that set them apart from others (Lamont 1992)....

Lamont's (1992, 2000) work—which attends to how people draw class boundaries—suggests that both affluent and working-class Americans construct a sense of superiority in relation to each other. She finds that working-class Americans often perceive the affluent as superficial and lacking integrity....

Women's deployment of slut discourse may be partly about negotiating class differences. It may define moral boundaries around class that also organize sexual behavior (i.e., how much and what kinds of sexual activity women engage in and with whom) and performances of femininity. The positions women take, and the success they experience when definitions conflict, may be influenced by prior social advantage. This perspective suggests that no group is entirely subject to, or in control of, slut discourse: all actively constitute it in interaction.

... We argue that women use sexual stigma to distance themselves from other women, but not primarily on the basis of actual sexual activity. Women use slut discourse to maintain status distinctions that are, in this case, linked closely to social class. Both low- and high-status women define their own performances of femininity as exempt from sexual stigma while labeling other groups as "slutty." It is only high-status women, though, who experience what we refer to as sexual privilege—the ability to define acceptable sexuality in high-status spaces.

METHODS

Our awareness of women's use of slut discourse emerged inductively from a longitudinal ethnographic and interview study of a cohort of 53 women who began college in the 2004–2005 academic year at Midwest University.[2] We supplement these data with individual and group interviews conducted outside the residence hall sample....

Ethnography and Longitudinal Interviews

A research team of nine, including two authors, occupied a room on a residence hall floor. When data collection commenced, the first author was an assistant professor in her late thirties and the second author a graduate student in her early twenties. The team included a male graduate student, an undergraduate sorority member, and an undergraduate from a working-class family. Variation in age, approach, and self-presentation facilitated different types of relationships with women on the floor (Erickson and Stull 1998).

The floor we studied was located in one of several "party dorms." Affluent students often requested this residence hall if they were interested in drinking, hooking up, and joining the Greek system. Few identified as feminist and all presented as traditionally feminine.

Floor residents were similar in many ways. They started college together, on the same floor, at the same school.[3] All were white, a result of low racial diversity on campus and segregation in campus housing. All but two identified as heterosexual and only one woman was not born in the United States. This homogeneity, though a limitation, allowed us to isolate ways that social class shaped women's positions on campus and moral boundaries they drew with respect to sexuality and gender presentation. Assessment of class background was based on parental education and occupation, student employment during the school year, and student loans. Of the sample, 54 percent came from upper-middle or upper-class backgrounds; we refer to these women as affluent. The remainder grew up in working, lower-middle, or middle-class families; we refer to these women as less affluent.

Women were told that we were there to study the college experience, and indeed, we attended to all facets of their lives. We observed throughout the academic year, interacting with participants as they did with each other (Corsaro 1997). We let women guide conversations and tried to avoid revealing our attitudes. This made it difficult for them to determine what we were studying, which behaviors interested us, and how we might judge them—minimizing the effects of social desirability.

We also conducted five waves of interviews—from women's first year of college to the year after most graduated. We include data from 189 interviews with the 44 heterosexual women (83% of the floor) who participated in the final interview. The interviews ranged from 45 minutes to 2.5 hours.

All waves covered a broad range of topics, including partying, sexuality, relationships, friendships, classes, employment, religion, and relationships with parents. The first wave included a question about how women might view "a girl who is known for having sex with a lot of guys." This wording reveals our early assumption that the slut label was about sexual activity and generated little discussion when women stayed close to the prompt. Later we realized that this, too, provided data. Aware that we were attempting to ask about "sluts," many women offered a definition of a "real" slut, as if to educate us. We also draw on the frequent, unsolicited use of slut discourse emerging from discussions of college sexuality, peers, and partying. Women were most concerned with the slut label during the first year of college, as status hierarchies were being established.

Classification into Status Groups

We classified women according to participation in the Greek party scene, which was the most widely accepted signal of peer status on campus. We categorized 23 women as high status and 21 as low status.

High-status women exhibited a particular style of femininity valued in sororities. The accomplishment of "cuteness"—a slender but fit, blonde, tan, fashionable look—required class resources. Women also gained admission on the basis of "good personalities"—indicated by extroversion, interest in high-end fashion, and familiarity with brand names. Sorority membership was almost a requirement for high status: only four women managed to pursue alternative paths into the party scene. One benefited from her relationship with an athlete, another from residence in a luxurious apartment complex with a party reputation, and two capitalized on dense high school networks.

Status fell largely, although not entirely, along class lines: the 23 high-status women were primarily upper class and upper middle class, in part because they had time and money to participate. Most were from out of state, which corresponded with wealth due to the high cost of out-of-state tuition. Some middle-class women who successfully emulated affluent social and sexual styles were also classified as high status.

The remaining 21 women were excluded from the Greek party scene. Fifteen lower-middle-class and working-class women lacked the economic and cultural resources necessary for regular participation and were low status by default. They shared this designation with six middle-class to upper-class women who did not join sororities. These women had few friends on campus and expressed attitudes critical of the Greek party scene. They did not perform the gender style that would have increased their status. Two identified as lesbian, and the others viewed themselves as alternative or nerdy. For these women, compliance would have been challenging and uncomfortable....

SLUT BOUNDARY WORK

The results are organized in three sections. First, we discuss how women simultaneously produce and evade slut stigma through interaction and their investment in this cultural work. We then show that status on campus, organized largely by social class, shapes how women define sluttiness. High- and low-status women draw moral boundaries consistent with their own classed styles of femininity, effectively segregating the groups. As we discuss in a final section, low-status women sometimes attempt to enter the dominant social scene. There they find themselves classified according to high-status standards, which places them at risk of public sexual stigma. In contrast, high-status women are exempt from public slut shaming. This, we argue, is a form of sexual privilege.

Producing Slut Stigma Through Discourse

Years after high school, two young women became angry as they revisited instances when abstinence failed to protect them from slut stigma:

WOMAN 1: I was a virgin the first time I was called a slut.

WOMAN 2: I was too.

OTHER WOMAN: Really?

WOMAN 1: Yeah, because no one knew [I was really a virgin].

WOMAN 2: They all thought I slept with people. That's what my volleyball coach said to all my friends, that I was the one that was going to be causing trouble when I get older, and now every one of my friends has had sex with like a hundred people!

WOMAN 1: Or are pregnant or have been pregnant.

WOMAN 2: Yeah, exactly.

FIRST AUTHOR: What were they responding to?

WOMAN 1: Like, if masturbation were to come up ... I wouldn't be afraid to talk about it. I think people got the wrong idea from that.

WOMAN 2: In high school, they called me a cocktease. I didn't do anything but ... I have always been the open one. (Off-Campus Group)

As was often the case, slut stigma was disconnected from sexual behavior. Yet rather than challenge the use of this label, these women, like others, endorsed it. They argued that the accusations were problematic because they were inaccurate. They even suggested that their friends who had sex with "like a hundred people" or "have been pregnant" were more appropriate targets—deflecting stigma onto someone else.

Conversations in which women discussed and demarcated the line between good and bad girls—labeling others negatively while positioning themselves favorably—were common. All but three women, or 93 percent, revealed familiarity with terms like *slut, whore, skank,* or *ho. Good girl, virgin,* or *classy* were used to indicate sexual or moral superiority. Women drew hierarchical distinctions within groups as well as between ingroup and outgroup members. Friends were easy targets, as women believed that they knew more about their sexual behavior than that of other women. As we discuss later, though, public slut shaming was commonly directed at members of the opposing status group.

These cases might be seen as textbook examples of defensive othering—a common strategy for managing stigma. Yet aspects of slut stigma differ from what social psychological models of stigma predict. The criteria for assigning stigma were unclear and continually constructed through interaction. Women were both potential recipients of sexual stigma and producers of it—simultaneously engaged in both defensive and oppressive othering. As one insightful woman put it, "I feel like you're more likely to say [slut] if you maybe feel like you could potentially be called that" (Abby Y1). There was no stable division between stigmatized and normal individuals.

It was rare for the slut label to stick to any given woman, a requirement for status loss and persistent discrimination across situations. Most labeling occurred in private and was directed at targets unaware of their stigmatization. As one woman reported about her friend's sexual relationship,

> She just keeps going over there because she wants his attention because she likes him. That's disgusting. That to me, if you want to talk about slutty, that to me is whoring yourself out. And, I mean, I hate to say that because she is one of my best friends, but good God, it's like how stupid can you be? (S06)

Often the labeled were women viewed as sexual competition. As Becky told us,

> My boyfriend, girls hit on him all the time, and during Halloween he told me this story about a girl who was wearing practically nothing.... She went up to him [and he asked,] "What are you supposed to be?" And she said, "I'm a cherry. Do you want to pop my cherry?" She lifts up her skirt and she's wearing a thong that had a cherry on it. That's skanky. That's so skanky. (Y1)

Whether friends, enemies, or as detailed below, women in the other status group, targets served as foils for women's claims of virtue.

The labeled woman did not even need to exist. Women sometimes referred to others who were so generic, interchangeable, or socially distant as to be apocryphal—the "mythical slut." For instance, sorority women in a group interview explained how serenading, a common Greek practice, was "ruined" by a "complete slut" who purportedly "had sex with a guy in front of everybody." The connection to the "slut" was tenuous: no one actually knew her—only of her. Her behavior, being particularly public in nature, was used to delimit the acceptable.

... Women feared public exposure as sluts. Virtually all expressed the desire to avoid a "bad reputation. I know that I wouldn't want that reputation" (Olivia Y1). At times they seemed to be assuring us (and themselves) of their virtue. As one anxiously reported, "I'm not a fast-paced girl. I'm a good girl" (Naomi Y1). In the context of a feminist group interview, one woman came close to positively claiming a slut identity: she proclaimed that she was done with her "secret life of being promiscuous" and was "coming out to people now.... I'm promiscuous, dammit!" Yet she proceeded to admit that she was really only "out" to her friends, noting, "I don't tell some of my friends—a lot of my friends. That's why I really love my feminist thing. I reserve it, as people aren't going to judge me." Even she feared public censure.

Class and Status Differences in Moral Boundaries

… For affluent women, a primary risk of sex in college was its potential to derail professional advancement and/or class-appropriate marriage. Hooking up, particularly without intercourse, was viewed as relatively low risk because it did not require costly commitment (Hamilton and Armstrong 2009). When asked who hooked up the most on campus, Nicole responded, "All … the people who came to college to have a good time and party" (Y1). Women even creatively reframed sexual exploration as a necessary precondition for a successful marriage. As Alicia explained, "I'm glad that I've had my one-night stands … because now I know what it's supposed to feel like when I'm with someone that I want to be with…. I feel bad for some of my friends…. They're still virgins" (Y1).

High-status women rejected the view that all sexual activity outside of relationships was bad. They viewed sexual activity along a continuum, with hooking up falling conveniently in the middle. Becky's nuanced definition of hooking up is illustrative. She argued that "kissing [is] excluded"—minimizing this favorite activity of hers in seriousness. As she continued, "You have kissing over here [motions to one side] and sex over here [motions to the other]…. Anything from making out to right before you hit sex is hooking up…. I think sex is in its own class" (Y1).

This view hinged on defining a range of sexual activities—such as "hardcore making out, heavy petting" (Becky Y1), mutual masturbation, and oral sex—as not "sex." "Sex," as women defined it, referred only to vaginal intercourse. Hannah described herself as a virgin to both researchers and her mother, despite admitting to oral sex with a hookup partner. She joked with her mother about a missed period, "Must be from all the sex I've been having. And she's like, uhhhh…. I was like, Mom, I'm just kidding. I'm still a virgin" (Y2). Hannah was not alone. Research suggests that many young Americans do not define oral-genital contact as "having sex" (Backstrom, Armstrong, and Puentes 2012; Vannier and Byers 2013).

Vaginal intercourse outside of relationships was viewed as more problematic. Becky, for example, judged those who engaged in extrarelational intercourse. When asked how often she hooked up, Becky emphasized participation in low- to middle-range activities: "I mean, I wasn't like a slut or anything. There'd be weekends I wouldn't want to do anything except make out with someone, and there's weekends I wouldn't want to do anything, like maybe a little bit of a kiss" (Y1). When the discussion turned to vaginal intercourse she—like most women—mentioned only sex with her boyfriend.

Yet having vaginal intercourse in a hookup was sometimes permissible—as long as women did not do so "too many" times or "too easily." As Tara claimed, "I think when people have sex with *a lot* of guys that aren't their boy-friends that's really a slut" (Y1, emphasis added). She was vague about the number, unable to articulate whether one, five, or 50 hookups with intercourse made a woman a slut. Another woman, who had more sexual partners than her friends, claimed that the number of partners was irrelevant. She noted, "Slutty doesn't mean how many people [you slept with]. It just means how easy you are. Like, if a guy wants it, are you gonna give it to him?" (Abby Y1).

To high-status women, looking "trashy" was more indicative of sluttiness than any amount of sexual activity. Women spent hours trying to perfect a high-status sexy look without crossing the line into sluttiness. This was often a social exercise: women crowded in front of a mirror, trying on outfits and accessories until everyone assembled approved. As Blair described, "A lot of the girls when we we're going out ... they're asking, 'I don't look slutty, do I?'" The process was designed to protect against judgment by others, although it also provided personal affirmation. For Blair, the fact that she and her sorority sisters asked these critical questions signaled that they were "classier.... That's important" (Y1).

Blair was not the only woman to contrast a desirable, classy appearance with an undesirable, trashy appearance. For instance, Alicia noted, "If my house is considered the trashy, slutty house and I didn't know that and someone said that [it] would hurt my feelings. [Especially] when I'm thinking ... it's the classy house" (Y1). *Classy* denoted sophisticated style, while *trashy* suggested exclusion from the upper rungs of society, as captured in the phrase "white trash" (Kusenbach 2009). They rarely referred to actual less-affluent women—who, by virtue of their exclusion from social life, were invisible (see Fiske 2011). Instead, women used labels to mark gradations of status in their bounded social world. By closely aligning economic advantage and moral purity, women who pulled off a classy femininity were beyond reproach.

The most successful women were those who constructed a seamless upper-middle-class gender presentation. Sororities actively recruited these women. As Alicia continued,

> Let's say I'm president of the house or something and I [want to] keep the classy [sorority] name that we've had from the previous year then [we need] more people with that classy [sorority] look.... The preppy, classy, good girl that likes to have fun and be friendly. You know, the perfect girl. (Y1)

Similarly, when asked to define her sorority's reputation, one sorority woman responded with a single word, "classy," on which another focus group member elaborated: "I think we would be the girl next door."

The "perfect girl" or "girl next door" indexed the wholesome, demure, and polite—but fun-loving—interactional style characteristic of affluent white women (Bettie 2003; Trautner 2005). Alicia's use of the word *preppy* offered another class clue: this style originated on elite Eastern college campuses and was exemplified by fashion designers like Ralph Lauren, known for selling not only clothing but an advantaged lifestyle (Banks and Chapelle 2011). The preppy female student displayed confidence in elite social settings and could afford the trappings necessary to make a good impression.

Accomplishing a classy presentation required considerable resources. Parent-funded credit cards allowed women to signal affluent tastes in clothing and makeup. Several purchased expensive MAC-brand purple eye shadow that read as classy rather than the drugstore eye shadow worn by at least one working-class woman. As Naomi told us, "I'm high maintenance.... I like nice things [laughs]. I guess in a sense, I like things brand name" (Y1). Without jobs, they had time to go tanning, get their hair done, do their nails, shop, and keep up with fashion trends. By college, these women were well versed in classed interactional styles and bodywork. Many had cultivated these skills in high-school peer cultures as cheerleaders, prom queens, and dance squad members.

High-status women also knew the nuanced rules of the party scene before arrival. Most had previous party experience and brought advice from college-savvy friends and family with them. Becky described one such rule, about attire:

> [Halloween is] the night that girls can dress skanky. Me and my friends do it. [And] in the summer, I'm not gonna lie, I wear itty bitty skirts.... Then there are the sluts that just dress slutty, and sure they could be actual sluts. I don't get girls that go to fraternity parties in the dead of winter wearing skirts that you can see their asses in. (Y1)

As she noted, good girls do not wear short skirts or revealing shirts without social permission. She was aware that women who dressed provocatively were not necessarily "actual sluts," but her language suggested belief in such women's existence, necessitating efforts to avoid being placed in this category. Another woman highlighted ways that dress and deportment could be played off each other. She noted that it was acceptable for women to "have a short skirt on" if "they're being cool" but "if they're dancing really gross with a short skirt on, then like, oh slut. You've got to have the combination" (Lydia Y1). Women lacking familiarity with these unstated rules started at a disadvantage.

In general, classy girls did not get in trouble, draw inappropriate attention, or do anything "weird." For instance, one supposed slut was "involved with drugs, and she stole a lot of stuff, and her parents sent her to boarding school" (Nicole Y1). Others were described as having "problems at home with their families and stuff" (Nicole Y4). In one case, a slut was remarkable for "eat[ing] ketchup for dinner [laughter]. [First Author: Like, only ketchup?] Right, she has some issues" (Erica and Taylor Y1). These activities were not sexual. Instead, they represented failure to successfully perform an affluent femininity, with sexual stigma applied as the penalty.

The low-status view: nice versus bitchy. The notion that youth should participate in hookups was foreign to less-affluent women, whose expectations about appropriate relationship timelines were shaped by a different social world. Many of their friends back home were already married or had children. Amanda, a working-class woman, recalled, "I thought I'd get married in college.... I wanted to have kids before I was 25" (Y4). Hooking up made little sense uncoupled from the desire to postpone commitment. As one less-affluent woman noted,

Who would be interested in just meeting somebody and then doing something that night? And then never talking to them again? ... I'm supposed to do this; I'm supposed to get drunk every weekend. I'm supposed to go to parties every weekend ... and I'm supposed to enjoy it like everyone else. But it just doesn't appeal to me. (Valerie Y1)

Lacking access to classed beliefs supporting sexual exploration, less-affluent women treated sexual activity outside of relationships as morally suspect. As lower-middle-class Olivia explained,

I have really strong feelings about the whole sex thing.... I know that some people have boyfriends and they've been with them for a long time, and I understand that. But I listen to some people when they talk about [hooking up].... I know that personally for me, I would rather be a virgin for as much as I can than go out and do God knows who and do whatever. (Y1)

As discussed in the Methods section, not all low-status women lacked class advantage, but even low-status women from affluent families opposed hooking up. As upper-middle-class Madison noted, "I just don't [hook up].... I'm not really into that kinda thing, I guess. I just don't like getting with random people" (Y1). Similarly, upper-middle-class Linda described herself as "very sexually conservative" in contrast to her "liberal" floor, in part due to their participation in hooking up (Y1).

Some low-status women were confused about hooking up, as they were excluded from the social networks where the practice made sense. When asked for a definition, Mary, a middle-class woman, responded, "Good question. I honestly, I couldn't tell you what some of their... I mean I've heard them use [the word] and I'm kind of like, well what does that mean? Did you have sex with them or did you just make out with them or ... ?" (Y1). Working-class Megan had not even heard of hooking up until we asked her about it. She equated hooking up with an alleged sorority hazing ritual in which "they would tie the girls up naked on a bed and then a guy would come in and they would have sex with them" (Y1).

Without insider cultural knowledge, low-status women did not make the same fine-grained distinctions between types of sexual activity outside of relationships. For these women, the relevant divide was whether the activity occurred in a relationship or not. They assumed that hookups, like most committed relationships, involved vaginal intercourse. A roommate pair explained:

> *HEATHER:* A lot of the girls ... they're always like oh you hooked up.
>
> *STACEY:* We're not used to that. Hooking up means you guys fucked.... I'd be like omigod and everyone else's like what? And I'm like you guys hooked up? They'd be like so?
>
> *SECOND AUTHOR:* You thought everyone was having random sex?
>
> *STACEY:* [I felt like saying] you slut.
>
> *HEATHER:* At first we were like, what is this place? (Y1)

These two women would briefly (and unsuccessfully) attempt to befriend affluent partiers on the floor. This provided them with more information about the complexities of hooking up, although they did not alter their own sexual practices.

Low-status women maintained a distinction between themselves and those who hooked up. As Olivia noted,

> My friends are similar when it comes to things like [sex]. We don't think of it as doing whatever with who knows who.... I'm sure there's more people that are like me, but I know there are people who just do it casually. They don't think of it as anything 'cause a lot of them have done it before. For them it's different. (Y1)

Her explanation, using us-versus-them language, divided college women into two groups and implied her group was superior.

The judgment low-status women passed on their high-status peers was about more than sexuality. They often derided sorority women and those who attended parties. As Carrie described, "[My sister] who goes to [private college] is in [a sorority]. Umm, hello. All those girls are sluts. Sorry, they were. All they did was drink and go to parties. She's not like that so she deactivated" (Y1). Linda referred to women in the Greek system as "the party sluts" (Y4).

Underlying this disapproval was a rejection of their partying peers' interactional style. Madison, right after she transferred to a regional college, explained what she disliked about many women on the floor:

> Sorority girls are kinda whorish and unfriendly and very cliquey. If you weren't Greek, then you didn't really matter.... I feel like most, if not all, the sorority girls I met at MU were bitches and stuck up. [In response to the indignation of a friend from another school, who was present during this segment of the interview:] I met [Sasha's sorority] sisters and they're really nice. (Y3)

Madison equated sluttiness with exclusivity—being bitchy, stuck up, cliquey, and unfriendly. She contrasted this with the desirable trait, "niceness," which she was obligated to attribute to Sasha and her friends.

Niceness, also described as being "friendly," "laid back," or "down home," referenced a classed femininity in which social climbing, expensive consumption patterns, and efforts to distinguish oneself as "better than" others were disparaged. Madison rejected high-status femininity, despite her own affluence. She explained,

> Most of the girls ... they seem to be snotty. There were a few girls that are just like [my friend's and my] level, where we aren't gonna be, oh we have money, we're gonna live better than you. But there are a few that definitely you could tell they had like an unlimited income. They went shopping all the time. (Y3)

Similarly, Stacey—who was from a lower-middle-class family—remarked bitterly, "There's a lot of rich bitches in sororities, and they have everything that their daddy gives them.... I mean, they probably saw on TV we're the number one party school, like, four years ago and they're like, 'Daddy, Mommy, I wanna go there!'" (Y3/Y4). Sluttiness and wealth were often conflated. As Alana reported, "Some people think [this dorm is where] the whores are. You know, oh those 'Macslutts in MacAdams.... ' People think [it's] like the rich people.... Their stereotypes might be true" (Y1).

These women expressed considerable class and status vulnerability—even animosity. Their private commentary was pointed, directed at specific high-status women. As Fiske (2011) suggests, those at the bottom of a hierarchy tend to be excruciatingly aware of those above them, whereas those with status attend less to those below them. Lacking language to make sense of the class differences that permeated social life at Midwest University, the slut label did cultural work. Low-status women conflated unkindness and perceived promiscuity when they called high-status women "slutty." Their use of the term captured both their reactions to poor treatment and the unfairness of others' getting away with sexual behavior they viewed as inappropriate (and for which they would have been penalized). Slut discourse was thus employed in privately waged battles of class revenge. As we discuss below, this animosity had few consequences for high-status women.

Status, Affluence, and Competing Boundaries. Slut discourse helped establish and maintain boundaries between high- and low-status women. Midway through college there were no friendships crossing this line, despite the cross-group interactions necessitated by living on the same floor. Women enforced moral boundaries on uneven ground. Most cases of conflict occurred when low-status women—lured by the promise of fun, status, and belonging—attempted to interact with high-status women, especially in the party scene. There was not much movement in the other direction: high-status women had little to gain by associating with low-status women.

Women rarely labeled others publicly. We recorded only five instances in our first-year residence hall observations. None of the women carried a negative reputation outside the situations where labeling occurred. These interactions, however, were among the most explosive and painful we witnessed. Targets were low-status—and, in four cases, less-affluent—women who attempted to make inroads with high-status women.

High-status women valued a muted, polite, and demure femininity. This contrasted with the louder, cruder, overtly sexual femininity exhibited by Stacey and Heather, a working-class roommate pair who, early in the year, attempted to associate with partiers on the floor. As field notes recount,

> Whitney ... came out into the hall as Heather and Stacey (applying finishing touches to her tube top) came out. Both were in tight pants (one black one brown?) and tight tops. They had plenty of makeup on (this was clear from far away) and tall heels.... They were headed for another dorm to say "hey" to a guy that Stacey had met. Whitney made a comment about how dressed up they were to just say "hey." [The girls] laughed it off and very loudly yelled something about going to "whore around." (FN 9-15-04)

In this incident, high-status Whitney implicitly passed judgment on Heather and Stacey, whose clothing and demeanor violated high-status norms of self-presentation. The two women immediately understood that their behavior was being coded as sexually deviant. Ironically, their attempt at saving face—by joking about "whoring around"—likely made Whitney's comment seem even more warranted in the eyes of their affluent peers.

Several months after the hallway incident, Stacey was watching a television show with several high-status women who lived near her:

> One of the characters was hooking up with somebody new and Stacey said, "Slut-bag!" Chelsea said, "Stacey?" as if to imply jokingly that she had no right to call this woman a slut. Stacey was clearly offended by this and said indignantly, "I am NOT a slut." Chelsea, seeing her take it so badly, said that she really didn't mean it that way and that she was joking but Stacey stormed off anyway. (FN 1-13-05)

Stacey attempted to apply her own definition of slut to the actions of the television character, calling her out for hooking up. Chelsea rejected this, turning the label back on Stacey, who was offended. Later, a lower-middle-class woman attempted to defend Stacey. She remarked, "It's not like Stacey sleeps around anyway." The damage had already been done though. None of the other women in the room chimed in to confirm Stacey's virtue.

In another instance, the "wrong" choice of an erotic partner landed working-class Monica a label. As we recorded,

> Monica's really open flirting and sexuality with Heather's brother was looked down on by people on the floor. Many rolled eyes and insinuate[d] that she was being slutty or inappropriate. This guy (both because he was someone's brother and because he was clearly working-class—not in a frat or middle-class) was the wrong object. (FN 2-10-05)

From the perspective of high-status women, good girls only flirted with affluent men who had high status on campus. This disadvantaged less-affluent women, who were often drawn to men sharing their class background. These men were not in fraternities or necessarily even in college.

Monica's dalliance with Heather's brother might have escaped notice had she not also made brief forays into the party scene. Monica and her middle-class roommate Karen—who worked her way into the high-status group—ended the year in a vicious battle, flinging the slut label back and forth behind each other's backs. Monica, however, was singled out for judgment by shared acquaintances. Prior to their dramatic split, Monica and Karen often kissed each other at parties—a form of same-sex eroticism often intended to appeal to men (Hamilton 2007). Several floormates decided in conversation that Monica was "somewhat weird and 'slutty'... [while] Karen's sexuality or sluttiness never came up.... It wasn't even a question" (FN 3-8-05). Monica lacked friends positioned to spread similar rumors about Karen. Unexpectedly, Monica left shortly before the end of the year and did not return to Midwest University.

Monica's, Stacey's, and Heather's experiences illustrate the challenges women from less-advantaged backgrounds faced if they attempted to break into the party scene. They were also at risk of acquiring sexual stigma back home, where they were judged for associating with rich partiers. For instance, Monica's hometown acquaintances started a virulent rumor that she had an abortion while at Midwest University. This suggests that people in her hometown shared the construction of sluttiness we described earlier, viewing affluent college girls as sluts in contrast with down-to-earth, small-town girls. Monica had been tainted by association.

In contrast, the only affluent woman to be publicly shamed was from the low-status group. She had angered many of her floormates with her blatant and public homophobia. They retaliated by writing derogatory comments, including the slut label, on the whiteboard posted on her door. Aside from this case, affluent women were virtually exempt from public shaming by other women, whether at school or at home, where their friends' definitions were roughly in sync with their own.

This freedom from stigma is particularly remarkable considering what we ascertained about women's sexual activities. All but one high-status woman hooked up during college in between committed relationships. Some low-status women also hooked up, but usually only once or twice before deciding it was not for them. Nearly two thirds of this group did not hook up at all. A few low-status women left college without having had vaginal intercourse, but no high-status women refrained from intercourse entirely. Most low-status women limited their sexual activity to relationships. Low-status women reported to us, on average, roughly 1.5 fewer sexual partners (for oral sex or intercourse) during college than high-status women. These patterns underscore the disconnect between vulnerability to slut stigma and sexual activity.

From the perspective of low-status women, the sexual activities of high-status peers were riskier than their own strategy of restricting sex to relationships (or avoiding it altogether)—yet high-status women evaded the most damaging kind of labeling. As long as they were discreet and did not, as one put it, "go bragging about the guys I've hooked up with," high-status women experienced minimal threat of judgment by others (Lydia Y1). Upper-middle-class Rory, who with more than 60 partners was the most sexually active woman we interviewed, explained, "I'm the kind a girl that everybody would like talk shit about if they knew.... I have this really good image. Hah. And people don't think of me that way. They think I'm like nice and smart, and I'm like yeah" (S07). Casual sexual activity posed little reputational risk for savvy, affluent women who maintained a classy image.

DISCUSSION

Slut discourse was ubiquitous among the women we studied. Sexual labels were exchanged fluidly but rarely became stably attached to particular women. Stigma was instead produced in interaction, as women defined their virtue against real or imagined bad girls. The boundaries women drew were shaped by status on campus, which was closely linked to class background. High-status women considered the performance of a classy femininity—which relied on economic advantage—as proof that one was not trashy. In contrast, low-status women, mostly from less-affluent backgrounds, emphasized niceness and viewed partying as evidence of sluttiness.

Both groups actively reconstituted the slut label to their advantage. Despite this, they were not equally situated to enforce their moral boundaries. High-status women operated within a discursive system allowing greater space for sexual experimentation. When low-status women attempted to participate in high-status social worlds, they risked public slut shaming. At the same time, their more restrictive definitions lacked social consequences for higher-status women. This, as we argue below, is a form of sexual privilege. Low-status women resented the class and sexual advantages of their affluent peers and unsuccessfully used sexual stigma in an attempt to level differences.

Class, Race, and Moral Boundaries

The behaviors of women and girls are often viewed through the lens of sexual and gender inequality, particularly where sexual practices are concerned (Bettie 2003; Wilkins 2008). Certainly, sexual double standards are real and may guide men's use of the slut label against women (Crawford and Popp 2003). But equalizing sexual standards—while undoubtedly an important goal—would not necessarily eliminate slut shaming, which assists women in drawing class boundaries.

… The white women in this study operated in racially homogeneous social worlds, making it easier for us to see class-based processes. Race is not absent from their accounts, however. The notion of the "girl next door" and even the "nice" down-home girl are both racialized. Had we also studied the small nonwhite student population on campus—who, like less-affluent women, were excluded from the predominately white Greek system—it is likely that we would have recognized moral boundaries drawn around race. Indeed, Garcia's (2012) Latina participants viewed "sluttiness" as primarily white (also see Espiritu 2001).

Sexual Privilege

Classed resources provided affluent white women with more room to maneuver sexually. They drew on the notion that young adulthood should be about exploration to justify sexual experimentation in noncommitted sexual contexts (Hamilton and Armstrong 2009). Slut discourse, rather than constraining their sexual options, ensured that they could safely enjoy the sexual opportunities of the party scene. Those without the time, money, and knowledge needed to effect a "classy" appearance lacked similar protections. It is thus unsurprising that women who hook up on residential college campuses are more likely to be affluent and white (Owen et al. 2010; Paula England, personal communication with second author, 2013).[4]

The definition of sluttiness offered by the low-status women in our study does, however, have a place in youth culture. See, for example, this definition of "sorostitute" (a play on prostitute) from Urban Dictionary:

> You can find me on campus in the SUV my daddy bought for me....
> I never leave my sorority house without my letters somewhere on me.
> I date a fratdaddy. I don't care that he cheats on me with other sorostitutes because I cheat on him too.... Looks are all that matter to me.
> I spent money that was supposed to be for books on tanning and manicures. I have had plastic surgery. I'm always well dressed. I pop my collar and all of my handbags—my Louis [Vuitton], my Kate Spade, my Prada—are real. If I look like this, frat boys will want me and other sororities will be jealous. I look better than you, I act better than you, I AM better than you.[5]

The circulation of this term suggests that our participants are not alone in attempting to label affluent sorority women as slutty.

Sexual privilege, however, involves the ability to define acceptable sexuality in ways that apply in high-status spaces. High-status women in our study were deeply embedded in the dominant social scene on campus. Over the years, they moved into positions of greater influence—for instance, later selecting the women who joined them in elite sorority houses. They did not care what marginalized individuals thought of them as these opinions were inconsequential both during college and beyond. As gatekeepers to the party scene, however, high-status women had considerable power over low-status women who wished to belong. It is in this context that the sexual activity of advantaged women becomes invisible.

This is not to downplay men's power in sexualized interactions or deny the gendered sexual double standard faced by women. Yet we differ from the classic framework posed by Connell (1987), in which no femininity holds a position of power equivalent to that of hegemonic masculinity among men (but see Schippers 2007). We argue that women are actively invested in slut shaming because they have something to gain. They are not simply unwitting victims of men's sexual dominance. The winners—those whose femininities are valued—enjoy sexual privilege. This is a benefit also extended to men who display a hegemonic masculinity (DeSantis 2007; Sweeney 2013). It indicates the importance of attending to dynamics within—not only across—gender.

Stigma at the Discursive Level

The questions generally answered by social psychological research on stigma—who the labeled and labelers are, how deviants are labeled and respond to stigma—are indeed important. A focus on the individual level does not, however, provide a complete picture of stigma processes. Our work, building on that of gender scholars and cultural sociologists, points to the value of examining how stigma is constituted and circulated.

A discursive approach suggests that the social psychological model of "othering" might be constructively reworked (Jones et al. 1984; Crocker et al. 1998; Crocker 1999). Subordinates may succeed in generating alternative public classification systems or subtly reworking dominant ones. For example, the actions of low-status women are not exclusively devoted to adapting to meaning systems established by high-status, socially dominant women on campus. Instead, they produce their own discursive system demarcating the line between good and bad girls in a way that benefits them.

The process of othering may thus provide ongoing opportunities for reclassification, potentially along entirely different dimensions than designated by oppressors—even if alternative frames are difficult to sustain. Othering may be not only oppressive or defensive but also confrontational or challenging. Indeed, the example of the sorostitute suggests cultural resistance to classification systems exempting affluent, high-status college women's sexual behavior from stigma.

To see this process, stigma research must be explicitly intersectional, looking at how dominants and subordinates draw on dimensions of stratification to define within-group hierarchies. Here, for instance, women draw on classed understandings of femininity and acceptable sexuality to deflect sexual stigma and define themselves as morally superior. Without a classed lens, it is easy to miss the competition among women that motivates women's participation in slut shaming.

Attention to how sets of categories are constructed and organized also generates questions for future research. We might ask why and when some discursive systems—not others—are in play. This focus introduces room for multiple, competing ways of constituting stigma. It raises questions of power and status in the successful application of stigma—that is, whose definitions of deviance are more influential? At the level of discourse, it is also easier to see variation across types of stigma. Why are some forms particularly rigid and likely to stick, while others—like the slut or fag labels—more fluid and able to constrain the actions of all individuals, not just a recognizable group of deviants? Attention to the discursive level makes it easier to detect additional, subtler bases for stigma and better ascertain its operation.

These questions may be difficult to answer in the laboratories where much social psychological research on stigma is conducted (Hebl and Dovidio 2005; Trautner and Collett 2010). An expanded focus necessitates a parallel openness to ethnography, interviews, and other qualitative methods, alongside conventional approaches. Qualitative techniques are often ideal for studying interactions within and across social groups and capturing the processes through which discourse is created and circulates.

As we noted in the introduction, some research—notably Pascoe's (2007) analysis of the circulation of the fag epithet—pushes in this direction. Yet research traditions often develop separately, even when similar concepts are explored. For example, Pascoe's research neither cites nor is cited by scholars studying stigma. This limits production of knowledge across subfields—for example, social psychology, cultural theory, and gender theory—that would benefit from greater dialogue. Our research highlights the potential of cross-fertilization and calls for more work in this vein.

NOTES

1. "Slut." *Urban Dictionary*. Retrieved December 18, 2013 (http://www.urbandiction-ary.com/define.php?term=slut).
2. We refer to the university with a pseudonym.
3. At the start of the study, 51 women were freshmen, and 2 were sophomores.
4. Paula England's Online College Social Life Survey of 21 four-year colleges and universities includes maternal education as the measure of social class. These data indicate that women whose mothers have either a BA or an advanced degree report significantly higher numbers of hookups than those whose mothers have a high school degree or less. White women also report significantly greater numbers of hookups than women in all other racial/ethnic categories.
5. "Sorostitute." *Urban Dictionary*. Retrieved December 18, 2013 (http://www.urban-dictionary.com/define.php?term=sorostitute).

REFERENCES

Adler, Patricia A. and Peter Adler. 1998. *Peer Power: Preadolescent Culture and Identity*. New Brunswick, NJ: Rutgers University Press.

Alexander, Michele G. and Terri D. Fisher. 2003. "Truth and Consequences: Using the Bogus Pipeline to Examine Sex Differences in Self-Reported Sexuality." *Journal of Sex Research* 40(1):27–35.

Armstrong, Elizabeth A. and Laura T. Hamilton. 2013. *Paying for the Party: How College Maintains Inequality*. Cambridge, MA: Harvard University Press.

Backstrom, Laura, Elizabeth A. Armstrong, and Jennifer Puentes. 2012. "Women's Negotiation of Cunnilingus in College Hookups and Relationships." *Journal of Sex Research* 49(1):1–12.

Banks, Jeffrey and Doria de la Chapelle. 2011. *Preppy: Cultivating Ivy Style*. New York: Random House.

Bell, Leslie C. 2013. *Hard to Get: Twenty-Something Women and the Paradox of Sexual Freedom*. Berkeley, CA: University of California Press.

Berger, Joseph, Cecilia L. Ridgeway, and Morris Zelditch. 2002. "Construction of Status and Referential Structures." *Sociological Theory* 20(2):157–79.

Bettie, Julie. 2003. *Women Without Class: Girls, Race, and Identity*. Berkeley, CA: University of California Press.

Blinde, Elaine M. and Diane E. Taub. 1992. "Women Athletes as Falsely Accused Deviants: Managing the Lesbian Stigma." *Sociological Quarterly* 33(4):521–33.

Bourdieu, Pierre. 1984. *Distinction: A Social Critique of the Judgment of Taste*. Cambridge, MA: Harvard University Press.

Butler, Judith. 1990. *Gender Trouble: Feminism and the Subversion of Identity*. New York: Routledge.

Coleman, James S. 1961. *The Adolescent Society: The Social Life of the Teenager and Its Impact on Education*. New York: Free Press.

Collins, Patricia Hill. 2004. *Black Sexual Politics: African Americans, Gender, and the New Racism*. New York: Routledge.

Connell, Raewyn. 1987. *Gender and Power: Society, the Person, and Sexual Politics*. Stanford, CA: Stanford University Press.

Corsaro, William A. 1997. *The Sociology of Childhood*. Thousand Oaks, CA: Pine Forge Press.

Crawford, Mary and Danielle Popp. 2003. "Sexual Double Standards: A Review and Methodological Critique of Two Decades of Research." *Journal of Sex Research* 40(1):13–26.

Crocker, Jennifer. 1999. "Social Stigma and Self-Esteem: Situational Construction of Self-Worth." *Journal of Experimental Social Psychology* 35(1):89–107.

Crocker, Jennifer, Brenda Major, and Claude M. Steele. 1998. "Social Stigma." Pp. 504–53 in *The Handbook of Social Psychology*, ed. Daniel Todd Gilbert, Susan T. Fiske, and Gardner Lindzey, Fourth ed. Boston, MA: McGraw-Hill.

DeSantis, Alan D. 2007. *Inside Greek U.* Lexington: University of Kentucky Press.

Eder, Donna, Catherine Colleen Evans, and Stephen Parker. 1995. *School Talk: Gender and Adolescent Culture*. New Brunswick, NJ: Rutgers University Press.

Erickson, Ken and Donald D. Stull. 1998. *Doing Team Ethnography: Warnings and Advice*. Thousand Oaks, CA: Sage.

Espiritu, Yen Le. 2001. "'We Don't Sleep around Like White Girls Do': Family, Culture, and Gender in Filipina American Lives." *Signs* 26(2):415–40.

Fine, Gary Alan. 1992. *Manufacturing Tales: Sex and Money in Contemporary Legends*. Knoxville, TN: University of Tennessee Press.

Fiske, Susan T. 2011. *Envy Up, Scorn Down: How Status Divides Us*. New York: Russell Sage.

Foucault, Michel. 1978. *The History of Sexuality. Vol. 1, An Introduction*. Translated by R. Hurley. New York: Vintage.

Garcia, Lorena. 2012. *Respect Yourself, Protect Yourself: Latina Girls and Sexual Identity*. New York: New York University Press.

Gieryn, Thomas. 1983. "Boundary-Work and the Demarcation of Science from Non-Science: Strains and Interests in Professional Ideologies of Scientists." *American Sociological Review* 48(6):781–95.

Goffman, Erving. 1963. *Stigma: Notes on the Management of Spoiled Identity*. Englewood Cliffs, NJ: Prentice Hall.

Gorman, Thomas J. 2000. "Cross-Class Perceptions of Social Class." *Sociological Spectrum* 20(1):93–120.

Hamilton, Laura. 2007. "Trading on Heterosexuality: College Women's Gender Strategies and Homophobia." *Gender & Society* 21(2):145–72.

Hamilton, Laura and Elizabeth A. Armstrong. 2009. "Gendered Sexuality in Young Adulthood: Double Binds and Flawed Options." *Gender & Society* 23(5):589–616.

Hebl, Michelle and John F. Dovidio. 2005. "Promoting the 'Social' in the Examination of Social Stigmas." *Personality and Social Psychology Review* 9(2):156–82.

Hurtado, Sylvia, Jeffrey Milem, Alma Clayton-Pedersen, and Walter Allen. 1999. *Enacting Diverse Learning Environments: Improving the Climate for Racial/Ethnic Diversity in Higher Education.* Washington, DC: Graduate School of Education and Human Development, George Washington.

Jones, Edward E., Amerigo Farina, Albert H. Hastorf, and Rita French. 1984. *Social Stigma: The Psychology of Marked Relationships.* New York: W.H. *Freeman.*

Jones, Nikki. 2010. *Between Good and Ghetto: African American Girls and Inner City Violence.* New Brunswick, NJ: Rutgers University Press.

Killian, Caitlin and Cathryn Johnson. 2006. "'I'm Not an Immigrant!': Resistance, Redefinition, and the Role of Resources in Identity Work." *Social Psychology Quarterly* 69(1):60–80.

Kusenbach, Margarethe. 2009. "Salvaging Decency: Mobile Home Residents' Strategies of Managing the Stigma of 'Trailer' Living." *Qualitative Sociology* 32(4):399–428.

Lamont, Michèle. 1992. *Money, Morals, and Manners: The Culture of the French and the American Upper-Middle Class.* Chicago, IL: University of Chicago Press.

Lamont, Michèle. 2000. *The Dignity of Working Men: Morality and the Boundaries of Race, Class, and Immigration.* Cambridge, MA: Harvard University Press.

Lamont, Michèle and Virág Molnár. 2002. "The Study of Boundaries in the Social Sciences." *Annual Review of Sociology* 28:167–95.

Laumann, Edward, John Gagnon, Robert Michael, and Stuart Michaels. 1994. *The Social Organization of Sexuality: Sexual Practices in the United States.* Chicago, IL: University of Chicago Press.

Link, Bruce G. and Jo C. Phelan. 2001. "Conceptualizing Stigma." *Annual Review of Sociology* 27:363–85.

Lucas, Jeffrey W. and Jo C. Phelan. 2012. "Stigma and Status: The Interrelation of Two Theoretical Perspectives." *Social Psychology Quarterly* 75(4):310–33.

Major, Brenda and Laurie T. O'Brien. 2005. "The Social Psychology of Stigma." *Annual Review of Psychology* 56(1):393–421.

McCall, Leslie. 1992. "Does Gender Fit? Bourdieu, Feminism, and Conceptions of Social Order." *Theory and Society* 21(6):837–67.

Milner, Murray, Jr. 2006. *Freaks, Geeks, and Cool Kids: American Teenagers, Schools, and the Culture of Consumption.* New York: Routledge.

Nack, Adina. 2002. "Bad Girls and Fallen Women: Chronic STD Diagnoses as Gateways to Tribal Stigma." *Symbolic Interaction* 25(4):463–85.

Ortner, Sherry B. 1991. "Reading America: Preliminary Notes on Class and Culture." Pp. 163–89 in *Recapturing Anthropology: Working in the Present*, edited by R. G. Fox. Santa Fe, NM: School of American Research.

Owen, Jesse, Galena Rhoades, Scott Stanley, and Frank Fincham. 2010. "'Hooking Up' among College Students: Demographic and Psychosocial Correlates." *Archives of Sexual Behavior* 39(3):653–63.

Pascoe, C. J. 2007. *Dude, You're a Fag: Masculinity and Sexuality in High School.* Berkeley, CA: University of California Press.

Payne, Elizabethe. 2010. "Sluts: Heteronormative Policing in the Stories of Lesbian Youth." *Educational Studies* 46(3):317–36.

Phillips, Lynn M. 2000. *Flirting with Danger: Young Women's Reflections on Sexuality and Domination.* New York: New York University.

Pyke, Karen and Tran Dang. 2003. "'FOB' and 'Whitewashed': Identity and Internalized Racism among Second Generation Asian Americans." *Qualitative Sociology* 27(2): 147–72.

Remez, Lisa. 2000. "Oral Sex among Adolescents: Is It Sex or Is It Abstinence?" *Family Planning Perspectives* 32(6):298–304.

Ridgeway, Cecilia. 2011. *Framed by Gender: How Gender Inequality Persists in the Modern World.* New York: Oxford University.

Ringrose, Jessica and Emma Renold. 2012. "Slut-Shaming, Girl Power and 'Sexualisation': Thinking through the Politics of the International SlutWalks with Teen Girls." *Gender and Education* 24(3):333–43.

Saguy, Abigail C. and Anna Ward. 2011. "Coming Out as Fat: Rethinking Stigma." *Social Psychology Quarterly* 74(1):53–75.

Schalet, Amy. 2011. *Not Under My Roof: Parents, Teens, and the Culture of Sex.* Chicago, IL: University of Chicago Press.

Schippers, Mimi. 2007. "Recovering the Feminine Other: Masculinity, Femininity, and Gender Hegemony." *Theory and Society* 36(1):85–102.

Schwalbe, Michael, Daphne Holden, Douglas Schrock, Sandra Godwin, Shealy Thompson, and Michele Wolkomir. 2000. "Generic Processes in the Reproduction of Inequality: An Interactionist Analysis." *Social Forces* 79(2):419–52.

Skeggs, Beverley. 1997. *Formations of Class and Gender: Becoming Respectable.* Thousand Oaks, CA: Sage.

Stuber, Jenny M. 2006. "Talk of Class: The Discursive Repertoires of White Working- and Upper-Middle-Class College Students." *Journal of Contemporary Ethnography* 35(3):285–318.

Sweeney, Brian. Forthcoming. "Sorting Women Sexually: Masculine Status, Sexual Performance, and the Sexual Stigmatization of Women." *Symbolic Interaction.*

Tanenbaum, Leora. 1999. *Slut! Growing up Female with a Bad Reputation.* New York: Seven Stories.

Thoits, Peggy A. 2011. "Resisting the Stigma of Mental Illness." *Social Psychology Quarterly* 74(1):6–28.

Trautner, Mary Nell. 2005. "Doing Gender, Doing Class: The Performance of Sexuality in Exotic Dance Clubs." *Gender & Society* 19(6):771–88.

Trautner, Mary Nell and Jessica L. Collett. 2010. "Students Who Strip: The Benefits of Alternate Identities for Managing Stigma." *Symbolic Interaction* 33(2):257–79.

Vannier, Sarah A. and E. Sandra Byers. 2013. "A Qualitative Study of University Students' Perceptions of Oral Sex, Intercourse, and Intimacy." *Archives of Sexual Behavior* 42(8):1573–81.

West, Candace and Don H. Zimmerman. 1987. "Doing Gender." *Gender & Society* 1(2):125–51.

White, Emily. 2002. *Fast Girls: Teenage Tribes and the Myth of the Slut*. New York: Scribner.

Wilkins, Amy C. 2008. *Wannabes, Goths, and Christians: The Boundaries of Sex, Style, and Status*. Chicago, IL: University of Chicago Press.

Wolf, Naomi. 1997. *Promiscuities: The Secret Struggle for Womanhood*. New York: Random House.

24

Queering the Sexual and Racial Politics of Urban Revitalization

DONOVAN LESSARD

If we look at the results of national surveys measuring societal acceptance of gays and lesbians, we get a rosy picture of increasing tolerance. For example, 48% of Americans in 2012 supported gay marriage, as opposed to 11% in 1988 (Smith and Son 2013). A 2013 poll even found that 52% of Republicans and conservative independents between ages 18 and 49 support same-sex marriage (Cohen 2013). Additionally, visibility and representations of gays and lesbians in U.S. media have steadily increased from "almost no gay characters" before 1970 (Fisher et al. 2007) to 64 LGBT characters on scripted prime time television (Gay and Lesbian Alliance Against Defamation 2014). Gay marriage is now legal in the United States. It appears that we have arrived in an unprecedented era of tolerance for gays and lesbians.

But is this the reality for all LGBTQ people? Are some LGBTQ groups accepted more than others, valued more? Which LGBTQ groups are less likely to be accepted? And how does that acceptance/intolerance manifest itself in urban space? Taking a spatial, intersectional approach complicates this notion of a queer golden age—a productive complication, in my opinion.

SEXUALITY, SPACE, AND INTERSECTIONALITY

Sexual identity does not exist in a spatial vacuum. This becomes especially evident when we look at the historical importance of how the establishment of gay neighborhoods in cities throughout North America created economic and political bases for gay identities and communities (Levine 1979; D'Emilio 1998; Knopp 1987; Castells 1983). Numerous gay and queer scholars and novelists have written about the role of "sites of male-centered sexual opportunity"— gay bars, parks, truck stops, bathhouses, and bookstores—in the formation of gay collective identity in the mid-20th century (Higgins 1999; Leap 1999; Chauncey 1994; White 1982). During the same period, lesbians also claimed space in urban centers through running and operating bookstores, coffee shops, and bars that created networks that enabled the development of a political identity, created a sense of community, and challenged heterosexism (Enke 2007).

Space is also important because it is where the actual operation of hierarchies and power relations take place. As opposed to an approach that is just focused on specific identities (gay/lesbian, Black, white, etc.) queer and feminist theorists who take a place-based approach to studying social movements and identity are able to see the operation of multiple identities and power relations and how they interact with each other on the ground and change over time; this is one of the strengths of a place-based approach. This is what is meant by an *intersectional analysis*—an analysis that considers the interplay of multiple identities, forms of power, and oppression, including race, ethnicity, sexuality, gender, and ability. This intersectional perspective is especially important when we begin to think about the current state of LGBTQ acceptance in communities around the United States.

URBAN REVITALIZATION AND GLBTQ COMMUNITIES

Following white flight from urban centers in the 1950s and 1960s, to deindustrialization in the 1970s and 1980s, and the recent foreclosure crisis, city planners, tourist agencies and real estate developers are strategizing ways to "revitalize" inner cities around the United States. Curiously, gays and lesbians are increasingly looked to as a niche market of "urban pioneers" who "turns neighborhoods around" (Murphy 2010; Manalansan 2005; Chasin 2001). "Urban pioneers" is a term used to describe the middle-class people who initially move into a working-class or poor neighborhood. Historically, these are people with middle-class cultural capital (high education attainment and middle-class origins) but lower income than their wealthier middle-class peers. Artists and students are two social groups that have historically been "urban pioneers." Another is middle-class, often white, gay people.... Indeed, many researchers and theorists on urban revitalization have argued that gay populations are often "urban pioneers" who raise housing prices in poor neighborhoods (Castells 1983; Zukin 1995; Smith 1996).

But this begs several questions: *which* groups of gay people become valued citizens courted by developers as an expanding niche market and are hailed by urbanists as "adventurous creatives" who may bolster neighborhood housing values? In contrast, which gays and queers become labeled violent deviants thought to be jeopardizing neighborhood safety and well-being?

In the summers of 2010 and 2011, I performed seven months of ethnographic research. I interviewed 35 GLBTQ activists, organizers, and frequenters of various queer spaces in Minneapolis, Minnesota. I argue that the gay people that urban studies scholars, city planners, developers, and many lay people hail as "urban pioneers" become visible as "gay" because of their class status and race. This is supported by my ethnography of Athena, a majority white queer art space in a neighborhood targeted for "urban revitalization." I then present the story of Minnemen, a majority GLBTQ youth of color space across town in a different neighborhood, a neighborhood experiencing urban blight. I describe how the poor and working-class GBLTQ youth of color of Minnemen experience disproportionate policing. In this way, they are unfairly rendered as indicators of "urban blight"; their claims to space vigorously opposed by local media, neighbors and police. It should be noted that both of Athena and Minnemen spaces are attempting to form underground, unlicensed spaces for the people who frequent them. Both are incredibly meaningful for those who frequent them. However, their differing class and racial make-up and differing neighborhoods lead to very different perceptions of their value to their neighborhoods by neighbors, media, police, and the city.

RELUCTANT QUEER ARTS "URBAN PIONEERS"

In the summers of 2010 and 2011, I helped organize, conduct, and clean up after events at a queer community arts center in Minneapolis called Athena. Athena is located on the second floor of an apartment building on Chicago Avenue, a busy thoroughfare two miles south of downtown Minneapolis that serves as the boundary between the Central and Powderhorn Park neighborhoods. At the time of this research, Athena did not have a mission, a board of directors, or any type of licensing with the city—it was a technically illegal, underground space. The rent for the space was generated through several art studio spaces used by artists during the day and through a wide breadth of events programming at night—ranging from foods swaps (which were shut down by the Food and Drug Administration shortly after this research was concluded), life-drawing classes, a feminist film series, and a particularly wild performance art and dance night called Divine. In interviews, many patrons emphasized the benefits that Athena's space had brought them, as a place for meeting other radical queers, creating a base of support in the face of increasing HIV infection rates, and dealing with gay bashings, among other things. The organizers of the space meet on a monthly basis and understand it to be an affinity space for queer community groups needing a place to meet and hold events. While there is a degree of diversity of age, race, and class among attendees of events, a common topic of conversation and concern was the overwhelming whiteness of Athena. The majority of those I interviewed from Athena were white, in their early 20s or 30s, and had attended at least some college.

Athena became located where it is largely because of the timing of their search for space, available money, and its central location. In 2010, when the organizers signed the lease on the space, the neighborhood was still relatively affordable for renters, with a median rental unit rate of U.S. $844 a month. However, average housing values in the neighborhood have almost doubled, from U.S. $118,150 in 2000 to U.S. $203,543 in 2011. This rapid increase in housing values occurred in the latter part of the decade, following the conversion of a formerly empty catalogue distribution center into a mixed-income development with low-income rental housing, 50 shops, restaurants, and bars. Next to the converted factory a new transit hub has been put in, which allows for bus access to nearly every part of the city and many suburbs. New organic restaurants and coffee shops are opening at both the north and south ends of the neighborhood. Despite a poverty rate that still hovers around the 40% mark, the neighborhood is undergoing noticeable change as students, artists, and young professionals increasingly see the neighborhood as a place to live and socialize.

These changes are the actualization of a "Framework Plan" for "neighborhood revitalization" drafted in 2007 by neighborhood associations, the business council and adopted by the City Council in 2008 (Community Design Group 2007). The Plan calls for the repurposing of existing buildings for mixed-income housing, artist live-work studios, rezoning for new development, reducing visible indicators of "blight" through strict enforcement of zoning and land use regulations, as well as increasing police presence and changing "perceptions of lack of safety in the project area."

Although the goal of these changes is to revitalize the neighborhood, it may also work in tandem with the beginning of the gentrification of the neighborhood. Gentrification is not a uniform process and varies according to the city and neighborhood it takes place within, but it typically includes middle- or upper-middle class people moving into formerly poor or working class neighborhoods, housing development corporations knocking down, rebuilding, and/or renovating the housing stock, and increasing housing values that can displace the poor or working-class residents of the neighborhood. The new organic coffee shops and restaurants and bicycle lanes in the neighborhood work as symbolic markers for the changes occurring in the neighborhood.

These changes are not lost on the organizers of Athena, and they strive to be a community institution that does not contribute to the changes in the neighborhood. However, that is challenging. For instance, Connor, a white queer man organizer in his late 30s at Athena, told me that at an early open house for the space a white woman "who positioned herself as a recent arrival" to the neighborhood walked up to him and told him "thank you for helping get the gangs out of the neighborhood." Connor says that he told her,

> You know, we're actually trying to keep a low profile, and we don't
> really necessarily want to change the composition of anything. Our goal
> is just to try to make the space work without having an intervention in
> the way that anyone else lives their life in the neighborhood.

He then said, "But of course that isn't possible."

This does not mean that the queers running the art space do not try, however. The organizers at Athena were very aware of the impact they *could* have on the neighborhood. For instance, Brian, a queer man in his early 30s, told me, "The people across the street ... they're Latino and there are quite a few people living there. We don't want to have these blaring loud parties keeping them up— maybe they're undocumented and can't call the police." Early in the summer of

2010 someone donated air-conditioning units to the space. At first glance, it seemed that this was for the benefit of the patrons of the space. However, Tim, a bearded, white queer man in his late 20s, told me that the air-conditioning units were there to keep the windows shut so that the neighbors would not be bothered and the police would not come. Similarly, when a party in the summer of 2011 filled the space far over capacity, with people outside on the sidewalk, drinking and smoking cigarettes and talking loudly around midnight, the organizers of the event decided to shut it down and asked everyone to leave.

These vignettes illustrate how the queers of Athena are aware of and seek to avoid abusing their privilege or playing into the "urban pioneers" narrative that newly arriving, higher socioeconomic status residents (such as the one who thanked Connor for moving to the neighborhood), the police, and developers may ascribe to them. Despite their intentions, several structures of power interpret them as "urban pioneers" who are helping to "revitalize" the neighborhood.

CONSTRUCTING BLACK QUEER SPACE AS A VIOLENT THREAT TO NEIGHBORHOOD SAFETY

Located about three miles east of Athena in a different neighborhood is the Minnemen space, an unlicensed, underground warehouse space that draws a mostly queer youth of color crowd. I first heard about Minnemen when I attended a "ball culture" primer event at a nonprofit office building next door to Athena that hosts several nonprofit organizations that serve trans people of color. Ball culture is a primarily youth of color subculture and balls usually consist of drag performers competing in several categories. I first saw Michael, an energetic Black gay man in his early 30s who runs the Minnemen space, speaking at a rally in defense of CeCe McDonald (a Black Minneapolis trans woman who defended herself against a white supremacist attacker, killing him in the process, and faced several years of prison). Michael began the space for the stated purpose of promoting "HIV and safe sex awareness in our GLBT community" and aiming to bring together "people of all colors, men and women between the ages of 15 to adult." Jillian, the head of a nonprofit organization providing services to trans youth of color, told me that Minnemen is, "a good place to meet folks and just be making sure that flyers are being passed out, and folks know what else is happening in the community." The events typically include DJs, dancing and performing dancers and sometimes run into the early hours of the morning. The Minnemen space has become an important space of sociability for queer youth of color in Minneapolis, many of whom are homeless or have tentative housing situations and cannot afford to go to gay bars or clubs in downtown.

In the summer of 2011, Minnemen came under fire from neighbors, the police, and the city. In a sensationalistic article about Minnemen in *The Star Tribune* in September of 2011, the author wrote that a recent party was, "an alcohol-fueled orgy with male strippers." The article emphasizes the violence of the space, with McKinney (2011) writing that in six months prior to the article, "29 police calls were made to the building for loud music, fights and weapons." A (white) neighborhood resident is quoted as saying, "The neighborhood has been trying to stop them for six months now." A local city council member assures that, "city staff is pursuing every possible angle to deal with this illegal business."

Key to understanding the crackdown on Minnemen is an attention to the larger demographic and economic forces that are changing the neighborhood, dynamics that are quite different from the neighborhood that houses Athena. Census and American Community Survey data show that the neighborhood's median household income fell from U.S. $40,390 (in 2010 inflation adjusted dollars) to U.S. $37,382 from 2000 to 2010. Furthermore, whereas the neighborhood was only composed of 6% African Americans in 2000, in 2010 the neighborhood was 16% African Americans, with double and triple the percentages of Native Americans and Asians moving to the neighborhood by 2010, respectively. The disparities in poverty rates by race are stark: while 20% of whites in the neighborhood live below the poverty line, more than 60% of African Americans in the neighborhood are living below the poverty level. A wave of poor and working-class African Americans, perhaps displaced from other neighborhoods, had recently moved into this majority white working-to-middle class neighborhood. Minnemen, and the increasing visibility of Black and Brown bodies—especially at night—appears to be a lightning rod for anxieties that the predominantly white residents have about changes in their neighborhood. This has resulted in increased neighborhood and police surveillance, with the effect of criminalizing queer youth of color.

RACE, SEXUALITY, AND URBAN REVITALIZATION

While Athena organizers actively try to avoid participating in potentially gentrifying the neighborhood (thus aiding developers or the police), they nonetheless benefit from their majority status of whiteness and their arts orientation. In the view of the police, other neighborhood residents, and the City government, their queerness, sexuality, and edgy performances can be folded into a freedom-of-the-arts and arts-as-urban-revival narratives that shield them from the types of repercussions that Minnemen faced. Similarly, even though they may not agree with being lumped in with actors and processes gentrifying the neighborhood, they sometimes are: this is especially evident from the response of the new arrival who thanked Connor for "helping to get the gangs out of the neighborhood."

Athena is contending with a history of white, middle- and upper-class gay people both actively and passively participating in gentrification and aiding police in purging poor people of color from neighborhoods—taming the supposed wild, forgotten inner-city for the manifest destiny of development, commerce, and "renewal." For instance, white gay activists actively participated in "neighborhood safety patrols" in New York and San Francisco in the 1970s and 1980s that originally began with the good intentions of protecting gay people from being bashed, but this quickly dovetailed with state and private developer interests to target poor people of color as threats to "public safety" in need of being "cleaned up" (Hanhardt 2013). Additionally, Manalansan (2005) documents how homeless queer youth of color in New York City have been systematically targeted for removal from the Christopher Street Pier, historically a "gay" haven, after high-end condominiums were built in the early 2000s. Interestingly, the piers were marketed heavily to gay and lesbian professionals.

Many of the same things that the media accused Minnemen of—overt sexual behavior, loudness, weapons—I had seen occur at Athena. I had seen creative performances that included nudity and an informed, queer vulgarity that was of a quality that could be at art museum events, art shows and theater festivals. I had observed people drinking illegally, loudly yelling at each other outside. I knew people that carried pepper spray, batons, or other self-defense weapons because they had been attacked before and needed to be able to defend themselves if it happened again. During the time of this research, the only police censure that Athena received was a warning issued to the landlord for posting a flyer for an event that advertised cheap alcohol.

Through a philosophical commitment and several practical measures to respect neighbors, Athena limited their exposure to neighbors and the police. However, this could only be done so much. There were times when over 50 people were on the sidewalk drinking, shouting, and smoking cigarettes and police drove by but did not stop. The same actions by Athena's mostly white patrons and Minneman's mostly Black and Latino patrons are viewed quite differently by police and neighbors. It seems likely that the race and class of these queer neighborhood organizations and how they interact with the demographic characteristics of the neighborhood shape the interpretation through larger narratives of neighborhood change (whether revitalization or decline).

Indeed, Athena is contending with a common urban ideology that represents "gay people" as middle- or upper middle-class, white people. These gay folk are the "urban pioneers" who supposedly help restore blighted neighborhoods. This urban ideology holds that blighted neighborhoods become rundown as a result of the irresponsibility of the poor people of color who lived there. It crucially ignores the many federal, state, and local policies that segregated people of color into specific urban neighborhoods and encouraged white flight into the suburbs after World War II, the practices of banks refusing to give housing loans to people of color in majority white neighborhoods throughout the 20th century (commonly referred to as "red lining"), and a century of failed public housing policies that concentrated Black poverty in inner cities.

CONCLUSIONS

City officials and police have a vested interest in claiming victory in the "renewal" of the inner-city, reductions in crime and increased perceptions of safety. The comparison of the situations of Athena and of Minnemen shows how neighborhood-level processes of urban revitalization and attendant demographic changes, along with city planning and policing objectives, lead to differential perceptions of the value of queerness to a neighborhood.

This case study also reveals the power of a space-based, intersectional lens in disentangling the complicated ways that race and class shape interpretations of sexuality and the differential experiences of GLBTQ groups, based on their race and class. The simplistic argument that "gays" raise housing values and propel gentrification relies on an extremely narrow, racialized and class-specific definition of "gays." Problematically, this narrow definition whitewashes gay identity with a middle-class sheen and renders GLBTQ people of color, working-class, and poor people—those more likely to be *displaced* by later phases of gentrification rather than propelling it—invisible. This exacerbates the already existing race and class divides between GLBTQ people and makes it even harder for these divides to be crossed, much less bridged.

Queer space matters. It matters because it gives GLBTQ people a place to be creative, to create art, to create new ways of doing gender and sexuality, to challenge heteronormativity, and to create friendships and families and share resources. I would venture to argue that queer space is even more important to social groups with less resources, such as the Black, young, working-class, poor, and homeless patrons of Minnemen. This is because they cannot afford access to many gay clubs and bars and they may be aggressively carded or kicked out of bars and clubs for using the "wrong" bathroom if they are trans. Even Athena is in a tenuous place: they are threatened by being priced out of their location as the value of the housing stock in their neighborhood rises. Through an intersectional lens, then, the common narrative that gay people raise housing values and propel gentrification (Florida and Mellander 2007) can be flipped. As gentrification renders many formerly affordable neighborhoods unaffordable and racialized policing heightens to sanitize these areas for middle-class consumption, the structural conditions for queer space evaporate. And the world becomes just that much less queer.

BIBLIOGRAPHY

Castells, Manuel. 1983. "Cultural Identity, Sexual Liberation and Urban Structure: The Gay Community in San Francisco." In *The City and the Grassroots: A Cross-Cultural Theory of Urban Social Movements*, pp. 138–170. London: Edward Arnold.

Chasin, Alexandra. 2001. *Selling Out: The Gay and Lesbian Movement Goes to Market*. New York: Palgrave.

Chauncey, George. 1994. *Gay New York*. New York: Basic Books.

Cohen, Jon. "Gay Marriage Support Hits New High in Post-ABC Poll." *The Washington Post*. 18 March 2013. Available at: www.washingtonpost.com/news/the-fix/wp/2013/03/18/gay-marriage-support-hits-new-high-in-post-abc-poll/. Accessed April 13, 2018.

Community Design Group, LLC. 2007. The 38th Street and Chicago Avenue Small Area/Corridor Framework Plan. Available at: www.ci.minneapolis.mn.us/www/groups/public/@cped/documents/webcontent/convert_274471.pdf. Accessed April 13, 2018.

D'Emilio, John. 1998. "Capitalism and Gay Identity." In *The Lesbian and Gay Studies Reader*, eds H. Abelon, M. A. Borak and D. Halperin, pp. 467–476. New York: Routledge.

Enke, Anne. 2007. *Finding the Movement: Sexuality, Contested Space, and Feminist Activism.* Durham: Duke University Press.

Fisher, D. A., Hill, D. L., Grube, J. W. and Gruber, E. L. 2007. "Gay, Lesbian, and Bisexual Content on Television: A Quantitative Analysis Across Two Seasons." *Journal of Homosexuality* 52(3–4): 167–188. doi: 10.1300/J082v52n03_08.

Florida, Richard and Mellander, Charlotte. 2007. "There Goes the Neighborhood: How and Why Bohemians, Artists, and Gays Effect Regional Housing Values." Available at: http://creativeclass.typepad.com/thecreativityexchange/files/Florida_Mellander_Housing_Values_1.pdf. Accessed April 13, 2018.

Gay and Lesbian Alliance Against Defamation. 2014. "2014: Where are we on TV?" Washington: DC. Available at: www.glaad.org/files/GLAAD-2014-WWAT.pdf. Accessed April 13, 2018.

Hanhardt, Christina. 2013. *Safe Space: Gay Neighborhood History and the Politics of Violence.* Durham: Duke University Press.

Higgins, Ross. 1999. "Baths, Bushes, and Belonging: Public Sex and Gay Community in Pre-Stonewall Montreal." In *Public Sex/Gay Space*, ed. William Leap, pp. 240–268. New York City: Columbia University Press.

Knopp, L. 1987. "Social Theory, Social Movements, and Public Policy: Recent Accomplishments of the Gay and Lesbian Movements in Minneapolis, MN." *International Journal of Urban and Rural Research*, 11(2): 243–261.

Leap, William. 1999. "Introduction." In *Public Sex/Gay Space*, ed. William Leap, pp. 1–22. New York City: Columbia University Press.

Levine, M. P. 1979. "Gay Ghetto." In *Social Perspectives in Lesbian and Gay Studies: A Reader*, eds Peter M. Nardi and Beth E. Schneider, pp. 194–206. London: Routledge.

Manalansan, Martin. 2005. "Race, Violence, and Neoliberal Spatial Politics in the Global City," *Social Text* 23(3–4 84–85): 141–155.

McKinney, Matt. 2011. "That's Not Me: Man at Center of City Investigation Says He's Unfairly Targeted." *Star Tribune*, September 28, 2011. Available at: www.startribune.com/man-accused-ofrunning-wild-minneapolis-club-says-he-s-unfairly-targeted/130710943/. Accessed April 13, 2018.

Murphy, Kevin P. 2010. "Gay Was Good: Progress, Homonormativity, and Oral History." In *Queer Twin Cities*, eds. Twin Cities GLBT History Project, pp. 305–318. Minneapolis, MN: University of Minnesota Press.

Smith, Neil. 1996. *The New Urban Frontier: Gentrification and the Revanchist City.* London: Routledge.

Smith, Tom and Son, Jaseok. 2013. *Trends in Attitudes Toward Sexual Morality.* Chicago, IL: NORC.

White, Edmund. 1982. *A Boy's Own Story.* New York: Random House.

Wyatt, Edward. 2013. "Most of U.S. is Wired, But Millions Aren't Plugged In." *The New York Times*. August 18, 2013. Available at: https://www.nytimes.com/2013/08/19/technology/a-push-to-connect-millions-who-live-offline-to-the-internet .html. Accessed April 13, 2018.

Zukin, Sharon. 1995. *The Cultures of Cities*. Cambridge, MA and Oxford: Blackwell Publishers.

PART III

Social Institutions and Social Issues

MARGARET L. ANDERSEN
AND PATRICIA HILL COLLINS

In this section of the book, we look at how systems of intersecting inequalities shape social institutions and contemporary social issues. Topics such as jobs, health care, educational reform, immigration policy, and violence populate the daily news and, perhaps, arise in conversations with friends. All of these contemporary topics have their origins in the structure of our social institutions—institutions that are structured through the very systems of inequality that we examined in Part II.

Indeed, the major social institutions in the United States have been historically founded on practices that were quite specifically organized around race and gender exclusion. Early in the nation's founding, Black Americans provided free labor that built the nation's economy and enriched certain White people, even while Black people were being denied the full rights of citizenship. Women, too, were formally excluded from various institutions, until quite recently in the nation's development: denied the right to vote, excluded from many jobs, and unable to enroll in prestigious educational institutions. Such practices, although in some regard things of the past, nonetheless shape how social institutions have developed, thus advantaging or disadvantaging people based on the intersecting forces of race, class, gender—and the other systems of social power that we have been examining, including sexual identity, ethnicity, disability, and regional residence.

These institutional patterns are *systemic*. That is, they are part of everyday life but are based on social forces that, although abstract, have real and lasting consequences in how different people live and die.

Social institutions are abstract entities—nothing more than collective actions that accumulate over an extended period of time and come to be thought of as "just the way things are." Economic institutions, for example, are patterns of how work is organized and how people are rewarded (or not) for their labor. Those with the most power and the most resources then guide how the institution operates through rules, laws, and regulations that shape the behavior of people within the institution.

The dominant American ideology portrays institutions as neutral in their treatment of different groups. Indeed, the liberal framework of the law makes access to public institutions (such as education and work) allegedly gender- and race-blind. Still, institutions routinely differentiate on the basis of race, class, and gender. The lens of intersectionality, however, debunks the notion that social institutions are neutral. Rather, intersectionality teaches you that social institutions have underlying structures that are specifically framed by the various axes of inequality examined in this book.

As institutions are established, they become channels for allocating societal privileges and penalties. The type of work you do, the structure of your family, the relationships you have, whether your religion will be recognized or suppressed, the kind of education and health care you receive, the rights you hold (or do not), even your likelihood of being killed by the police: All of these are shaped by institutional systems in society. People rely on social institutions to meet their needs, but how well institutions serve particular groups is shaped by systemic inequality.

Institutions can and do change over time because societal conditions evolve and because people challenge them. Yet, institutions are resistant to change. It often takes a movement from outside the institution to make substantial change.

Sometimes a social issue develops because of the failure of a given institution, such as when the absence of a sound immigration policy leaves people at risk of deportation, even if they have been working and paying taxes for years. Other times an issue arises because the institution is working exactly as intended, but with negative consequences for particular groups. An example is the joblessness that Black men and, increasingly, White, working-class men are experiencing because of technological change and globalization (discussed further below).

Capitalism is the very core of U.S. economic institutions. Capitalism depends upon the exploitation of certain groups of workers so that profits accrue to others.

Further, capitalism organizes the labor market along lines of race, class, and gender. The result is the vast and growing economic inequality that the nation is currently witnessing and that has a specific impact on you depending on your race, class, and gender location in this institutional system.

Fully understanding social issues thus requires the perspective that intersectionality provides. Without that view, the causes and consequences of any given social issue are less likely to be known. Take the issue of health. It is well documented that health disparities because of race are substantial. But, your particular risk of poor health and/or poor health care is also shaped by your gender, your social class, your age, your sexual identity, *and* your citizenship status (López and Gadsden 2016, this volume). Together, these social facts determine your likelihood of disease, your access to good health care, and your likelihood of an early death.

Yet, people in the United States tend to understand most social issues in individualistic terms. A violent encounter, for example, might result in the victim being blamed for being in the wrong place at the wrong time or, in the case of sexual violence, as somehow "asking for it." Whether particular individuals are perceived as somehow blameworthy is also related to the systems of power we examine here. Women receiving public assistance, especially women of color, are scorned for allegedly not wanting to work, while White, middle-class women are glorified if they are "stay-at-home moms." Asian American women who encourage their children's success are ridiculed as "tiger Moms," while wealthy White families who invest thousands of dollars in their children's private education are envied. When you take an intersectional perspective, you see the entangled dynamics of race, ethnicity, class, gender, and sexuality in shaping our most pressing national problems.

Here we examine five topics: work, jobs, and the labor market; families and relationships; education and health care; citizenship and nationality; and violence and criminalization. Hopefully, the perspective you develop will also help you analyze other social issues as they arise.

WORK, JOBS, AND THE LABOR MARKET

Economic transformation is changing who works and where; who is being displaced; how wealth and income are distributed; whether one faces discrimination at work; and how work is organized, among other changes. You cannot understand these changes without the intersectional lens of race, class, gender, and other systems of power.

Economic restructuring refers to the changes taking place in people's lives as the result of three major factors. First, *globalization* means that jobs in the United States are increasingly tied to an international division of labor and the global flow of capital (that is, money and other financial resources). The experiences of U.S. workers, including those who are out of work, are integrally connected to global economic forces that encourage business owners to export jobs while also importing cheaper labor. This is one way owners cut their costs and increase their profits.

Second, economic restructuring refers to the *change from a manufacturing-based economy to a service-based economy.* In a manufacturing-based economy, workers make goods. Traditionally, manufacturing jobs have been manual, blue-collar jobs—skilled labor in many cases, but not work that requires higher education. Think of coal miners, automobile assembly plant workers, and farming, as examples. In a service-based economy, the majority of jobs provide some sort of service—either directly or indirectly. Some service sector jobs are highly paid and require professional training (teachers, physicians, and IT developers, as examples), but the majority of jobs in this sector are low-wage, require limited skill or training, and are subject to high rates of turnover (fast-food workers, retail workers, cleaners, and hotel maids, as examples). The result of this form of restructuring is a highly bifurcated labor force.

Third, economic restructuring is the result of rapid *technological change,* especially the information revolution. Some jobs have simply become obsolete because machines, perhaps even robots, have replaced the work once done by human hands. When was the last time you saw an elevator operator? If you are young enough, you may not even know what that is! Even the jobs that were once skilled, manual labor are, in many cases, now done by machines. Technological innovation means that in the restructured economy, good jobs (even in the manufacturing sector) require some form of higher education and strong technological skills. Without such credentials, workers get stuck in the bottom rung of the labor market, if they are employed at all. Some simply become unemployable except in the most menial and unstable jobs.

In the restructured economy, race, class, gender—and immigration status—strongly predict who ends up where in this bifurcated labor market. A dual economy operates with high-paying, stable, and benefitted jobs for some (investment bankers, software engineers, and biotech researchers, as examples) and low-paying service work for everyone else (food-service workers, nursing home aides, and child care workers, as examples). Altogether, these changes have resulted in a very different workforce than one would have seen forty or fifty years ago.

The labor market is then structured with: (1) a *primary labor market* where there are relatively high wages, opportunities for advancement, employee benefits, and rules of due process that protect workers' rights; and, (2) a *secondary labor market* characterized by low wages, little opportunity for advancement, few benefits, and little protection for workers. Perhaps not surprisingly, women, people of color, and immigrants are most likely to be found in the secondary labor market. In other words, gender, class, and race segregation are structured into the dual labor market.

You can see this for yourself by just looking around you and seeing who is doing what jobs. Women of color are most likely to be working in occupations where most of the other workers are also women of color. At the same time, men of color are also segregated into particular segments of the market. Indeed, there is a direct connection between gender and race segregation and wages because wages are lowest in occupations where women of color predominate (U.S. Department of Labor 2018). This is what it means to say that institutions are structured by the intersection of race, class, and gender.

Economic restructuring has been underway for some time, but it is now having dramatic effects on jobs and thus on people's economic and personal well-being. Anyone who cannot adapt to these changes by having the necessary education and training may face a lifetime of dead-end work or no work at all—a fact that has plagued many racial-ethnic minorities and is now increasingly affecting White blue-collar workers as well. When jobs disappear, whole communities are disrupted, as shown in the opening essay here by sociologist William Julius Wilson ("Jobless Ghettoes: The Social Implications of the Disappearance of Work in Segregated Neighborhoods"). Wilson's essay was written based on research done in Chicago in the 1990s, but the conditions he identifies have not changed and may be even worse.

The worker who years ago found a decent-paying job in an automobile assembly plant or a steel mill—perhaps even without a high school education—is unlikely today to do any better than a minimum-wage job. For those in the White working-class, work that in the past usually led to a lifetime of steady employment with decent job benefits had a decent chance of economic security. Such jobs were also the route to social mobility for many people of color and earlier immigrants. For many, especially from the manufacturing sector of the economy, that promise no longer exists. The result is the displacement of workers that many would say is driving some of our national politics (Hochschild 2016).

Joblessness affects not only individuals but, as Wilson shows, high rates of joblessness can decimate whole communities. Being loosened from the social bonds that work provides lies at the root of many social problems that even employed people in such neighborhoods, experience. This is happening not only in inner city neighborhoods, where Black men in particular face very high rates of joblessness, but also in whole swaths of the nation where White working-class people have lost manufacturing jobs in mining, the automobile industry, and textile mills as work has gone overseas or become automated. Research shows that when these workers are displaced, if they become re-employed, they are likely to end up in lower-paying, less stable service work. Simply put, whether in Appalachia, the Midwest Rust Belt, urban ghettoes, or farming areas of the South and West, economic disruption is causing mass upheaval for those who cannot compete in this new economy.

These economic transformations reverberate now in the fortunes—or misfortunes—of young people. As Jennifer Silva points out ("Working-Class Growing Pains"), the path to adulthood used to follow a particular route: education, job, marriage, then childbearing—and in that order. Now, young people find that the traditional paths to adulthood are blocked, and they may then experience an "extended adolescence" through the lack of work and stable marriage partners. Young people are then stereotyped as lazy, shiftless, and self-absorbed, taking the blame for problems that lie in the society that has changed before them. Silva's focus on working-class youth shows how easily blaming the victim falls upon these young people.

Sandra Weissinger's study of women workers in Walmart ("Gender Matters. So Do Race and Class: Experiences of Gendered Racism on the Walmart Shop Floor") is a good illustration of how race, class, and gender are structured in the labor market. Her discussion of the experiences of women workers in the retail industry shows how the division of labor in the world of work is very much conditioned by the interaction of race, class, and gender.

The social dynamics of race, gender, and class in the labor market also appear in the biases, sometimes unconscious, that can limit people's opportunities for work. Something as simple as one's name can result in discriminatory behavior that shapes labor market outcomes for particular groups. This has been cleverly studied by Marianne Bertrand and Sendhil Mullainathan ("Are Emily and Greg More Employable than Lakisha and Jamal? A Field Experiment on Labor Market Discrimination"). This research used a controlled experiment to document the fact that discrimination by race continues to exist, as does gender discrimination (Nunley et al. 2015), even in an age where discrimination is illegal. Together, these articles show the institutional structure of work and the labor market, and how much they are shaped by the overlapping structures of race, class, and gender.

FAMILIES AND RELATIONSHIPS

The family is another institution profoundly influenced by intersecting systems of race, class, gender, and sexuality. Families are often idealized as sites for love and support—a private world where people are nurtured and protected from the forces of the "outside" world. But families—and people's experiences within them—are produced by the same social forces that shape other social institutions.

In reality, families are highly diverse—in their form, the resources available to them, and how well they are able to nurture and protect family members, especially family members who are the most vulnerable, that is the young and the old. Yet, whether one's family is perceived as somehow "legitimate" depends on how much one's family conforms to the dominant ideal. As the essays in this section will show, families often sustain and support people, as they navigate the terrain of an unequal society. But families are also buffeted by the inequities of the systems of power examined in this book, even while being commonly blamed for the perpetuation of these inequalities.

Bonnie Thornton Dill's essay, "Our Mothers' Grief: Racial-Ethnic Women and the Maintenance of Families," provides a historical portrait of African American, Chinese American, Mexican American, and White families, and how women's roles within them have been shaped by their specific histories. She especially examines the *reproductive labor* of women, meaning both the paid and unpaid work that sustains human life, including cooking, cleaning, child care, elder care, and the emotional labor of caring. Dill shows how family organization and women's reproductive labor, particularly that of women of color, have been directly influenced by each group's placement in the larger society, especially by the intersecting dynamics of race, class, and gender—and we would now add sexual identity.

Families have historically been a site of contestation, meaning that some families have been perceived as illegitimate and have consequently been denied the institutional resources and public support that other families enjoy. Mignon R. Moore and Michael Stambolis-Ruhstorfer ("LGBT Sexuality and Families at the Start of the Twenty-First Century") show that LGBTQ families still face social and legal obstacles to family formation, even while contemporary attitudes have become more accepting of LGBTQ families. Who counts as a family? Even with greater acceptance of diversity in families, people who deviate from traditional family form still face prejudice and the denial of certain rights. Moore and Ruhstorfer review current data on LGBTQ families, issues of parenting and adoption in LGBTQ families, and new research on transgender families.

They conclude that research on families would be further enriched through greater attention to the intersections of race, class, gender, and sexuality.

Roberta Espinoza ("The Good Daughter Dilemma: Latinas Managing Family and School Demands") focuses on how Latinas in higher education manage what she calls the "good daughter dilemma." The good daughter dilemma is the pressure daughters feel in pursuing education when they are encouraged not to stray too far from home and family responsibilities. Her research is on Latina doctoral students, but the "good daughter dilemma" is common among first-generation and working-class students as well. Espinoza finds that Latinas negotiate the competing demands of family responsibilities and education by becoming either "integrators" or "separators"—strategies they employ to manage the two sets of commitments. Her conclusions show how important it is to change the race, ethnic, and gender structures of institutions because, as she says, educational institutions have failed to legitimate family commitments. Were they to do so, the institution would work better for all concerned.

Finally, in this section, Amy Steinbugler ("Loving across Racial Divides") explains that interracial couples have always faced intense public scrutiny, but now face more subtle forms of exclusion and hostility—even though interracial relationships are more widely accepted. Within such relationships, both partners have to confront racism in their everyday lives—a fact further complicated by institutional patterns of racial residential segregation. Given the prevalence of segregated neighborhoods, one partner in an interracial relationship is likely to feel conspicuous as an "only" in the very place where he or she lives.

These articles collectively show that although the past injustices of racism may seem to be gone, racism still shapes, the everyday life of people in families and intimate relationships. Intersectional thinking that connects the realities of race, class, gender, and sexuality reveals these new understandings about families and can shatter the myths that continue to blame families for a host of social problems.

EDUCATION AND HEALTH

Education and health care are also institutions that face significant challenges in today's world. Indeed, some would argue that both are institutions in crisis. People worry that schools are failing our children. The lack of adequate health care populates the daily news and is at the core of current political wrangling. Great disparities exist in how well people are educated and how well they are cared for—disparities that are largely accounted for by the intersections of race, class, and gender, along with age, sexuality, ethnicity, and disability. Simply put, education and health care institutions reflect and reproduce the inequalities that characterize society as a whole.

Even with laws in place that promise school desegregation, schools remain highly segregated by race. Despite the fact that desegregation policies were generated to attack schooling inequalities more than sixty years ago, racial isolation in schools is actually growing. In fact, public schools are resegregating such that segregation in schools is now higher than it was in 1980. Furthermore, significant numbers of African American and Latinx students attend schools that are over 90 percent minority students—so called "majority-minority" schools (Orfield and Frankenberg 2014). This would not in itself be a problem were it not for the fact that such schools are so underresourced (Darling-Hammond 2010). How can we expect a more racially just society when young people grow up in isolation from other groups?

Persistent inequality in schools has generated much attention and research on the causes and consequences of inequality in schooling—notably in the so-called achievement gap. But, as Gloria Ladson-Billings argues ("From the Achievement Gap to the Education Debt: Understanding Achievement in U.S. Schools"), we should be thinking not just about the race and gender gap in achievement, but also about the debt owed to those who suffer from poor education and social neglect. Ladson-Billings asks whether we can afford such disparate experiences in a society where democracy depends on an educated citizenry.

Students in the undocumented population face their own challenges within educational institutions. William Perez, Roberta Espinoza, Karina Ramos, Heidi M. Coronado, and Richard Cortes ("Academic Resilience among Undocumented Latino Students") have studied undocumented Latino students and find that, although they experience the same stresses as other Latinos, they have additional stresses by virtue of their immigration status. Undocumented students are ineligible for scholarships and live under the constant fear of family separation and/or deportation. Parents may not have the educational background to help them navigate educational institutions that were not designed for them. Also, like other students who hold jobs while in school, undocumented students have to balance the multiple demands of work and schooling.

Still, Perez and his colleagues report resilience among undocumented students, especially when they have personal and environmental resources such as supportive adults and strong peer relationships. Remember that, although most of the public attention to undocumented immigrants has focused on Latinas/os, Asians are estimated to be about 14 percent of the undocumented population and should not be forgotten (Ramakrishnan and Shah 2017).

Students with learning and other disabilities also face unique circumstances within educational institutions, as shown in the personal narrative of Michael, a Black working-class student with a learning disability. David J. Connor ("Michael's Story: 'I Get into So Much Trouble Just by Walking': Narrative Knowing and Life at the Intersections of Learning Disability, Race, and Class") discusses Michael's location at the intersections of race, class, and disability. Michael also faces a vicious cycle of segregated housing, limited educational options, and restricted employment, which in turn leads to segregated housing. O'Connor's article also examines the different levels where institutional power operates—the structural, interpersonal, and ideological levels of society.

In the final article of this section, Nancy López and Vivian L. Gadsden ("Health Inequalities, Social Determinants, and Intersectionality") argue that an intersectional perspective is essential for understanding persistent health disparities. They show that an intersectional perspective is necessary not only for documenting such disparities, but also for reducing them and promoting a healthier society.

Is the promise of education as a path for social mobility a fading dream? Does the nation have the will to tackle persistent health disparities and deliver better health care to all of its citizens? As López and Gadsden conclude, an intersectional framework is needed to illuminate and change the conditions that stifle the health and well-being of people in communities marginalized by virtue of race, class, and gender.

CITIZENSHIP AND NATIONAL IDENTITY

Citizenship and national identity have become increasingly important to understand as the United States grapples with the increasing diversity of its population. Many of the issues of the day are entangled with questions about who can be a citizen and how national identity is defined. Citizenship is conferred by the *state*—that is, the organized system of power and authority in society. The state is supposed to protect all citizens, regardless of their race, class, and gender (as well as other characteristics such as disability and age), yet state policies routinely privilege some groups while denying basic rights of citizenship to others.

The very concept of citizenship in the United States has been implicitly—and sometimes explicitly—defined by casting particular immigrant groups as "other." Even a cursory look at our nation's history shows how the full rights of citizenship were denied to many—Black men and women toiling under slavery; Asian immigrants having their rights curtailed at various points in the nation's history; women being denied the right to vote; Chicanos having their land taken away by war and a subsequent treaty (The Treaty of Hidalgo). Now, in the post-9/11 context, Arab Americans are suspect citizens who are routinely profiled as terrorists and perceived as threats to "real" Americans. The very meaning of "being an American" is fraught with patterns of certain groups being excluded from the rights of citizenship—both formal and informal.

The earliest example comes from the history of Native people. C. Matthew Snipp ("The First Americans: American Indians") details phases in U.S. history when American Indians have been removed, restricted, relocated, and robbed of resources. His historical account shows that the founding of the United States entailed the settling and seizing of Native lands and culturally annihilating Native peoples.

Indeed, despite the ideology of the "melting pot," national identity in the United States has been closely linked to a history of White privilege. As Lillian Rubin points out in "Is This a White Country, or What?" the term *American* is usually assumed to mean White. People of color such as African Americans and Asian Americans become distinguished from the "real" Americans by virtue of their race. More importantly, certain benefits are reserved for those deemed to be "deserving Americans." You might think of Rubin's article as you hear contemporary references to so-called "illegal aliens." What different connotation emerges from the phrase "undocumented workers?" Which nomenclature makes people seem to be "other," "distant," and "un-American?"

As Min Zhou argues ("Are Asian Americans Becoming White?"), being defined as "white" is a social and historical process. As some groups achieve material success, they may become perceived as "white," even when, as was the case of Jewish and Irish immigrants, they have earlier been labeled as "nonwhite." Zhou shows how Asian Americans—an identity that merges different ethnic backgrounds—are simultaneously stereotyped as the "model minority" while still being perceived as outsiders or alien to American society. Her work shows how groups can become racialized depending on the historic circumstances they encounter.

Being perceived as "alien" or "other" has many risks. Tonya Golash-Boza ("Feeling Like a Citizen, Living as a Denizen: Deportees' Sense of Belonging") relates the experience of legal permanent residents who run the constant risk of deportation with even a minor infraction—a traffic stop, a petty crime, or simply being in the wrong place at the wrong time. Her research shows that in addition to the formal rights of citizenship, there is a social dimension to citizenship, namely the sense of belonging that citizenship bestows. As we can see through the articles in this section, that sense of belonging is very much conditioned by the intersections of race, ethnicity, class, gender, and sexuality.

VIOLENCE AND CRIMINALIZATION

In the last section of this part, we examine violence and the criminalization of people who are also targets of racist, sexist, and homophobic violence. Police shootings of Black men and women are all too common. Hate crimes against Muslims, Jewish people, and LGBTQ people are on the rise (Potok 2017). School shootings wrench our hearts on an all-too often basis. Usually violence is depicted as the action of crazed individuals who are socially maladjusted, angry, or "sick." Although this may be the case for some, from an intersectional perspective violence originates in the social structures that then guide individual action.

Who is perceived as violent also shifts. Within a system of race, class, and gender inequality, some groups become criminalized—suspect by their very existence, and thus more likely to be targeted by social control agents such as the police, border patrol officials, and court officers. An intersectional perspective teaches you that violence and criminalization takes place within institutions that are structured by the systems of power we have been examining. Thus, whereas for those with racial and gender privilege, institutions such as the police and the courts work to protect them, for others the criminal justice system itself is a source of violence.

Victor M. Rios, now a distinguished university professor, reveals this in reporting his experience as a young man of color ("Policed, Punished, Dehumanized: The Reality for Young Men of Color Living in America"). He and his brother were routinely harassed by the police. He now refers to such harassment (and worse) as state-sanctioned violence against marginalized populations. Punishment of Black and Latino youth, as he explains, starts early and continues through young adulthood, resulting in what Rios calls *social death*—the process by which Black and Latino youth lose their humanity by being policed, punished, and dehumanized by institutions that are part of the *youth control complex*—that is, schools, community centers, social workers, merchants, community members, and sometimes even family members who systematically single out minority youth for constant oversight and surveillance.

Criminalization also targets immigrants who, despite the actual facts, are routinely suspected of criminal behavior. Rubén G. Rumbaut and Walter Ewing ("The Myth of Immigrant Criminality and the Paradox of Assimilation") debunk the widespread myths that immigrants are criminals. Through a review of the most careful research, they show that immigrants are less likely to commit crime than either native-born citizens or second and third-generation immigrants. The paradox of assimilation is that crime is more likely the longer one is in the United States.

Eileen Pittaway and Linda Bartolomei ("Refugees, Race, and Gender: The Multiple Discrimination against Refugee Women") look at violence in a different context: the vulnerability of migrant women to rape and sexual violence. They document the high rates of gendered violence against migrant women and show the connection of such violence to armed conflict. As they point out, women of color, indigenous women, and Dalit women (Indian women of the lowest caste) are especially at risk, showing the need for an intersectional analysis that understands the confluence of gender, race, and class in explaining and stopping sexual violence.

Natalie Sokoloff ("The Intersectional Paradigm and Alternative Visions to Stopping Domestic Violence") provides an illustration of how such an understanding matters in organized efforts to stop domestic violence. Sokoloff shows that the most marginalized women—that is poor women, women of color, immigrant women, and lesbians—are subjected to violence, but can be empowered to tackle this issue within diverse communities. She demonstrates that those working to stop domestic violence must not focus solely on dominant group experiences. Using an inclusive framework has enabled poor, immigrant, lesbian, and racial-ethnic women to have a voice in the understanding of and support systems for domestic violence.

Altogether, the articles in this section show how embracing an intersectional analysis of race, class, gender, and sexuality is important to creating an awareness of how society can become more safe, secure, and just.

REFERENCES

Darling-Hammond, Linda. 2010. *The Flat World and Education*. New York: Teachers College Press.

Hochschild, Arlie Russell. 2016. *Strangers in Their Own Land: Anger and Mourning on the American Right*. New York: The New Press.

Nunley, John M., Adam Pugh, Nicholas Romero, and R. Alan Seals. 2015. "Racial Discrimination in the Labor Market for Recent College Graduates: Evidence from a Field Experiment." *The B.E. Journal of Economic Analysis & Policy* 15 (3): 1093–1125.

Orfield, Gary, and Erica Frankenberg. 2014. *Brown at 60: Great Progress, A Long Retreat, and an Uncertain Future*. Los Angeles: Civil Rights Project/Proyecto Derechos Civiles, University of California, Los Angeles.

Potok, Mark. 2017. "The Year in Hate and Extremism." *Intelligence Report* (February), Montgomery, AL: Southern Poverty Law Center. www.splcenter.org

Ramakrishnan, Karthick, and Sono Shah. 2017. "One Out of Every Seven Asian Immigrants is Undocumented." Data from Center for Migration Studies. http://data.cmsny.org and http://aapidata.com/blog/asian-undoc-1in7/. Accessed June 1, 2017.

U.S. Department of Labor. January 2018. *Employment and Earnings*. Washington, D.C.: U.S. Government Printing Office. www.dol.gov

25

Jobless Ghettos

The Social Implications of the Disappearance of Work in Segregated Neighborhoods

WILLIAM JULIUS WILSON

In 1950, a substantial portion of the urban black population was poor but working. Urban poverty was quite extensive, but people held jobs. However, in many inner-city ghetto neighborhoods in 1990, most adults were not working in a typical week. For example, in 1950, 69 percent of all males 14 and over held jobs in a typical week in the three neighborhoods that represent the historic core of the Black Belt in Chicago—Douglas, Grand Boulevard, and Washington Park. But by 1990, only four in ten in Douglas worked in a typical week, one in three in Washington Park, and one in four in Grand Boulevard. In all, only 37 percent of all males 16 and over held jobs in a typical week in these neighborhoods.

The disappearance of work has had devastating effects not only on individuals and families kit also on the social life of neighborhoods as well. Inner-city joblessness is a severe problem that is often overlooked or obscured when the focus is mainly on poverty and its consequences. Despite increases in the concentration of poverty since 1970, inner cities have always featured high levels of poverty, but the levels of inner-city joblessness reached in 1990 were unprecedented.[1]

It should be noted that when I refer to "joblessness" I am not solely referring to official unemployment. The unemployment rate represents only the *official* labor force—that is, those who are actively looking for work. It does not include those who are outside of or have dropped out of the labor market, including the nearly 6 million males age 25–60 who appear in the census statistics but do not show up in the labor statistics.

These uncounted males in the labor market are disproportionately represented in the inner-city ghettos....

The consequences of high neighborhood joblessness are more devastating than those of high neighborhood poverty. A neighborhood in which people are poor, but employed, is much different from a neighborhood in which people are poor and jobless. Many of today's problems in the inner-city ghetto neighborhoods—crime, family dissolution, welfare, low levels of social organization, and so on—are fundamentally a consequence of the disappearance of work.

It should be clear that when I speak of the disappearance of work, I am referring to the declining involvement in or lack of attachment to the formal labor market. It could be argued that the general sense of the term "joblessness" does not necessarily mean "non-work." Many people who are officially jobless are nonetheless involved in informal activities, ranging from unpaid housework to income from work in the informal or illegal economies.

Housework is work; baby-sitting is work; even drug dealing is work. However, what contrasts work in the formal economy with work activity in the informal and illegal economies is that work in the formal economy has greater regularity and consistency in schedules and hours. The demands for discipline are greater. It is true that some work activities outside the formal economy also call for discipline and regular schedules. Several studies reveal that the social organization of the drug industry is driven by discipline and a work ethic, however perverse. However, as a general rule, work in the informal and illegal economies is far less governed by norms or expectations that place a premium on discipline and regularity. For all these reasons, when I speak of the disappearance of work, I mean work in the formal economy, work that provides a framework for daily behavior because of the discipline and regularity that it imposes.

Thus, a youngster who grows up in a family with a steady breadwinner and in a neighborhood in which most of the adults are employed will tend to develop some of the disciplined habits associated with stable or steady employment—habits that are reflected in the behavior of his or her parents and of other neighborhood adults. These might include attachment to a routine, a recognition of the hierarchy found in most work situations, a sense of personal efficacy attained through the routine management of financial affairs, endorsement of a system of personal and material rewards associated with dependability and responsibility, and so on. Accordingly, when this youngster enters the labor market, he or she has a distinct advantage over the youngsters who grow up in households without a steady breadwinner and in neighborhoods that are not organized around work—in other words, a milieu in which one is more exposed to the less-disciplined habits associated with casual or infrequent work.

In the absence of regular employment, a person lacks not only a place in which to work and the receipt of regular income but also a coherent organization of the present—that is, a system of concrete expectations and goals. Regular employment provides the anchor for the spatial and temporal aspects of daily life. It determines where you are going to be and when you are going to be there. In the absence of regular employment, life, including family life, becomes less coherent. Persistent unemployment and irregular employment hinder rational planning in daily life, a necessary condition of adaptation to an industrial economy.

EXPLANATIONS OF THE GROWTH OF JOBLESS GHETTOS

What accounts for the growing proportion of jobless adults in inner-city communities? An easy explanation would be racial segregation. However, a race-specific argument is not sufficient to explain recent changes in such neighborhoods. After all, these historical Black Belt neighborhoods were just as segregated by skin color in 1950 as they are today, yet the level of employment was much higher then. One has to account for the ways in which racial segregation interacts with other changes in society to produce the recent escalating rates of joblessness.

The disappearance of work in many inner-city neighborhoods is in part related to the nationwide decline in the fortunes of low-skilled workers. Over the past two decades, wage inequality has increased sharply and gaps in labor market outcomes between the less- and more- skilled workers have risen substantially. Research suggests that these changes are the result of "a substantial decline in the relative demand for the less-educated and those doing more routinized tasks compared to the relative supply of such workers."[2] Two factors appear to have reduced the relative demand for less-skilled workers—the computer revolution (i.e., skill-based technological change) and the internationalization of economic activity. Inner-city workers face an additional problem—the growing sub urbanization of jobs. Most ghetto residents cannot afford cars and therefore rely on public transit systems that make the connection between inner-city neighborhoods and suburban job locations difficult and time consuming.

Although the relative importance of the different underlying causes of the growing jobs problems of the less-skilled, including those in inner city, continues to be debated, there is little disagreement about the underlying trends. They are unlikely to reverse themselves.

Changes in the class, racial, and demographic composition of inner-city neighborhoods have also contributed to the high percentage of jobless adults in these neighborhoods. Because of the steady outmigration of more advantaged families, the proportion of non-poor families and prime-age working adults has decreased sharply in the typical inner-city ghetto since 1970. These changes have made it increasingly difficult to sustain basic neighborhood institutions or to achieve adequate levels of social organization. The declining presence of working- and middle-class blacks has also deprived ghetto neighborhoods of key resources, including structural resources, such as residents with income to sustain neighborhood services, and cultural resources, such as conventional role models for neighborhood children.

It is not surprising, therefore, that our research in Chicago revealed that inner-city ghetto residents share a feeling of little informal social control of their children. A primary reason is the absence of a strong organizational capacity or an institutional resource base that would provide an extra layer of social organization in their neighborhoods. It is easier for parents to control the behavior of the children in their neighborhoods when a strong institutional resource base exists and when the links between community institutions such a churches, schools, political organizations, businesses, and civic clubs are strong or secure. The higher the density and stability of formal organizations, the less illicit activities such as drug trafficking, crime, prostitution, and the formation of gangs can take root in the neighborhood.

It is within this context that the public policy discussion on welfare reform and family values should be couched. Our Chicago research suggests that, as employment prospects recede, the foundation for stable relationships becomes weaker over time. More permanent relationships such as marriage give way to temporary liaisons that result in broken unions, out-of-wedlock pregnancies and births, and, to a lesser extent, separation and divorce. The changing norms concerning marriage in the larger society reinforce the movement toward temporary liaisons in the inner city, and therefore economic considerations in marital decisions take on even greater weight. The evolving cultural patterns are seen in the sharing of negative outlooks toward marriage and toward the relationships between males and females in the inner city, outlooks that are developed in and influenced by an environment featuring persistent joblessness. This combination of factors has increased out-of-wedlock births, weakened the family structure, expanded the welfare rolls, and, as a result, caused poor inner-city blacks to be even more disconnected from the job market and discouraged about their role in the labor force. The economic marginality of the ghetto poor is cruelly reinforced, therefore, by conditions in the neighborhoods in which they live.

In the eyes of employers in metropolitan Chicago, the social conditions in the ghetto render inner-city blacks less desirable as workers, and therefore many employers are reluctant to hire them. One of the three studies that provided the empirical foundation for *When Work Disappears* included a representative sample of employers in the greater Chicago area who provided entry-level jobs. An overwhelming majority of these employers, both white and black, expressed negative views about inner-city ghetto workers, and many stated that they were reluctant to hire them. For example, a president of an inner-city manufacturing firm expressed a concern about employing residents from certain inner-city neighborhoods:

> If somebody gave me their address, uh, Cabrini Green, I might unavoidably have some concerns. [*Interviewer:* What would your concerns be?] That the poor guy probably would be frequently unable to get to work and... I probably would watch him more carefully, even if it wasn't fair, than I would with somebody else. I know what I should do though is recognize that here's a guy that is trying to get out of his situation and probably will work harder than somebody else who's already out of there and he might be the best one around here. But I, I think I would have to struggle accepting that premise at the beginning.

In addition to qualms about the neighborhood milieu, employers frequently mentioned concerns about applicants' language skills and educational training. An employer from a computer software firm expressed the view "that in many businesses the ability to meet the public is paramount and you do not talk street talk to the buying public. Almost all your black welfare people talk street talk. And who's going to sit them down and change their speech patterns?" A Chicago real estate broker made a similar point:

> A lot of times I will interview applicants who are black, who are sort of lower class.... They'll come to me and I cannot hire them because their language skills are so poor. Their speaking voice for one thing is poor....They have no verbal facility with the language... and these... you know, they just don't know how to speak and they'll say "sales-mens" instead of "salesmen" and that's a problem.... They don't know punctuation, they don't know how to use correct grammar, and they cannot spell. And I can't hire them. And I feel bad about that and I think they're being very disadvantaged by the Chicago public school system.

Another respondent defended his method of screening out most job appli-cants on the telephone on the basis of their use of "grammar and English."

> I have every right to say that that's a requirement for this job. I don't care if you're pink, black, green, yellow, or orange, I demand someone who speaks well. You want to tell me that I'm a bigot, fine, call me a bigot.

Finally, an inner-city banker claimed that many blacks in the ghetto "simply cannot read. When you're talking our type of business, that disqualifies them immediately. We don't have a job here that doesn't require that somebody have minimum reading and writing skills."

How should we interpret the negative attitudes and actions of employers? To what extent do they represent an aversion to blacks per se and to what degree do they reflect judgments based on the job-related skills and training of inner-city blacks in a changing labor market? I should point out that the state-ments made by the African-American employers concerning the qualifications of inner-city black workers do not differ significantly from those of the white employers. Whereas 74 percent of all the white employers who responded to the open-ended questions expressed negative views of the job-related traits of inner-city blacks, 80 percent of black employers did so as well.

This raises a question about the meaning and significance of race in certain situations—in other words, how race intersects with other factors. A key hypoth-esis in this connection is that, given the recent shifts in the economy, employers are looking for workers with a broad range of abilities: "hard" skills (literacy, numeracy, basic mechanical ability, and other testable attributes) and "soft" skills (personalities suitable to the work environment, good grooming, group-oriented work behaviors, etc.). While hard skills are the product of education and raining—benefits that are apparently in short supply in inner-city schools—soft skills are strongly tied to culture and are therefore shaped by the harsh envi-ronment of the inner-city ghetto. If employers are indeed reacting to the differ-ence in skills between white and black applicants, it becomes increasingly difficult to discuss the motives of employers: are they rejecting inner-city black applicants out of overt racial discrimination or on the basis of qualifications?

Nonetheless, many of the selective recruitment practices do represent what economists call statistical discrimination: employers make assumptions about the inner-city black workers *in general* and reach decisions based on those assumptions before they have had a chance to review systematically the qualifications of an individual applicant. The net effect is that many black inner-city applicants are never given the chance to prove their qualifications on an individual level because they are systematically screened out by the selective recruitment process. Statistical discrimination, although representing elements of class bias against poor workers in the inner city, is clearly a matter of race. The selective recruitment patterns effectively screen out far more black workers from the inner city than Hispanic or white workers from the same types of backgrounds. But race is also a factor, even in those decisions to deny employment to inner-city black workers on the basis of objective and thorough evaluations of their qualifications. The hard and soft skills among inner-city blacks that do not match the current needs of the labor market are products of racially segregated communities, communities that have historically featured widespread social constraints and restricted opportunities.

Thus, the job prospects of inner-city workers have diminished not only because of the decreasing relative demand for low-skilled labor, the suburbanization of jobs, and the social deterioration of ghetto neighborhoods, but also because of negative employer attitudes. This combination of factors presents a real challenge to policy-makers. Indeed, considering the narrow range of social policy options in the "balance-the-budget" political climate, how can we immediately alleviate the inner-city jobs problem—a problem that will undoubtedly grow when the new welfare reform bill takes full effect.

PUBLIC POLICY DILEMMAS

To what extent will the inner-city jobs problem respond to macroeconomic levers that can act to enhance growth and reduce unemployment? I include here fiscal policies that regulate government spending and taxation and monetary policies that influence interest rates and control the money supply. If jobs are plentiful even for less-skilled workers during periods of economic expansion, then labor shortages reduce the likelihood that hiring decisions will be determined by subjective negative judgments concerning a group's job-related traits.

But given the fundamental structural decline in the demand for low-skilled workers, fiscal and monetary policies designed to enhance economic growth will have their greatest impact in the higher-wage sectors of the economy. Many low-wage workers, especially those in high-jobless inner-city neighborhoods who are not in or have dropped out of the labor force and who also face the problem of negative employer attitudes, will not experience any improvement in their job prospects because of such policies.

If firms in the private sector cannot use or refuse to hire low-skilled adults who are willing to take minimum-wage jobs, then the jobs problem for inner-city workers cannot be adequately addressed without considering a policy of public-sector employment of last resort. Indeed, until current changes in the labor market are reversed or until the skills of the next generation can be upgraded before it enters the labor market, many workers, especially those who are not in the official labor force, will not be able to find jobs unless the government becomes an employer of last resort. This argument applies especially to low-skilled inner-city black workers. It is bad enough that they face the problem of shifts in labor-market demand shared by all low-skilled workers; it is even worse that they confront negative employer perceptions about their work-related skills and attitudes.

... We could face a real catastrophe in many urban areas if steps are not taken soon to enhance the job prospects of hundreds of thousands of inner-city youths and adults.

NOTES

1. Parts of this essay are based on my latest book. *When Work Disappears: The World of the New Urban Poor* (New York: Alfred A. Knopf, 1996), which included three research studies conducted in Chicago between 1986 and 1993. The first of these included a random survey of nearly 2,500 poor and non-poor African-American, Latino, and white residents in Chicago's poor neighborhoods; a subsample of 175 participants from this survey who were reinterviewed and answered open-ended questions; a survey of 179 employers selected to reflect the distribution of employment and firm sizes in the metropolitan area; and comprehensive ethnographic research, including participant-observation research and life-history interviews in a representative sample of inner-city neighborhoods.

   The second study included a survey of a representative sample of 546 black mothers and up to two of their adolescent children (ages 11 to 16—or 887 adolescents), in working- and middle-class neighborhoods and high-poverty neighborhoods. Finally, the third study featured a survey of a representative sample of 500 respondents from two high-joblessness neighborhoods on the South Side of Chicago and six focus-group discussions involving the residents and former residents of these neighborhoods.

2. Lawrence Katz, "Wage Subsidies for the Disadvantaged," Working Paper 5679, National Bureau of Economic Research, Cambridge, MA, 1996, p. 2.

26

Working Class Growing Pains

JENNIFER M. SILVA

In a working-class neighborhood in Lowell, Massachusetts, I sat across the kitchen table from a 24-year-old white woman named Diana. The daughter of a dry cleaner and a cashier, Diana graduated from high school and was accepted into a private university in Boston. She embarked on a criminal justice degree while working part-time at a local Dunkin' Donuts, taking out loans to pay for her tuition and room and board. But after two years, Diana began to doubt whether the benefits of college would ever outweigh the costs, so she dropped out of school to be a full-time cashier.

She explained, "When I work, I get paid at the end of the week. But in college, I would have had to wait five years to get a degree, and once I got that, who knows if I would be working or find something I wanted to be." Now, close to a hundred thousand dollars in debt, Diana has forged new dreams of getting married, buying a home with a pool in a wealthy suburb of Boston, and having five children, a cat, and a dog—by the time she is 30.

But Diana admitted that she can't even find a man with a steady job to date, let alone marry, and that she will likely regret her decision to leave school: "Everyone says you can't really go anywhere unless you have a degree. I don't think I am going to make it anywhere past Dunkin' when I am older, and that scares me to say. Like it's not enough to support me now."

Living with her mother and bringing home under $275 per week, Diana is stuck in an extended adolescence with no end in sight. Her yardsticks for adulthood—owning her own home, getting married, finishing her education, having children, and finding a job that pays her bills—remain spectacularly out of reach. "Your grandparents would get married out of high school, first go steady, then get married, like they had a house," she reflected. "Since I was 16, I have asked my mother when I would be an adult, and she recently started saying I'm an adult now that I'm working and paying rent, but I don't feel any different."

What does it mean to "grow up" today? Even just a few decades ago, the transition to adulthood would probably not have caused Diana so much confusion, anxiety, or uncertainty. In 1960, the vast majority of women married before they turned 21 and had their first child before 23. By 30, most men and women had moved out of their parents' homes, completed school, gotten married, and begun having children. As over a decade of scholarly and popular

SOURCE: Republished with permission of Sage publications, from Working class growing pains by Jennifer M. Silva, *Contexts*, Vol. 13, No. 2, pp. 26–31, © 2014 American Sociological Association; permission conveyed through Copyright Clearance Center, Inc.

literature has revealed, however, in the latter half of the twentieth century traditional markers of adulthood have become increasingly delayed, disorderly, reversible—or have been entirely abandoned. Unlike their 1950s counterparts, who followed a well-worn path from school to work, and courtship to marriage to childbearing, men and women today are more likely to remain unmarried; to live at home and stay in school for longer periods of time; to switch from job to job; to have children out of wedlock; to divorce; or not have children at all.

LONG AND WINDING JOURNEY

Growing up, in essence, has shifted from a clear-cut, stable, and normative set of transitions to a long and winding journey. This shift has been greeted with alarm, and the Millennial Generation has often been cast as entitled, self-absorbed, and lazy. In 2013, for example, *Time* magazine's cover story on "The Me Me Me Generation" headlined: "Millennials are lazy, entitled narcissists who still live with their parents." And a poll conducted in 2011 by the consulting firm Workplace Options found that the vast majority of Americans believe that Millennials don't work as hard as the generations before them. The overriding conclusion is that things have gotten worse—and that young people are to blame.

But this longing to return to the past obscures the restrictions—and inequalities—that characterized traditional adult milestones for many young people in generations past. As the historian Stephanie Coontz reminds us, in the 1950s and 60s women couldn't serve on juries or own property or take out lines of credit in their own names; alcoholism and physical and sexual abuse within families went ignored; factory workers, despite their rising wages and generous social benefits, reported feeling imprisoned by monotonous work and merciless supervision; and African Americans were denied access to voting, pensions, and healthcare.

The social movements for civil rights, feminism, and gay pride that emerged during subsequent decades erased many of these barriers, granting newfound freedoms to young adults in their wake. In many ways, young people today have a great deal more freedom and opportunity than their 1950s counterparts: women, especially, can pursue higher education, advance in professional careers, choose if and when to have children, and leave abusive marriages. And all young adults have more freedom to choose a partner regardless of sex or race.

As psychologist Jeffrey Arnett argues: "More than ever before, coming of age in the twenty-first century means learning to stand alone as a self-sufficient person, capable of making choices and decisions independently from among a wide range of possibilities." But that's not the whole story. Just as many social freedoms for young people have expanded, economic security—stable, well-paid jobs, access to health insurance and pensions, and affordable education—has contracted for the working class. Meanwhile, the growing fragility of American families and communities over the same time period has placed the responsibility for launching young adults into the future solely on the shoulders of themselves and their parents.

For the more affluent young adults of this "Peter Pan Generation"—those with a college fund, a parent-subsidized, unpaid internship, or an SAT coach, the freedom to delay marriage and childbearing, experiment with flexible career paths, and pursue higher education grants them the luxury to define adulthood in their own terms. But working-class men and women like Diana have to figure out what it means to be a worthy adult in a world of disappearing jobs, soaring education costs, shrinking social support networks, and fragile families.

From 2008–2010, I interviewed 100 working-class men and women between the ages of 24 and 34—people who have long ago reached the legal age of adulthood but still do not feel "grown up." I went from gas stations to fast food chains, community colleges to temp agencies, tracking down working-class young people, African Americans and whites, men and women, and documenting the myriad obstacles that stand in their way. And what I heard was profoundly alarming: caught in the throes of a merciless job market and lacking the social support, skills, and knowledge necessary for success, working-class young adults are relinquishing the hope for traditional markers of adulthood—a home, a job, a family—at the heart of the American Dream.

My conversations with these men and women uncovered the contours of a new definition of working-class adulthood: one characterized by low expectations of loyalty in work, wariness toward romantic commitment, widespread distrust of social institutions, and profound isolation from and hostility toward others who can't make it on their own. Simply put, growing up today means learning to depend on no one but yourself.

WORK AND LOVE AMID INEQUALITY

Pervasive economic insecurity, fear of commitment, and confusion within institutions make the achievement of traditional markers of adulthood impossible and sometimes undesirable. The majority of the young people I spoke with bounce from one unstable service job to the next, racking up credit card debt to make ends meet and fearing the day when economic shocks—an illness, a school loan coming out of deferment—will erode what little stability they have.

Upon leaving high school, they quickly learned that they shouldn't expect loyalty or respect from their jobs. Jillian, a 26-year-old white woman, started out as a line cook, making $5.50 an hour the year she graduated from high school. Under the guidance of her manager, she worked her way up the line until she was his "right hand man," running the line by herself and making sure everyone cleaned up their stations at the end of a long day. When Bill died suddenly from a heart attack, the owner waited to hire a new manager, causing a year of skeleton crews, chaos, and backbreaking 70-hour work weeks.

Jillian knew that she was lucky to have all those hours a week to work, especially in the recession, and she didn't complain: "... you basically worshipped the ground they walked on because they gave you a job. You had to keep your mouth shut." But when Jillian pushed for changes and the owner snapped, "You won't get respect anywhere else, so why expect it here?" She quit. "I thought I had it going good for a while there. But everything really came to a screeching halt, and I bought a car, and now not having a job...I feel like I'm starting over."

Indeed, growing up means learning that trusting others, whether at school, home, or work, will only hurt them in the end. Rob is a 26-year-old white man whom I met while recruiting at a National Guard training weekend in Massachusetts. Rob told me his story in an empty office at the armory because he was currently "crashing" on his cousin's couch. When he graduated from his vocational high school, he planned to use his training in metals to build a career as a machinist: "Manufacturing technology, working with metal, I loved that stuff," he recalled longingly. As he attempted to enter the labor market, however, he quickly learned that his newly forged skills were obsolete.

"I was the last class at my school to learn to manufacture tools by hand," he explained. "Now they use CNC [computer numerical controlled] machine programs, so they just draw the part in the computer and plug it into the machine, and the machine cuts it... I haven't learned to do that, because I was the last class before they implemented that in the program at school, and now if you want to get a job as a machinist without CNC, they want five years' experience. My skills are useless."

Over the last five years, Rob has stacked lumber, installed hardwood floors, landscaped, and poured steel at a motorcycle factory. His only steady source of income since high school graduation has been his National Guard pay, and although he recently returned from his second 18-month deployment in Afghanistan, he is already considering a third: "I am looking for a new place. I don't have a job. My car is broken. It's like, what exactly can you do when your car is broken and you have no job, no real source of income, and you are making four or five hundred dollars a month in [military] drills." He explains his economic predicament: "Where are you going to live, get your car fixed, on $500 a month? I can't save making 500 bucks a month. That just covers my bills. I have no savings to put down first and last on an apartment, no car to get a job. I find myself being like, oh what the hell? Can't it just be over? Can't I just go to Iraq right now? Send me two weeks ago so I got a pay-check already!"

Insecurity seeps into the institution of family, leaving respondents uncertain about both the feasibility and desirability of commitment. Deeply forged cultural connections between economic viability, manhood, and marriage prove devastating, as men's falling wages and rising job instability leave them uncertain about the meanings of masculinity in the twenty-first century. Brandon, a 34-year-old black man who manages the night shift at a women's clothing chain, explained matter-of-factly, "No woman wants to sit on the couch all the time and watch TV and eat at Burger King. I can only take care of myself now. I am missing out on life but making do with what I have."

For working-class women who have grown up shouldering immense social and economic burdens on their own, being responsible for another person who may ultimately let them down doesn't feel worth the risk. Lauren, a 24-year-old barista who was kicked out of her father's house when she came out as a lesbian, has weathered years of addiction, homelessness, and depression, finally emerging as a survivor, sober and able to pay her own rent. She has chosen to remain single because she fears having to take care of someone else.

"I mean, everybody's life sucks, get over it! My mom's an alcoholic, my dad kicked me out of the house. It's not a handicap; it has made me stronger. And I want someone who has you know similarly overcome their respective obstacle and learned and grown from them, rather than someone who is bogged down by it and is always the victim." As Lauren suggests, since intimacy carries with it the threat of self-destruction, young working-class men and women forego the benefits of lasting commitment, including pooled material resources, mutual support, and love itself.

Children symbolize the one remaining source of trust, love, and commitment; while pregnancies are usually accidental, becoming a parent provides motivation, dignity, and self-worth. As Sherrie, whose pregnancy gave her the courage to break up with her abusive boyfriend, explained: "You have a baby to take care of! My daughter is the reason why I am the way I am today. If I didn't have her, I think I might be a crackhead or an alcoholic or in an abusive relationship!" Yet the social institutions in which young adults create families can work against their desire to nurture and protect their children.

Rachel, a young black single mother, joined the National Guard in order to go to college for free through the GI Bill. However, working 40 hours a week at her customer service job, attending weekend army drills, and parenting has left her with little time for taking college classes. Hearing rumors that her National Guard unit may deploy to Iraq for a third time in January, she is tempted to put in for discharge so that she is not separated from her son again. However, her desire to give her son everything she possibly can—including the things she can buy with the higher, tax-free combat pay she receives when she deploys—keeps her from signing the papers: "I am kinda half and half with the deployment coming up. I could use it for the money. I could do more for my son. But I missed the first two years of my son's life and now I might have to leave again. It's just rough. You can't win."

DISTRUST AND ISOLATION

Common celebrations of adulthood—whether weddings, graduations, housewarming's, birthdays—are more than just parties; they are rituals for marking community membership and shared, public expressions of commitment, obligation, rights, and belonging. But for the young men and women I spoke with, there was little sense of shared joy or belonging in their accounts of coming of age. Instead, I heard story after story of isolation and distrust experienced within a vast array of social institutions, including higher education, the criminal justice system, the government, and the military. While we may think of the life course as a process of social integration, marked by public celebrations of transitions, young working-class men and women depend on others at their peril.

They believe that a college education will provide the tools for success. Jay, a 28-year-old black man, struggled through seven years of college. He failed several classes after his mother suffered a severe mental breakdown. After being expelled from college and working for a year, helping his mom get back on her feet, he went before the college administration and petitioned to be reinstated. He described them as "a panel of five people who were not nice." As Jay saw it, "It's their job to hear all these sob stories, you know I understand that, but they just had this attitude, like you know what I mean, 'oh your mom had a breakdown and you couldn't turn to anyone?' I just wanted to be like, fuck you, but I wanted to go to college, so I didn't say fuck you." When he eventually graduated, when he was 25, he "was so disillusioned by the end of it, my attitude toward college was like, I just want to get out and get it over with, you know what I mean, and just like, put it behind me, really." He shrugged: "I felt like it wasn't anything to celebrate. I mean I graduated with a degree. Which ultimately I'm not even sure if that was what I wanted, but there was a point where I was like I have to pick some bullshit I can fly through and just get through. I didn't find it at all worthwhile."

Since graduating three years ago, with a communications major, Jay has worked in a series of food service and coffee shop jobs. Reflecting on where his life has taken him, he fumed: "They were just blowing smoke up my ass—the world is at my fingertips, you can rule the world, be whatever you want, all this stuff. When I was 15, 16, I would not have envisioned the life I am living now. Whatever I imagined, I figured I would wear a suit every day, that I would own things. I don't own anything. I don't own a car. If I had a car, I wouldn't be able to afford my daily life. I'm coasting and cruising and not sure about what I should be doing."

Christopher, a 24-year-old, who has been unemployed for nine months, further illustrates how distrust and isolation is intensified by bewildering interactions with institutions. As he put it, "I have this problem of being tricked...Like I will get a phone call that says, you won a free supply of magazines. And they will start coming to my house. Then all of a sudden I am getting calls from bill collectors for the subscriptions to *Maxim* and *ESPN*. It's a run around: I can't figure out who to call. Now I don't even pick up the phone, like I almost didn't pick up when you called me."

Recently, Christopher was taxed $400 for not purchasing mandatory health insurance in Massachusetts, which he could not afford because he was unemployed, and did not know how to access for free. Like many of my respondents, he lacks the skills and know-how to navigate the institutions that frame the transition to adulthood. He tells his coming of age story as one incident of deception after another—each of which incurs a heavy emotional and financial cost. But while he acknowledges that he has not achieved the traditional markers of adulthood, he still believes that he is at least partially an adult because of the way he has learned to manage his feelings of betrayal: "I ended up the way I am because of my experiences. I have seen crazy shit. Like now if I see someone beating someone up in the street, I don't scream. I don't care. I have no emotions or feelings." Growing up hardened against and detached from the world, and dependent on no one, Christopher protects himself from the possibility of trickery and betrayal.

REMAKING WORKING-CLASS ADULTHOOD

The working-class men and women I spoke with lack the necessary knowledge, skills, credentials, and money to launch themselves into a secure adult future, as well as the social support and guidance to protect themselves from economic and social turmoil. But despite their profound anger, betrayal, and loss, they do not want pity—and they do not expect a handout. On the contrary, at a time when individual solutions to collective structural problems is a requirement for survival, they believe that adulthood means taking responsibility for one's own successes and failures. Emma, who works as a waitress, praised her grandfather who worked his way up digging ditches for a gas company; she says it is now up to her to "take what you are given and utilize it correctly." Similarly, Kelly, a line cook who has lived on and off in her car, explains, "Life doesn't owe me any favors. I can have a sense of my own specialness and individuality, but that doesn't mean that anybody else has to recognize that or help me accomplish my goals."

This bootstrap mentality, while highly praised in our culture, has a darker side: blaming those who can't make it on their own. Wanda, the daughter of a tow-truck driver who wants to go to college but can't afford the tuition, expresses anger at her parents' lack of economic support: "I feel like it's their fault they don't have nothing." Working-class youth have little trust even in those closest to them and—despite the social and economic forces that work against their efforts—they blame themselves for their shortcomings.

Julian, a young black man, is a disabled vet who is unemployed, divorced, and living with his mother. Describing his inability to find a steady job and lasting relationship, he tells me: "…Every day I look in the mirror, and I could bullshit you right now and tell you that race has something to do with it. But at the end of the day looking in the mirror, I know where all my shortcomings come from. From the things that I either did not do or I did and I just happen to fail at them." They believe that understanding their shortcomings in terms of structural barriers to mobility is a crutch; both blacks and whites are hostile toward others who do not take sole responsibility for their own failures.

John, a 27-year-old black man who sells shoes, explained: "Society lets it [race] affect me. It's not what I want to do, but society puts tags on everybody. You gotta be presentable, take care of yourself. It's about how a man looks at himself and how people look at him. Some people use it as a crutch, but it's not gonna be my crutch." That is, while black men and women acknowledge that discrimination persists, they see navigating racism as an individual game of cunning. All make a virtue out of not asking for help, out of rejecting dependence and surviving completely on their own, mapping these traits onto their definitions of adulthood. Those who fail to "fix themselves" are met with disdain and disgust—they are not worthy adults.

This hardening against oneself and others could have profound personal and political consequences for the future of the American working class. Its youngest members embrace self-sufficiency, blame those who are unsuccessful in the labor market, and choose distrust and isolation as the only way to survive. Rather than target the vast social, economic, and cultural changes that have disrupted the transition to adulthood—the decline of good jobs, the weakening of unions, the shrinking of communities—they target themselves. In the end, if they have to go it alone, then everyone else should, too. And it is hard to find even a glimmer of hope for their futures.

Their coming of age stories are still unfolding, their futures not yet written. To tell a different kind of story—one that promises hope, dignity, and connection—they must begin their journeys to adulthood with a living wage and the skills and knowledge to confront the future. They need neighborhoods and communities that share responsibility for launching them into the future. And they need new definitions of dignity that do not make a virtue out of isolation, self-reliance, and distrust. The health and vibrancy of all our communities depend on the creation and nurturance of definitions of adulthood that foster connection and interdependence.

27

Are Emily and Greg More Employable Than Lakisha and Jamal?

A Field Experiment on Labor Market Discrimination

MARIANNE BERTRAND AND SENDHIL MULLAINATHAN

Every measure of economic success reveals significant racial inequality in the U.S. labor market. Compared to Whites, African-Americans are twice as likely to be unemployed and earn nearly 25 percent less when they are employed (Council of Economic Advisers, 1998). This inequality has sparked a debate as to whether employers treat members of different races differentially. When faced with observably similar African-American and White applicants, do they favor the White one? Some argue yes, citing either employer prejudice or employer perception that race signals lower productivity. Others argue that differential treatment by race is a relic of the past, eliminated by some combination of employer enlightenment, affirmative action programs and the profit-maximization motive. In fact, many in this latter camp even feel that stringent enforcement of affirmative action programs has produced an environment of reverse discrimination. They would argue that faced with identical candidates, employers might favor the African-American one. Data limitations make it difficult to empirically test these views. Since researchers possess far less data than employers do, White and African-American workers that appear similar to researchers may look very different to employers. So any racial difference in labor market outcomes could just as easily be attributed to differences that are observable to employers but unobservable to researchers.

To circumvent this difficulty, we conduct a field experiment that builds on the correspondence testing methodology that has been primarily used in the past to study minority outcomes in the United Kingdom. We send resumes in response to help-wanted ads in Chicago and Boston newspapers and measure callback for interview for each sent resume. We experimentally manipulate perception of race via the name of the fictitious job applicant. We randomly assign very White-sounding names (such as Emily Walsh or Greg Baker) to half the resumes and very African-American-sounding names (such as Lakisha Washington or Jamal Jones) to the other half. Because we are also interested in how credentials affect the racial gap in callback, we experimentally vary the quality of the resumes used in response

SOURCE: Bertrand, Marianne, and Sendhil Mullainathan. 2004. "Are Emily and Greg More Employable than Lakisha and Jamal: A Field Experiment on Labor Market Discrimination." *American Economic Review,* 94(September): 991–1013. Copyright © 2004 American Economic Review. Reprinted by permission.

to a given ad. Higher-quality applicants have on average a little more labor market experience and fewer holes in their employment history; they are also more likely to have an e-mail address, have completed some certification degree, possess foreign language skills, or have been awarded some honors. In practice, we typically send four resumes in response to each ad: two higher-quality and two lower-quality ones. We randomly assign to one of the higher- and one of the lower-quality resumes an African-American-sounding name. In total, we respond to over 1,300 employment ads in the sales, administrative support, clerical, and customer services job categories and send nearly 5,000 resumes. The ads we respond to cover a large spectrum of job quality, from cashier work at retail establishments and clerical work in a mail room, to office and sales management positions.

We find large racial differences in callback rates. Applicants with White names need to send about 10 resumes to get one callback whereas applicants with African-American names need to send about 15 resumes. This 50-percent gap in callback is statistically significant. A White name yields as many more callbacks as an additional eight years of experience on a resume. Since applicants' names are randomly assigned, this gap can only be attributed to the name manipulation.

Race also affects the reward to having a better resume. Whites with higher-quality resumes receive nearly 30 percent more callbacks than Whites with lower-quality resumes. On the other hand, having a higher-quality resume has a smaller effect for African-Americans. In other words, the gap between Whites and African-Americans widens with resume quality. While one may have expected improved credentials to alleviate employers' fear that African-American applicants are deficient in some unobservable skills, this is not the case in our data.

The experiment also reveals several other aspects of the differential treatment by race. First, since we randomly assign applicants' postal addresses to the resumes, we can study the effect of neighborhood of residence on the likelihood of callback. We find that living in a wealthier (or more educated or Whiter) neighborhood increases callback rates. But, interestingly, African-Americans are not helped more than Whites by living in a "better" neighborhood. Second, the racial gap we measure in different industries does not appear correlated to Census-based measures of the racial gap in wages. The same is true for the racial gap we measure in different occupations. In fact, we find that the racial gaps in callback are statistically indistinguishable across all the occupation and industry categories covered in the experiment. Federal contractors, who are thought to be more severely constrained by affirmative action laws, do not treat the African-American resumes more preferentially; neither do larger employers or employers who explicitly state that they are "Equal Opportunity Employers." In Chicago, we find a slightly smaller racial gap when employers are located in more African-American neighborhoods.

I. PREVIOUS RESEARCH

With conventional labor force and household surveys, it is difficult to study whether differential treatment occurs in the labor market. Armed only with survey data, researchers usually measure differential treatment by comparing the labor market performance of Whites and African-Americans (or men and women) for which they observe similar sets of skills. But such comparisons can be quite misleading. Standard labor force surveys do not contain all the characteristics that employers observe when hiring, promoting, or setting wages. So one can never be sure that the minority and nonminority workers being compared are truly similar from the employers' perspective. As a consequence, any measured differences in outcomes could be attributed to these unobserved (to the researcher) factors.

This difficulty with conventional data has led some authors to instead rely on pseudo-experiments. Claudia Goldin and Cecilia Rouse (2000), for example, examine the effect of blind auditioning on the hiring process of orchestras. By observing the treatment of female candidates before and after the introduction of blind auditions, they try to measure the amount of sex discrimination. When such pseudo-experiments can be found, the resulting study can be very informative; but finding such experiments has proven to be extremely challenging.

A different set of studies, known as audit studies, attempts to place comparable minority and White actors into actual social and economic settings and measure how each group fares in these settings. Labor market audit studies send comparable minority (African-American or Hispanic) and White auditors in for interviews and measure whether one is more likely to get the job than the other. While the results vary somewhat across studies, minority auditors tend to perform worse on average: they are less likely to get called back for a second interview and, conditional on getting called back, less likely to get hired.

These audit studies provide some of the cleanest nonlaboratory evidence of differential treatment by race. But they also have weaknesses.... First, these studies require that both members of the auditor pair are identical in all dimensions that might affect productivity in employers' eyes, except for race. To accomplish this, researchers typically match auditors on several characteristics (height, weight, age, dialect, dressing style, hairdo) and train them for several days to coordinate interviewing styles. Yet, critics note that this is unlikely to erase the numerous differences that exist between the auditors in a pair.

Another weakness of the audit studies is that they are not double-blind. Auditors know the purpose of the study.... This may generate conscious or subconscious motives among auditors to generate data consistent or inconsistent with their beliefs about race issues in America. As psychologists know very well, these demand effects can be quite strong. It is very difficult to insure that auditors will not want to do "a good job." Since they know the goal of the experiment, they can alter their behavior in front of employers to express (indirectly) their own views. Even a small belief by auditors that employers treat minorities differently can result in measured differences in treatment. This effect is further magnified by the fact that auditors are not in fact seeking jobs and are therefore more free to let their beliefs affect the interview process.

Finally, audit studies are extremely expensive, making it difficult to generate large enough samples to understand nuances and possible mitigating factors. Also, these budgetary constraints worsen the problem of mismatched auditor pairs. Cost considerations force the use of a limited number of pairs of auditors, meaning that any one mismatched pair can easily drive the results. In fact, these studies generally tend to find significant differences in outcomes across pairs.

Our study circumvents these problems. First, because we only rely on resumes and not people, we can be sure to generate comparability across race. In fact, since race is randomly assigned to each resume, the same resume will sometimes be associated with an African-American name and sometimes with a White name. This guarantees that any differences we find are caused solely by the race manipulation. Second, the use of paper resumes insulates us from demand effects. While the research assistants know the purpose of the study, our protocol allows little room for conscious or subconscious deviations from the set procedures. Moreover, we can objectively measure whether the randomization occurred as expected. This kind of objective measurement is impossible in the case of the previous audit studies. Finally, because of relatively low marginal cost, we can send out a large number of resumes. Besides giving us more precise estimates, this larger sample size also allows us to examine the nature of the differential treatment from many more angles.

...

Our results indicate that for two identical individuals engaging in an identical job search, the one with an African-American name would receive fewer interviews. Does differential treatment within our experiment imply that employers are discriminating against African-Americans (whether it is rational, prejudice-based, or other form of discrimination)? In other words, could the lower callback rate we record for African-American resumes *within our experiment* be consistent with a racially neutral review of the *entire pool* of resumes the surveyed employers receive?

In a racially neutral review process, employers would rank order resumes based on their quality and call back all applicants that are above a certain threshold. Because names are randomized, the White and African-American resumes we send should rank similarly on average. So, irrespective of the skill and racial composition of the applicant pool, a race-blind selection rule would generate equal treatment of Whites and African-Americans. So our results must imply that employers use race as a factor when reviewing resumes, which matches the legal definition of discrimination....

II. CONCLUSION

This paper suggests that African-Americans face differential treatment when searching for jobs and this may still be a factor in why they do poorly in the labor market. Job applicants with African-American names get far fewer callbacks for each resume they send out. Equally importantly, applicants with African-American names find it hard to overcome this hurdle in callbacks by improving their observable skills or credentials.

Taken at face value, our results on differential returns to skill have possibly important policy implications. They suggest that training programs alone may not be enough to alleviate the racial gap in labor market outcomes. For training to work, some general-equilibrium force outside the context of our experiment would have to be at play. In fact, if African-Americans recognize how employers reward their skills, they may rationally be less willing than Whites to even participate in these programs.

REFERENCES

Council of Economic Advisers. *Changing America: Indicators of social and economic well-being by race and Hispanic origin.* September 1998, http://www.gpoaccess.gov/eop/ca/pdfs/ca.pdf.

Goldin, Claudia and Rouse, Cecilia. "Orchestrating Impartiality: The Impact of Blind Auditions on Female Musicians." *American Economic Review*, September 2000, *90*(4), pp. 715–41.

28

Gender Matters. So Do Race and Class

Experiences of Gendered Racism on the Wal-Mart Shop Floor*

SANDRA E. WEISSINGER

I n this case study, inequitable access to power found within one particular institution, Wal-Mart stores, is examined. As the largest corporation in the world—
with revenues larger than those accumulated by Switzerland and outselling Target,
Home Depot, Sears, Kmart, Safeway, and Kroger combined—the company is
quite powerful both in the United States and abroad (Litchenstein 2006). In the
U.S., 1.3 million are employed (Rathke 2006) within the four thousand stores
across the nation (Litchenstein 2006). Because the lives of so many converge
there, Wal-Mart has the potential to mirror many of the oppressive practices
observed in social relationships outside of the stores. For this reason, the statements
given by Wal-Mart employees are telling of the robust and adaptive nature of
discriminatory practices, regardless of geographic location. Within Wal-Mart, the
multidimensional nature of inequality can be illuminated as gender, race, and class
intersect and shape the experiences of the women that work at these stores.

CASE BACKGROUND

In 2004, 1.6 million plaintiffs were granted class action status, making theirs the
largest sex discrimination case seen in U.S. courts.** The suit, Dukes v. Wal-
Mart Stores Inc., began when current and former female Wal-Mart employees
from across the United States claimed that the retailer discriminated against them
in terms of access to promotions, wages similar to their male colleagues, and
their job tasks. Individual employees examined their personal issues and developed a sociological imagination (Mills 2000 [1959])—the ability to see that the
treatment they were subject to was shared, at least in part, by those with similar

---

*This essay is dedicated to Ms. Betty Dukes and the nearly two million women who were
courageous enough to stand in the gap for other employees experiencing similar treatment.
**Editors' note: In June 2011, the U.S. Supreme Court ruled against the women who
filed a class-action suit against Wal-Mart.

SOURCE: Weissinger, Sandra, E. 2009. "Gender Matters, So Do Race and Class:
Experiences of Gendered Racism on the Wal-Mart Shop Floor." *Humanity and Society*,
33(4): 341–362. Copyright © 2009 by the Author. Reprinted by permission of SAGE
Publications, Inc.

biographical characteristics. For example, when her supervisor referred to her as a "Mexican princess" in front of her peers, Gina Espinoza-Price (a plaintiff) was humiliated. She also realized that others were being treated in a similar unprofessional fashion.

Wal-Mart's defense lawyers posed several challenges to the plaintiffs; the most important was whether their case qualified as a class action suit. If the case had not qualified, plaintiffs would have had to file individual suits against the corporation. At best, this could produce rulings that benefited an individual plaintiff but failed to bring about large-scale changes to Wal-Mart's employment and promotions procedures that would benefit all employees. This suit is the impetus behind changes in Wal-Mart's human resources departments nationwide. All employees can now view and apply for all positions. They are also given the opportunity to officially declare their career goals in their computerized personnel file (Featherstone 2004).

The plaintiff for whom the case is named, Betty Dukes, is still employed at Wal-Mart. Dukes has stated that her motivation to continue speaking about employment practices at Wal-Mart comes from her religious background as a Christian minister. According to Dukes, "I am participating in this case in order to insure that young women such as my nieces and other women are treated fairly at every Wal-Mart store. The time has surely come for equality for women" (Rosen 2004). Dukes's statement is unique in that she sees her actions today as connected to the lives of others who will come after her. Although the other plaintiffs did not specifically list similar motivations, all of the plaintiffs made claims concerning inequality. These claims have been bolstered by the statistical findings of Drogin (2003), who found that across geographic locations, women working for Wal-Mart earn less than male employees who occupy the same positions. In addition, Drogin found that women were promoted into management positions at lower and slower rates than male employees.

The Dukes suit rightly identifies the effects of sex-based discrimination that are generalizable across women's experiences within Wal-Mart stores. To provide another vantage point from which to understand the effects of the discrimination experienced by these employees (and perhaps those at corporations that seek to model their businesses after Wal-Mart), I contend that discrimination due to biological sex differences alone does not explain the range of the plaintiffs' experiences or properly gauge individual and social damages caused when one must navigate treacherous environments on a daily basis. Rather, I argue, those individuals who are targeted for mistreatment experience such treatment as equally raced, classed, and gendered people whose lives exist within a web of intersecting and relational inequalities. Therefore, a reexamination of what plaintiffs of the class action suit said about their employment experiences (when guided by a gender primacy framework) is necessary so that the multiple and differing daily work experiences of the plaintiffs can be illuminated, adding to sociological knowledge concerning the work lives of women across standpoints. Reexamination is needed not simply to see difference in lives, but to reveal the persistence of inequality and the multiple ways discriminatory work atmospheres are maintained.

According to the files Wal-Mart made available to the court, women worked for the company longer and received better reviews from supervisors, yet earned less money and received fewer promotions when compared to male employees.

Women's Experiences across Race and Class

... Scholarship about the social locations of women has addressed not only the lived experiences of women of color, but also those of white women across class and geographic location. Enlarging our understanding of the differences between women, sociologists have documented the ways in which white women act as oppressors as well as the oppressed—benefiting from certain race and/or class positions. As explained by Blee (2002) and Frankenberg (1993), one experience of marginalization does not automatically inspire empathy in white women toward other oppressed communities. Whiteness, as all categories of difference, is socially constructed through carrying real consequences that bear upon individuals' everyday lives. "In a white-dominated society, whiteness is invisible. Race is something that adheres to others as a mark of otherness, a stigma of difference. It is not perceived by those sheltered under the cloak of privilege that masks its own existence" (Blee 2002:56). In short, white women can be marginalized due to gender and class, but experience certain benefits because of their race.

Methodology
... In this work I identify six biographical characteristics most plaintiffs mentioned in their declarations: gender, race, class, family makeup, geographic location, and age. To be clear, I do not argue that one area of oppression or facet of one's biography is more important than another. My argument is not one that posits that there is a monolithic or authentic experience either. Certainly, lived experiences can be similar and should be juxtaposed against one another for analysis. Rather, I argue that each employee had rich and unique work experiences resulting from the intersecting characteristics that make up their biographies.

... Past studies have failed to examine women's work experiences outside of the gender lens. Questioning how the multiple features of each plaintiff's biography influence and shape the stories they tell about work life at Wal-Mart is at the heart of this work.

Findings

Insider Outsider Statuses
Very few white plaintiffs mentioned their race but when they did, it was in relation to the race of others. These statements were made when the plaintiff reflected on "backstage" (Picca and Feagin 2007), private conversations they engaged in with white male supervisors. The failure of white women plaintiffs to mention race, except when noting the difference between themselves and people of color, illustrates that "whiteness is unmarked because of the pervasive nature of white domination" (Blee 2002:56). To illuminate this argument, the following excerpts show how two white women who held positions of authority at Wal-Mart stores addressed race.

The first quote is from Ms. Lorie Williams, the youngest declarant in this sample (twenty-six years old) and a single mother who supports her child [working] as a "Front End Manager," the lowest ranking management position in the occupational hierarchy at the stores.

> In 1996, I became the front-end manager. With no training, I was single-handedly responsible for hiring door greeters, cart pushers, over sixty new cashiers, and preparing the entire front end in order to transition the Collierville Wal-Mart into a Supercenter. Almost immediately, Co-Manager Doug Ayerst and new Store Manager Robert Hayes (he replaced Wes Grab in 1996) began criticizing the way in which I was managing the front end. They repeatedly complained to me about problems in the front end but did not give me any practical management advice and often gave inconsistent instructions. Store Manager Hayes once told me that the problem was that there were "too many damn women in the front end." On another occasion, Store Manager Hayes and Co-Manager Jim Belcoff pulled me aside to tell me that I needed to "whiten up [the staff of] the front end." Both of these statements seemed to indicate that they wanted me to stop hiring women and African-American cashiers. When I asked why, they indicated that the staff was "intimidating" to the clientele. I tried to explain that I did not believe this was true and that I was hiring the individuals whom I believed were the most qualified applicants. They were unresponsive.

… In the second example, the words of Melissa Howard, a 35-year-old white woman who had risen to the rank of Store Manager while raising her biracial child in the Midwest, are examined.

> … In July 1997, I was promoted to the position of Store Manager in Marysville, Kansas. I drove with my daughter's father, who is African-American, to find a place to live there. When we arrived in town, we were treated hostilely by a clerk in the local Wal-Mart store, the town realtor, and by a number of prospective landlords because we were a mixed-race family. At the last rental we visited, the landlord told us that my daughter was not safe and that we needed to be out of town by dark. I knew then that I could not move my family to this place. I called the Regional Personnel Manager Gary Coward, explained what had happened and asked to be placed as a store manager anywhere else. He refused and told me that I would have to go to Marysville as planned or accept a demotion and return as an assistant manager to the 86th Store in Indianapolis, one of the worst stores in the area. I took the demotion…. This was particularly humiliating because my employees in Plainfield had just thrown me a big going-away party to celebrate my promotion.

Melissa Howard was chastised by a male supervisor not just because of her gender, but because she broke a racialized stereotype of white women's chasteness (Collins 2000:132–134). By breaking this stereotype, her colleagues and supervisors saw her as a lower class white. As a result of this construction, the unequal ways individuals have access to power are produced and upheld (Brown 1995; DeVault 1999:27). As a white woman in an interracial relationship, Melissa Howard was punished and seen as an outsider by the white people with whom she interacted. It was as though she created the problems she and her family experienced in Kansas.

In both examples, these white women were treated as incompetent and at fault for the problems they experienced. In turn, this affected their work. In these examples it is clear that while these women faced oppression due to their gender, rules and relationships to power are established for individuals based on the multiple, and differently valued, statuses they hold. Melissa Howard seemingly rose above the occupational glass ceiling noted by many plaintiffs. However, her failure to adhere to dating rules held by those with power, including landlords, realtor, and the Wal-Mart Regional Manager, created experiences of discrimination. As an insider due to her race, but a subordinate due to gender and class, Lorie Williams was able to attain a lower-level management position, but was chastised and belittled because she hired men of color and women.

These stories illustrate the lived experiences of white women who work at Wal-Mart. To succeed, they were socialized to accept and replicate discrimination against people of color. Similarly, workers are trained to follow rules of "respectability" concerning their actions. Insiders like Lorie Williams who hold beliefs that make them outsiders, need support so they will continue to believe that the long-term benefits of their actions will outweigh the short-term punishments they endure for failing to replicate discriminatory practices.

The Continuing Significance of Race

… In the following statement, it is clear subordinates also draw on racial and gendered privileges to gain power, even in instances in which they occupy lower class positions. In her statement, Ms. Jennifer Johnson, a Black woman with a community college degree, describes her interpersonal relationships with store staff.

> In 1991, Mr. Pshek was promoted to District Manager in a different district. He told me that if I was willing to relocate to a store in his district, that he would promote me to Assistant Manager, a position that I had sought for some time. I agreed, although I knew that Scott Schwalback, Mark Melatesta, and a male named Rick had been promoted to Assistant Manager without having to switch districts. I was transferred to the Eustis, Florida store as an hourly employee. I worked in that store as a Department Manager for approximately five months without receiving the promotion to Assistant Manager that Mr. Pshek had promised. Finally, I spoke with Bob Hart, Regional Vice President, about the situation, and he told me I had to wait another three to fourth months to be promoted to Assistant Manager. I was finally promoted in February 1992. Both Scott Schwalback and Rick (a Department Manager in Furniture) became Department Managers after I had, since

I had worked for Wal-Mart for longer, but they were both promoted to Assistant Manager positions before I was. Thus, I worked much longer as a Department Manager before being promoted to Assistant Manager than similarly qualified men had. Shortly after I was promoted to Co-Manager, one of the male Assistant Managers who reported to me was disrespectful and avoided doing work I assigned him. I spoke to my Store Manager Kevin Robinson about it. He told me that the man was upset that I had been promoted instead of him, and said "you have two strikes against you: 1) you're a woman; 2) you're black."

Jennifer Johnson described the years of obstacles she navigated to gain promotions. Even with her experience and dedication, others reduced her accomplishments and used them to treat her with disdain. For her and other women of color, working in these stores means that they must not only do their jobs well, but they must also navigate a workplace where others use them as scapegoats for their own frustrations, block them from training opportunities, or sabotage them by not doing work these women of color assign them. In addition, those with the power to chastise offenders and influence workplace discussions of inequality did nothing. Therefore, women of color had to find their own ways to overcome inequality.

The above example illustrates blatant, hostile racism. But racism is not always aggressively expressed. Feagin (1991) argues after a lifetime of experiencing discrimination, people of color develop a special lens through which to recognize even subtle discrimination. For example, joking can be a subtle medium through which people of color experience belittlement and struggles for dominance. Ms. Gina Espinoza-Price, an energetic, innovative woman, worked her way up to become the District Manager of Mexico. She provided an example of how a white man used joking to force others to acknowledge the labels he created for them and to show dominance. Because he knew he would not face a penalty for his actions, he was able to set a behavioral example for other men under his supervision.

Male Photo Division Management behaved in ways that demeaned and belittled women and minorities. In fall 1996, there was a Photo District Manager meeting in Valencia, California. Wal-Mart had just hired a second female Photo District Manager for the western region, Linda Palmer. During dinner, Jeff Gwartney introduced all of the District Managers to Ms. Palmer using nicknames for the minorities and women. I was introduced as Gina, "the little Mexican princess." I was very offended by Mr. Gwartney's comment and left the dinner early. Throughout the meeting, men made sexual statements and jokes that I thought were very offensive. For example, a flyer with an offensive joke about women being stupid was left on my belongings. In February 1997, during an evaluation, I complained to One-hour Photo Divisional Manager Joe Lisuzzo about harassment based on gender at the previous Photo District Manager meeting. He replied that he would take care of it. I knew from trainings on Wal-Mart's sex harassment policy given by Wal-Mart Legal Department employee Canetta Ivy that company policy mandates that when someone complains of sexual harassment, an investigation must begin within twenty-four hours. Therefore, I expected to be interviewed as a part of an investigation. I was never called. A couple of weeks later, in March 1997, I saw Mr. Lisuzzo at a meeting. I asked him if he had been conducting an investigation of my sexual harassment complaint. He replied that it was being taken care of. I was never aware of any action taken in response to my complaint. Six weeks after complaining about sexual harassment, I was terminated.

Gina Espinoza-Price's statement highlights blatant examples of racism and sexism as well as subtle, institutional examples of discrimination as carried out by individuals who occupied similar class positions, but different status levels. Mr. Lisuzzo's handling of the sexual harassment claim filed by the plaintiff is an example of how the concerns of people are ignored, requiring victims of harassment and discrimination to waste emotional energy rationalizing their own reactions and creating coping tactics to help them interact with hostile colleagues. In addition, it shows how racist joking is often coupled with class and gender.

Class Matters

... Just as some women enjoy privilege and opportunities based on their race, class also shapes their experiences and actions. The following excerpt demonstrates the difficulties and frustrations of women of color who are single mothers trying to support their families with Wal-Mart wages.

... Ms. Uma Jean Minor, a single mother living in Alabama, described the need to remain gainfully employed.

As a single mother, I could not support my family on this wage [paid by Wal-Mart] and was forced to take a second, full-time day job at Food Fair. At the time, I planned to work this second job only until I was able to move to a higher paying position at Wal-Mart. I had no idea that it would take me another seven years to obtain a management position at Wal-Mart and that I would be working two full-time jobs for this entire period.

In this example, Uma Jean Minor notes that her Wal-Mart wages were not sufficient to lift her out of poverty, an argument echoed by living wage supporters throughout the United States (see McCarthy and Ciokajlo 2006; Talbott and Dolby 2003; Warren 2005). Though she shows a tremendous amount of will power, motivation, and agency, her actions would not be sustainable for most women in her position as a single mother of five. Missing from her declaration are the stories of fear, if not hardship, she lived through raising five children and working two full time jobs. In addition, we do not know if she had a network of kin or fictive kin on which she could rely. Therefore, we can only guess at the psychological and material effects discrimination had on her and her children.

Clearly, the effects of sex-based discrimination at Wal-Mart shape the lives of women and their families in different and important ways.

… For female heads of household whose families depend on their earnings, the wages, difficult decisions about how to navigate work, and workplace stress factor into whether it is worthwhile to work for poverty wages. Studies like that of Edin and Lein (1997) have demonstrated that the benefits lost by taking a low-wage job are detrimental to these women's children, who lose state funded health care, housing assistance, and food support. For women like the defendants, work provides another burden; because they need these jobs, they must sometimes sacrifice their dignity and morals in the face of oppressive supervisors (like Mary Crawford) and glass ceilings (like Uma Jean Minor) for the sake of their children.

DISCUSSION AND CONCLUSION

… These plaintiffs did not experience the waste of their talent, energy, and potential in the same way. Rather, it can be observed that the Dukes case is a collection of varying vignettes through which social scientists can observe how some women have more access to power and resources, leaving others to struggle because of their marginalized positions in the power hierarchy. To be clear, although some of the women at Wal-Mart have suffered because of sexism, gender abuse takes multiple forms allowing certain women privilege (even in their oppression) because power is not distributed equally.

REFERENCES

Blee, Kathleen M. 2002. *Inside Organized Racism: Women in the Hate Movement.* Berkeley: University of California Press.

Brown, Elsa B. 1995. " 'What has Happened Here': The Politics of Difference in Women's History and Feminist Politics." In D. C. Hine, W. King and L. Reed (eds). *We Specialize in the Wholly Impossible: A Reader in Black Women's History,* pp. 39–54. Brooklyn, NY: Carlson Publishing.

Collins, Patricia H. 2000. *Black Feminist Thought: Knowledge, Consciousness and the Politics of Empowerment.* New York: Routledge.

DeVault, Marjorie L. 1999. *Liberating Method: Feminism and Social Research.* Philadelphia: Temple University Press.

Drogin, Richard. 2003. *Statistical Analysis of Gender Patterns in Wal-Mart Workforce.* Retrieved from www.walmartclass.com on 22 Sept 2008.

Edin, Kathryn, and Lein, Laura. 1997. *Making Ends Meet: How Single Mothers Survive Welfare and Low-Wage Work.* New York: Russell Sage Foundation.

Feagin, Joe R. 1991. "The Continuing Significance of Race: Antiblack Discrimination in Public Places." *American Sociological Review* 56(1): 101–116.

Featherstone, Liza. 2004. *Selling Women Short*. New York: Basic Books.

Ferber, Abby L. 1999. "What White Supremacists Taught a Jewish Scholar about Identity." *The Chronicle of Higher Education* May 7: B6–B7.

Frankenberg, Ruth. 1993. *White Women, Race Matters: The Social Construction of Whiteness*. Minneapolis: University of Minnesota Press.

Litchenstein, Nelson. 2006. "Wal-Mart: A Template for Twenty-First Century Capitalism." Chapter 1 in Nelson Litchenstein (ed.). *Wal-Mart: The Face of Twenty-First Century Capitalism*. New York: The New Press.

McCarthy, Brendan, and Mickey Ciokajlo. 2006. "Clerics Slam Big-Box Wage Law: Ordinance Would Chase Jobs from City, They Contend." *Chicago Tribune*, July 18, 2006. Accessed from Lexis Nexis August 5, 2007.

Mills, C. Wright. 2000 [1959]. *The Sociological Imagination*. Oxford: Oxford University Press.

Picca, Leslie Houts, and Feagin, Joe R. 2007 *Two-Faced Racism: Whites in the Backstage and Frontstage*. New York: Routledge.

Rathke, Wade. 2006. "A Wal-Mart Workers Association? An Organizing Plan" in Litchenstein, Nelson (ed.). *Wal-Mart: The Face of Twenty-First Century Capitalism*. New York: The New Press.

Rosen, Ruth. 2004. *Big-Box Battle: A Review of Selling Women Short: The Landmark Battle for Workers' Rights at Wal-Mart, by Liza Featherstone*. Retrieved from http://www.longviewinstitute.org/research/rosen/walmart/sellingwomenshort on October 14, 2008.

Talbott, Madeline, and Doby, Michael. 2003. "Using the Big Box Living Wage Ordinance to Keep Wal-Mart Out of the Cities." *Social Policy* 34(2/3): 23–28.

Warren, Dorian T. 2005. "Wal-Mart Surrounded: Community Alliances and Labor Politics in Chicago." *New Labor Forum* 14(3): 17–23.

29

Our Mothers' Grief

Racial-Ethnic Women and the Maintenance of Families

BONNIE THORNTON DILL

REPRODUCTIVE LABOR[1] FOR WHITE WOMEN
IN EARLY AMERICA

In eighteenth- and nineteenth-century America, the lives of white[2] women in the United States were circumscribed within a legal and social system based on patriarchal authority. This authority took two forms: public and private. The social, legal, and economic position of women in this society was controlled through the private aspects of patriarchy and defined in terms of their relationship to families headed by men. The society was structured to confine white wives to reproductive labor within the domestic sphere. At the same time, the formation, preservation, and protection of families among white settlers was seen as crucial to the growth and development of American society. Building, maintaining, and supporting families was a concern of the State and of those organizations that prefigured the State. Thus, while white women had few legal rights as women, they were protected through public forms of patriarchy that acknowledged and supported their family roles of wives, mothers, and daughters because they were vital instruments for building American society....

In colonial America, white women were seen as vital contributors to the stabilization and growth of society. They were therefore accorded some legal and economic recognition through a patriarchal family structure....

Throughout the colonial period, women's reproductive labor in the family was an integral part of the daily operation of small-scale family farms or artisan's shops. According to Kessler-Harris (1981), a gender-based division of labor was common, but not rigid. The participation of women in work that was essential to family survival reinforced the importance of their contributions to both the protection of the family and the growth of society.

Between the end of the eighteenth and mid-nineteenth century, what is labeled the "modern American family" developed. The growth of industrialization and an urban middle class, along with the accumulation of agrarian wealth among Southern planters, had two results that are particularly pertinent to this

SOURCE: Dill, Bonnie Thornton. 1988. "Our Mothers' Grief: Racial-Ethnic Women and the Maintenance of Families." *Journal of Family History*, 13(4): 415–431. Copyright © 1988 by JAI Press Inc. Reprinted by permission of SAGE Publications, Inc.

discussion. First, class differentiation increased and sharpened, and with it, distinctions in the content and nature of women's family lives. Second, the organization of industrial labor resulted in the separation of home and family and the assignment to women of a separate sphere of activity focused on childcare and home maintenance. Whereas men's activities became increasingly focused upon the industrial competitive sphere of work, "women's activities were increasingly confined to the care of children, the nurturing of the husband, and the physical maintenance of the home" (Degler 1980, p. 26).

This separate sphere of domesticity and piety became both an ideal for all white women as well as a source of important distinctions between them. As Matthaei (1982) points out, tied to the notion of wife as homemaker is a definition of masculinity in which the husband's successful role performance was measured by his ability to keep his wife in the homemaker role. The entry of white women into the labor force came to be linked with the husband's assumed inability to fulfill his provider role.

For wealthy and middle-class women, the growth of the domestic sphere offered a potential for creative development as homemakers and mothers. Given ample financial support from their husband's earnings, some of these women were able to concentrate their energies on the development and elaboration of the more intangible elements of this separate sphere. They were also able to hire other women to perform the daily tasks such as cleaning, laundry, cooking, and ironing. Kessler-Harris cautions, however, that the separation of productive labor from the home did not seriously diminish the amount of physical drudgery associated with housework, even for middle-class women.... In effect, household labor was transformed from economic productivity done by members of the family group to home maintenance; childcare and moral uplift done by an isolated woman who perhaps supervised some servants.

Working-class white women experienced this same transformation but their families' acceptance of the domestic code meant that their labor in the home intensified. Given the meager earnings of working-class men, working-class families had to develop alternative strategies to both survive and keep the wives at home. The result was that working-class women's reproductive labor increased to fill the gap between family need and family income. Women increased their own production of household goods through things such as canning and sewing; and by developing other sources of income, including boarders and homework. A final and very important source of other income was wages earned by the participation of sons and daughters in the labor force. In fact, Matthaei argues that "the domestic homemaking of married women was supported by the labors of their daughters" (1982, p. 130)....

Another way in which white women's family roles were socially acknowledged and protected was through the existence of a separate sphere for women. The code of domesticity, attainable for affluent women, became an ideal toward which nonaffluent women aspired. Notwithstanding the personal constraints placed on women's development, the notion of separate spheres promoted the growth and stability of family life among the white middle class and became the basis for working-class men's efforts to achieve a family wage, so that they could keep their wives at home. Also, women gained a distinct sphere of authority and expertise that yielded them special recognition.

During the eighteenth and nineteenth centuries, American society accorded considerable importance to the development and sustenance of European immigrant families. As primary laborers in the reproduction and maintenance of family life, women were acknowledged and accorded the privileges and protections deemed socially appropriate to their family roles. This argument acknowledges the fact that the family structure denied these women many rights and privileges and seriously constrained their individual growth and development. Because women gained social recognition primarily through their membership in families, their personal rights were few and privileges were subject to the will of the male head of the household. Nevertheless, the recognition of women's reproductive labor as an essential building block of the family, combined with a view of the family as the cornerstone of the nation, distinguished the experiences of the white, dominant culture from those of racial ethnics.

Thus, in its founding, American society initiated legal, economic, and social practices designed to promote the growth of family life among European colonists. The reception colonial families found in the United States contrasts sharply with the lack of attention given to the families of racial-ethnics. Although the presence of racial-ethnics was equally as important for the growth of the nation, their political, economic, legal, and social status was quite different.

REPRODUCTIVE LABOR AMONG RACIAL-ETHNICS IN EARLY AMERICA

Unlike white women, racial-ethnic women experienced the oppressions of a patriarchal society but were denied the protections and buffering of a patriarchal family. Their families suffered as a direct result of the organization of the labor systems in which they participated.

Racial-ethnics were brought to this country to meet the need for a cheap and exploitable labor force. Little attention was given to their family and community life except as it related to their economic productivity. Labor, and not the existence or maintenance of families, was the critical aspect of their role in building the nation. Thus they were denied the social structural supports necessary to make *their* families a vital element in the social order. Family membership was not a key means of access to participation in the wider society. The lack of social, legal, and economic support for racial-ethnic families intensified and extended women's reproductive labor, created tensions and strains in family relationships, and set the stage for a variety of creative and adaptive forms of resistance.

African-American Slaves

Among students of slavery, there has been considerable debate over the relative "harshness" of American slavery, and the degree to which slaves were permitted or encouraged to form families. It is generally acknowledged that many slave-owners found it economically advantageous to encourage family formation as a way of reproducing and perpetuating the slave labor force. This became increasingly true after 1807 when the importation of African slaves was explicitly prohibited. The existence of these families and many aspects of their functioning, however, were directly controlled by the master. In other words, slaves married and formed families but these groupings were completely subject to the master's decision to let them remain intact. One study has estimated that about 32 percent of all recorded slave marriages were disrupted by sale, about 45 percent by death of a spouse, about 10 percent by choice, with the remaining 13 percent not disrupted at all (Blassingame 1972, pp. 90–92). African slaves thus quickly learned that they had a limited degree of control over the formation and maintenance of their marriages and could not be assured of keeping their children with them. The threat of disruption was perhaps the most direct and pervasive cultural assault[3] on families that slaves encountered. Yet there were a number of other aspects of the slave system which reinforced the precariousness of slave family life.

In contrast to some African traditions and the Euro-American patterns of the period, slave men were not the main provider or authority figure in the family. The mother-child tie was basic and of greatest interest to the slaveowner because it was critical in the reproduction of the labor force.

In addition to the lack of authority and economic autonomy experienced by the husband-father in the slave family, use of the rape of women slaves as a weapon of terror and control further undermined the integrity of the slave family.... The slave family, therefore, was at the heart of a peculiar tension in the master-slave relationship. On the one hand, slaveowners sought to encourage familial ties among slaves because, as Matthaei (1982) states: "... these provided the basis of the development of the slave into a self-conscious socialized human being" (p. 81). They also hoped and believed that this socialization process would help children learn to accept their place in society as slaves. Yet the master's need to control and intervene in the familial life of the slaves is indicative of the other side of this tension. Family ties had the potential for becoming a competing and more potent source of allegiance than the slavemaster himself. Also, kin were as likely to socialize children in forms of resistance as in acts of compliance.

It was within this context of surveillance, assault, and ambivalence that slave women's reproductive labor took place. She and her menfolk had the task of preserving the human and family ties that could ultimately give them a reason for living. They had to socialize their children to believe in the possibility of a life in which they were not enslaved. The slave woman's labor on behalf of the family was, as Davis (1971) has pointed out, the only labor the slave engaged in that could not be directly appropriated by the slaveowner for his own profit. Yet, its indirect appropriation, as labor crucial to the reproduction of the slaveowner's labor force, was the source of strong ambivalence for many slave women. Whereas some mothers murdered their babies to keep them from being slaves, many sought within the family sphere a degree of autonomy and creativity denied them in other realms of the society. The maintenance of a distinct African-American culture is testimony to the ways in which slaves maintained a degree of cultural autonomy and resisted the creation of a slave family that only served the needs of the master.

Gutman (1976) provides evidence of the ways in which slaves expressed a unique Afro-American culture through their family practices. He provides data on naming patterns and kinship ties among slaves that flies in the face of the dominant ideology of the period. That ideology argued that slaves were immoral and had little concern for or appreciation of family life.

Yet Gutman demonstrated that within a system which denied the father authority over his family, slave boys were frequently named after their fathers, and many children were named after blood relatives as a way of maintaining family ties. Gutman also suggested that after emancipation a number of slaves took the names of former owners in order to reestablish family ties that had been disrupted earlier. On plantation after plantation, Gutman found considerable evidence of the building and maintenance of extensive kinship ties among slaves. In instances where slave families had been disrupted, slaves in new communities reconstituted the kinds of family and kin ties that came to characterize Black family life throughout the South. These patterns included, but were not limited to, a belief in the importance of marriage as a long-term commitment, rules of exogamy that included marriage between first cousins, and acceptance of women who had children outside of marriage. Kinship networks were an important source of resistance to the organization of labor that treated the individual slave, and not the family, as the unit of labor (Caulfield 1974).

Another interesting indicator of the slaves' maintenance of some degree of cultural autonomy has been pointed out by Wright (1981) in her discussion of slave housing. Until the early 1800s, slaves were often permitted to build their housing according to their own design and taste. During that period, housing built in an African style was quite common in the slave quarters. By 1830, however, slaveowners had begun to control the design and arrangement of slave housing and had introduced a degree of conformity and regularity to it that left little room for the slave's personalization of the home. Nevertheless, slaves did use some of their own techniques in construction and often hid it from their masters....

Housing is important in discussions of family because its design reflects sociocultural attitudes about family life. The housing that slaveowners provided for their slaves reflected a view of Black family life consistent with the stereotypes of the period. While the existence of slave families was acknowledged, it certainly was not nurtured. Thus, cabins were crowded, often containing more than one family, and there were no provisions for privacy. Slaves had to create their own....

Perhaps most critical in developing an understanding of slave women's reproductive labor is the gender-based division of labor in the domestic sphere. The organization of slave labor enforced considerable equality among men and women. The ways in which equality in the labor force was translated into the family sphere is somewhat speculative....

We know, for example, that slave women experienced what has recently been called the "double day" before most other women in this society. Slave narratives (Jones 1985; White 1985; Blassingame 1977) reveal that women had primary responsibility for their family's domestic chores. They cooked (although on some plantations meals were prepared for all of the slaves), sewed, cared for their children, and cleaned house, all after completing a full day of labor for the master. Blassingame (1972) and others have pointed out that slave men engaged in hunting, trapping, perhaps some gardening, and furniture making as ways of contributing to the maintenance of their families. Clearly, a gender-based division of labor did exist within the family and it appears that women bore the larger share of the burden for housekeeping and child care....

Black men were denied the male resources of a patriarchal society and therefore were unable to turn gender distinctions into female subordination, even if that had been their desire. Black women, on the other hand, were denied support and protection for their roles as mothers and wives and thus had to modify and structure those roles around the demands of their labor. Thus, reproductive labor for slave women was intensified in several ways: by the demands of slave labor that forced them into the double-day of work; by the desire and need to maintain family ties in the face of a system that gave them only limited recognition; by the stresses of building a family with men who were denied the standard social privileges of manhood; and by the struggle to raise children who could survive in a hostile environment.

This intensification of reproductive labor made networks of kin and quasi-kin important instruments in carrying out the reproductive tasks of the slave community. Given an African cultural heritage where kinship ties formed the basis of social relations, it is not at all surprising that African American slaves developed an extensive system of kinship ties and obligations (Gutman 1976; Sudarkasa 1981). Research on Black families in slavery provides considerable documentation of participation of extended kin in child rearing, childbirth, and other domestic, social, and economic activities (Gutman 1976; Blassingame 1972; Genovese and Miller 1974)....

With individual households, the gender-based division of labor experienced some important shifts during emancipation. In their first real opportunity to establish family life beyond the controls and constraints imposed by a slavemaster, family life among Black sharecroppers changed radically. Most women, at least those who were wives and daughters of able-bodied men, withdrew from field labor and concentrated on their domestic duties in the home. Husbands took primary responsibility for the fieldwork and for relations with the owners, such as signing contracts on behalf of the family. Black women were severely criticized by whites for removing themselves from field labor because they were seen to be aspiring to a model of womanhood that was considered inappropriate for them. This reorganization of female labor, however, represented an attempt on the part of Blacks to protect women from some of the abuses of the slave system and to thus secure their family life. It was more likely a response to the particular set of circumstances that the newly freed slaves faced than a reaction to the lives of their former masters. Jones (1985) argues that these patterns were "particularly significant" because at a time when industrial development was introducing a labor system that divided male and female labor, the freed Black family was establishing a pattern of joint work and complementary tasks between males and females that was reminiscent of the preindustrial American families. Unfortunately, these former slaves had to do this without the institutional supports that white farm families had in the midst of a sharecropping system that deprived them of economic independence.

Chinese Sojourners

An increase in the African slave population was a desired goal. Therefore, Africans were permitted and even encouraged at times to form families subject to the authority and whim of the master. By sharp contrast, Chinese people were explicitly denied the right to form families in the United States through both law and social practice. Although male laborers began coming to the United States in sizable numbers in the middle of the nineteenth century, it was more than a century before an appreciable number of children of Chinese parents were born in America. Tom, a respondent in Nee and Nee's (1973) book, *Longtime Californ',* says: "One thing about Chinese men in America was you had to be either a merchant or a big gambler, have lot of side money to have a family here. A working man, an ordinary man, just can't!" (p. 80).

Working in the United States was a means of gaining support for one's family with an end of obtaining sufficient capital to return to China and purchase land. The practice of sojourning was reinforced by laws preventing Chinese laborers from becoming citizens, and by restrictions on their entry into this country. Chinese laborers who arrived before 1882 could not bring their wives and were prevented by law from marrying whites. Thus, it is likely that the number of Chinese-American families might have been negligible had it not been for two things: the San Francisco earthquake and fire in 1906, which destroyed all municipal records; and the ingenuity and persistence of the Chinese people who used the opportunity created by the earthquake to increase their numbers in the United States. Since relatives of citizens were permitted entry, American-born Chinese (real and claimed) would visit China, report the birth of a son, and thus create an entry slot. Years later the slot could be used by a relative or purchased. The purchasers were called "paper sons." Paper sons became a major mechanism for increasing the Chinese population, but it was a slow process and the sojourner community remained predominantly male for decades.

The high concentration of males in the Chinese community before 1920 resulted in a split-household form of family....

The women who were in the United States during this period consisted of a small number who were wives and daughters of merchants and a larger percentage who were prostitutes. Hirata (1979) has suggested that Chinese prostitution was an important element in helping to maintain the split-household family. In conjunction with laws prohibiting intermarriage, Chinese prostitution helped men avoid long-term relationships with women in the United States and ensured that the bulk of their meager earnings would continue to support the family at home.

The reproductive labor of Chinese women, therefore, took on two dimensions primarily because of the split-household family form. Wives who remained in China were forced to raise children and care for in-laws on the meager remittances of their sojourning husband. Although we know few details about their lives, it is clear that the everyday work of bearing and maintaining children and a household fell entirely on their shoulders. Those women who immigrated and worked as prostitutes performed the more nurturant aspects of reproductive labor, that is, providing emotional and sexual companionship for men who were far from home. Yet their role as prostitute was more likely a means of supporting their families at home in China than a chosen vocation.

The Chinese family system during the nineteenth century was a patriarchal one wherein girls had little value. In fact, they were considered only temporary members of their father's family because when they married, they became members of their husband's families. They also had little social value: girls were sold by some poor parents to work as prostitutes, concubines, or servants. This saved the family the expense of raising them, and their earnings also became a source of family income. For most girls, however, marriages were arranged and families sought useful connections through this process.

With the development of a sojourning pattern in the United States, some Chinese women in those regions of China where this pattern was more prevalent would be sold to become prostitutes in the United States. Most, however, were married off to men whom they saw only once or twice in the 20- or 30-year period during which he was sojourning in the United States. Her status as wife ensured that a portion of the meager wages he earned would be returned to his family in China. This arrangement required considerable sacrifice and adjustment on the part of wives who remained in China and those who joined their husbands after a long separation....

Despite these handicaps, Chinese people collaborated to establish the opportunity to form families and settle in the United States. In some cases it took as long as three generations for a child to be born on United States soil....

Chicanos

Africans were uprooted from their native lands and encouraged to have families in order to increase the slave labor force. Chinese people were immigrant laborers whose "permanent" presence in the country was denied. By contrast, Mexican-Americans were colonized and their traditional family life was disrupted by war and the imposition of a new set of laws and conditions of labor. The hardships faced by Chicano families, therefore, were the result of the United States colonization of the indigenous Mexican population, accompanied by the beginnings of industrial development in the region. The treaty of Guadalupe Hidalgo, signed in 1848, granted American citizenship to Mexicans living in what is now called the Southwest. The American takeover, however, resulted in the gradual displacement of Mexicans from the land and their incorporation into a colonial labor force (Barrera 1979). In addition, Mexicans who immigrated into the United States after 1848 were also absorbed into the labor force.

Whether natives of Northern Mexico (which became the United States after 1848) or immigrants from Southern Mexico, Chicanos were a largely peasant population whose lives were defined by a feudal economy and a daily struggle on the land for economic survival. Patriarchal families were important instruments of community life and nuclear family units were linked together through an elaborate system of kinship and godparenting. Traditional life was characterized by hard work and a fairly distinct pattern of sex-role segregation....

As the primary caretakers of hearth and home in a rural environment, *Las Chicanas* labor made a vital and important contribution to family survival....

Although some scholars have argued that family rituals and community life showed little change before World War I (Saragoza 1983), the American conquest of Mexican lands, the introduction of a new system of labor, the loss of Mexican-owned land through the inability to document ownership, plus the transient nature of most of the jobs in which Chicanos were employed, resulted in the gradual erosion of this pastoral way of life. Families were uprooted as the economic basis for family life changed. Some immigrated from Mexico in search of a better standard of living and worked in the mines and railroads. Others who were native to the Southwest faced a job market that no longer required their skills and moved into mining, railroad, and agricultural labor in search of a means of earning a living. According to Camarillo (1979), the influx of Anglo[4] capital into the pastoral economy of Santa Barbara rendered obsolete the skills of many Chicano males who had worked as ranchhands and farmers prior to the urbanization of that economy. While some women and children accompanied their husbands to the railroad and mine camps, they often did so despite prohibitions against it. Initially many of these camps discouraged or prohibited family settlement.

The American period (post-1848) was characterized by considerable transiency for the Chicano population. Its impact on families is seen in the growth of female-headed households, which was reflected in the data as early as 1860. Griswold del Castillo (1979) found a sharp increase in female-headed households in Los Angeles, from a low of 13 percent in 1844 to 31 percent in 1880. Camarillo (1979, p. 120) documents a similar increase in Santa Barbara from 15 percent in 1844 to 30 percent by 1880. These increases appear to be due not so much to divorce, which was infrequent in this Catholic population, but to widowhood and temporary abandonment in search of work. Given the hazardous nature of work in the mines and railroad camps, the death of a husband, father or son who was laboring in these sites was not uncommon. Griswold del Castillo (1979) reports a higher death rate among men than women in Los Angeles. The rise in female-headed households, therefore, reflects the instabilities and insecurities introduced into women's lives as a result of the changing social organization of work.

One outcome, the increasing participation of women and children in the labor force was primarily a response to economic factors that required the modification of traditional values....

Slowly, entire families were encouraged to go to railroad workcamps and were eventually incorporated into the agricultural labor market. This was a response both to the extremely low wages paid to Chicano laborers and to the preferences of employers who saw family labor as a way of stabilizing the workforce. For Chicanos, engaging all family members in agricultural work was a means of increasing their earnings to a level close to subsistence for the entire group and of keeping the family unit together....

While the extended family has remained an important element of Chicano life, it was eroded in the American period in several ways. Griswold del Castillo (1979), for example, points out that in 1845 about 71 percent of Angelenos lived in extended families and that by 1880, fewer than half did. This decrease in extended families appears to be a response to the changed economic conditions and to the instabilities generated by the new sociopolitical structure. Additionally, the imposition of American law and custom ignored and ultimately undermined some aspects of the extended family. The extended family in traditional Mexican life consisted of an important set of familial, religious, and community obligations. Women, while valued primarily for their domesticity, had certain legal and property rights that acknowledged the importance of their work, their families of origin and their children....

In the face of the legal, social, and economic changes that occurred during the American period, Chicanas were forced to cope with a series of dislocations in traditional life. They were caught between conflicting pressures to maintain traditional women's roles and family customs and the need to participate in the economic support of their families by working outside the home. During this period the preservation of some traditional customs became an important force for resisting complete disarray....

Of vital importance to the integrity of traditional culture was the perpetuation of the Spanish language. Factors that aided in the maintenance of other aspects of Mexican culture also helped in sustaining the language. However, entry into English-language public schools introduced the children and their families to systematic efforts to erase their native tongue....

Another key factor in conserving Chicano culture was the extended family network, particularly the system of *compadrazgo* or godparenting. Although the full extent of the impact of the American period on the Chicano extended family is not known, it is generally acknowledged that this family system, though lacking many legal and social sanctions, played an important role in the preservation of the Mexican community (Camarillo 1979, p. 13). In Mexican society, godparents were an important way of linking family and community through respected friends or authorities. Named at the important rites of passage in a child's life, such as birth, confirmation, first communion, and marriage, *compadrazgo* created a moral obligation for godparents to act as guardians, to provide financial assistance in times of need, and to substitute in case of the death of a parent. Camarillo (1979) points out that in traditional society these bonds cut across class and racial lines....

The extended family network—which included godparents—expanded the support groups for women who were widowed or temporarily abandoned and for those who were in seasonal, part-, or full-time work. It suggests, therefore, the potential for an exchange of services among poor people whose income did not provide the basis for family subsistence.... This family form is important to the continued cultural autonomy of the Chicano community.

CONCLUSION: OUR MOTHERS' GRIEF

Reproductive labor for Afro-American, Chinese-American, and Mexican-American women in the nineteenth century centered on the struggle to maintain family units in the face of a variety of cultural assaults. Treated primarily as individual units of labor rather than as members of family groups, these women labored to maintain, sustain, stabilize, and reproduce their families while working in both the public (productive) and private (reproductive) spheres. Thus, the concept of reproductive labor, when applied to women of color, must be modified to account for the fact that labor in the productive sphere was required to achieve even minimal levels of family subsistence. Long after industrialization had begun to reshape family roles among middle-class white families, driving white women into a cult of domesticity, women of color were coping with an extended day. This day included subsistence labor outside the family and domestic labor within the family. For slaves, domestics, migrant farm laborers, seasonal factory-workers, and prostitutes, the distinctions between labor that reproduced family life and which economically sustained it were minimized. The expanded workday was one of the primary ways in which reproductive labor increased.

Racial-ethnic families were sustained and maintained in the face of various forms of disruption. Yet racial-ethnic women and their families paid a high price in the process. High rates of infant mortality, a shortened life span, the early onset of crippling and debilitating disease provided some insight into the costs of survival.

The poor quality of housing and the neglect of communities further increased reproductive labor. Not only did racial-ethnic women work hard outside the home for a mere subsistence, they worked very hard inside the home to achieve even minimal standards of privacy and cleanliness. They were continually faced with disease and illness that directly resulted from the absence of basic sanitation. The fact that some African women murdered their children to prevent them from becoming slaves is an indication of the emotional strain associated with bearing and raising children while participating in the colonial labor system.

We have uncovered little information about the use of birth control, the prevalence of infanticide, or the motivations that may have generated these or other behaviors. We can surmise, however, that no matter how much children were accepted, loved, or valued among any of these groups of people, their futures in a colonial labor system were a source of grief for their mothers. For those children who were born, the task of keeping them alive, of helping them to understand and participate in a system that exploited them, and the challenge of maintaining a measure—no matter how small—of cultural integrity, intensified reproductive labor.

Being a racial-ethnic woman in nineteenth-century American society meant having extra work both inside and outside the home. It meant having a contradictory relationship to the norms and values about women that were being generated in the dominant white culture. As pointed out earlier, the notion of separate spheres of male and female labor had contradictory outcomes for the nineteenth-century whites. It was the basis for the confinement of women to the household and for much of the protective legislation that subsequently developed. At the same time, it sustained white families by providing social acknowledgment and support to women in the performance of their family roles. For racial-ethnic women, however, the notion of separate spheres served to reinforce their subordinate status and became, in effect, another assault. As they increased their work outside the home, they were forced into a productive labor sphere that was organized for men and "desperate" women who were so unfortunate or immoral that they could not confine their work to the domestic sphere. In the productive sphere, racial-ethnic women faced exploitative jobs and depressed wages. In the reproductive sphere, however, they were denied the opportunity to embrace the dominant ideological definition of "good" wife or mother. In essence, they were faced with a double-bind situation, one that required their participation in the labor force to sustain family life but damned them as women, wives, and mothers because they did not confine their labor to the home. Thus, the conflict between ideology and reality in the lives of racial-ethnic women during the nineteenth century sets the stage for stereotypes, issues of self-esteem, and conflicts around gender-role prescriptions that surface more fully in the twentieth century. Further, the tensions and conflicts that characterized their lives during this period provided the impulse for community activism to jointly address the inequities, which they and their children and families faced.

NOTES

1. The term *reproductive labor* is used to refer to all of the work of women in the home. This includes but is not limited to: the buying and preparation of food and clothing, provision of emotional support and nurturance for all family members, bearing children, and planning, organizing, and carrying out a wide variety of tasks associated with their socialization. All of these activities are necessary for the growth of patriarchal capitalism because they maintain, sustain, stabilize, and *reproduce* (both biologically and socially) the labor force.

2. The term *white* is a global construct used to characterize peoples of European descent who migrated to and helped colonize America. In the seventeenth century, most of these immigrants were from the British Isles. However, during the time period covered by this article, European immigrants became increasingly diverse. It is a limitation of this article that time and space does not permit a fuller discussion of the variations in the white European immigrant experience. For the purposes of the argument made herein and of the contrast it seeks to draw between the experiences of mainstream (European) cultural groups and that of racial/ethnic minorities, the differences among European settlers are joined and the broad similarities emphasized.

3. Cultural assaults, according to Caulfield (1974), are benign and systematic attacks on the institutions and forms of social organization that are fundamental to the maintenance and flourishing of a group's culture.

4. This term is used to refer to white Americans of European ancestry.

REFERENCES

Barrera, Mario. 1979. *Race and Class in the Southwest*. South Bend, IN: Notre Dame University Press.

Blassingame, John. 1972. *The Slave Community: Plantation Life in the Antebellum South*. New York: Oxford University Press.

Blassingame, John. 1977. *Slave Testimony: Two Centuries of Letters, Speeches, Interviews, and Autobiographies*. Baton Rouge, LA: Louisiana State University Press.

Camarillo, Albert. 1979. *Chicanos in a Changing Society*. Cambridge, MA: Harvard University Press.

Caulfield, Mina Davis. 1974. "Imperialism, the Family, and Cultures of Resistance." *Socialist Review* 4(2)(October): 67–85.

Davis, Angela. 1971. "The Black Woman's Role in the Community of Slaves." *Black Scholar* 3(4)(December): 2–15.

Degler, Carl. 1980. *At Odds*. New York: Oxford University Press.

Genovese, Eugene D., and Elinor Miller, eds. 1974. *Plantation, Town, and County: Essays on the Local History of American Slave Society*. Urbana: University of Illinois Press.

Griswold del Castillo, Richard. 1979. *The Los Angeles Barrio: 1850–1890*. Los Angeles: The University of California Press.

Gutman, Herbert. 1976. *The Black Family in Slavery and Freedom: 1750–1925*. New York: Pantheon.

Hirata, Lucie Cheng. 1979. "Free, Indentured, Enslaved: Chinese Prostitutes in Nineteenth-Century America." *Signs* 5 (Autumn): 3–29.

Jones, Jacqueline. 1985. *Labor of Love, Labor of Sorrow*. New York: Basic Books.

Kessler-Harris, Alice. 1981. *Women Have Always Worked*. Old Westbury: The Feminist Press.

Matthaei, Julie. 1982. *An Economic History of Women in America*. New York: Schocken Books.

Nee, Victor G., and Brett de Bary Nee. 1973. *Longtime Californ'*. New York: Pantheon Books.

Saragoza, Alex M. 1983. "The Conceptualization of the History of the Chicano Family: Work, Family, and Migration in Chicanos." Research Proceedings of the Symposium on Chicano Research and Public Policy. Stanford, CA: Stanford University, Center for Chicano Research.

Sudarkasa, Niara. 1981. "Interpreting the African Heritage in Afro-American Family Organization." Pp. 37–53 in *Black Families*, edited by Harriette Pipes McAdoo. Beverly Hills, CA: Sage Publications.

White, Deborah Gray. 1985. *Ar'n't I a Woman?: Female Slaves in the Plantation South*. New York: W. W. Norton.

Wright, Gwendolyn. 1981. *Building the Dream: A Social History of Housing in America*. New York: Pantheon Books.

LGBT Sexuality and Families at the Start of the Twenty-First Century*

MIGNON R. MOORE AND MICHAEL STAMBOLIS-RUHSTORFER

INTRODUCTION

Whether it is the study of the relationship between sexual satisfaction and marital quality or dating and relationship patterns of young adults, sexuality has often been a component of research on families. This interest in sexuality often took heterosexuality for granted when studying family formation and family processes. However, the literature on same-sex couple relationships and families headed by single parents who identify as lesbian or gay has grown exponentially. Social science research published since the start of the twenty-first century tackles many new questions about sexual minority families. Psychologists are writing about family processes, relationship quality between partners, and children's socioemotional outcomes and development. Sociologists have been considering these issues while also grappling with questions regarding how these families relate to social institutions such as school and legal systems and how these systems shape and are affected by lesbian and gay families. Sociological studies have also revealed the strategies adults in these households adopt to construct meaning so that they are perceived by outsiders as family.*...

Some caveats are necessary. First, most of the sociological research in this area focuses on same-sex couple households and not on single-parent

---

*A NOTE ON DEFINITIONS

In this review, we use terminology that represents the varying scope of populations in the research we review from tight to broad according to the sample and description in the cited study. Lesbian and gay refers to women and men who identify themselves as attracted, usually exclusively, to members of the same sex/gender. As an adjective, gay may sometimes refer to both gay men and lesbians. LGB (lesbian, gay, and bisexual) includes individuals who identify themselves as attracted to both sexes/genders. LGBT (lesbian, gay, bisexual, and transgender) includes individuals who have changed or are in the process of changing sexes or gender identities. Finally, sexual minority refers broadly to individuals whose sexual identity/behavior is marginalized by heterosexually prescribed norms.

SOURCE: Moore, Mignon R., and Michael Stambolis-Ruhstorfer. 2013. "LGBT Sexuality and Families at the Start of the Twenty-First Century." Modified with permission from the Annual Review of Sociology, Volume 39 © 2013 by Annual Reviews, http://www.annualreviews.org.

households. Second, most of the research focuses on households in which the partners identify as lesbian or gay and not bisexual or transgender (see sidebar). Third, much of the work reviewed is U.S.-based. An international and cross-cultural framework for the study of LGBT-parent families would be ideal, given that changes are happening around the world, in part independently and in part informed by one another, in how same-sex parent families are defined and understood. However, space limitations compel us to emphasize work done in the United States, though we make an effort to include research in other national contexts whenever possible. We recognize that by not including material from social systems in which same-sex marriage is more firmly institutionalized than in the United States (e.g., in Scandinavia and the Netherlands), as well as non-Western cultural contexts where sexual minority parenthood often takes place in the context of heterosexual unions while individuals also maintain a lesbian or gay identity, we may lose some purchase on the trends we identify. Finally, much of this work has been interdisciplinary rather than purely sociological, so we draw from the work that most closely speaks to sociological understandings.

WHAT MAKES A FAMILY?

The ideological debates on sexuality and family situate gay men and lesbians at the heart of broader discussions of family politics. Bernstein and Reimann (2001) argue that by making themselves visible as families, same-sex couple households reveal a subversive power that challenges dominant conceptions of gender. In her analysis of pro– and anti–gay marriage movements, Lehr (1999, p. 140) says that "gays and lesbians stand in a unique location from which to view family and from which to define a politics of family and private life" and should seek to undo marriage altogether in favor of new and creative ways of organizing their families beyond the liberal democratic discourse of the rights of couples. For Lehr, the recognition of same-sex marriage inadvertently supports politics that reinforce the disciplinary power of marriage in general. This perspective suggests that marriage should not bestow special legal privileges upon couples, regardless of sexual orientation. It advocates for a broader approach to thinking about what constitutes familial commitment and how it should be legally protected....

Other sociologists have found that even when same-sex couples want to marry and have society view their households as families, they are met with resistance. Stein's (2005) work shows the continuing ambivalence about the normalization of homosexuality and says that as a nation we remain divided over whether lesbians and gay men are the moral equivalent of heterosexuals. In *Counted Out: Same-Sex Relations and Americans' Definitions of Family*, Powell et al. (2010) analyze two waves of survey data, collected in 2003 and 2006, from a nationally representative sample of Americans about a variety of family-related topics, including their opinions about same-sex couple households. They find that a large segment of the U.S. population is ambivalent or resistant to the inclusion of same-sex couples in their definition of family because they believe these relationships threaten the heterosexual family form and undermine traditional gender and sexuality norms (Powell et al. 2010, p. 103). These ideologies not only represent disapproval or discomfort around gay sexuality but also speak to broader understandings of gender and sexuality.

DEFINING LESBIAN- AND GAY-PARENT FAMILIES

... From 2000 to 2010, the number of these households increased by about 80%, from 358,390 to 646,464. Different-sex unmarried couples increased by about 40% during this same period, whereas different-sex married couples increased by a much smaller rate of 3.7%. Nineteen percent of same-sex couples were raising children under age 18. These data suggest that currently there are approximately 125,000 same-sex couples raising nearly 220,000 children. The most recent report on LGBT parenting issued by the Williams Institute analyzes data from the Gallup Daily Tracking Survey collected from June to September 2012, the 2008–2010 General Social Survey, and the National Transgender Discrimination Survey. It estimates that as many as 6 million American children and adults have an LGBT parent (Gates 2013). It also finds that same-sex couples are more likely than different-sex couples to be raising an adopted or foster child. And these numbers reflect only those individuals who were in same-sex partnerships at the time of data collection.

... Same-sex couple households with children are not evenly spread across geographic regions or socioeconomic class. Same-sex parenting is more common in the South, where more than a quarter of same-sex couples are raising children (Gates 2011b). One finding across demographic studies is the greater level of interraciality among same-sex couples compared with different-sex married couples (Jepsen & Jepsen 2002, Rosenfeld & Kim 2005). Relative to Whites, child-rearing among same-sex couples is higher among African Americans, Latinos, and American Indian/Alaskan Natives and lower among Asian and Pacific Islanders. Forty percent of African American same-sex couples, 28% of Latino/a, 24% of American Indian/Alaskan Native, 12% of Asian and Pacific Islander, and 16% of White same-sex couples have children under age 18 living with them in the home (Gates 2011b, p. F3).

Although same-sex couples have relatively high educational attainment and earnings, same-sex couples with children tend to come from the working class and are more likely to be from racial minority groups.... Same-sex couple families are significantly more likely to be poor than are heterosexual married couple families. For example, African American same-sex couples have poverty rates that are significantly higher than both African American different-sex married couples and White same-sex couples. Children in lesbian and gay couple households have poverty rates twice those of children in heterosexual married couple households, and their families are more likely to receive government cash supports intended for low-income families.... These differences reveal race, ethnicity, and social class as mutually constitutive in the lives of sexual minority parents and their children (Moore & Brainer 2013).

Of the nearly 650,000 same-sex couples counted in Census 2010, 79,200 or about 12% have at least one partner who either is not a U.S. citizen or is a naturalized citizen, and these couples are raising more than 25,000 children (Konnoth & Gates 2011). Existing immigration law has made it difficult for these families to legalize through marriage and bars couples who do not share the same home country from pursuing permanent residency as a couple in either partner's country of origin....

PATHWAYS TO PARENTING

There are four dominant ways individuals in same-sex partner households come to parent children: through a prior relationship with a different-sex partner that resulted in the birth of a child/children, through adoption, through the use of assisted reproductive technologies, or by becoming a partner to someone who has done one or more of these things....

The most common route to parenthood for lesbians and gay men is one in which a person has children through heterosexual contact before taking on a gay identity (Telingator & Patterson 2008). However, the increasing social and political acceptance of lesbian and gay relationships over the past 20 years has created new opportunities for people to create families within the context of their unions or as single parents through donor insemination, surrogacy, and/or adoption. These intentional families, also referred to by Biblarz & Savci (2010) as "planned families," represent a generational shift in lesbian and gay parenting. They have generated sociological interest for several reasons. First, they reflect the desire of sexual minorities to have children outside of heteronormative circumstances and, to some scholars (Dunne 2000, Stacey 2006), represent a radical shift in the meaning of parenthood and family. Second, sexual minorities, and particularly gay men, must navigate complicated and sometimes hostile institutions to have children and then raise them (Berkowitz & Marsiglio 2007). Third, differences in gender, race, class, and region constrain and enable how such families overcome these barriers and accentuate the inequalities among them. Finally, as these intentional families pursue greater legal recognition, judicial and legislative scrutiny has pushed scholars to interrogate the details of these families and, in particular, the outcomes for children raised in lesbian- and gay-parent households.

Adoption

The adoption of children by lesbians and gay men has been a controversial issue in the United States and in many other countries, primarily centering around the capabilities of lesbian and gay adults to effectively parent and the question of whether a heterosexual mother and father are the most appropriate models for children's development and gender socialization. In the early 1980s, the number of children in foster care increased dramatically, and the practice of screening adoption applicants changed as child welfare professionals began to rethink their notions of what constituted acceptable mothers and fathers for waiting children. The concept of a so-called suitable family expanded to include single parents, grandparents, parents of different ethnicities, lower-income families, and, since the 1990s in many states, lesbians and gay men (Esposito & Biafora 2007). Drawing from three national data sources, Gates (2013) estimates that same-sex couples are currently raising more than 25,000 adopted and foster children in the United States. Same-sex couples raising children are four times more likely than their different-sex counterparts to be raising an adopted child and six times more likely to be raising foster children. These figures significantly underestimate LGBT parent adoption because they do not account for single-parent sexual minorities who have adopted or are fostering a child. National survey data suggest that lesbian and gay adoptive parents share several demographic characteristics with heterosexual adoptive parents. Gates's (2011b) analysis of Census 2010 showed that although racial and ethnic minority same-sex couples were significantly more likely to be raising children, White same-sex couples were significantly more likely to have used adoption to have children. Same-sex couples with adopted children have the highest average annual household income of all adoptive family types, including heterosexual married adoptive couples (Gates et al. 2007).

Shapiro (2013) writes about the ways state laws govern LGBT-parent families. She argues that the key distinction in the law for parental rights in same-sex couple households is whether one or both adults are the legal parent of the child. Once recognition as a legal parent is attained, the rights of parents do not vary based on sexual orientation, but legal differences across states and varying judge's adjudications, especially in the case of custody determinations, can create uncertainty for families. Given that in some jurisdictions only the biological parent in a gay or lesbian intentional family is automatically recognized as a legal guardian of the child, many parents seek out a second-parent adoption in which the nonbiological partner is fully recognized as an equal parent (Polikoff 2008)....

Although second-parent adoption offers a more secure legal environment for families to raise children, it also affects custody when couples end their relationships and is not generally used in lesbian stepfamilies. Work by Gartrell and colleagues (2011) shows that second-parent adoption makes custody decisions more complicated for judges but results in a more equitable sharing of time between parents than in instances when only one parent was recognized as the legal guardian. Furthermore, the likelihood that a child reported feeling close to both mothers was significantly higher in families that had second-parent adoptions. Evidence from the law and society literature suggests that the granting of second-parent adoption is subject to the individual appreciation of judges and is highly variable across jurisdictions, which can exaggerate inequalities among sexual minority families (S.G. Mezey 2008).

Among couples seeking to adopt children through the foster care system or through private adoption, Shapiro says the variance in state law regarding the right of same-sex couples to legally marry affects the ability of these partners to attain legal parent status....

Transgender Couple Households

Analyzing relationships in which one person is transgender reveals a complex and nuanced dynamic of sexual and gender identity that distinguishes them from lesbian, gay, or heterosexual couples (Sanger 2010). Interviews with partners of transgender individuals show that they perform a great deal of "gendered labor" (Ward 2010) to help their partners achieve their desired gender both during and after transition. Women in relationships with transmen report doing more of the housework and emotional work such as nurturing their partner or managing his medical care and health advocacy (Pfeffer 2010). They justify the inequalities in the division of labor in terms of individual preference and choice that allow them to create a "family myth" of gender equality and maintain a feminist identity (Pfeffer 2010).

Transgender families are uniquely situated socially and institutionally because they can subvert or maintain legal and social norms. If they are perceived by others as ordinary heterosexual couples despite their own desire to transgress, they can use "normative resistance" to work against this "queer invisibility" by rejecting expectations of marriage and monogamy (Pfeffer 2012, p. 580). At the same time, they can also use "inventive pragmatism" to take advantage of the social and material resources of existing heteronormative structures such as legal marriage and parenthood (p. 587)....

Confronting Stigma

Despite the social changes for lesbian and gay couple households in recent decades, they still face obstacles in the legal system that pose barriers to their full recognition as families. In interviews with young adults raised by lesbian and gay parents, Robitaille & Saint-Jacques (2009) found that youths experienced both direct and indirect forms of stigmatization. They were more likely to have been teased or belittled because they had same-sex parents. They also reported feeling stigmatized when teachers and other adults discussed same-sex marriage or homosexuality in negative ways. Most of the adult children queried by Leddy et al. (2012) felt that lesbian families face both structural and interpersonal barriers to successful integration in society. Notwithstanding these impediments, they also reported receiving mostly positive responses from their peers about their families (though some reported negative or indifferent reactions). Of those who had experienced bullying because of their mothers' sexual orientation, they reported feeling hurt and angry, as well as embarrassed.

Faced with such experiences, lesbian and gay families create strategies to confront stigma. In addition to communicating closely with their children and teaching them how to talk about their family structures to others, they also create dense and diverse social support networks with families, friends, and especially other lesbian and gay parents (Peplau & Fingerhut 2007, Bos & van Balen 2008). These relationships allow them to have both material and psychological support to counteract the negative effects of homophobia. Children of lesbian and gay parents, especially adolescents and adults, develop their own techniques of dealing with the stigma associated with their parents' sexual orientation. For example, Leddy et al. (2012) found that young adults pursued several strategies including keeping quiet about their lesbian parents, speaking up and educating their peers when faced with negative comments, and, to a lesser degree, directly confronting instances of homophobia or seeking out support groups. Lick et al. (2012) linked county-level indices of social climate to psychological adjustment in individuals raised by sexual minority parents. Better outcomes were reported for children living in areas that were more supportive of LGBT populations and had antidiscrimination laws to protect sexual minority populations. These findings suggest a role that public policies and laws can play in reducing the effects of stigma for youth raised in LGBT-parent families.

… Finally, we stress that more attention needs to be paid to families on the margins: families of color, working-class families, transgender families, and households whose structures are outside the couple norm (i.e., poly-parent families). Yet beyond increasing the volume and centrality of empirical studies dedicated to these families, we suggest that the theoretical insights of scholars already studying them be expanded and developed. The understandings we gain from learning about Black lesbian families or families with transgender members, for instance, require us to rethink how we address the intersections of race, class, gender, and sexuality and have ramifications for the field of family studies more broadly. By interrogating the very categories of analysis that tend to dominate research on lesbian and gay families, this groundbreaking work gives scholars the opportunity to push forward a rich research agenda that avoids taking the meanings of family for granted.

LITERATURE CITED

Berkowitz D, Marsiglio W. 2007. Gay men: negotiating procreative, father, and family identities. *J. Marriage Fam.* 69:366–81.

Bernstein M, Reimann R, eds. 2001. *Queer Families, Queer Politics: Challenging Culture and the State.* New York: Columbia Univ. Press.

Biblarz TJ, Savci E. 2010. Lesbian, gay, bisexual, and transgender families. *J. Marriage Fam.* 72:480–97.

Bos HMW, van Balen F. 2008. Children in planned lesbian families: stigmatisation, psychological adjustment and protective factors. *Cult. Health Sex.* 10:221–36.

Dunne GA. 2000. Opting into motherhood: lesbians blurring the boundaries and transforming the meaning of parenthood and kinship. *Gend. Soc.* 14:11–35.

Esposito D, Biafora FA. 2007. Toward a sociology of adoption: historical deconstruction. In *Handbook of Adoption: Implications for Researchers, Practitioners, and Families*, ed. RA Javier, AL Baden, FA Biafora, A Camacho-Gingerich, pp. 17–31. Thousand Oaks, CA: Sage.

Gartrell N, Bos HMW, Goldberg N. 2011. Adolescents of the U.S. National Longitudinal Lesbian Family Study: sexual orientation, sexual behavior, and sexual risk exposure. *Arch. Sex. Behav.* 40:1199–209.

Gates GJ. 2011b. Family formation and raising children among same-sex couples. *Natl. Counc. Fam. Relat.* FF51:F2–F4.

Gates GJ. 2013. *LGBT parenting in the United States.* Rep., Williams Inst., Los Angeles. http://williamsinstitute.law.ucla.edu/wp-content/uploads/LGBT-Parenting.pdf.

Jepsen LK, Jepsen CA. 2002. An empirical analysis of the matching patterns of same-sex and opposite-sex couples. *Demography* 39:435–53.

Konnoth CJ, Gates GJ. 2011. *Same-sex couples and immigration in the United States.* Rep., Williams Inst., Los Angeles. http://williamsinstitute.law.ucla.edu/wp-content/uploads/Gates-Konnoth-Binational-Report-Nov-2011.pdf.

Leddy A, Gartrell N, Bos H. 2012. Growing up in a lesbian family: the life experiences of the adult daughters and sons of lesbian mothers. *J. GLBT Fam. Stud.* 8:243–57.

Lehr V. 1999. *Queer Family Values.* Philadelphia, PA: Temple Univ. Press.

Lick D, Tornello S, Riskind R, Schmidt K, Patterson CJ. 2012. Social climate for sexual minorities predicts well-being among heterosexual offspring of lesbian and gay parents. *Sex. Res. Soc. Policy* 9:99–112.

Mezey SG. 2008. *Gay Families and the Courts: The Quest for Equal Rights.* Lanham, MD: Rowman & Littlefield.

Moore MR, Brainer A. 2013. LGBT-parent families: innovations in research and implications for practice. See *Goldberg & Allen* 2013, pp. 133–48.

Peplau LA, Fingerhut AW. 2007. The close relationships of lesbians and gay men. *Annu. Rev. Psychol.* 58:405–24.

Pfeffer CA. 2010. "Women's work"? Women partners of transgender men doing housework and emotion work. *J. Marriage Fam.* 72:165–83.

Polikoff N. 2008. *Beyond (Straight and Gay) Marriage: Valuing All Families Under the Law.* Boston, MA: Beacon.

Powell B, Bolzendahl C, Geist C, Carr Steelman L. 2010. *Counted Out: Same-Sex Relations and Americans' Definitions of Family.* Ithaca, NY: CUP Serv.

Robitaille C, Saint-Jacques M. 2009. Social stigma and the situation of young people in lesbian and gay families. *J. Homosex.* 56:421–42.

Rosenfeld MJ, Kim B. 2005. The independence of young adults and the rise of interracial and same-sex unions. *Am. Sociol. Rev.* 70:541–62.

Sanger T. 2010. *Trans People's Partnerships: Towards an Ethics of Intimacy.* Basingstoke, UK: Palgrave Macmillan.

Shapiro J. 2013. The law governing LGBT-parent families. See *Goldberg & Allen* 2013, pp. 291–304.

Stacey J. 2006. Gay parenthood and the decline of paternity as we knew it. *Sexualities* 9:27–55.

Stein A. 2005. Make room for daddy: anxious masculinity and emergent homophobias in neopatriarchal politics. *Gend. Soc.* 19:601–20.

Telingator CJ, Patterson CJ. 2008. Children and adolescents of lesbian and gay parents. *J. Am. Acad. Child Adolesc. Psychiatry* 47:1364–68.

31

The Good Daughter Dilemma
Latinas Managing Family and School Demands

ROBERTA ESPINOZA

Understanding the "good daughter dilemma" that Latinas pursuing higher education face requires a discussion of the cultural value of *familismo*, which emphasizes loyalty, reciprocity, and solidarity (Vega, 1990). *Familismo* includes strong identification and attachment to the family, both nuclear and extended, and requires members to prioritize family over individual interests. Unlike dominant U.S. culture that values independence and self-sufficiency, *familismo* emphasizes cooperation and interdependence.

Latinas with a strong sense of *familismo* have a schema of obligations and reciprocities that is important for them to fulfill in their youth and adult lives. The cultural template of expected obligations in Latina/o families most often cited include language/cultural brokering, sibling caretaking and financial contributions. Latina/o family obligations include spending time with family and staying close to home. These obligations may hold special significance for females who are more often socialized into a caretaker role within Latina/o families with a strong *familismo* orientation. Studies examining the obligations Latina/o children fulfill consistently find that the responsibilities are more likely to fall on girls than boys. The cultural value of *marianismo*, for example, modeled on the Catholic Virgin Madonna, prescribes dependence, subordination, responsibility for domestic chores, and selfless devotion to family. This creates an expectation that the "good Latina woman" will always prioritize family needs above her own individual needs. However, there is a significant contradiction embedded in this cultural template for Latinas with a strong sense of *familismo* who pursue higher education because the time dedicated to school directly competes with time available for family. Parents who still expect daughters to continue contributing to the family while pursuing higher education places a demand on her to fulfill multiple, and often competing, obligations at the same time. As a result, Latinas find themselves caught in a cultural bind between meeting the demands of their individualistic-oriented school culture and their collectivist-oriented family culture.

The double-edged sword Latinas with a strong sense of *familismo* face is that their connections to family, which undoubtedly compete with school, gives them a sense of belonging which they draw on to do well academically. Latinas with

SOURCE: Espinoza, R. 2010. "The Good Daughter Dilemma: Latinas Managing Family and School Demands." *Journal of Hispanic Higher Education*, 9(4), 317–330.

a high collectivist orientation often find the individualistic culture of school alienating prompting them to maintain strong ties to family. Although their obligations to family may conflict with school, those connections are very important to their ability to get through successfully. In fact, research has found that family obligations which tie ethnic minority youths to their families have positive academic consequences. Hardway and Fuligni (2006) found that Latina/o adolescents were more likely to persist in postsecondary education if they had a stronger sense of family obligation. Latina/o students who placed a great value on their role in the family performed better in school indicating that their connection to family helped them focus on their academics. Other studies have found that strong family ties are important not only for academic success but also in the adjustment of Latinas/os to higher education environments.

Although feeling strongly connected to family may provide a sense of emotional well-being for Latinas, it is not clear how the obligation to regularly engage in behaviors that support the family affect Latina students' adjustment to and success in higher education. The requirement to fulfill such family obligations competes with the time demands of their educational pursuits, making the transition to higher education more challenging....

BICULTURALISM

Biculturalism theory helps us understand how Latinas manage the conflicts and tensions between two different cultures. Accordingly, a bicultural person is competent in two cultures, engages in typical behaviors of both cultures, embraces the opportunity to remain involved in practices and lifestyles of both cultures, and feels a sense of belonging to both cultural communities. Levels of biculturalism, however, can vary from individual to individual. Berry (1990) has suggested that the position of ethnic minorities can best be described in terms of two independent dimensions: (a) retention of one's cultural traditions and (b) establishment and maintenance of relationships with mainstream society. Those who retain the traditions and values of their culture of origin and also develop and maintain identification with the larger society are said to be "bicultural."

In behavioral terms, bicultural individuals are competent and active within the contexts of both their native and new culture in which they are immersed. These individuals have an extensive role repertoire referring to the range of cultural or situational appropriate behaviors they have developed from being exposed to two cultures (LaFromboise, Coleman, & Gerton, 1993). At the high end of the bicultural scale, an individual is called a "blended bicultural," and at the low end of the scale, an individual is considered an "alternating bicultural" (Phinney & Devich-Navarro, 1997). A blended bicultural blends or integrates their two cultures, whereas an alternating bicultural switches between their two cultures keeping them separate. Although the biculturalism framework helps understand how Latinas negotiate the macro structures of culture, Chicana Feminism informs the daily social ethnic identity management from a more micro perspective.

CHICANA FEMINISM: MULTIPLE IDENTITIES

Chicana feminist theory highlights that a necessary survival skill when living between two cultures is learning how to maintain a distinct ethnic or cultural identity while at the same time learning to adapt to the dominant mainstream culture (Vera & de los Santos, 2005). In her theoretically groundbreaking book entitled *Borderlands: La Frontera*, Chicana feminist Gloria Anzaldúa (1987) proposed a third, hybrid identity that develops from the process of constantly straddling two cultures in everyday life called the *mestiza* identity, "the new mestiza copes by developing a tolerance for contradictions, a tolerance for ambiguity... she learns to juggle cultures." (p. 79). According to Anzaldúa (1987), the mestiza identity, which can take on many unique forms, is developed to manage two cultures that are always in direct conflict with one another. The term *mestiza* has come to mean living with ambivalence while balancing opposing powers (Delgado Bernal, 2001).

Anzaldúa (1987) describes the conflict and tension generated by the polarities experienced when one is torn between the needs of the home or ethnic culture and the demands of the Anglo world such as that which resides within educational institutions; the dual nature of identity in the border culture means learning two ways of thinking, speaking, and sometimes two distinct languages. The duality emerges when Chicanas exist in the all-too familiar states called "'belonging' and 'not belonging'" (Elenes, 1997, p. 363). The mestiza identity is characterized by a resiliency that allows the Chicana to shift in and out of habitual formations and movement from a single goal to divergent thinking (Anzaldúa, 1987). Cherrie Moraga (1986) also explores the issues of multiple and fluid identities that are based on self-knowledge by asserting the possibility of changing or transforming one's identity and actively maintaining more than one at any point in time.

These important feminist theories elucidate the borders Chicana/Latina women are forced to cross on a daily basis, constantly shifting in and out of different social contexts with diverse gender expectations to which they adjust.... Informed by the bicultural and *mestiza* identity frameworks, this study examines the integrator and separator strategies Latina doctoral students develop to manage and balance family relationships with school demands.

METHOD

Participants

Fifteen Latina doctoral graduate students who attend universities in Northern California were interviewed. Only women who were at least in their second year of full-time graduate study were selected. The average age of participants was 27 years. Eleven of the women were in the Social Sciences and four were in the Humanities. Twelve participants were Mexican American/Chicana, one was Puerto Rican, one was Ecuadoran, and one was Salvadoran. Twelve participants were either first- or second-generation in the United States (i.e., either they themselves or both of their parents were born in a Spanish-speaking country) and three were fifth- or above generation (i.e., both parents were born in the United States). Thirteen participants were first-generation in their families to attend college and two had at least one parent with a bachelor of arts degree. None of the participants were married nor had children at the time of the interview. All participants reported growing up in families with strong *familismo* orientations which included high levels of familial household responsibilities, expectations of spending time with family over friends, sibling caretaking, doing household chores, and language/cultural brokering.

Procedure

Participants were recruited from various university-sponsored social events including students of color receptions and student list serves. Once an interest in participating in the study was expressed via email, a time and location for the interview was scheduled. A snowball sampling methodology was utilized by asking participants to refer women they knew matched the research criteria on completion of their interview. Interviews lasted on average 80 minutes and were audiorecorded and fully transcribed. After each transcript was reviewed, a few participants were selected for a follow-up interview to get clarification on responses that needed further elaboration or were not clear due to poor audio quality. All participants have been assigned a pseudonym to protect their identity.

Interview Protocol

Participants were first asked questions about early childhood experiences growing up, then about relationships with family as graduate students. Many of the interview questions focused on family responsibilities and care (e.g., caring for others and being cared for). Specific questions that were asked included, "When you were growing up, what messages did you receive from your family about how you should care for them? What was expected of you when you were young? What was your role in your family growing up?" Participants were also asked about their relationship with their parents in graduate school compared to their childhood, with questions including, "What is your role in your family now? How has your relationship with your family changed since you started graduate school? What has been the biggest change? Has the contact with your family (phone or physical) decreased, increased, or stayed the same since you started graduate school?"

A careful systematic analysis of the transcribed interviews was conducted by two different researchers for validity. Each interview transcript was read in-depth for emerging themes that fell under two main topics: family relationships and responsibilities growing up, and family relationships and responsibilities during graduate school. After all themes under each topic were identified, they were compared across interviews for commonalities, then combined into major thematic categories from which two emerged; integrators and separators.

FINDINGS

Latina doctoral students balanced the demands of school and family in graduate school in two different ways. One group of nine women, the *integrators*, managed family expectations and obligations by explicitly communicating with family members about their school responsibilities. Integrators blended family and school by first explaining the nature of their school demands, then enlisting their family's support to enhance their academic success. The second group of six women, the *separators*, actively organized their daily lives to keep family and school separate to minimize tension and conflict. Although the separators prioritized family similar to their integrator counterparts, they often felt they had to keep their schooling experiences separate to protect their relationships with family members. Both patterns demonstrate high levels of biculturalism and fluid *mestiza* identities that helped them juggle the contradictions of their two distinct social worlds as they pursued educational advancement.

The Integrators

Veronica's decision to attend graduate school 500 miles away from home defied an important expectation of being a good daughter—to stay close to home. The way she explained her decision to her parents reflects the specific strategies many integrators employed. Veronica sat her parents down one Saturday afternoon and explained to them that she had been offered an amazing opportunity to attend one of the most prestigious schools in the nation for her doctorate. In her explanation, she mentioned key details that she knew her parents needed to hear. She told them that the university was giving her a multiple year fellowship to pay for her studies and that earning an advanced level education was going to benefit the entire family. Veronica also explained that taking this next step in her education was going to allow her to fulfill their dreams of a highly educated daughter. After all, the main reason her parents immigrated to the United States was to provide to their children educational opportunities they never had in El Salvador. Veronica skillfully clarified the process to her parents and related it to their expectations of her as a good daughter. Although she knew her parents "did not fully understand what doctoral studies entailed," after her explanation they trusted she was making a good decision.

Veronica continued to successfully manage her relationship with her family throughout graduate school. The strategy of communicating to parents detailed information about heavy school workloads and deadlines was typical of the women who fell into this category. Integrators were very explicit with their families about the demands placed on them at school when they knew it directly conflicted with family expectations that might lead to the perception of being a bad daughter. Veronica recounted a situation her third year of graduate school when a critical school deadline kept her from returning home for the holidays:

> My parents always expect me to come home for the holidays even the small ones like Thanksgiving, but last year I couldn't go because

I needed the time to finish my MA paper which was due by the end of the term. When my mom called and asked if I would be coming home, I had to tell her that I couldn't come home for Thanksgiving because I had school work to do. She was very disappointed and kept trying to convince me to come home. I explained to her that the reason I couldn't come home was because I needed the extra time to work on my paper. I also explained to her that if I didn't finish the paper by the end of the term, I would not get my MA degree until May. It was a hard situation for both of us, but eventually she understood.

Veronica explained to her mother why in this one instance she was prioritizing school over family. She told her mother that although she could not make it home for Thanksgiving, she would be home in 3 weeks for winter break which she felt eased her mother's concerns. Veronica, like other women in this category, made concerted efforts to have discussions with their families about school to introduce them to their school lives which is consistent with a highly bicultural orientation and strong *mestiza* identity.

For integrators, having good communication with their families allowed them to more easily manage their good daughter reputations which were of great importance to them. Dolores described how she tried to be up-front with her parents who often expect her to come home every weekend by saying things such as, "I would love to come home right now, but instead of coming home Friday, Saturday, and Sunday, can I just come home on Sunday and we'll spend the whole day together?" Doing this consistently over time has allowed her to establish "really strong communication" with her parents in which "… they trust that I know what I'm doing and that I want good things for myself and will only do things that are healthy." Learning to balance family expectations and school demands by explaining her educational world to her parents, however, has been a learning process for Dolores. When she was an under-graduate student, she did not explain school to her parents, which resulted in being "cut-off from the family for six *long* months." Since her parents did not understand her school schedule, and because at that time she did not explain it to them, her parents jumped to conclusions calling her a "*muchacha librada*," meaning "a woman with loose morals," which equated to being a bad daughter. Dolores' experience highlights the fluidity of her *mestiza* identity, and overtime, found ways to reconcile her two social spheres of school and family by drawing on her bicultural orientation.…

Integrators explained school demands to their families to not only negotiate family expectations but also enlist their families' support in their school endeavors. Kathy, for example, recalled how when she was studying for her oral qualifying exams, she explained to her mother how stressful the process was and, as it got closer to taking her exams, called on her mother for emotional support. After doing this, her mother came up from Southern California to take care of her the weekend before her exams:

I've learned how to ask for support from my family. I've been able to call and let them know that I need their support, that I need them to call me. Like I asked my mom to come the weekend before my orals just to be here with me and that was just very revolutionary for me… it was just an amazing turning point for us… and she came and made me like little cupcakes and made me sopa de pollo (chicken soup) for the week so I would not have to cook and it was so wonderful… it was like okay there is a different way of being a graduate student and it could incorporate your mom.

Getting support from her family is how Kathy integrated family with being a doctoral graduate student. Similar to Dolores, being an integrator is something Kathy has learned to do over time as she pursues higher education. In college, she explained that her mother would see the "aftermath of a stressful semester" when she would come home and sleep for days and her mother "didn't understand what got her there," which always caused many misunderstandings. Now she realizes, "learning how to get support during the process [of graduate school] is important." Thus explaining school to her parents has allowed Kathy to merge her identities as a good student and a good daughter; illuminating the unique *mestiza* identity she has developed to integrate family and school.

The Latinas who used this strategy are highly bicultural in their ability to integrate family and school demonstrating mastery of appropriate role behaviors in both their collectivist Latino home culture and their individualistic Anglo-school culture. Not only do they negotiate conflicts when they occur but they also manage their ethnic identity within each cultural context. As noted by *mestiza* identity theorists, integrators negotiate cultural contradictions by actively using new knowledge from their everyday experiences to transform and fuse their social identities. The integrator strategy often blends being a good daughter with being a good student.

The Separators

Another strategy Latina doctoral students described to balance family relationships with graduate school demands is by keeping these two social worlds away from one another. Rosa is an example of an individual who managed family and school roles exceptionally well by keeping them completely separate. One of the reasons Rosa chose to keep family and school separate is because she felt they clashed as she eloquently explained,

Maybe you can think of graduate school in terms of like a culture clash, right where you've got this American individualism thing where you're supposed to care about yourself and do what you can to move yourself forward in school, and family kind of disappears from that, as well as family obligations and needs. I think grad school is very, I feel like grad school fits into that right, you're kind of seen as an individual and you're not supposed to have too many outside forces pulling at you, but then I've also got this family obligation and I don't feel it's a bad thing to say it's obligation, I mean it's a good thing right, because we support each other, but I feel kind of like this pull between doing what most

Americans do and just kind of doing what I need to do to get ahead, and then also at the same time dealing with something that's been instilled in me always, you know your family comes first and they're very important.

Rosa felt that family responsibilities were not legitimated by institutions of higher education since the expectation to prioritize her studies and research above all else was very evident. However, family is an important part of Rosa's life and they often count on her to meet family obligations such as attending life events (i.e., birthdays) and lending support when needed (i.e., helping siblings with homework). She explained a situation in her first year of graduate school where she was forced to choose between family and school:

So if my two younger siblings have trouble with their homework and my mom does not understand and my dad is not home, my mom will call me and she'll be like last year when I was doing Professor Smith's midterm I got a call at like about 10:00 o'clock at night and she [her mother] was like "your brother can't put adjectives in these sentences, can you help him?" and I was like okay… and then I did, I helped him.

The decision for Rosa in this instance was easy, family before school, a value tied to her sense of *familismo*. In fact, she did not even explain to her mother that she was right in the middle of working on a midterm exam due the next day. Instead, she just put her school work aside, helped her brother, and resumed working on her midterm when she got off the phone with him. This example illustrates how Rosa keeps family separate from school, but still manages to actively prioritize family. Rosa felt that doing this would minimize conflict, which was consistent with her notion of being a good daughter.

Other women who fell into this category did not explain school to their families out of concern and to protect them from worrying. Although Celia has a strong relationship with her mother, she did not ever mention to her in their weekly phone conversations the many stressful aspects of graduate school. Part of the reason Celia did not talk to her mother about school is because she felt guilty for moving far away to attend her graduate program. Celia moved across the country for graduate school and felt that she had already defied being a good daughter because she did not stay close to home. She stated, "There's a lot of sort of guilt feelings because of the distance and sort of thinking about… I think that my family is sort of confused about why I need to be so far away for school." Unlike Veronica who explained to her parents why she was moving away for graduate school, Celia did not discuss the decision with her mother to minimize the stress and conflict. Instead, Celia managed her good daughter reputation by keeping her roles within the contexts of family and school separate so as not to worry her family further.

As a separator, Luciana also alternated between school and family to maintain her status of a good daughter. When her family needed her at home, she did not hesitate to interrupt her school work to make herself available to her family. Recently, when her father had surgery she stopped working on her dissertation and made time to go down to Los Angeles to be with him. She explains, "My father just had surgery this past summer and I went down for the surgery. I went down to be with him during and after the surgery and sometimes it is hard to find time to get down there, but I do." Family is Luciana's first priority and she behaved in ways that demonstrated this to her family. She always made a concerted effort to be physically present when needed at home for critical life events (i.e., health issues) to provide support (i.e., emotional, financial) and show her commitment to strong collectivist family values. Like all separators, Luciana sees the clear incongruence between her graduate program and family expectations. As such, she actively employs a strategy for balancing them by keeping them in separate spheres of her life.

Unlike the integrators, the separators' *mestiza* identity variant is characterized by a "good daughter" role that is separate from the "good student" role. The separators manage the cultural ambiguities of their social contexts by adapting to role behaviors differently in each environment as a result of their high levels of biculturalism. While the integrators communicate and explain to family the expectations of them in the two worlds in which they are immersed—family and school—the separators employ behaviors that keep them compartmentalized and separate. Each approach reflects two variations of their distinct cultural and ethnic identity location along the Latino and American cultural continuums.

IMPLICATIONS FOR POLICY, PRACTICE, AND RESEARCH

The findings from this study have implications for institutions of higher education to support the educational pursuits of Latinas.... To enhance academic success and persistence, support efforts must include institutionally sanctioned and formalized ways to educate faculty and administrators about Latinas family–school dilemmas.

First, universities should improve their outreach efforts to inform Latina students of the various support services available to them even before they start graduate school. It is imperative that they are immediately connected with supportive organizations, educators, and peers that can help them adjust to their new school environment. Institutions and departments need to be proactive in providing Latinas with information and experiences that make them feel a sense of belonging and connectedness to their new academic homes. These efforts will greatly diminish the balancing act that Latinas engage in, thus freeing them to better engage in school and their departments.

Second, faculty who work the closest with Latina students need to attend informational workshops that highlight the challenges they face on entering the university. It is important for educational agents to understand the positive impact they can have as academic role models by helping Latinas through the various hurdles of graduate school while simultaneously legitimating their *familismo*. Faculty often overlooks having a life outside of school, which alienates students who have other obligations and responsibilities outside the university....

CONCLUSIONS

Latina doctoral candidates who grow up with a strong *familismo* cultural orientation highly value being a good daughter throughout their lives while at the same time they aspire for academic success at the highest educational level. In their pursuit of higher education they may experience the pressure of fulfilling multiple, and often competing roles. In an effort to maintain their statuses of both a good student and good daughter, some Latinas draw on their *mestiza* identity and bicultural orientation to navigate the cultural bind they face. To effectively navigate the two social spheres of family and school, Latinas may employ strategies of integration and separation as described in this study. Educational institutions must consider these processes as they develop and implement better support programs for Latinas to ensure their overall educational success.

REFERENCES

Anzaldúa, G. E. (1987). *Borderlands/la frontera: The new mestiza.* San Francisco: Aunt Lute.

Berry, J. (1990). Psychology of acculturation. In J. Berman (Ed.), *Cross-cultural perspectives: Nebraska symposium on motivation* (pp. 201–234). Lincoln: University of Nebraska Press.

Delgado Bernal, D. (2001). Learning and living pedagogies of the home: The Mestiza consciousness of Chicana students. *International Journal of Qualitative Studies in Education, 14,* 623–639.

Elenes, C. A. (1997). Reclaiming the borderlands: Chicana/o identity, difference, and critical pedagogy. *Educational Theory, 47,* 359–375.

Hardway, C., & Fuligni, A. J. (2006). Dimensions of family connectedness among adolescents with Mexican, Chinese, and European backgrounds. *Developmental Psychology, 42,* 1246–1258.

LaFromboise, T. D., Coleman, H. L. K., & Gerton, J. (1993). Psychological impact of biculturalism: Evidence and theory. *Psychological Bulletin, 114,* 489–535.

Moraga, C. (1986). From a long line of vendidas: Chicanas and feminism. In T. De Lauretis (Ed.), *Feminist studies/critical studies* (pp. 27–34). New York: Women of Color Press.

Phinney, J. S., & Devich-Navarro, M. (1997). Variations in bicultural identification among African American and Mexican American adolescents. *Journal of Research on Adolescence, 7*(1), 3–32.

Vega, W. A. (1990). Hispanic families in the 1980s: A decade of research. *Journal of Marriage and the Family, 52,* 1015–1024.

Vera, H., & de los Santos, E. (2005). Chicana identity construction: Pushing the boundaries. *Journal of Hispanic Higher Education, 4*(2), 102–113.

32

Loving Across Racial Divides

AMY STEINBUGLER

"I think that people in interracial relationships give up something, you know, give up an ease about living, in some ways," said Leslie Cobbs, a 30-something white woman, surrounded by novels, textbooks, and old photographs, in the Brooklyn apartment she shares with her black partner Sylvia Chabot. Leslie acknowledges that African Americans have to "think about race all the time," but insists that there are unique racial issues that stem from being in an interracial relationship.

Sylvia and Leslie face challenges that are usually less overt than those black/white couples would have confronted 50 years ago. There is little chance that they might lose their jobs, get kicked out of their church, or be denied housing simply because one of them is black and the other is white. While injustices like these still occur, when they do they are noteworthy. Undisguised discrimination against interracial couples is no longer typical. Nor do relationships like theirs inspire the raw disbelief that Sidney Poitier and Katharine Houghton famously elicited when they portrayed a young couple in the 1967 film, *Guess Who's Coming to Dinner*.

Still, Sylvia says, "We're always looking for spaces where we can be together as a couple, [be] validated, feel comfortable." Compared to straight interracial pairs, same-sex partners like Sylvia and Leslie can count on far fewer legal protections, and also are vulnerable to homophobia from coworkers, family members, and strangers on the street.

Interracial couples in the United States have always attracted public scrutiny. Until the mid-twentieth century, this attention was almost entirely negative. Legal sanctions prevented people of African, Asian and, sometimes, Native American descent from marrying whites. Black/white relationships, especially those between white women and black men, drew the harshest condemnation. Black communities treated such couples as disreputable; white communities often threatened, physically harmed, or ostracized them.

In recent years, interracial couples are more likely to encounter hope than censure, at least in terms of public discourse. Some observers liken current legal prohibitions against same-sex marriage to anti-miscegenation laws before the 1967 Supreme Court ruling in *Loving v. The State of Virginia*. Social commentators paint contemporary interracial marriage as a victory for equality and freedom. A 2001 *Time* magazine article celebrated interracial unions as representing

SOURCE: Steinberger, A. 2014. "Loving Across Racial Divides." *Contexts*, 13(2), 32–37. Copyright © 2014 by the American Sociological Association. Reprinted by permission of SAGE Publications, Inc.

an intimate "vanguard" who "work on narrowing the divisions between groups in America, one couple at a time."

More recently, a Pew Research Center report released in 2012 suggests a positive shift in public attitudes towards intermarriage. Forty-three percent of Americans now view the trend for more people of different races to marry each other as a change for the better. About two-thirds say it would be fine with them if a family member "married out" of their racial or ethnic group. Even black/white relationships, which have long elicited the fiercest disapproval and the strongest legal sanctions, are becoming more acceptable.

Despite the supposed acceptance of dating and marrying across racial lines, only a small percentage of people in the United States—according to the 2010 Census, less than 7% of all heterosexual married couples—actually do so. Among gay and lesbian couples, approximately 14% are interracial—about the same proportion as among heterosexual unmarried partners.

Low intermarriage rates notwithstanding, many people embrace the popular notion that Americans have truly become "colorblind." But racism is more than just a matter of prejudice; liberalizing racial attitudes coexist with the stubborn persistence of racism. For the past 12 years, I have studied couples who love across racial difference in Philadelphia, New York, and Washington, D.C. What I've found is that while hostility toward interracial pairs, like racism itself, has become more subtle, race continues to powerfully impact everyday life for Sylvia, Leslie, and the other 39 interracial couples I interviewed. Racism, manifested in neighborhood segregation and racial self-understandings, shapes everyday life, creeping up in the most ordinary circumstances, like walking through their neighborhood, or deciding where to get a drink.

RACED SPACES

Mary Chambers, a heterosexual woman of Afro-Caribbean descent, knew that her husband Neil was sometimes uncomfortable in their middle-class, majority-black neighborhood. The neighborhood, which includes many sprawling, three-story, Tudor houses, feels suburban, though it is located in a small city less than 30 miles from Manhattan. Mary thinks that Neil "would prefer to live in a community where he's more comfortable with the people, in a community where he can look around and see his own race. [One] with more white people." She finds this discouraging. "I wish that I could change his perspective on it and really make him see that the community we live in is valuable."

In the years Neil spent in this neighborhood, he became very aware of his own whiteness. It was impossible for him not to think about race in everyday social interactions. It was typical, he says, to go into the grocery store or the post office and be "the only white guy there." He continues, "It's not a bad thing, you know. It's not like I feel like I'm going to get mugged or, um, I'm going to get hurt. But you have to understand, it's kind of like when you deal with white people, they have their prejudice—they have what they're used to. And then when you deal with black people, it's really the same thing, you know?"

The residential racial segregation of blacks and whites has been slowly declining nationwide. But in northeastern cities like New York and Philadelphia, highly segregated neighborhoods remain the norm. (In New York, the racial composition of neighborhoods is so lopsided that 79% of blacks would have to move in order to achieve a balanced distribution—in which the percentage of blacks in every neighborhood mirrors their share of the city's total population.)

Neighborhoods that are black or white often pose problems for interracial couples because they set the stage for situations in which one partner feels uncomfortable or conspicuous. "No matter where we live," one white woman lamented, "one of us is not going to be in the right neighborhood." A black partner agreed, "As diverse as [New York] city is, to me it's still pretty segregated." This sense of belonging or not belonging is something interracial partners often brought up as we talked.

Neighborhood divisions are stressful when one partner feels conspicuous and has to look out for racial undercurrents in everyday social interactions. Such divisions create racial fatigue, though they affect black and white partners differently. For Neil Chambers and other whites in my study, being in the racial minority feels awkward because it happens so rarely. Noticing one's own whiteness is a new experience that can prompt an unsettled feeling. One of the taken for granted privileges of being white is the tendency to think of yourself not as a *white* person—*just* as a person. I asked one white woman who is married to a black man how often she thinks of herself as white. "I don't," she said. "Well, maybe if I were in an all-black environment, and I'm the only white person there. That's the only time."

Black interracial partners also noticed when they were among only a handful of blacks in a neighborhood or social gathering. But for these middle-class Americans, that experience was not uncommon. Many worked or had gone to school in majority-white environments. Compared to the whites in my study, black partners tended to be much more accustomed to being in the numerical minority. In contrast to the unease of white partners, who sometimes felt intimidated in black neighborhoods, the discomfort of black partners was linked to a history of violence against their racial group. It was one of many instances in which black and white partners perceived race very differently.

RACIAL ORIENTATIONS

Tamara is white and Scott is black. When this 30-something unmarried straight couple decided to move in together, they needed to transport countless boxes of books and clothes from Tamara's place in Philadelphia to a nearby city. Tamara wanted Scott to drive the SUV she had borrowed from a friend. But like many other black men in the United States, Scott was concerned about racial profiling.

He didn't want to get pulled over driving a borrowed car, especially given that his cell phone wasn't working properly.

Scott recalls telling Tamara, "'I'd really rather you drive... because when—if [I] get pulled over... I can't dial [the woman who owns the car]. Now it's just me and some cop and he's probably going to treat you better than he's going to treat me.'" For Scott, this was a routine calculation. For Tamara, it didn't seem like a big deal. This reflects a broader disjuncture in how—and how often—each of them thinks about race. Scott, continuing the story of the move, tells me, "It's my job to consider that, whereas...I don't think [me getting pulled over] is automatically something that she considers—and you know what? It doesn't bother me that it isn't, because how could it be? Someone [who is] Black can really, really think, like could have their mind go to that."

For Scott, anticipating everyday acts of prejudice and discrimination is second nature, ever since his grandmother told him about the lynching of Emmett Till. But Tamara, whose whiteness has shielded her from being the target of racial animus, is only now learning to consider how the accumulation of a lifetime of racial experiences informs even small decisions, like who will drive an SUV full of books and clothes 30 minutes away.

Daniel, who is black, and Shawn, who is white, thought their racial orientations were very similar until they became the adoptive parents of two black boys. When Daniel and Shawn began to talk about their sons' future schooling, they soon discovered that they conceptualize racism differently. "We both share the same basic political views," Shawn explained. But Daniel, he believed, "subscribes to a kind of conspiracy theory that white America has banded together to exclude black America." Shawn sees the racism, he says, but "I don't see it as being organized in the same way. Because I'm white and nobody ever came to me and said, 'Hey, lets get together and do this thing to the black people.' So that's a difference of philosophy." Daniel sees "[racism] as institutionalized and I see it just as kind of widespread."

Daniel and Shawn's perspectives lead to important differences in dealing with racism. While Shawn has no intention of letting his sons get hurt, he is more comfortable taking a wait-and-see approach. Daniel feels more strongly about the need to be proactive about racial discrimination at school. "I'm not looking for trouble," says Daniel. It's just that "I don't want to be asleep when that stuff happens."

For Daniel, there are limits to what a white person can understand about discrimination. Shawn, he says, "doesn't expect that kind of behavior and it's hard [for him] to believe that in fact that can happen." He "just doesn't believe that a teacher would look at an eight-year-old and actually treat one eight-year-old differently from another simply because of the color of their skin." While both men are deeply invested in protecting their sons from racism, their conflicting racial orientations to the subject remain unresolved.

Shawn's strategies for parenting black sons and Tamara's skepticism about racial profiling reflect the attitudes of many white Americans who question the scope and severity of contemporary racism. Surprisingly, white partners' intimacy with black people did not substantially challenge their racial perspectives, casting doubt on the notion that interracial partners represent an enlightened, "post-racial" vanguard.

STEREOTYPES—AND EXCEPTIONS

Gary, 54, and his wife Soonja, 58, met in Korea. They have been married for 20 years. Gary chose to marry a Korean woman because, he said, a "good" wife should be loyal and subservient, and to him, Soonja's race signifies these traits. Soonja, too, believes that her choice of husband may reflect upon the kind of person she is. She chafes at being associated with what Koreans regard as stereotypical "international marriages," temporary sexual relationships between local, uneducated Korean women and American military men.

Gary's whiteness and American citizenship did not hold any special appeal for Soonja or her family. But over time, part of what has made their marriage work is that Soonja believes there are distinct cultural differences that make American husbands better partners than Korean husbands. "I'm glad that I didn't marry a Korean, who ignores his wife, drinks a lot, and comes home late." Gary confides that many of his friends "actually say that they envy me because they understand that Asian women are very good wives and…good mothers."

Sociologist Kumiko Nemoto has researched marriages between whites and Asians in the United States. White men, according to Nemoto, commonly associate Asian and Asian American women with family and domesticity. Younger Asian women in interracial relationships, she found, are more likely to define themselves as egalitarian, ambitious, and aesthetically (as opposed to domestically) feminine, challenging these stereotypes.

Vivian, 25, who grew up in a Chinese family is attracted to Peter, 27, in part because she sees them as equals. He is intelligent, got good grades in college, and is economically mobile. "We have a mental connection," she says. Peter's professional ambitions and work ethic help Vivian think of herself as a modern Asian American woman. Peter, for his part, is proud to appreciate beauty that falls outside of normative white femininity: "[I'm] more attracted to ideas and people who are more exotic…I think Asian features are prettier than white features." He likes "darker-skinned women" and the "shapes of Asian eyes."

In the U.S. racial order, particular Asian groups (such as Japanese and Chinese) are positioned much closer to whites than blacks. Asian Americans certainly experience racial discrimination and the false presumptions embedded within the idea of the "model minority." Even so, some Asian groups are increasingly seen as what sociologist Eduardo Bonilla-Silva calls "honorary Whites." As the 2010 Census shows, intermarriages between Asian Americans and whites are far more common than those between blacks and whites.

Couples in my study also used racial-gender stereotypes about their partners to describe themselves. Still, some whites were careful to portray their partner as exceptional, rather than a typical example of their racial group. As Neil said of Mary, who is of African American descent: "She's not someone who would curse or, you know, say anything that's inappropriate or off color. Not that she's a saint but—she has a certain background—she's not offensive to you. She's very pleasant."

What this and other examples suggest is that the prejudice interracial couples encounter from strangers is only one small part of how race shapes their everyday lives. Race is a social system that shapes neighborhoods, orientations, and identities, and plays a critical role in intimate relationships. Despite the gains of the Civil Rights movement and the historic election of the first black president, the racial categories we are assigned to at birth have tremendous material consequences for how our lives unfold. Racial inequalities affect the wealth which we have access to, the neighborhoods we live in, the type and amount of the healthcare we are able to get, and the quality of our children's schools. Even in this supposed post-racial moment, our position in the racial system shapes the way we see the world.

33

From the Achievement Gap to the Education Debt
Understanding Achievement in U.S. Schools

GLORIA LADSON-BILLINGS

One of the most common phrases in today's education literature is "the ... achievement gap." The term produces more than 11 million citations on Google. "Achievement gap," much like certain popular culture music stars, has become a crossover hit. It has made its way into common parlance and everyday usage. The term is invoked by people on both ends of the political spectrum, and few argue over its meaning or its import. According to the National Governors' Association, the achievement gap is "a matter of race and class. Across the U.S., a gap in academic achievement persists between minority and disadvantaged students and their white counterparts." It further states: "This is one of the most pressing education-policy challenges that states currently face" (2005). The story of the achievement gap is a familiar one. The numbers speak for themselves. In the 2005 National Assessment of Educational Progress results, the gap between Black and Latina/o fourth graders and their White counterparts in reading scaled scores was more than 26 points. In fourth-grade mathematics the gap was more than 20 points (Education Commission of the States, 2005). In eighth-grade reading, the gap was more than 23 points, and in eighth-grade mathematics the gap was more than 26 points. We can also see that these gaps persist over time (Education Commission of the States).

Even when we compare African Americans and Latina/os with incomes comparable to those of Whites, there is still an achievement gap as measured by standardized testing (National Center for Education Statistics, 2001). While I have focused primarily on showing this gap by means of standardized test scores, it also exists when we compare dropout rates and relative numbers of students who take advanced placement examinations; enroll in honors, advanced placement, and "gifted" classes; and are admitted to colleges and graduate and professional programs.

Scholars have offered a variety of explanations for the existence of the gap. In the 1960s, scholars identified cultural deficit theories to suggest that children of color were victims of pathological lifestyles that hindered their ability to benefit from schooling (Hess & Shipman, 1965; Bereiter & Engleman, 1966; Deutsch, 1963). The 1966 Coleman Report, *Equality of Educational Opportunity* (Coleman et al.), touted the importance of placing students in racially integrated classrooms. Some scholars took that report to further endorse the cultural deficit theories and to suggest that there was not much that could be done by schools to improve the achievement of African American children. But Coleman et al. were subtler than that. They argued that, more than material resources alone, a combination of factors was heavily correlated with academic achievement. Their work indicated that the composition of a school (who attends it), the students' sense of control of the environments and their futures, the teachers' verbal skills, and their students' family background all contribute to student achievement. Unfortunately, it was the last factor—family background—that became the primary point of interest for many school and social policies.

Social psychologist Claude Steele (1999) argues that a "stereotype threat" contributes to the gap. Sociolinguists such as Kathryn Au (1980), Lisa Delpit (1995), Michèle Foster (1996), and Shirley Brice Heath (1983), and education researchers such as Jacqueline Jordan Irvine (2003) and Carol Lee (2004), have focused on the culture mismatch that contributes to the gap. Multicultural education researchers such as James Banks (2004), Geneva Gay (2004), and Carl Grant (2003), and curriculum theorists such as Michael Apple (1990), Catherine Cornbleth (and Dexter Waugh; 1995), and Thomas Popkewitz (1998) have focused on the nature of the curriculum and the school as sources of the gap. And teacher educators such as Christine Sleeter (2001), Marilyn Cochran-Smith (2004), Kenneth Zeichner (2002), and I (1994) have focused on the pedagogical practices of teachers as contributing to either the exacerbation or the narrowing of the gap.

But I want to use this opportunity to call into question the wisdom of focusing on the achievement gap as a way of explaining and understanding the persistent inequality that exists (and has always existed) in our nation's schools. I want to argue that this all-out focus on the "Achievement Gap" moves us toward short-term solutions that are unlikely to address the long-term underlying problem.

DOWN THE RABBIT-HOLE

Let me begin the next section of this discussion with a strange transition from a familiar piece of children's literature:

> *Alice started to her feet, for it flashed across her mind that she had never before seen a rabbit with either a waistcoat-pocket, or a watch to take out of it, and burning with curiosity, she ran across the field after it, and fortunately was just in time to see it pop down a large rabbit-hole under the hedge. In another moment down went Alice after it, never once considering how in the world she was to get out again.*

> Lewis Carroll, *Alice's Adventures in Wonderland*

The relevance of this passage is that I, like Alice, saw a rabbit with a watch and waistcoat-pocket when I came across a book by economist Robert Margo entitled *Race and Schooling in the American South, 1880–1950* (1990). And, like Alice, I chased the rabbit called "economics" down a rabbit-hole, where the world looked very different to me. Fortunately, I traveled with my trusty copy of Lakoff and Johnson's (1980) *Metaphors We Live By* as away to make sense of my sojourn there. So, before making my way back to the challenge of school inequality, I must beg your indulgence as I give you a brief tour of my time down there.

NATIONAL DEBT VERSUS NATIONAL DEFICIT

Most people hear or read news of the economy every day and rarely give it a second thought. We hear that the Federal Reserve Bank is raising interest rates, or that the unemployment numbers look good. Our ears may perk up when we hear the latest gasoline prices or that we can get a good rate on a mortgage refinance loan. But busy professionals rarely have time to delve deeply into all things economic. Two economic terms—"national deficit" and "national debt"—seem to befuddle us. A deficit is the amount by which a government's, company's, or individual's spending exceeds income over a particular period of time. Thus, for each budget cycle, the government must determine whether it has a balanced budget, a budget surplus, or a deficit. The debt, however is the sum of all previously incurred annual federal deficits. Since the deficits are financed by government borrowing, national debt is equal to all government debt.

Most fiscal conservatives warn against deficit budgets and urge the government to decrease spending to balance the budget. Fiscal liberals do not necessarily embrace deficits but would rather see the budget balanced by increasing tax revenues from those most able to pay. The debt is a sum that has been accumulating since 1791, when the U.S. Treasury recorded it as $75,463,476.52 (Gordon, 1998). Thomas Jefferson (1816) said, "I ... place economy among the first and most important virtues, and public debt as the greatest of dangers to be feared. To preserve our independence, we must not let our rulers load us with perpetual debt."...

But the debt has not merely been going up.... Even in those years when the United States has had a balanced budget, that is, no deficits, the national debt continued to grow. It may have grown at a slower rate, but it did continue to grow....

THE DEBT AND EDUCATION DISPARITY

By now, readers might assume that I have made myself firmly at home at the Mad Hatter's Tea Party. What does a discussion about national deficits and national debt have to do with education, education research, and continued education disparities? It is here where I began to see some metaphorical concurrences between our national fiscal situation and our education situation. I am arguing that our focus on the achievement gap is akin to a focus on the budget deficit, but what is actually happening to African American and Latina/o students is really more like the national debt. We do not have an achievement gap; we have an education debt.…

… I am arguing that the historical, economic, sociopolitical, and moral decisions and policies that characterize our society have created an education debt. So, at this point, I want to briefly describe each of those aspects of the debt.

THE HISTORICAL DEBT

Scholars in the history of education, such as James Anderson (1989), Michael Fultz (1995), and David Tyack (2004), have documented the legacy of educational inequities in the United States. Those inequities initially were formed around race, class, and gender. Gradually, some of the inequities began to recede, but clearly they persist in the realm of race. In the case of African Americans, education was initially forbidden during the period of enslavement. After emancipation we saw the development of freedmen's schools whose purpose was the maintenance of a servant class. During the long period of legal apartheid, African Americans attended schools where they received cast-off textbooks and materials from White schools. In the South, the need for farm labor meant that the typical school year for rural Black students was about 4 months long. Indeed, Black students in the South did not experience universal secondary schooling until 1968 (Anderson, 2002). Why, then, would we not expect there to be an achievement gap?

The history of American Indian education is equally egregious. It began with mission schools to convert and use Indian labor to further the cause of the church. Later, boarding schools were developed as General George Pratt asserted the need "to kill the Indian in order to save the man." This strategy of deliberate and forced assimilation created a group of people, according to Pulitzer Prize writer N. Scott Momaday, who belonged nowhere (Lesiak, 1992). The assimilated Indian could not fit comfortably into reservation life or the stratified mainstream. No predominately White colleges welcomed the few Indians who successfully completed the early boarding schools. Only historically Black colleges, such as Hampton Institute, opened their doors to them. There, the Indians studied vocational and trade curricula.

Latina/o students also experienced huge disparities in their education. In Ferg-Cadima's report *Black, White, and Brown: Latino School Desegregation Efforts in the Pre– and Post*–Brown v. Board of Education *Era* (2004), we discover the longstanding practice of denial experienced by Latina/os dating back to 1848. Historic desegregation cases such as *Mendez v. Westminster* (1946) and the Lemon Grove Incident detail the ways that Brown children were (and continue to be) excluded from equitable and high-quality education.

It is important to point out that the historical debt was not merely imposed by ignorant masses that were xenophobic and virulently racist. The major leaders of the nation endorsed ideas about the inferiority of Black, Latina/o, and Native peoples. Thomas Jefferson (1816), who advocated for the education of the American citizen, simultaneously decried the notion that Blacks were capable of education. George Washington, while deeply conflicted about slavery, maintained a substantial number of slaves on his Mount Vernon Plantation and gave no thought to educating enslaved children.

A brief perusal of some of the history of public schooling in the United States documents the way that we have accumulated an education debt over time. In 1827 Massachusetts passed a law making all grades of public school open to all pupils free of charge. At about the same time, most Southern states already had laws forbidding the teaching of enslaved Africans to read. By 1837, when Horace Mann had become head of the newly formed Massachusetts State Board of Education, Edmund Dwight, a wealthy Boston industrialist, felt that the state board was crucial to factory owners and offered to supplement the state salary with his own money. What is omitted from this history is that the major raw material of those textile factories, which drove the economy of the East, was cotton—the crop that depended primarily on the labor of enslaved Africans (Farrow, Lang, & Frank, 2005). Thus one of the ironies of the historical debt is that while African Americans were enslaved and prohibited from schooling, the product of their labor was used to profit Northern industrialists who already had the benefits of education....

This pattern of debt affected other groups as well. In 1864 the U.S. Congress made it illegal for Native Americans to be taught in their native languages. After the Civil War, African Americans worked with Republicans to rewrite state constitutions to guarantee free public education for all students. Unfortunately, their efforts benefited White children more than Black children. The landmark *Plessy v. Ferguson* (1896) decision meant that the segregation that the South had been practicing was officially recognized as legal by the federal government.

Although the historical debt is a heavy one, it is important not to overlook the ways that communities of color always have worked to educate themselves. Between 1865 and 1877, African Americans mobilized to bring public education to the South for the first time. Carter G. Woodson (1933/1972) was a primary critic of the kind of education that African Americans received, and he challenged African Americans to develop schools and curricula that met the unique needs of a population only a few generations out of chattel slavery.

THE ECONOMIC DEBT

As is often true in social research, the numbers present a startling picture of reality. The economics of the education debt are sobering. The funding disparities that currently exist between schools serving White students and those serving students of color are not recent phenomena. Separate schooling always allows for differential funding. In present-day dollars, the funding disparities between urban schools and their suburban counterparts present a telling story about the value we place on the education of different groups of students.

The Chicago public schools spend about $8,482 annually per pupil, while nearby Highland Park spends $17,291 per pupil. The Chicago public schools have an 87% Black and Latina/o population, while Highland Park has a 90% White population. Per pupil expenditures in Philadelphia are $9,299 per pupil for the city's 79% Black and Latina/o population, while across City Line Avenue in Lower Merion, the per pupil expenditure is $17,261 for a 91% White population. The New York City public schools spend $11,627 per pupil for a student population that is 72% Black and Latina/o, while suburban Manhasset spends $22,311 for a student population that is 91% White (figures from Kozol, 2005).

One of the earliest things one learns in statistics is that correlation does not prove causation, but we must ask ourselves why the funding inequities map so neatly and regularly onto the racial and ethnic realities of our schools. Even if we cannot prove that schools are poorly funded *because* Black and Latina/o students attend them, we can demonstrate that the amount of funding rises with the rise in White students. This pattern of inequitable funding has occurred over centuries. For many of these populations, schooling was nonexistent during the early history of the nation; and, clearly, Whites were not prepared to invest their fiscal resources in these strange "others."

Another important part of the economic component of the education debt is the earning ratios related to years of schooling. The empirical data suggest that more schooling is associated with higher earnings; that is, high school graduates earn more money than high school dropouts, and college graduates earn more than high school graduates. Margo (1990) pointed out that in 1940 the average annual earnings of Black men were about 48% of those of White men, but by 1980 the earning ratio had risen to 61%. By 1993, the median Black male earned 74% as much as the median White male....

THE SOCIOPOLITICAL DEBT

The sociopolitical debt reflects the degree to which communities of color are excluded from the civic process. Black, Latina/o, and Native communities had little or no access to the franchise, so they had no true legislative representation. According to the Civil Rights Division of the U.S. Department of Justice, African Americans and other persons of color were substantially disenfranchised in many Southern states despite the enactment of the Fifteenth Amendment in 1870 (U.S. Department of Justice, Civil Rights Division, 2006).

The Voting Rights Act of 1965 is touted as the most successful piece of civil rights legislation ever adopted by the U.S. Congress (Grofman, Handley, & Niemi, 1992). This act represents a proactive attempt to eradicate the sociopolitical debt that had been accumulating since the founding of the nation.

… The dramatic changes in voter registration are a result of Congress's bold action. In upholding the constitutionality of the act, the Supreme Court ruled as follows:

> Congress has found that case-by-case litigation was inadequate to combat wide-spread and persistent discrimination in voting, because of the inordinate amount of time and energy required to overcome the obstructionist tactics invariably encountered in these lawsuits. After enduring nearly a century of systematic resistance to the Fifteenth Amendment, Congress might well decide to shift the advantage of time and inertia from the perpetrators of the evil to its victims. (*South Carolina v. Katzenbach*, 1966; U.S. Department of Justice, Civil Rights Division, 2006)

It is hard to imagine such a similarly drastic action on behalf of African American, Latina/o, and Native American children in schools. For example, imagine that an examination of the achievement performance of children of color provoked an immediate reassignment of the nation's best teachers to the schools serving the most needy students. Imagine that those same students were guaranteed places in state and regional colleges and universities. Imagine that within one generation we lift those students out of poverty.

The closest example that we have of such a dramatic policy move is that of affirmative action. Rather than wait for students of color to meet predetermined standards, the society decided to recognize that historically denied groups should be given a preference in admission to schools and colleges. Ultimately, the major beneficiaries of this policy were White women. However, Bowen and Bok (1999) found that in the case of African Americans this proactive policy helped create what we now know as the Black middle class.

As a result of the sociopolitical component of the education debt, families of color have regularly been excluded from the decision-making mechanisms that should ensure that their children receive quality education. The parent—teacher organizations, school site councils, and other possibilities for democratic participation have not been available for many of these families. However, for a brief moment in 1968, Black parents in the Ocean Hill-Brownsville section of New York exercised community control over the public schools (Podair, 2003). African American, Latina/o, Native American, and Asian American parents have often advocated for improvements in schooling, but their advocacy often has been muted and marginalized. This quest for control of schools was powerfully captured in the voice of an African American mother during the fight for school desegregation in Boston. She declared: "When we fight about schools, we're fighting for our lives" (Hampton, 1986).

Indeed, a major aspect of the modern civil rights movement was the quest for quality schooling. From the activism of Benjamin Rushing in 1849 to the struggles of parents in rural South Carolina in 1999, families of color have been fighting for quality education for their children (Ladson-Billings, 2004). Their more limited access to lawyers and legislators has kept them from accumulating the kinds of political capital that their White, middle-class counterparts have.

THE MORAL DEBT

A final component of the education debt is what I term the "moral debt." I find this concept difficult to explain because social science rarely talks in these terms....

... [A] moral dent reflects the disparity between what we know is right and what we actually do. Saint Thomas Aquinas saw the moral debt as what human beings owe to each other in the giving of, or failure to give, honor to another when honor is due. This honor comes as a result of people's excellence or because of what they have done for another. We have no trouble recognizing that we have a moral debt to Rosa Parks, Martin Luther King, Cesar Chavez, Elie Wiesel, or Mahatma Gandhi. But how do we recognize the moral debt that we owe to entire groups of people? How do we calculate such a debt?...

What is that we might owe to citizens who historically have been excluded from social benefits and opportunities? Randall Robinson (2000) states:

> No nation can enslave a race of people for hundreds of years, set them free bedraggled and penniless, pit them, without assistance in a hostile environment, against privileged victimizers, and then reasonably expect the gap between the heirs of the two groups to narrow. Lines, begun parallel and left alone, can never touch, (p. 74)

Robinson's sentiments were not unlike those of President Lyndon B. Johnson, who stated in a 1965 address at Howard University: "You cannot take a man who has been in chains for 300 years, remove the chains, take him to the starting line and tell him to run the race, and think that you are being fair"' (Miller, 2005)....

... Taken together, the historic, economic, sociopolitical, and moral debt that we have amassed toward Black, Brown, Yellow, and Red children seems insurmountable, and attempts at addressing it seem futile. Indeed, it appears like a task for Sisyphus. But as legal scholar Derrick Bell (1994) indicated, just because something is impossible does not mean it is not worth doing.

WHY WE MUST ADDRESS THE DEBT

... On the face of it, we must address it because it is the equitable and just thing to do. As Americans we pride ourselves on maintaining those ideal qualities as hallmarks of our democracy. That represents the highest motivation for paying this debt. But we do not always work from our highest motivations.

Most of us live in the world of the pragmatic and practical. So we must address the education debt because it has implications for the kinds of lives we can live and the kind of education the society can expect for most of its children. I want to suggest that there are three primary reasons for addressing the debt—(a) the impact the debt has on present education progress, (b) the value of understanding the debt in relation to past education research findings, and (c) the potential for forging a better educational future.

The Impact of the Debt on Present Education Progress

... As I was attempting to make sense of the deficit/debt metaphor, educational economist Doug Harris (personal communication, November 19, 2005) reminded me that when nations operate with a large debt, some part of their current budget goes to service that debt. I mentioned earlier that interest payments on our national debt represent the third largest expenditure of our national budget. In the case of education, each effort we make toward improving education is counterbalanced by the ongoing and mounting debt that we have accumulated. That debt service manifests itself in the distrust and suspicion about what schools can and will do in communities serving the poor and children of color. Bryk and Schneider (2002) identified "relational trust" as a key component in school reform. I argue that the magnitude of the education debt erodes that trust and represents a portion of the debt service that teachers and administrators pay each year against what they might rightfully invest in helping students advance academically.

The Value of Understanding the Debt in Relation to Past Research Findings

The second reason that we must address the debt is somewhat selfish from an education research perspective. Much of our scholarly effort has gone into looking at educational inequality and how we might mitigate it. Despite how hard we try, there are two interventions that have never received full and sustained hypothesis testing—school desegregation and funding equity. Orfield and Lee (2006) point out that not only has school segregation persisted, but it has been transformed by the changing demographics of the nation. They also point out that "there has not been a serious discussion of the costs of segregation or the advantages of integration for our most segregated population, white students" (p. 5). So, although we may have recently celebrated the 50th anniversary of the *Brown* decision, we can point to little evidence that we really gave *Brown* a chance. According to Frankenberg, Lee, and Orfield (2003) and Orfield and Lee (2004), America's public schools are more than a decade into a process of resegregation. Almost three-fourths of Black and Latina/o students attend schools that are predominately non-White. More than 2 million Black and Latina/o students—a quarter of the Black students in the Northeast and Midwest—attend what the researchers call apartheid schools. The four most segregated states for Black students are New York, Michigan, Illinois, and California.

The funding equity problem, as I illustrated earlier in this discussion, also has been intractable. In its report entitled *The Funding Gap 2005*, the Education Trust tells us that "in 27 of the 49 states studied, the highest-poverty school districts receive fewer resources than the lowest-poverty districts.... Even more states shortchange their highest minority districts. In 30 states, high minority districts receive less money for each child than low minority districts" (p. 2). If we are unwilling to desegregate our schools *and* unwilling to fund them equitably, we find ourselves not only backing away from the promise of the *Brown* decision but literally refusing even to take *Plessy* seriously. At least a serious consideration of *Plessy* would make us look at funding inequities....

The Potential for Forging a Better Educational Future

Finally, we need to address what implications this mounting debt has for our future. In one scenario, we might determine that our debt is so high that the only thing we can do is declare bankruptcy. Perhaps, like our airline industry, we could use the protection of the bankruptcy laws to reorganize and design more streamlined, more efficient schooling options. Or perhaps we could be like developing nations that owe huge sums to the IMF and apply for 100% debt relief. But what would such a catastrophic collapse of our education system look like? Where could we go to begin from the ground up to build the kind of education system that would aggressively address the debt? Might we find a setting where a catastrophic occurrence, perhaps a natural disaster—a hurricane—has completely obliterated the schools? Of course, it would need to be a place where the schools weren't very good to begin with. It would have to be a place where our Institutional Review Board and human subject concerns would not keep us from proposing aggressive and cutting-edge research. It would have to be a place where people were so desperate for the expertise of education researchers that we could conduct multiple projects using multiple approaches. It would be a place so hungry for solutions that it would not matter if some projects were quantitative and others were qualitative. It would not matter if some were large-scale and some were small-scale. It would not matter if some paradigms were psychological, some were social, some were economic, and some were cultural. The only thing that would matter in an environment like this would be that education researchers were bringing their expertise to bear on education problems that spoke to pressing concerns of the public. I wonder where we might find such a place?...

... [T]he cumulative effect of poor education, poor housing, poor health care, and poor government services creates a bifurcated society that leaves more than its children behind. The images should compel us to deploy our knowledge, skills, and expertise to alleviate the suffering of the least of these. They are the images that compelled our attention during Hurricane Katrina. Here, for the first time in a very long time, the nation—indeed the world—was confronted with the magnitude of poverty that exists in America.

In a recent book, Michael Apple and Kristen Buras (2006) suggest that the subaltern can and do speak. In this country they speak from the barrios of Los Angeles and the ghettos of New York. They speak from the reservations of New Mexico and the Chinatown of San Francisco. They speak from the levee breaks of New Orleans where they remind us, as education researchers, that we do not merely have an achievement gap—we have an education debt.

REFERENCES

Anderson, J. D. (1989). *The education of Blacks in the South, 1860–1935.* Chapel Hill, NC: University of North Carolina Press.

Anderson, J. D. (2002, February 28). *Historical perspectives on Black academic achievement.* Paper presented for the Visiting Minority Scholars Series Lecture. Wisconsin Center for Educational Research, University of Wisconsin, Madison.

Apple, M. (1990). *Ideology and curriculum* (2nd ed.). New York: Routledge.

Apple, M., & Buras, K. (Eds.). (2006). *The subaltern speak: Curriculum, power and education struggles.* New York: Routledge.

Au, K. (1980). Participation structures in a reading lesson with Hawaiian children. *Anthropology and Education Quarterly, 11*(2), 91–115.

Banks, J. A. (2004). Multicultural education: Historical development, dimensions, and practices. In J. A. Banks & C. M. Banks (Eds.), *Handbook of research in multicultural education* (2nd ed., pp. 3–29). San Francisco: Jossey-Bass.

Bell, D. (1994). *Confronting authority: Reflections of an ardent protester.* Boston: Beacon Press.

Bereiter, C., & Engleman, S. (1966). *Teaching disadvantaged children in preschool.* Englewood Cliffs, NJ: Prentice Hall.

Bowen, W., & Bok, D. (1999). *The shape of the river.* Princeton, NJ: Princeton University Press.

Brice Heath, S. (1983). *Ways with words: Language, life and work in communities and classrooms.* Cambridge, UK: Cambridge University Press.

Brown v. Board of Education 347 U.S. 483 (1954).

Bryk, A., & Schneider, S. (2002). *Trust in schools: A core resource for improvement.* New York: Russell Sage Foundation.

Cochran-Smith, M. (2004). Multicultural teacher education: Research, practice and policy. In J. A. Banks & C. M. Banks (Eds.), *Handbook of research in multicultural education* (2nd ed., pp. 931–975). San Francisco: Jossey-Bass.

Coleman, J., Campbell, E., Hobson, C., McPartland, J., Mood, A., Weinfeld, F. D., et al. (1966). *Equality of educational opportunity.* Washington, DC: Department of Health, Education and Welfare.

Cornbleth, C., & Waugh, D. (1995). *The great speckled bird: Multicultural politics and education.* Mahwah, NJ: Lawrence Erlbaum.

Delpit, L. (1995). *Other people's children: Cultural conflict in the classroom.* New York: Free Press.

Deutsch, M. (1963). The disadvantaged child and the learning process. In A. H. Passow (Ed.), *Education in depressed areas* (pp. 163–179). New York: New York Bureau of Publications, Teachers College, Columbia University.

Education Commission of the States. (2005). *The nation's report card.* Retrieved January 2, 2006, from http://nces.ed.gov/nationsreportcard

Education Trust. (2005). *The funding gap 2005.* Washington, DC: Author.

Farrow, A., Lang, J., & Frank, J. (2005). *Complicity: How the North promoted, prolonged and profited from slavery.* New York: Ballantine Books.

Ferg-Cadima, J. (2004, May). *Black, White, and Brown: Latino school desegregation efforts in the pre– and post–* Brown v. Board of Education *era.* Washington, DC: Mexican-American Legal Defense and Education Fund.

Foster, M. (1996). *Black teachers on teaching.* New York: New Press.

Frankenberg, E., Lee, C., & Orfield, G. (2003, January). *A multiracial society with segregated schools: Are we losing the dream?* Cambridge, MA: The Civil Rights Project, Harvard University.

Fultz, M. (1995). African American teachers in the South, 1890–1940: Powerlessness and the ironies of expectations and protests. *History of Education Quarterly, 35*(4), 401–422.

Gay, G. (2004). Multicultural curriculum theory and multicultural education. In J. A. Banks & C. M. Banks (Eds.), *Handbook of research in multicultural education* (2nd ed., pp. 30–49). San Francisco: Jossey-Bass.

Gordon, J. S. (1998). *Hamilton's blessing: The extraordinary life and times of our national debt.* New York: Penguin Books.

Grant, C. A. (2003). *An education guide to diversity in the classroom.* Boston: Houghton Mifflin.

Grofman, B., Handley, L., & Niemi, R. G. (1992). *Minority representation and the quest for voting equality.* New York: Cambridge University Press.

Hampton, H. (Director). (1986). *Eyes on the prize* [Television video series], Blackside Productions (Producer). New York: Public Broadcasting Service.

Hess, R. D., & Shipman, V. C. (1965). Early experience and socialization of cognitive modes in children. *Child Development, 36,* 869–886.

Irvine, J. J. (2003). *Educating teachers for diversity: Seeing with a cultural eye.* New York: Teachers College Press.

Jefferson, T. (1816, July 21). *Letter to William Plumer. The Thomas Jefferson Paper Series. 1. General correspondence,* 1651–1827. Retrieved September 11, 2006, from http://rs6.loc.gov/cgi-bin/ampage

Kozol, J. (2005). *The shame of the nation: The restoration of apartheid schooling in America.* New York: Crown Publishing.

Ladson-Billings, G. (1994). *The dreamkeepers: Successful teachers of African American children.* San Francisco: Jossey-Bass.

Ladson-Billings, G. (2004). Landing on the wrong note: The price we paid for *Brown. Educational Researcher, 33*(7), 3–13.

Lakoff, G., & Johnson, M. (1980). *Metaphors we live by.* Chicago: University of Chicago Press.

Lee, C. D. (2004). African American students and literacy. In D. Alvermann & D. Strickland (Eds.), *Bridging the gap: Improving literacy learning for pre-adolescent and adolescent learners, Grades 4–12.* New York: Teachers College Press.

Lesiak, C. (Director). (1992). *In the White man's image* [Television broadcast]. New York: Public Broadcasting Corporation.

Margo, R. (1990). *Race and schooling in the American South, 1880–1950.* Chicago: University of Chicago Press.

Mendez v. Westminster 64F. Supp. 544 (1946).

Miller, J. (2005, September 22). New Orleans unmasks apartheid American style [Electronic version]. *Black Commentator, 151.* Retrieved September 11, 2006, from http://www.blackcommentator.com/151/151_miller_new_orleans.html

National Center for Education Statistics. (2001). *Education achievement and Black-White inequality.* Washington, DC: Department of Education.

National Governors' Association. (2005). *Closing the achievement gap.* Retrieved October 27, 2005, from http://www.subnet.nga.org/educlear/achievement!

Orfield, G., & Lee, C. (2004, January). *Brown at 50: King's dream or Plessy's nightmare?* Cambridge, MA: The Civil Rights Project, Harvard University.

Orfield, G., & Lee, C. (2006, January). *Racial transformation and the changing nature of segregation.* Cambridge, MA: The Civil Rights Project, Harvard University.

Plessy v. Ferguson 163 U.S. 537 (1896).

Podair, J. (2003). *The strike that changed New York: Blacks, Whites and the Ocean Hill-Brownsville Crisis.* New Haven, CT: Yale University Press.

Popkewitz, T. S. (1998). *Struggling for the soul: The politics of schooling and the construction of the teacher.* New York: Teachers College Press.

Robinson, R. (2000). *The debt: What America owes to Blacks.* New York: Dutton Books.

Sleeter, C. (2001). *Culture, difference and power.* New York: Teachers College Press.

South Carolina v. Katzenbach 383 U.S. 301, 327–328 (1966).

Steele, C. M. (1999, August). Thin ice: "Stereotype threat" and Black college students. *Atlantic Monthly, 284,* 44–47, 50–54.

Tyack, D. (2004). *Seeking common ground: Public schools in a diverse society.* Cambridge, MA: Harvard University Press.

U.S. Department of Justice, Civil Rights Division. (2006, September 7). *Introduction to federal voting rights laws.* Retrieved September 11, 2006, from http://www.usdoj.gov/crt/voting/intro/intro.htm

Woodson, C. G. (1972). *The mis-education of the Negro.* Trenton, NJ: Africa World Press. (Original work published 1933).

Zeichner, K. M. (2002). The adequacies and inadequacies of three current strategies to recruit, prepare, and retain the best teachers for all students. *Teachers College Record, 105*(3), 490–511.

34

Academic Resilience Among Undocumented Latino Students

WILLIAM PEREZ, ROBERTA ESPINOZA, KARINA RAMOS,
HEIDI M. CORONADO, AND RICHARD CORTES

In 2005, there were 1.8 million undocumented youth under the age of 18 living in the United States. Latinos represent approximately 78% of this undocumented population. Each year, approximately 80,000 undocumented students reach high school graduation age. Of these high school graduates, approximately 13,000 enroll in public colleges and universities across the country (Passel, 2006). Not only do these students endure the same stressors and risk factors as other Latino and immigrant youth, but they also face constant institutional and societal exclusion and rejection due to their undocumented status. They are not eligible for most scholarships, do not qualify for any form of government sponsored financial assistance, are not eligible to apply for a driver's license, are legally barred from formal employment, and may be deported at any time. The social, educational, and psychological experiences of these immigrant youth raise a number of important questions: What specific social and environmental characteristics mediate their school success in the presence of numerous factors that place them at risk for low achievement? And how can the risk and resilience framework help us better understand the academic achievement patterns of undocumented immigrant Latino youth?

IMMIGRANT YOUTH

Migration is one of the most radical transitions and life changes an individual or family can endure. For immigrant children, the migration experience fundamentally reshapes their lives as familiar patterns and ways of relating to other people dramatically change. Some potential stressors related to migration include loss of close relationships, housing problems, a sense of isolation, obtaining legal documentation, going through the acculturation process, learning the English language, negotiating their ethnic identity, changing family roles, and adjusting to the schooling experience....

SOURCE: Perez, W., et al. 2009. "Academic Resilience Among Undocumented Latino Students." *Hispanic Journal of Behavioral Sciences*, 31(2), 149–181. Copyright © 2009 by SAGE Publications. Reprinted by permission of SAGE Publications, Inc.

With respect to Latino immigrant youth, research suggests a host of socio-cultural experiences related to the acculturation process are extremely stressful (Cervantes & Castro, 1985).... Discrepancies in the values and practices of Hispanic children and their parents may create pressure in selecting which set of cultural norms and expectations to adhere to, those of their culture of origin or those of mainstream culture.

UNDOCUMENTED IMMIGRANT STUDENTS

Although literature exists on first and second generation immigrants, there is a lack of research on the undocumented immigrant student population. In one of only a handful of studies, Dozier (1993) found three central emotional concerns for undocumented college students: fear of deportation, loneliness, and depression. Dozier found that students' fear of deportation was so central to undocumented students, it influenced almost every aspect of their lives. Some students, reported being afraid of going to hospitals because they worried that their immigration status would be questioned. Because their legal status made it impossible to obtain work authorization, they were sometimes forced to stay in bad work conditions because they feared not being able to find another job. In addition, undocumented students were often reluctant to develop close emotional relationships with others for fear of their undocumented status being discovered. Despite these stressors, the undocumented students in Dozier's study managed to accumulate the necessary academic record to be accepted into college. How did they manage such accomplishments in the face of numerous obstacles?

... The findings from the few studies focusing on undocumented Latino youth suggest that while both documented and undocumented immigrant Latino youth face similar educational and psychological risks, undocumented youth's precarious legal status translates into additional risk factors and sources of stress. However, the psychological and academic effects of legal marginalization have not been fully studied or addressed by researchers.

PSYCHOLOGICAL RESILIENCE

Resilience is the process of overcoming the negative effects of risk exposure, coping successfully with traumatic experiences, and avoiding the negative trajectories associated with those risks. A key requirement of resilience is the presence of both risk and protective factors that either help bring about a positive outcome or reduce and avoid a negative outcome. Resilience theory, though it is concerned with risk exposure among adolescents, is focused more on strengths rather than deficits and understanding healthy development in spite of high-risk exposure. Personality characteristics and environmental social resources are thought to moderate the negative effects of stress and promote positive outcomes despite risks....

The purpose of this study was to examine the role of protective resources in mediating the academic achievement of undocumented Latino youth. Our study uses three main indicators of academic success: high grade point average (GPA), high number of academic awards, and high number of academically rigorous Honors and AP courses. The assumption is that these represent significant accomplishments for undocumented Latino students who must surmount a multitude of obstacles to attain them. The primary hypothesis of this study was that academic resilience is at least partially explained by the extent to which personal and environmental resources are available to them. The study's focus on student personal and environmental resources is based on the premise that these factors are important antecedents of school achievement. Thus, we sought to answer the following questions: What is the relationship between risk, protective factors (personal and environmental), and academic achievement among undocumented immigrant Latino students? And how do undocumented students with different configurations of risk and protective factors differ in their academic performance? It was hypothesized that undocumented students with high levels of risk factors, but also high levels of both personal and environmental protective factors, would have higher academic outcomes than students who have similar levels of risk factors, but lower levels of both personal and environmental protective factors.

METHOD

... One hundred and ten undocumented Latino high school, community college, and university students from across the United States participated in this study. The average age of participants was 19.97 years ($SD = 2.15$). A total of 62% of subjects were female. The male to female ratio in this study is similar to college enrollment rates for Latinos.... The high school group in this study was gender balanced with 50% female. The average number of years living in the United States was 13.03 ($SD = 4.80$) for males and 13.90 ($SD = 4.15$) for females. Male participants came to the United States when they were 7.25 ($SD = 5.25$) years old while female participants had an average arrival age of 6.79 ($SD = 3.89$) years old. A total of 18% participants were high school seniors, 34% were community college students, and 48% were students at a B.A.-granting university.

Participants were selected from a convenience sample recruited using email and flyer advertisements to various Latino student organizations at colleges and high schools in Southern California. Information flyers were also passed out in several high school and college classrooms. We also asked participants to forward our information to other students who met our criteria of being undocumented. The recruitment flyers and email invited students to participate in a research study that focused on "the educational experiences of undocumented students." This is the only detail that participants received regarding the purpose of the study. The email and printed flyer announcements contained a link to an online survey hosted by email The online survey did not collect names, e-mails, school names, or any other type of identifying information to protect the confidentiality of participants.

The first part of the online survey consisted of open-ended questions that asked participants to list their academic achievements, civic engagement experiences, extracurricular activities, leadership positions, and enrollment in advanced level academic courses. The second part of the survey consisted of school background and demographic information. The third and final part of the online questionnaire consisted of various Likert-type style, self-reported questions designed to assess distress levels, perceived societal rejection due to undocumented status, bilingualism, student valuing of school, parental valuing of school, and friends valuing of school. The survey took approximately 45 minutes to complete. The following is a description of the independent and dependent variables used in this study.

MEASURES

For the purpose of this study, we operationalized four key theoretical concepts: risk, environmental protective factors, personal protective factors, and academic outcomes. A detailed description of all key variables used in the analyses follows.

Risk factors. As Table 1 indicates, We used four measures of risk factors: employment during high school, sense of rejection related to undocumented status, low parental educational attainment, and large family size.

High school employment. Students were asked, "How many hours per week did you work in high school?" High school employment was considered a risk if students worked more than 20 hours per week....

Parental education. Students were asked how many years of schooling their mother and father had completed. We took the average of the mother's and father's years of education to create a parental education index. Parental education was considered a risk factor if the parental education index was less than a high school education.

Family size. Students were asked to report their total number of brothers and sisters. Family size was considered a risk factor if participants reported having three or more siblings....

Rejection due to undocumented status. This scale was developed specifically for this study and was composed of three statements such as "Because of my undocumented background I feel that I am not wanted in this country." Respondents indicated their responses ranging from 7 (*strongly agree*) to 1 (*strongly disagree*). Higher scores indicated higher feelings of alienation.

Personal protective factors. There are four measures of personal protective factors: having been identified as gifted or talented during their early education, valuing of schooling, bilingualism, and coping with distress.

Gifted. If the participant had been designated as gifted or talented during her or his elementary or middle school education, it was considered a personal protective factor.

Distress scale. The distress scale was composed of 12 statements such as "Lately do you feel sad?" and "Lately do you feel that you don't have much energy?" Respondents indicated their responses on a scale ranging from 7 (*always*) to 1 (*never*). Low distress was considered a protective factor, as those reporting lower distress scores demonstrate more adaptive coping with the high levels of stress that undocumented students contend with on a daily basis....

Bilingualism. The bilingualism scale was composed of eight statements such as "How well do you understand Spanish?" and "How well do you read English?" Respondents indicated their responses on a scale ranging from 4 (*very well*) to 1 (*not at all*). Students who reported understanding, speaking, reading, and writing both English and Spanish "very well" were considered having the protective factor of being highly bilingual.

Valuing of schooling. The valuing of schooling scale was composed of five statements such as "How important is it for you to do well in school?" and "How important is it to earn good grades?" Respondents indicated their responses on a scale ranging from 7 (*very important*) to 1 (*not important at all*). Higher scores on the valuing of school scale was the second personal protective factor.

Environmental protective factors. Five environmental protective factor measures were used: parental valuing of school, friends valuing of schooling, participation in extracurricular activities, participation in volunteer activities, and growing up with both parents.

Extracurricular activities. Students were asked to list all the extracurricular activities they participated in during high school. *Extracurricular participation* was defined as participation in the following activities: Student council, sports, band/music/choir, drama/theater, newspaper/magazine/ yearbook, cultural dance, clubs, YMCA/YWCA, Boys/Girls Club. After the extracurricular activities were coded, they were counted and summed to create an extracurricular participation score. Since extracurricular participation provides multiple opportunities for developing relationships with other academically engaged peers and school agents, higher counts of extracurricular activities was considered an environmental protective factor.

Volunteer activities. Students were asked in an open-ended format to list all volunteer activities they participated in during high school. *Volunteerism* was defined as participation in any one of the following activities: providing a social service, working for a cause and political activism, tutoring, and functionary work. *Performing social service* entailed interaction with people in need such as visiting, feeding, or caring for the homeless, poor, sick, elderly, or handicapped. *Working for a cause* and *political activism* were defined as having engaged in activities focused on a particular social issue or cause such as the environment, a political party, human rights, or other causes that did not entail direct interaction with the needy. *Tutoring* was defined as coaching, child care, and academic tutoring. *Functionary work* was defined as having participated in volunteer activities that entailed cleaning and maintenance work or organizing and administrative work such as beach cleanup. After the volunteer activities were coded, they were counted and summed to create a volunteer score. Higher scores were considered an environmental protective factor.

Family composition. Students were asked to indicate which parents or guardians they lived with growing up. Students were asked to select from the following 10 choices: "Both my mother and my father in the same house," "Only my mother," "My mother and stepfather," "Only my father," "My father and stepmother," "Some of the time in my mother's home and some in my father's," "Other relative (aunt, uncle, grandparents, etc.)," "Guardian or foster parent who is not a relative," "No parents or guardians (I lived alone or with friends)," and "Other." Student responses were then coded into one of two categories: (1) lived with both parents growing up, or (2) other. Growing up with both parents was considered an environmental protective factor.

Parent valuing of schooling. The parental valuing of school scale was composed of two statements: "For my parents, me getting good grades in school is?" and "For my parents, me going to college after high school is?" Respondents indicated their responses on a scale ranging from 4 (*very important*) to 1 (*not important*). High scores on the parental valuing of school was considered an environmental protective factor....

Friends valuing of schooling. The friends valuing of school scale was composed of three statements: "For my friends, getting good grades in school is?" and "For my friends, going to college after high school is?" Respondents indicated their responses on a scale ranging from 4 (*very important*) to 1 (*not important*). High scores on the friends valuing of school was considered an environmental protective factor....

Academic outcomes. The risk and protective factors described above were the independent variables used to understand the academic performance patterns of undocumented Latino students. The four academic outcomes used were high school GPA, number of school awards received during high school, and number of academically rigorous Honors and AP courses taken.

GPA. The GPA variable was calculated by asking students to report their overall high school GPA on a standard 4.0 scale....

School awards. Students were asked in an open-ended format to list all awards they received in high school. An *academic award* was defined by student of the month award, honor roll, attendance award, spelling bee/ writing/poetry contest award, subject award (i.e., science award), school sports award, band/music/choir award, community service award, citizenship award for good behavior, or student of the year award. After the awards were coded, they were counted and summed to create a total awards score.

Honors and AP courses. Students were asked to list all Honors and AP courses they took during high school. Examples of Honors courses included Honors English or Honors biology, while examples of AP courses included AP calculus or AP chemistry. After all courses were coded, they were counted and summed to create a total number of academically rigorous courses score....

RESULTS AND DISCUSSION

The results of this study support the premise that a constellation of protective resources can serve to buffer or protect students from the detrimental effects of psychosocial conditions that place them at risk of academic failure. Overall, three main conclusions can be drawn from the results: (a) academic success (resilience) was related to both personal and environmental resources, (b) when various resources were present, academic performance was generally positive, even in the presence of multiple sources of psychosocial risk, and (c) compared to the protected group, the high-risk and resilient groups suffered significantly higher levels of adversity. Resilient youth, however, had greater levels of environmental and personal resources than high-risk students, who in turn exhibited lower levels of academic success. These findings add to the growing evidence suggesting that personal and environmental resources facilitate academic success among youth growing up in environments where they are exposed to elevated psychosocial risks.

The psychosocial stressors examined in this study, such as undocumented status, socioeconomic hardship, and low parental education represent significant challenges for Latino immigrant adolescents. Thirtysix percent of participants we surveyed were found to be highly vulnerable, that is, reporting various risk factors, but lacking in personal and environmental resources. The results from this study suggest that when faced with the challenges of living in poverty, working long hours at a job during school, low levels of parental education, and feeling a high sense of rejection due to their legal status, resilient undocumented Latino youth draw on available personal and environmental resources. Results also suggest that not all undocumented Latino youth face high levels of risk factors. Some undocumented students have lower levels of risk exposure accompanied by a host of protective factors. In this study, 42% of participants reported low levels of risk accompanied by high levels of protective factors. The remaining 58% reported high levels of risk exposure. It should be noted that only 22% of participants met criteria for resilience, that is, high levels of personal and environmental resources to cope with high levels of risk.

The results indicate that protective factors cannot be ignored as important mediators of achievement, beyond the potentially detrimental effects that psychosocial risk factors have on academic performance.... The results of this study revealed three distinct profiles: high-risk students with low levels of protective factors (*high-risk*), low risk students with high levels of protective factors (*protected*), and high-risk students with high levels of protective factors (*resilient*). Compared to protected students, resilient and high-risk students worked longer hours at a job during high school, reported higher levels of feelings of societal rejection due to their undocumented status, and had parents with lower levels of schooling. The comparisons of academic outcomes indicate that resilient and protected students had significantly higher academic achievement levels than high-risk students. While all three groups had similar levels of personal protective factors, protected and resilient students had much higher levels of environmental protective factors. Both resilient and protected students reported higher levels of parental valuing of school, extracurricular participation, and volunteering.

Findings consistently indicated that giftedness, valuing of school, extracurricular participation, and volunteerism were significant predictors of academic achievement among undocumented youth. Giftedness was assigned a protective role, especially in high adversity situations. This result is consistent with other research findings emphasizing the significance of cognitive resources such as IQ, and general intellectual functioning on the development and maintenance of good adaptation in adversity (Luthar, 1991; Masten et al., 1999; Ripple & Luthar, 2000).

In a study with academically resilient Latino students at a highly selective university, Arellano and Padilla (1996) found that a significant percentage of them had been identified as gifted early in their schooling. In our study, bilingualism and distress did not seem to play an important role as a resource in the interplay between adversity and academic success. This finding may be the result of the measurement methods used in this study. A more comprehensive way to measure bilingualism and distress might yield more valid results.

Resilient and protected youth possessed a wider repertoire of environmental resources than their high-risk peers. They seemed to enjoy higher levels of parental valuing of school as well as greater integration in school communities through their extracurricular and volunteer activities. This combination of resources seemed to allow them to maintain a higher level of academic functioning even though they shared the same high levels of psychosocial risks as the high-risk students. Another important finding is the similarity in academic outcomes between the protected and resilient students. Given their lower levels of psychosocial risks and high levels of protective resources, protected youth will have higher academic outcomes than resilient youth, as predicted by resilience theory. This finding from our study possibly reflects the fact that not being particularly tried by adversity, the protected individuals' subjective sense of well-being remained unchallenged and largely left intact by external environmental threats. On the whole, the resilient group seemed to adapt well under adversity as indicated by their high academic achievement levels. The high-risk group in this study appeared to lack the necessary resources to maintain the same high levels of academic achievement as protected and resilient students.

The primacy of extracurricular participation and volunteerism as environmental resources available to resilient and protected students underscores the importance of environmental opportunities to develop relationships with supportive adults and peers engaged in prosocial activities. The results suggest that school opportunities to develop social support play a critical role in encouraging students to succeed in high school. Overall, extracurricular participation and volunteerism were the strongest predictors of academic achievement among undocumented Latino students with the resilient students reporting the highest levels of these two environmental protective resources.

If environmental factors can contribute to academic resilience within individuals, then those factors can be modified to increase the protection or assets in undocumented students' lives. Increasingly, schools are being explored for their potential to strengthen the resilience of children and youth. The number of children served by schools, and the amount of time in which students are influenced by their school environments from kindergarten through 12th grade, are primary reasons for such efforts. The role of the school in child development, the capacity of school personnel to develop competence in students, and the ability of the school to serve as an organizational base for mobilizing linkages with parents and community resources are other reasons for using schools to enhance resilience. Building developmental assets supports the academic mission of schools because higher levels of assets are associated with greater academic achievement and lower rates of school dropouts....

CONCLUSIONS AND IMPLICATIONS

The results of this study offer new insights into how resilient undocumented Latino youth draw on specific personal, family, and school resources to circumvent the effects of various stressors as well as social and institutional barriers to become academically successful. Although the research on undocumented resilient immigrant youth is only now beginning to emerge, it is important to continue studying this growing population. We need more quantitative and qualitative studies that help us better understand the psychosocial impact of immigration and immigration policies of receiving countries on immigrant children and adolescents.

REFERENCES

Cervantes, R. C., & Castro, F. G. (1985). Stress, coping, and Mexican American mental health: A systematic review. *Hispanic Journal of Behavioral Sciences, 1,* 1–73.

Dozier, S. B. (1993). Emotional concerns of undocumented and out-of-status foreign students. *Community Review, 13,* 33–29.

Luthar, S. S. (1991). Vulnerability and resilience: A study of high-risk adolescents. *Child Development, 62,* 600–616.

Masten, A. S., Hubbard, J. J., Gest, S. D., Tellegen, A., Garmezy, N., & Ramirez, M. (1999). Competence in the context of adversity: Pathways to resilience and maladaptation from childhood to late adolescence. *Development and Psychopathology, 11,* 143–169.

Passel, J. S. (2006). *The size and characteristics of the unauthorized migration population in the U.S.: Estimates based on the March 2005 current population survey.* Washington, D.C.: Pew Hispanic Center.

Ripple, C. H., & Luthar, S. S. (2000). Academic risk among inner-city adolescents: The role of personal attributes. *Journal of School Psychology, 8,* 277–298.

35

Michael's Story

"I get into so much trouble just by walking": Narrative Knowing and Life at the Intersections of Learning Disability, Race, and Class

DAVID J. CONNOR

I begin by contextualizing this research. As a former high school special educator working in New York City, I noticed patterns of entrenched racial segregation within public schools. I was always struck by the fact all students labeled learning disabled (LD) in "self-contained" classes were predominantly filled with black and Latino(a) adolescents, usually male. As I learned more about LD in college courses, I noticed that the label signified different outcomes for different people. What seemed to be a beneficial category of disability to middle-class, white students, by triggering various supports and services—served to disadvantage black and/or Latino(a) urban youngsters who were more likely to be placed in restrictive, segregated settings. While a small group of scholars have addressed racial and class issues within the field of special education, I feel that the actual students are still largely absent as people—living, thinking, knowing, being. With this in mind, I decided to create collaborative research with a selection of young, urban, black, working-class people who had been labeled LD at some point during their school career. This article details the lived experience of one young man.

A central premise of this study is to explore the phenomenon of intersectionality of learning disability, race, and class. The data presented were co-constructed with Michael, the participant-researcher, as part of a larger year-long study exploring young adults' understanding of their positionality within the discourse of LD. Data were culled from six semi-structured 1 1/2 hour interviews held at a local community college. During the first interview, Michael and I established mutual expectations for our meetings and began by focusing on how he understood his own LD. In subsequent meetings, the conversation flowed between his personal experience and observances of LD throughout his life span, as well as issues of race and class, both in and out of school. Between each session I transcribed the interviews; I shared all data with Michael. His

SOURCE: Connor, David J. 2006. Michael's Story: "I get into so much trouble just by walking": Narrative Knowing and Life at the Intersections of Learning Disability, Race, and Class. *Equity & Excellence in Education*, 39(2): 154–165, DOI: 10.1080/ 10665680500533942, copyright © University of Massachusetts Amherst, College of Education, www.umass.edu/education, reprinted by permission of Taylor & Francis Ltd, http://www.tandfonline.com on behalf of University of Massachusetts Amherst, College of Education.

transcribed ideas, thoughts, and experiences were arranged by me, merging into an ongoing portrait-in-progress that was always read by Michael, who in turn made suggestions about what to include and omit. He also granted permission to select his words to be arranged in poetic form.

VALUING NARRATIVE

Through the use of personal narrative, I foreground the experience of a person whose voice is not usually sought. Using narrative to know a person in a particular context is my prime interest; narrative elevates ordinary occurrences to important experiences, worthy of careful study and deliberation. Narrative stories are one of the most widely used ways of communicating and help "people make sense of their lives and the lives of others" (Richardson, 1990, p. 10). In many respects, a personal narrative is an authentic, political form of self-representation that holds power to promote change. As Richardson writes, personal narrative "is the closest to the human experience, and it rejuvenates the sociological imagination in the service of liberatory civic discourses and transformative social projects" (p. 65). Indeed, it can be argued that counter-storytelling brings new knowledge to both the minority who are speaking, and the majority who will listen:

> Members of the majority race should listen to stories, of all sorts, in order to enrich their own reality. Reality is not fixed, not a given. Rather, we construct it through our conversations, through our lives together. Racial and class-based isolation prevents the hearing of diverse storied and counter stories. It diminishes the conversation through which we create reality. (Delgado, 1989, 107)

... I chose narrative poetry to represent Michael because it "allows the Other to speak for ... himself" (Lincoln & Denzin, 2000, p. 1050).

MICHAEL'S STORY

I. I Write Backwards

I write backwards.
  My mother made sure, by the time I went to kindergarten, I'd be
    writing the right way.
She'd sit me at the table for hours.
She'd write the ABCs the correct way,
But I'd write it backwards all of the time.
The teachers wouldn't notice that stuff.
I was born in Brooklyn.
I'm nineteen going on twenty.
Dyslexia ... I still have it.
10 I think it is natural.

Everyone is born with something, so it's not like a disability.

People say, you have dyslexia, da-da-da, but to me it really didn't do anything.

I just had to work harder.

   If people understood, they wouldn't make fun of learning disabilities.

   People really won't know about what learning disabilities are.

I think in this society, a learning disability—

It's like a downfall, it makes you look bad.

They look at you like you're not like them,

And they're *"Yeah, you have a learning disability. What's wrong wichoo? You don't*

20   *Learn right? Oh man, how is it? How does it feel? Is it like*
AIDS or something?"

They distance themselves, like you're about to die.

And it's just a simple thing. I learn—it takes me a little while longer than you.

So if it takes you an hour, it might take me an hour and half, two hours.

That's all it is.

   If you went on a job interview and had a learning disability, they expect a slow kid:

   "He doesn't know how to track."

   "He's not gonna be good on the cash register."

   "He's not gonna be able to flip the burger properly."

   So, say you have a learning disability, if you have to put that on your application—

30   You check "yes,"... they probably won't wanna hire you.

You're normal just like everyone else.

Coz in life, people with learning disabilities *don't* have to work harder.

It doesn't take a genius to work in McDonald's.

   They let the students know, which kids is in special ed., *which I don't think is right,*

   When you're in regular ed. you hear all of the bad things about the special ed. kids.

   I'd hear, *"The retards, look at the retards ... Oh man, those kids are retarded."*

   You'd hear horrible things about them.

I think they should get rid of the title "special ed." ... it should be just forgotten about.

   Before I got into special ed.

40   I'd just sit there, playing around, lolly-gagging.

   Then in junior high school I was just falling down, so they was like
   "Michael's not catching up with the class.." da-da-da

   My resource room teacher said, *"Maybe you should get him tested."*

   Then I was tested some more, so I was being tested, tested, and they said,
   "You can either get left back ... or you gotta go to special ed."

   I hated it.

That's the most embarrassing thing to a kid.

Everyone thinks you're slower than everyone else.
When I was in there I was like *"Oh man, you have to get out ..."*
50   I really wanted to kill myself. One day, when they first put me in there,
I think I sat in my bath tub all day trying to drown myself.
I was so pissed off.
   The way they get your parent ...
   "If you don't sign it, your child's gonna be left back
   Coz he can't keep up with the rest of the children ...
   He'll be more embarrassed to be left back than to be in special ed."
   She ends up signing the paper.
   You just lost your rights right there.
Once you're in, it's just like Hell.
60   If they don't want to get you out, if they need a certain amount of numbers
   in that class,
Your behind is gonna stay in there until you graduate.
   The sad thing about it is that they put you in a separate part of the
   building.
   So basically you're labeled already by being put in that department, in
   that little section.
   You can never see the other kids.
My mother looked at the IEPs,
It'd just sit there, like junk mail.
From ninth, tenth, eleventh grade, nothing went on. I didn't get promoted,
I worked hard to get out, but I never got out.
   Then I stopped caring. I just didn't care.
70   I'd act like the rest of them, sit and play around, making jokes, throwing
   stuff.
   I thought if they're keeping me in here, I might as well do what I wanna
   do.
The system ... they really need to do it all over, coz the things they do.
Why put kids in special ed?
   From seventh grade through junior high school graduation I bust my ass.
   I did everything I had to do to get out of special ed.
   In ninth grade I did even better than I did in junior high school.
   I was maintaining about an eighty-five average. I was on the silver honor
   roll.
They never removed me but said, *"In high school you'll get out."*
They always promised me, *"The next year, the next year, you'll get out."*
80   The kids who are placed in that special ed. classroom, they don't want to
   learn.
Most of the kids act like they don't care.
They destroy the rooms, they play around, make jokes, throw stuff ...
But once you're placed in an environment with regular ed. kids, they sit
   there ...
A totally different person, coz they don't want to embarrass themselves.
When you're placed in a room with people who they claim that is your kind,

You don't care.
"Oh, we all special ed., so we can all act the same way."
"You're here, I'm here. So something's wrong with all of us."
"He stupid, I'm stupid, we're all in special ed."

90 When it starts at a young age, when it goes up, it just gets more corrupt.
"I'm special ed. I'm slow in all these classes. I don't need to do none of the work."
They don't think they're good in anything.
  It seems like teachers have pity on the special ed. kids.
  "He can't pass the test, but he's a good boy, so let's just pass him."
  I used to sit in the corner and like, *"This work, Oh my God, might as well
    go to sleep."*
When you're in regular ed., everyone loves you and adores you
In special ed, you're treated differently.
They have mercy on you.
When you do special ed. work, they try to help you too much.

100 Special class work was easy.
  Basically special ed. kids only stick with special ed. kids.
  Coz, at the lunch table if they find out that you're in special ed.,
  No one's gonna hang out with you—you might as well hang out with
    your own kind.
  You're just lost, labeled as a reject.
  So if we knew each other for ten years, and we was in high school,
  And they put me in special ed., we could hang out *after* school,
  But in school we can't associate with each other because you're a special
    ed. kid.
  They don't want to let their reputation go down.
In high school, no one ever really knew I was in special ed.

110 Coz I'd sit there and carry myself like a regular kid.
  I'd come to all my classes. Just to make sure the hall is clear,
  I'd go into the classroom and hide in the corner, coz you don't want people
    to know.
  Once they find out, girls don't want to date you ... no one wants to talk to
    you.
  I kept it to myself, I still do keep it to myself.
  All my friends, and my previous girlfriends never knew.
  I'm not embarrassed about it
  —It's just that something I'm not gonna sit there and go tell people.
    I was mad because I didn't get out.
    After a while I'd be there disrupting the class

120 In the eleventh grade they told me *"If you don't pass all your Regents.
    You're not going to get a diploma, You're going to get an IEP diploma."*
    *"So you're telling me all this time I've been sitting here.
    If I don't pass a test I'm going to get a special ed. diploma, and I can't go to
      college?"*
    I got pissed off.
My sister was like, *"What's an IEP diploma?"*

My mother was all complaining, *"They never told us this,*
They should of informed students of this when they were … juststarting high
school!"
That's when my mother started to get pissed off.
    What was I going to do, if I didn't pass my tests on time?
130   I was going to make sure I failed until I could graduate with my regular
        diploma.
    If I had to stay in high school my whole life to get a regular diploma, I
        would stay.
    I couldn't see myself going out with an IEP diploma—it made me look
        like I was weak.
When I got my regular ed. diploma, that was one of the most proudest
    times of my life,
Coz it showed I strived to do what I wanted to do and I got what I had to
    get.

II. They Say That Black People Only Stick With Black People

They say that black people only stick with black people.
Black kids hang … that's all that was in the school.
You're only allowed to go to your neighborhood.
They don't let you experience each other.
You was with black kids from K through eight,
140 You don't wanna go to school with the white kids.
You don't know how to operate with anything else but black kids.
    But then in junior high school, my mother beat the system.
    Instead of us going to 72,
    We went to Junior High School 210, in a white neighborhood.
    There was black kids in there and white kids and Chinese kids.
    When we was in that school, we actually got to mix and mingle with
        different crowds.
    When you hung out, it was like, *"Yo, these kids are just like us."*
People suspect us.
They stereotype black people as thieves, criminals and all that—you're
    gonna steal.
150 So people really don't trust you.
They'll take your bag before you go into the store.
It makes you not wanna shop.
    Being a black person is hard because you get stopped by the police
        constantly.
    I get stopped all of the time.
    I get into so much trouble just by walking.
I remember, the first time when I was walking from my old junior high
    school.
So I'm walking, chilling and … 10, 15 cops run up on me—
And throw me against the corner wall

"Put you hands on your head!" da-da-da.

160 They cuffed me up and all that, and one cop was talking a whole bunch of mess.

"You fit the description of this guy who was robbing this place."

"Sir, how do I fit the description of somebody when I'm like the only black person out right now with an S-curl?"

"Shut up before I throw you through the damn window,"

Once he found out it wasn't me, I was going to write his badge number down.

He let another lady hold me, got into the car and left.

When I asked her for his badge number, she said,

"Oh no, it's not necessary, just go home."

That's when I started to get real uptight about cops.

170 Ever since then, whenever I walk around ...

Like one day I'm driving—

I turn the corner and I stop my car coz it was raining hard and my car's overheated,

And this lady cop pulled up on me and says,

"Oh, get out the car" da-da-da,

And wanted to search my car. Saying I was trying to pick up prostitutes.

She made all of us get out of the car, and searched it when they didn't have the right.

That's another reason what made me cut my braids.

One day me and my girlfriend was sitting in a car,

She's about to get a permit, and I'm just letting her drive up and down the parking lot.

180 We stop driving, and we just parked in the corner.

So the cops go, *"Were you smoking weed?"*

I said, *"I never smoked a day in my life and I don't drink."*

Just because you're black, they expect that you smoke.

"Just shut up, and sit back."

She told me to open up my ashtray, da-da-da.

III. You Could Get Out of the Hood If You Wanna Get Out of the Projects

You could get out of the hood if you wanna get out of the projects.

It's really how hard you work for it.

I started in Brooklyn, and my mother worked her way up to Queens,

She wasn't going to stay in the projects for her whole life.

190 You just take your behind to school and get out of there.

My mother—she's a social worker—single parent, did everything on her own.

She works for *Coumantaros House.*

Instead of going to prison, you serve time there.

She has a masters degree and still works two jobs to pay off all the bills.
You're still working-class if you got more than one job.
When I was fourteen, I had a job.
Since then I never asked my mother for money.
Most of the time I was doing counselor jobs for smaller kids,
And then when I was 16, I had a job at the YMCA.
200 I helped the kids there with the after school centers.
I think I was bright enough so I started working with all ages.
Then when I was 16 I worked at McDonald's.
I left McDonald's just because ...
I'd work whenever they wanted me to work—
And they giving other people training shirts and management before me.
"Oh, your time will come, your time will come."
My time didn't come.
Three or four people come in after me and got promotions over me.
I left.
210 After McDonald's I went to Ringo's, a video store, trying to become management.
It never happened, I ended up leaving, taking too long.
But after that I worked at a construction site.
Now I'm at the university, as a porter.
Some people—they don't even know how it is to leave their neighborhoods.
"Oh, I'm scared to leave. I don't know what's gonna happen on the outside."
They *can* move out of there.
Finally.

ANALYZING AT THE INTERSECTIONS

In their research on race and gender, Crenshaw (1993) and Collins (1990, 2000) developed intersectional analyses that allowed for a more multi-dimensional understanding of racialized and gendered experiences of black women and, in particular, the workings of domination. Collins' (1990) matrix of domination was designed to illuminate the ways in which various forms of oppression intersect in everyday "reality." She explains,

> Whether viewed through the lens of a single system of power, or through that of intersecting oppressions, any particular matrix of domination is organized via four interrelated domains of power, namely, the structural, disciplinary, hegemonic, and interpersonal domains. Each domain serves a particular purpose. The structural domain organizes oppression, whereas the disciplinary domain manages it. The hegemonic domain justifies oppression, and the interpersonal domain influences every-day lived experience and the individual consciousness that ensues. (p. 276)

In some respects, systems of oppression can be viewed as a series of ropes and pulleys like the intricate rigging of a schooner, all involved in the escalation and de-escalation of sails. Each system is prone to the unpredictable influence of outside forces and, as they are all connected, a movement in one system potentially impacts all others, thereby simultaneously subjecting them to pressures from the inside as well as the outside.

Like Collins and Crenshaw, I recognize that embracing the intricacies of an intersectional approach allows the contemplation of what are usually perceived of as discrete discourses to be fused. In this research, the three merged discourses of disability, race, and class helps to "uncover what lies hidden between them" (Crenshaw, 1993, p. 114), offering opportunities to construct a better means of conceptualizing and politicizing LD as experienced by an urban, black, working-class student. Thus, such an approach is valuable because it helps illuminate Michael's understanding of life in terms of his positionality at the intersection of disability, race, and class, a position long repressed through dominant established modes of power. His story is an example of subjugated knowledge, replete with understandings "develop[ed] in cultural contexts controlled by oppressed groups" (Collins, 2000, p. 286). In the following sections, I will briefly elaborate upon each element within the matrix of domination used with the goal of understanding Michael's experience at the intersections of disability, race, and class.

The Structural Domain

In the structural domain, the organization of social institutions is viewed as contributing to the reproduction of subordination over time. It is important to note that social organizations and institutions may appear permanent and immutable but are—in postmodernist thought—the opposite, impermanent and mutable. Disability, race, and class can be viewed as constructs that serve the means of social organization, with each suggesting a preferred state of being (able-bodied, white, middle-class) over "others" (disabled, people of color, working-class). Thus, even though segregation according to disability, race, and class may be viewed by some people as "natural," or even "inevitable," they are actually lodged in binaric terms, forged and maintained by humans that elevate the worth of one group over the other. While somewhat useful, a binaric view proves reductionistic when "coupling" terms, oversimplifying the complexities involved in the hierarchical structuring within and among concepts such as disability, race, and class. In addition, traditional research attempts to isolate these concepts as separate markers of identity, extricating "one" from the others, thereby creating a simplistic and limited sense of understanding. Because all three markers of identity are experienced concurrently, oppressions (or privileges) within each discourse interlock, requiring all individuals to simultaneously negotiate multiple—sometimes competing—knowledges, thereby creating new epistemological understandings.

Michael holds ambivalent feelings toward his LD. On one hand he says, "It's not like a disability," (Line 11) thinks it "natural"(Line 10), and considers himself "normal" (Line 31). On the other hand, he is secretive and prefers not to disclose his status of LD, yet knows a difference exists in that he "it takes a while longer [to learn]" (Line 22). As a black male, he is acutely aware of largely low expectations placed upon him by the school system and society in general. In addition, he understands the material implications of having been literally positioned where many urban, working-class, black people are—housing projects that inhibit outward and/or upward mobility.

Michael's story reveals the interconnectedness of forces that serve to circumscribe his world: segregated housing patterns are enmeshed with limited schooling options; limited schooling options are entwined with restricted opportunities for employment; restricted opportunities for employment lead to continuing patterns of segregation in housing. In brief, expectancies for working-class black males predetermined by the majority of society are low and even lower if he is labeled disabled. Michael's circumstances clearly indicate his assumed place within patterns of social reproduction. Nonetheless, Michael's story also encompasses his "in progress" attempts to transcend the restrictive nature of racial and class divisions, operating together in his local environment. Noteworthy are the calculated moves of Michael's mother that allow her children to cross the "zoned" boundaries imposed by school districts, thereby enabling her children to partake in a far more racially and culturally diverse educational environment.

Michael is aware of how insular schooling patterns are, drawing attention to how they not only duplicate segregated housing patterns but also reflect segregated social interactions for people with disabilities. He notes, "Special ed. kids only stick with special ed. kids (Line 101) … if they find out that you're in special ed., no one's gonna hang out with you" (Line 103), and "Black people only stick with black people" (Line 135). Such comments indicate that there is segregation within segregation. More disturbing, perhaps, is the implication that Blacks in special education tend to favor other Blacks. Thus, social segregation is compounded by disability and race. Given that individuals come to understand the world through the social circles they inhabit, the Russian doll-like nesting of one "type" of person only communicating with the same "type," serves further to shrink their world, exacerbating oppressions.

Through his narrative, Michael conveys the "world" of an urban, working-class, black student labeled LD as significantly circumscribed; his exposure to, and interactions with, nondisabled and middle-class students are virtually nonexistent; his interactions with Whites and Asian American children are highly curtailed. To wit, the lack of exposure to diverse ways of being and thinking help create a highly contained comfort level, within the boundaries of "a ghetto" (socially, culturally, materially, and psychologically), a place from which an individual may be discomforted to leave, as evidenced by Michael's friends in Brooklyn.

Insecurity in managing the world outside of the neighborhood also has serious implications for obtaining potential jobs. Michael is industrious, gaining job experiences from an early age. On the other hand, he feels he was "passed over" for promotion in several jobs, inspiring him to quit. Interestingly, the causes of non-promotion are never mentioned. However, he does not think that a learning disability makes any difference outside of school in the world of work, explaining, "People with disabilities *don't* have to work harder. It doesn't take a genius to work in McDonalds" (Line 32). Based upon his personal experience, it is clear to understand Michael's rationale. Yet, it is worthy to note that Michael's frame of reference is based in working-class assumptions; he equates the world of work with demands placed upon workers in fast food franchises. Furthermore, he believes it best if individuals labeled LD do not self-identify as such on job applications for fear of being misjudged and stereotyped as *unable*.

In sum, the structural domain shows how Michael conceived of himself in terms of his positionality within three discourses: (1) acceptance of his own disability, but rejecting his public status of "special ed."; (2) acknowledgment that Blacks have far fewer opportunities and choices than Whites in terms of education, housing, and jobs—and the influence this phenomenon exerts upon self–expectation and the expectations of others; and (3) recognition and acceptance of being working-class, along with a wariness of the middle-class, balanced by a desire to change his current economic status to become upwardly mobile. In contemplating the intersectional positioning of Michael in terms of LD, race, and class, a recurring theme is one of containment. His comments reveal experiences and understandings of having his movements literally inhibited in terms of disability (segregated classes, IEP diplomas), race (workplaces, public places, school locations), and class (opportunities for further education, job expectations, neighborhoods in which to live). Thus, the limitations imposed upon Michael indicate very real forms of social restraint. The following section examines how these limitations are maintained.

The Disciplinary Domain

The disciplinary domain of power manages oppression. Collins (2000) notes the significance of bureaucracy as a form of social organization, believing that both "the structural and disciplinary domains of power operate through system wide social policies managed primarily through bureaucracies" (p. 285). Schools are prime examples of bureaucratic control and perhaps no more so than in special education. When students fail to achieve predetermined standards in academic, social, emotional, behavioral, and physical realms, they are often referred to evaluators who specialize in ascertaining whether a "disability" is present within the student. If one is "detected," the student enters the ultra-bureaucratic realm of special education. Many students who are labeled disabled are subsequently placed in separate facilities for all or most of their daily schooling. Thus, it can be argued that schools provide the necessary conditions for disciplinary power, allowing practices that manipulate and control individual behavior, self–identity, and expectations—thus *creating* special education students, and undermining their rights.

Michael lamented the loss of privacy in relation to his disability status. He states, by virtue of labeling and placement, school authorities "Let [general education] students know, *which I don't think is right*, which kids is in special ed." (Line 34). This standard procedure results in students who are labeled LD attempting to conceal the "disabled" aspect of identity. Here, we clearly see how schools discipline individual bodies through "commonplace" arrangements of spatialization and how such configurations serve to shape a reified "us" and "them" division. These divisions serve as physical, symbolic, and psychological segregation, a location Michael bitterly analogizes to Hell—it is easy to get in, it is hard to get out.

As Michael phrases it, "The way they get your parent .. ." (Line 53), is telling, in that it indicates the manipulative power operant within evaluation meetings. He understands that parents have to be involved in the initial decision making but once they have signed the papers, it is as if they have served their purpose in granting the label and acknowledging placement. Michael does not like the term LD because he claims it makes people "look at you like you're not them" (Line 18). Thus, his solution would be to "get rid of the title 'special education'" (Line 38). With the label of LD and placement in special education, Michael believes he has lost his rights. It is no wonder that he urges stopping the use of labels and the dissolution of special education departments. He paints a picture of "merciful" teachers, overzealous in their desire to help, and unable or unwilling to give challenging work in an atmosphere infused with low expectations. The sacred cow of special education—the IEP—is regarded as "junk" (Line 66) although it purportedly "speaks" for Michael. His experiences exemplify how the everyday apparatus of special education bureaucracy works, leaving him to conclude, "The system ... they really need to do it all over, coz the things they do. Like why put kids in special ed?" (Lines 72–73).

In terms of the disciplinary domain, Michael views those in ability, race, and class positions different from his as having a much easier life. General education students are "loved and adored"; Whites are assumed to have middle-class status and therefore more economic power. Contrastingly, as per his self-description: In school he is the special ed. pariah; on the street he is the innocent youth guilty of being black; and in the workplace he is the laborer passed over. On all fronts, he is always aware of being underestimated by educators, law enforcement, and employers—a positionality unfamiliar to nondisabled, white, middle-class professionals, who overwhelmingly constitute the fields of education, law, and private industry.

As previously mentioned, while segregation by race and class are not officially sanctioned, separation according to disability is. The segregation of black students is not perceived as an issue of race. Schools, therefore, are organizations that can significantly limit educational opportunities and contribute to social reproduction in terms of disability *and* race. Furthermore, such reproduction is made possible by the very nature of schooling being middle-class both in its basic values and middle-class teaching force. Thus, by contemplating the interstices of disability, race, and socioeconomics in public schools, it becomes apparent how patterns of social inequalities are repeated, ultimately maintaining the status quo. For example, students are often prevented from fully interacting with each other if they are deemed disabled and/or are from different races. These schooling arrangements are due to institutional structures and requirements according to disability (testing, IEPs, etc.) interacting with estab-lished social patterns according to race (housing, work, etc.). Such arrangements are rarely challenged by professionals who stand to benefit from working within them. Indeed, the justification of many, particularly urban professionals in special education, is predicated upon the clinical hunt for disability by day to feed existing segregated programs, before personal retreat into "safe" neighbor-hoods by night (read: middle-class, and largely racially segregated). The "us and them" division so apparent in Michael's narrative between able-bodied and disabled is clearly marked within school organizations, and the apparatus that sifts and separates does not recognize its complicity in reproducing entrenched social patterns.

The Hegemonic Domain

The hegemonic domain rationalizes oppression. Collins (2000) describes how it supports the other domains, because "the hegemonic domain of power aims to *justify* [my emphasis] practices in these domains of power" (p. 284). In every cul-ture there exists a dominant ideology—namely, a system of ideas and beliefs that underlie common practices that permeate all levels of society. The roots of such ideologies are embedded in longstanding and deeply entrenched ways of think-ing that posit one group of people (less valued) as inferior to others (more val-ued). The hegemonic domain is able to manipulate ideology and culture by acting as a mediating link "between social institutions (structural domain), their organizational practices (disciplinary domain), and the level of everyday social interaction (inter-personal domain)" (Collins, 2000, p. 284). In postmodern terms, the discursive interactions between all domains of influence serve to posi-tion individuals in terms of power. Thus, the hegemonic domain is enormously influential in how people are constituted through being disabled, raced, and classed.

Foucault (1980) describes how the disciplining effect of power permeates "the very grain of individuals, touches their bodies and inserts itself into their actions and attitudes, their discourses, [their] learning processes and everyday lives" (p. 39). Thus, effective systems of oppression foster an individual's own subordination in structures through the cultivation of a consciousness that will-ingly self-inscribes into a classification system that is usually limiting and possi-bly detrimental to the person. Conveying the reaction of a student with LD who is placed in segregated classes, Michael says "He's gonna sit there like, *'I'm special ed. I'm slow in all these classes. I don't need to do none of the work,'*" (Line 91) explaining, "They don't think they're good in anything" (Line 92). By his account, even placid students become transformed by their placement in special education. Of course, Michael's initial fear of being placed in special education conveyed his sense of horror in being placed with "the retards" (Line 36). Interestingly, on several occasions Michael associates stigmatization with death. On being "pronounced" special ed., his fleeting bathtub death wish later symbolically resurfaces when describing how he feels others view him as contaminated, akin to having a condition as life-threatening as

HIV/AIDS. Thus, the label of LD believed as innocuous by those who distribute it, has a far deeper meaning for those so labeled: it is tantamount to educationally-imposed leprosy, a condition from which non-disabled students (and many teachers) retreat.

Michael's multiple incidents with the police demonstrate how "the long arm of the law" has positioned him—as a black male—in an adverse light. For example, when repeatedly stopped in his car, he is an assumed thief; when teaching his girlfriend to drive in an empty parking lot, he is presumed to be smoking marijuana; when waiting for his overheated car to cool down, he is accused of cruising for prostitutes. He recognizes these "charges" made by others and feels their expectations bear down upon him in his stating, "Just because you're black, they expect you to smoke [marijuana]" (Line 183) and openly acknowledges that "they stereo-type black people as thieves, criminals, and all that" (Line 149). In one telling instance, after being stopped for the umpteenth time by the law, Michael recalls, "That's another reason what made me cut my braids" (Line 177). While seemingly a small gesture, it symbolizes a larger phenomenon, that of yielding to the preferred images upheld by the dominant culture, images that deemphasize or erase obvious signs of "blackness," ostensibly making "it" (him) less easy to detect.

In terms of class, Michael is desirous of finishing community college and taking advantage of possibilities that completion of a degree will bring. His future looks optimistic in terms of opportunities often not afforded within working-class lifestyles. Obviously a hard worker with a plethora of experience, Michael believes in his own agency and sees education as a way out of the ghetto. On another note, it is evident that individuals similarly positioned to Michael in terms of disability, race, and class, are rarely represented in the mass media, and if so, portrayals are stereotypic and inaccurate. His very existence, however, defies such widespread invisibility, yet ironically also signifies hypervisibility in terms of surveillance by those in positions of power. Paradoxically, both instances—invisibility and hypersurveillance—are examples of regulation in that they promote and uphold social and cultural disempowerment of individuals at these particular intersections.

The Interpersonal Domain

The interpersonal domain is interconnected with other domains in that it functions through commonplace practices of how people interact with one another. It is at this microlevel of social organization that we must consider responses to the relentless nature of oppression because everyday experiences continually influence and shape the consciousness of each person, and how he or she functions within collectives.

There is always a degree of tension, then, between what individuals think, feel, and know, and what the "experts," "people in charge," "those in positions of authority," and "those in power" assert what knowledge is "correct" and therefore believed as real, or "the truth."

In contemplating disability, race, and class, it is useful to look at the everyday experiences for Michael, in terms of interactions, routines, and incidents. His desire to leave special education was summarily dismissed via a string of ongoing placebos, interminable placations that he came to understand as hollow. Michael's unhappiness within special education remained until he graduated with a "regular" diploma. In addition, he is guarded about the label of LD, believing it to be a private matter, regardless of it being part of the public domain as a "special education student." In terms of race, Michael comes to self-regulate his looks (cutting braids, lessening the degree of blackness) and exercise caution in deciding when and where to drive, realizing that being profiled according to gender, age, and skin color are commonplace occurrences. In terms of class, like his mother, he pushes for academic success as key to a better life.

Michael's interactions within the personal domain are evident through his participation in routines and recollection of specific incidents. Pervasive throughout Michael's story, when contemplating the education system, is his strong sense of "us and them," as if people are reified into "special" and "general" students. His routine of deliberately arriving to class late perhaps exemplifies the "choices" students feel they are forced to make. Michael also notes the vagaries of special education; sometimes it appears to be caring, while at other times he lists a litany of complaints about daily experiences: low-level repetitious work, the disintegration of student behavior (at times including his own), increased levels of violence, and being uninformed about educational options. On another note, an important and often unacknowledged aspect of school segregation is that students in special education are commonly subject to verbal abuse in the form of teasing and taunts. Michael describes how the barbs he once threw at others became directed to him, "stupid" (Line 89), "something wrong" (Line 88), "slow" (Line 25), and "retarded" (Line 36).

Life in the interpersonal domain at the interstices of being labeled LD, black, and working-class in an urban setting foregrounds an interesting issue: In being subject to discriminatory dialogue or actions, how does an individual know which aspects of his or her identity drives the discrimination? For example, when Michael is shadowed and demeaned by store attendants while shopping, is it because he is black and/or working-class and/or and young? Can this incident at the microlevel illuminate how oppressions often circulate, overlap, and merge without clearly drawn lines? This instance also illustrates how oppressions based on disability, race, and class move between the macro and the micro dimensions—indiscernible in source and unpredictable in strength as they flow through the structural, disciplinary, and hegemonic domains, flooding into everyday personal experience. Personal interactions at the microlevel reveal ways in which oppression impacts people within their everyday circumstance, yet it is arguably impossible for an individual to determine where oppression as a result of one marker of identity starts and the others stop. That said, perhaps specific instances of discrimination are perceived/felt to be as a result of one particular marker of identity (such as in Michael's case, being black) that is *compounded* by others (such as being working-class/perceived as poor, and/or young). Regardless, the interpersonal domain of everyday power plays manifest within relationships and interactions experienced by Michael both in and out of school leave him aware of how he has been positioned by others, influencing a degree of self-monitoring through word and gesture.

Empowerment

The interplay between oppression and activism within the matrix of domination recognizes that interconnected domains of power are responsive to human agency. In other words, "Such thought views the world as a dynamic place where the goal is not merely to survive, to fit, or to cope; rather, it becomes a place where we feel ownership and accountability" (Collins, 2000, p. 290). This notion is important to include because it reveals forms of resistance to dominant ways of thinking, ways in which young people seemingly positioned at "the bottom of the well" (Bell, 1992, p. xi) assert themselves.

Within Michael's narrative, his counterclaims are manifold. First, he understands dyslexia as natural, within the realm of human variation. Second, he appears to understand the workings—and therefore limitations—of special education. For example, he realizes that being "decertified" is not a real option for working-class and poor students whose parents are discouraged from being active participants. Indeed, keeping students in special education appears to serve the configurations of schools; he notes there has to be a certain number per class, and that is what holds most weight in decisions of placement.

Michael's hopes and dreams, his attempts to mobilize staff members to get him out, and his plot to fail classes until he passed the required state examinations to obtain a diploma, are examples of him "pushing back" at the system and can be viewed as a bittersweet experience. It is bitter in that his attempt to get out of special education was unsuccessful. It is, however, also sweet because he did graduate. Michael knows the odds against him to succeed were formidable. Staying in special education was high stakes because expectations placed upon him and his peers constituted obtaining, at most, an IEP diploma.

Michael's critique of special education is often quite scathing. His beliefs are molded by his own experiences and observations. He observes with great acuity that many students in his own disabled, raced, and classed position are not served well by the field of special education, for what the field appears to regard as benevolent and beneficial feels like the opposite. In addition to his critique of special education, Michael's counter-challenging spirit is evident in descriptions of interchanges with police, such as the request for a disrespectful officer's badge number. However, he is aware of how far he can protest in interchanges, and tempers his responses according to his understanding of the way things are. In terms of class, Michael believes the parameters placed upon black youth like himself can be transcended, yet must come from agency within. He believes "They *can* move" (Line 216) both literally and figuratively.

Summary

Michael's story illustrates how issues of disability, race, and class intersect. As a student labeled LD, he is stigmatized and ostracized by peers. Unlike race and class, LD does not trigger "pride," but rather shame. Stereotypic mainstream expectations of Michael position him as lazy and/or "unable" because of misperceptions of LD and long held assumptions of racial inferiority. His fate is "sealed" in special education because his working-class mother, while to some extent a class-straddler is still unable to "take on" the intricacies of special education bureaucracy. Unlike middle-class students, Michael does not have a sense of entitlement to attend college; he knows he has to work extra hard to transcend the expectations placed upon him. At the less valued side of three binaries, Michael understands the sense of restraint placed upon him as a black, "disabled" learner, from a working-class background. Michael's story is contrary to the meritocratic myth of individuals being on "an even playing field" in current society. Rather, he faces formidable challenges in order to break from the containment(s) imposed upon him. His everyday story reveals ingenuity and resilience in the face of adversity on many fronts.

IMPLICATIONS FOR THEORY, RESEARCH, AND PRACTICE

In terms of theory, when using an intersectional approach, traditionally compartmentalized conceptualizations (such as disability, race, and class) are intertwined, requiring the researcher to view a familiar arena (such as schools) and people (such as students) with a new lens. The matrix of domination is a useful analytical tool to assist in making connections between the macro and micro aspects of human existence. Ways in which discourses of disability, race, and class interlock become apparent through examining the arrangements of long term social structures, noticing the ways these cultural arrangements function to discipline individuals, identifying representations circulating throughout the mass media, and contemplating the impact of all these influences upon interpersonal experiences.

This study is an example of emancipatory research and, as such, allowed the participant-researcher the *act of naming*, usually reserved for the dominant group (Whites, males, the able-bodied, and so on). It is clear from Michael's portrait in which he "names" his position(s) in the world (both inside and outside of school) how his understanding significantly contrasts with the that of commonplace deficit-based models used by teachers and researchers (Kauffman & Hallahan, 1995). This study indicates that by doing research with those labeled as "others"—with the specific desire to use their own words to represent themselves—enables the co-construction of knowledge. Such knowledge, therefore, attempts to addresses widespread misrepresentation, as well as imbalances of power....

CONCLUSION

In designing this research, an intention was to deepen my understanding of a particular phenomenon—life at the intersections of being a black, working-class youth labeled LD, the largest group of students placed in special education in urban settings. I simultaneously sought to create an opportunity for Michael to further his own understanding, making him a researcher of his own life. It is my belief that in undertaking this project, I have worked with a person from a highly specific knowledge community who has acted as a bridge between theory and practice, helping to create a "seeding ground for competing stories that may lead to meaningful, enduring, educational change" (Anderson, 1997, p. 132).

REFERENCES

Anderson, L. W. (1997). The stories teachers tell and what they tell us. *Teaching and Teacher Education, 13*(1), 131–136.

Bell, D. A. (1992). *Faces at the bottom of the well: The permanence of racism.* New York: Basic Books.

Collins, P. H. (1990). *Black feminist thought: Knowledge, consciousness, and the politics of empowerment.* New York: Routledge.

Collins, P. H. (2000). *Black feminist thought: Knowledge, consciousness, and the politics of empowerment* (2nd ed.). New York: Routledge.

Crenshaw, K. W. (1993). Beyond racism and misogyny: Black feminism and 2 Live Crew. In M. J. Matsuda, C. R. Lawrence, R. Delgado, & K. W. Crenshaw (Eds.), *Critical race theory, assaultive speech, and the first amendment.* (pp. 111–132). Boulder, CO: Westview Press.

Kauffman, J. M., & Hallahan, D. P. (1995). *The illusion of full inclusion.* Austin, TX: Pro-Ed.

Lincoln, Y. S., & Denzin, N. K. (2000). The seventh movement: Out of the past. In N. K. Denzin & Y. S. Lincoln (Eds.), *Handbook of qualitative research* (pp. 1047–1065). Thousand Oaks, CA: Sage.

Richardson, L. (1990). *Writing strategies: Reaching diverse audiences* (Vol. 21). Newbury Park, CA: Sage.

36

Health Inequities, Social Determinants, and Intersectionality

NANCY LÓPEZ AND VIVIAN L. GADSDEN

Problems in health disrupt the human developmental process. They undermine the quality of life and opportunities for children, youth, and families, particularly those exposed to vulnerable circumstances. Despite incremental change within and across health-serving agencies and increased health education and scrutiny of patient care, we continue to see significant disparities in the quality of health and life options that children in racial and ethnic minority, low-income homes and neighborhoods experience (Bloche, 2001). Research has uncovered several interconnections between health and environmental and social factors (Chapman & Berggren, 2005; Thorpe & Kelley-Moore, 2013), but has not always shifted paradigms sufficiently to either disentangle intersecting inequalities or tease apart the ways in which social factors and structural barriers at once interlock to prevent meaningful and sustainable change.

In this essay, we focus on the potential and promise that intersectionality holds as a lens for studying the social determinants of health, reducing health disparities, and promoting health equity and social justice....

We ask: How do we engage in inquiry and praxis (action and reflection) that departs from the understanding that intersecting systems of oppression, including race/structural racism, class/capitalism, ethnicity/ethnocentrism, color/colorism, sex and gender/patriarchy, and sexual orientation/heterosexism, nationality and citizenship/nativism, disability/ableism and other systemic oppressions intersect and interact to produce major differences in embodied, lived race-gender that shape the social determinants of health? How can we as scholars, researchers, and practitioners concerned with child and family well-being take seriously the reality of intersecting systems of power intersecting to produce lived race-gender-class and other social locations of disadvantage and develop an intersectionality health equity lens for advancing health equity inquiry, knowledge projects and praxis?

We argue that the potential power of intersectionality as a transformational paradigm lies in two domains relevant to understanding social determinants. First, it is a critical knowledge project that questions the status quo and raises questions about the meaning and relationship between different social categories and

SOURCE: Reprinted with permission from Health Inequities, Social Determinants, and Intersectionality, Nancy López, PhD, University of New Mexico; Vivian L. Gadsden, EdD, University of Pennsylvania, by the National Academy of Sciences, Courtesy of the National Academies Press, Washington, D.C.

intersecting systems of privilege and oppression (Collins, 2008, 2015; Collins & Bilge, 2016; Bowleg, 2008; Yuval-Davis, 2011). It also pushes against the idea of "blaming the victim"—that is, the simplicity of explaining health or educational outcomes by attributing problems to individuals' genetics or cultural and social behaviors alone. Second, by focusing on power relations at the individual, institutional and global levels and the convergence of experiences in a given sociohistorical context and situational landscape, it serves as an anchor to advance equity and social justice aims for marginalized communities that have experienced and continue to experience structural inequalities (Crenshaw, 1993; Collins, 2008, 2009, 2015; Weber, 2010)....

There is growing evidence and professional wisdom to suggest that health disparities do not exist in isolation, but are part of a reciprocal and complex web of problems associated with inequality and inequity in education, housing, and employment (Schultz and Mullings, 2006; Weber, 2010; LaVeist and Isaac, 2013; Williams and Mohammed, 2013). These disparities affect the unborn child through social-emotional challenges such as maternal stress and diagnosed and undiagnosed medical problems, including higher prevalence of gestational and preexisting diabetes in some pregnant populations. In other cases, they are observable at birth, particularly pronounced when prenatal care is unavailable, when the importance of care is not understood fully, and when young children are not exposed to the cognitive and social-emotional stimulation needed to thrive. These and other problems are manifested in parental stress, for example, in mother-headed and two-parent, low-income, and immigrant households alike. Parent and family adversity may reduce the number and quality of resources available and life experiences for children and families in the early years and throughout the life course. Such adversity is exacerbated by structural barriers that limit employment opportunities, increase housing instability, and contribute to homelessness, and that constrain efforts by families to effect positive change....

Racism and discrimination are overwhelmingly significant factors, but are not the only critical dimensions related to identity to be considered (Williams & Mohammed, 2013). They are tied inextricably to multiple identities and social locations that children, youth, and adults assume, and define a context for health (Bauer et al., 2016; Brown et al., 2016). One might argue that there is no issue more important than ensuring health. How a person understands this point and is able to act upon it is determined by more than her or his cognitive ability to engage the idea. It is influenced as well by a range of dynamic and situational identities and social positions that are biologic, cultural, and epigenetic, by social determinants (i.e., where people are born, grow up, work, and age, and interact with their changing environments), and by a person's social experiences and encounters than her or his self-agency across a variety of social settings. Even individuals with the strongest work ethic and sense of agency, when faced with daily problems associated with intersectionality across any combination of racial, class, gender, sexual orientation, language, or disability systemic oppressions and discrimination, may find fighting against these inequalities daunting....

Our objective in the remainder of this essay is to provide a discussion of the possibilities for innovation in conceptualization, methodologies, and practices that can promote human development and health equity through an "intersectionality health equity lens." We employ Jones' (2016) definition of health equity. Jones defines health equity as "the [active] assurance of optimal conditions for all people." Dr. Jones explains that we can get there by "valuing everyone equally, rectifying historic inequities and distributing resources according to need." Jones invites us to think deeply and critically about equity as a never ending process that requires constant and ongoing vigilance and not just an outcome that once accomplished can be forgotten. Building on Jones' (2016) and Collins and Bilge's (2016) ideas about equity and intersectionality we define an intersectionality health equity lens, as ongoing critical knowledge projects, inquiry and praxis that can include research, teaching, and practice approaches that are attentive to the ways in which systems of inequality interlock to create conditions for either health equity or health inequities (Collins & Bilge, 2016; Collins, 2008, 2015; Crenshaw, 1993).

We also embrace Collins and Bilge (2016:25) core ideas of intersectionality, namely a focus on inequality, relationality and connectedness, power, social context, complexity and social justice. They use the analogy of "domains of power" to paint a picture of the way that power is visible at the "interpersonal" or individual level in terms of who is advantaged or disadvantaged at the level of social interactions. For example, individuals may experience privilege and or disadvantages when searching for a job, housing, interacting with law enforcement or even when accessing a voting booth. Collins and Bilge (2016:8) assert:

> *"Using intersectionality as an analytic lens highlights the multiple nature of individual identities and how varying combinations of class, gender, race, sexuality, and citizenship categories differentially position every individual."*

Collins and Bilge (2016) also underscore that we must always be attentive to the "disciplinary" level as a domain of power that organizes and regulates the lives of people in ways that echo our distinct social positions with regard to systems of oppression. For example, rules about who will or will not be seen at a medical office because of the ability to pay a co-pay, who will or will not be admitted to a domestic violence shelter based on their English Proficiency and who has access to gifted classroom, based on IQ test scores that are rooted in eugenicist origins, will inevitably impact the conditions for the advancement of health equity.

Collins and Bilge also invite us to reflect on how power is also visible at the "cultural" level or in the realm of ideas, norms and narratives. For Collins and Bilge (2016) ideas matter and how messages are manufactured creates explanations, justifications or challenges to the status quo vis-à-vis inequalities. For instance, if the idea that racialized health inequalities are simply a matter of individual behavior, food ways and choice, and that we life in a meritocracy, where your station in live is simply a matter of individual effort, then we are subscribing to what Bonilla-Silva refers to as "colorblind" racism or the belief that present day realities of race-gaps in health only mirror individual deficits of individuals or defective cultures.

The last arena where Collins and Bilge interrogate the dynamic of power, include the "structural" level or at the level of institutional arrangements, which interrogates how intersecting systems of institutionalized power, whether in the economy and labor market in terms of who's labor is valued and who is exploited, or at the political level in who is granted substantive citizenship rights and privileges and who is not, as well as at the level of who has access to structures of political power and influence, shapes the institutionalization of the conditions for health equity. For example, the struggle for sovereignty of indigenous people as evidenced in the Standing Rock movement to protect indigenous land and water for generations in South Dakota, provides a snapshot of the structural location of indigenous nations and capitalist neoliberal actors that are in a struggle to define the environmental context for current and future generations, which will have grave consequences for health justice for marginalized indigenous communities.

While an intersectionality health equity lens may inform or drive interdisciplinary or transdisciplinary research, it must also be considered as part of both the process of conceptualizing the problem and the product of research on the problem. Throughout this essay, readers should consider the potential applications of an intersectionality health equity lens, how its use enhances (or disrupts) our understanding of salient and longstanding issues, what might be learned from its use that will inform and deepen research and practice with children and families who are among the marginalized in society, and what types of intersectionality-focused approaches might lead to health access and equity. In the next section, we focus on the contributions of an intersectionality health equity lens for research and for promoting health equity.

AN "INTERSECTIONALITY HEALTH EQUITY LENS" FOR SOCIAL JUSTICE

When developing or applying an intersectionality health equity lens, the researcher engages in deep self-reflection that contextualizes and recognizes the ways in which race, gender, class, sexual orientation, disability, and other axes of inequality constitute intersecting systems of oppression. Such systems produce very different lived experiences for entire categories of people who are embedded within complex webs and social networks at different levels, for example family, neighborhood, and community as well as institutional and structural. These lived experiences can either enhance or challenge the developmental pathways of children through adulthood and the ability of parents and families to ensure a positive trajectory for their children. They affect both the individual child and the networks and communities in which children live and grow and that define their access to resources.

An intersectionality health equity lens for the purposes of our discussion takes on the broader, philosophical meaning attached to praxis as process involving health, educational, and social service researchers and practitioners in not only self-reflection but also action. Critical self-reflection allows researchers and practitioners to continually and closely examine their own race, gender, class, sexual orientation, disability, language, nativity/citizenship and social position, and their relationship to systems of inequality as part of intersecting systems of oppression and privilege. It argues for researchers and practitioners to draw upon their own experiences with health inequities and discrimination and to understand and respond to new or subtle forms of inequities and discrimination. These subtle forms of inequity and discrimination are sometimes so deeply embedded in and accepted as societal practices that they may be difficult to uncover, yet render many children and families hopeless. The interplay between and among relevant systems and the statuses accompanying power attributed to different ethnic, racial, cultural, and socioeconomic groups affect both individuals and their social networks (e.g., family, neighborhood, and community). They are tied directly to and within institutional and structural hierarchies.

Crenshaw (1993) points to the entrenched nature of inequity, underscoring the need for a useful paradigm in which to locate the issues faced by African American women and other racially stigmatized, visible minority women of color. Credited with creating a systematic analysis of the concept of intersectionality, Crenshaw (1993) urged readers to "map the margins" by focusing on those social locations that remain invisible. She argues that such invisibility results from a reliance on a mythical universal "black experience" (e.g., when we assume that the default category is the "black male experience" and by the same token when we speak about "'women's experiences" and assume that all women's experiences are represented in white women's experience). In each of these dominant conceptualizations of the black [male] and [white] woman's experience, heteronormativity is the invisible structure.

Crenshaw (1993) also illustrates how language, and potentially nativity and citizenship status, can serve as other axes of stratification that have received less attention than race and class. To illustrate her point, Crenshaw flexes her intersectional lens to bring into sharp relief the effects of "good intentions" on the real lives of women. She demonstrates that despite their good intentions, some domestic violence shelters may operate in ways that ignore the plight of immigrant women with children who may not speak English and are unable to access domestic violence shelters. It goes without saying that this would structurally exclude immigrant (both documented and undocumented) women and their children who do not speak English. "Nativism, English Only" categories are the invisible, yet real, structural barriers to addressing domestic violence in the aforementioned situation. By the same token, members of lesbian, gay, bisexual, transgender, queer, and in transition (LGBTQI) communities may not face explicit rules about being barred from these services because of their gender identity, but if counselors and other providers assume that their clients are in heterosexual, gender-conforming relationships, heteronormativity can operate as another type of an informal barrier.

One might well ask, given the complex relationships in addressing identity, whether it is possible to create intersectionality-grounded projects that integrate the issues of race, class, gender, disability, and other identities, statuses, and social locations in research on health and well-being for the range of issues facing marginalized children, youth, and families. Although we do not have a simple response, we highlight the need to address the real or perceived complexity of creating such projects and allowing time and resources for them to be developed well and to be refined. We similarly understand the limitations of relying on one-dimensional categories that are, at best, additive, for example, first race, then maybe class, then maybe gender, depending on the focus of the research. As the World Health Organization (WHO, 2015) and several health researchers before (e.g., LaVeist and Isaac, 2013; Williams & Mohammed, 2013) suggest, understanding the social determinants of health requires a broad reach to identify, and respond to, the embedded and entrenched inequities of policies that are situated in place and context.

Intersectionality health equity lenses help us understand that every person's experience is fundamentally different than the experience of others, based on their unique identity and structural positions within systems of inequality and structural impediments (Feagin & Sikes, 1994; Nakano Glenn, 2002; López, 2003, 2013a; Weber, 2010). More than just a theory or framework to be used selectively, it is a commitment to developing a relentlessly critical and self-reflective lens that begins with the premise that race, class, gender, and other axes of social identities are intertwined and mutually constitutive, and that such a lens can help advance health disparities research, practice, and leadership by making the invisible visible.

INTEGRATING RACE, GENDER, CLASS, AND SEXUALITY AS LIVED EXPERIENCES: A CASE EXAMPLE

In considering intersectionality projects, we must be aware of the overwhelming inequities associated with longstanding problems of race and gender and the added problems of poverty and class, problems that have narrowed in some cases over time but where inequality persists. It should come as no surprise that an intersectionality-focused project might appear opaque or obscure initially, despite its potential to uncover the breadth of issues faced in ensuring health and well being....

López (2013b) proposes the "racialized-gendered social determinant of health" as a heuristic device or framework for centering the lives of marginalized communities. This framework consists of two major concepts: (1) "lived race-gender" and (2) "racialized-gendered pathways of embodiment." López (2003) offers an example of the enactment of these concepts in the minds and experiences of both the observer and the observed. For example, she makes explicit the ways in which race-gender disparities are enacted and experienced in school and society by young Dominican and Caribbean men and women in what she calls "New York Immigration and Racialization." Consider Orfelia's narrative on the public's perceptions of blacks, hispanics, and whites and the differential result of their identities on these perceptions:

If you put on the news, anyone who does anything bad, if he's not Black, he's Hispanic .... You watch the news and you see that when any white guy does something, you won't see their face. They might just say it, and that's all. But if it's a Dominican, a Hispanic, a Black, they put him on for about two minutes, so that you can know him. (p. 23)

Orfelia points to the ways in which she has internalized race and gender stigma as dominant identity markers and their intersections with place (Queens in New York) and other intersectional identities such as immigrant and Spanish speaker. The mental health costs of feeling racially stigmatized may become embodied by many youth who also feel what sociologist W. E. B. DuBois coined in 1903 as the "double consciousness" experienced by blacks in the U.S. context or the sense of always being seen with contempt, pity, or disdain because of one's stigmatized status (DuBois, 1999; Vidal-Ortiz, 2005).

López also underscores the dominance of race and gender identities, along with other identities (e.g., social class, sexual orientation, age, ethnicity and nativity, and legal status) that form the basis for education and health frameworks. She draws upon a personal example to demonstrate connections among race, gender, sexuality, and social class and the significance attached to heteronormativity.

While race, gender, and class were overriding identities in the short narrative, heteronormativity was the silent but overpowering lens for López and her cousin. As López notes, the nature and type of her cousin's experiences in and out of school, within family and community contexts, and with stressors that were unnamed distinguished the two cousins. As she suggests through this anecdote, sexuality played only a small though apparently significant part in the everyday encounters that her cousin faced. What remains unanswered are questions about the ways race and gender (male and Dominican) played in her cousin's schooling, and the ways that gender nonconformance (what we now refer to as transgendered identity) produced barriers to health access, care, prevention, and maintenance; to employment; to housing; and to the daily acceptances that allow individuals to maintain not just a healthy personal racial, gendered, class, ethnic, or sexual identity but also an identity that can be embraced in full in all social domains and situations that López's cousin traversed throughout their short life.

Focusing on López's cousin's experiences from a health equity perspective, several additional questions are raised: Did the health system fail her cousin, or was it the larger social system that did not accept their intersectional identities? To what degree do our current systems of data collection make her cousin's intersecting lived oppressions vis-à-vis race, national origin, class, sexuality, gender identity, and nativity invisible? If we collect data only on gender identity and not class, nativity, citizenship, ethnicity, language, and/ or national origin, do we make some social locations invisible? Do we ignore the temporal element of identities across the life course? How would López's cousin's life experiences have been different if her cousin had been from an LGBTQI middle class, Dominican immigrant family that was light skinned, white-looking Latinx and not a visible minority? All of these data challenges are opportunities for establishing communities of practice committed to intersectionality praxis (action and reflection)....

DEVELOPING AN INTERSECTIONALITY HEALTH EQUITY LENS: CHANGING THE NARRATIVE FOR SOCIAL JUSTICE

What happens when health research takes an intersectional stance in producing and using knowledge to effect positive practice and social change and advance equity? In what ways do our personal and professional positionalities contribute to this intersectional stance, our research, and the opportunities afforded by our ways of seeing and knowing the world? How do we address the health inequalities and inequities that reduce these opportunities for children, youth, and families and redirect them to promote social justice?

We are aware that the answers to these questions require time, depth of inquiry, and breadth of analysis, and that they contribute to, rather than outline, a social justice framework. Throughout this essay, we argue that critical, self-reflexive intersectionality health equity lens and praxis depend upon a visceral commitment to uncovering the workings of the multiple systems of inequality in unpacking the social determinants of health. Such a lens might be expanded to become an "intersectionality equity" lens that questions further how our research, teaching, and practice can enact Crenshaw's (1993) idea of "mapping the margins." To achieve this, Crenshaw argues, we must center the lives of groups that remain often invisible when we talk about the generic working class "women" or "men" or "Latinos" or "LGBTQ" communities.

In moving forward, we also must be committed to enlarging and diversifying the pool of research scientists who study the issues. By diversity within an intersectionality health equity lens, we are referring to research scientists whose own awareness of their intersectional identities—that is, ethnicity, race, gender, class, sexuality, nativity, and disability—pushes them to design research that produces greater knowledge and clarity about the conceptualization of sound intersectionality-grounded studies and the range of methods to ensure new knowledge, better applications of knowledge, and effective uses of knowledge to guide our understanding of human development and health....

In developing our focus on intersectionality and social determinants of health, we attach our analysis to the goals of advancing social justice, where commitments to equality and equity reside and power is shared.... For all health and health policy researchers, scholars, practitioners, and community leaders who embrace a social justice framework, an intersectionality health equity lens could help to illuminate the often stifled issues that affect the health, development, and well-being of children and families in marginalized communities. This would mean that they would take seriously the ways in which institutional rights and duties allow people to participate and receive resources such as health, education, and social services in ways that are fundamentally shaped by intersecting inequalities. That would also mean promoting equal access to the fair distribution of wealth, equal opportunity, and equality of outcome by making the invisible visible through interrogating how race and class systems of oppression work together in shaping the social determinants of health.

REFERENCES

Bauer, G. (2014). Incorporating intersectionality theory into population health research methodology: Challenges and the potential to advance health equity. *Social Science and Medicine*, 110:10–17.

Bloche, M. G. (2001). Race and discretion in American medicine. *Yale Journal of Health Policy, Law, and Ethics* 1(1):5–5.

Bowleg, L. (2008). "The Methodological Challenges of Qualitative and Quantitative Intersectionality Research," *Sex Roles*, 59:312–325.

Brown, T. H., Richardson, L. J., Hargrove, T. W., & Thomas, C. S. (2016). Using Multiple-hierarchy Stratification and Life Course Approaches to Understand Health Inequalities The Intersecting Consequences of Race, Gender, SES, and Age. *Journal of Health and Social Behavior*, 57(2), 200–222.

Collins, P. H. (2008). *Black feminist thought: Knowledge, consciousness, and the politics of empowerment.* New York: Routledge.

Collins, P. H. (2009). *Another kind of public education: Race, schools, the media and democratic possibilities.* Boston, MA: Beacon Press.

Collins, P. H., and S. Blige. (2016). *Intersectionality.* Malden, MA: Polity Press.

Crenshaw, K. (1993). Mapping the margins: Intersectionality, identity politics, and violence against women of color. In *Critical race theory: The key writings that formed the movement,* edited by K. Crenshaw, N. Gotanda, G. Peller, K. Thomas New York: New Press, pp. 357–383.

DuBois, W. E. B. (1999). On spiritual strivings. In *The souls of black folk.* edited by H. L. Gates and T. H. Oliver. New York: W. W. Norton and Company.

Feagin, J. R., and M. P. Sikes. (1994). *Living with racism: The black middle-class experience.* Boston, MA: Beacon Press.

Jones, C. (2016). "Closing General Session: American Public Health Association 2016, Meeting Theme: Creating the Healthiest Nation: Ensuring the Right to Health," American Public Health Association Annual Meeting, Denver, CO.

LaVeist, T. A., and L. Isaac, eds. (2013). *Race, ethnicity, and health,* 2nd ed. San Francisco, CA: Jossey-Bass.

López, N. (2003). Hopeful girls, troubled boys: Race and gender disparity in urban education. New York: Routledge.

López, N. (2013a). Contextualizing lived race-gender and the racialized gendered social determinants of health. In *Mapping "race": Critical approaches to health disparities research,* edited by L. Gómez and N. López. New Brunswick, NJ: Rutgers University Press, pp. 179–211.

López, N. (2013b). Killing two birds with one stone? Why we need two separate questions on race and ethnicity in the 2020 census and beyond. *Latino Studies* 11(3):428–438.

Nakano Glenn, E. (2002). *Unequal freedom: How race and gender shaped American citizenship and labor.* Boston, MA: Harvard University Press.

Schultz, A., and L. Mullings. (2006). Intersectionality and health: An introduction. In *Gender, race, class and health: Intersectional approaches,* edited by A. Schultz and L. Mullings. San Francisco, CA: Jossey-Bass, pp. 3–20.

Thorpe, R., and J. Kelley-Moore. (2013). Life course theories of race disparities: A comparison of cumulative dis/advantage perspective and the weathering hypothesis. In *Race, Ethnicity, and Health*, edited by T. A. LaVeist and L. Isaac. San Francisco, CA: Jossey-Bass, pp. 355–375.

Vidal-Ortiz, S. (2005). Sexuality and gender in Santería: LGBT identities at the crossroads of Santería religious practices and beliefs. In *Gay religion*, edited by S. Thumma and E. Gray. Walnut Creek, CA: Altamira Press, pp. 115–138.

Weber, L. (2010). Defining contested concepts. In *Understanding race, class, gender and sexuality: A conceptual framework*. New York: Oxford University Press, pp. 23–43.

WHO (World Health Organization). (2006). *Constitution of the world health organization. Basic Documents*, 45th ed., Supplement. Available from: http://www.who.int/governance/eb/who_constitution_en.pdf.

WHO. (2015). *Health systems equity update*. Available from: http://www.who.int/healthsystems/topics/equity/en/.

Williams, D., and S. Mohammed. (2013). Racism and health, I: Pathways & scientific evidence. *American Behavioral Scientist* 57(8):1152–1173.

Yuval-Davis, N. (2011). *The politics of belonging: Intersected contestations*. London, UK: Sage.

Zuberi, T. (2001). *Thicker than blood: How racial statistics lie*. U of Minnesota Press.

37

The First Americans
American Indians

C. MATTHEW SNIPP

By the end of the nineteenth century, many observers predicted that American Indians were destined for extinction. Within a few generations, disease, warfare, famine, and outright genocide had reduced their numbers from millions to less than 250,000 in 1890. Once a self-governing, self-sufficient people, American Indians were forced to give up their homes and their land, and to subordinate themselves to an alien culture. The forced resettlement to reservation lands or the Indian Territory (now Oklahoma) frequently meant a life of destitution, hunger, and complete dependency on the federal government for material needs.

Today, American Indians are more numerous than they have been for several centuries. While still one of the most destitute groups in American society, tribes have more autonomy and are now more self-sufficient than at any time since the last century. In cities, modern pan-Indian organizations have been successful in making the presence of American Indians known to the larger community, and have mobilized to meet the needs of their people (Cornell 1988; Nagel 1986; Weibel-Orlando 1991). In many rural areas, American Indians and especially tribal governments have become increasingly more important and increasingly more visible by virtue of their growing political and economic power. The balance of this [reading] is devoted to explaining their unique place in American society.

THE INCORPORATION OF AMERICAN INDIANS

The current political and economic status of American Indians is the result of the process by which they were incorporated into Euro-American society (Hall 1989). This amounts to a long history of efforts aimed at subordinating an otherwise self-governing and self-sufficient people that eventually culminated in widespread economic dependency. The role of the U.S. government in this process can be seen in the five major historical periods of federal Indian

SOURCE: Snipp, C. Matthew. 1996. "American Indians." Pp 390–403 in *Origins and Destinies*, edited by S. Pedraza and R. Rumbaut. Belmont, CA: Wadsworth. © 1996 Cengage Learning.

relations: removal, assimilation, the Indian New Deal, termination and relocation, and self-determination.

Removal

In the early nineteenth century, the population of the United States expanded
rapidly at the same time that the federal government increased its political and
military capabilities. The character of Indian-American relations changed after
the War of 1812. The federal government increasingly pressured tribes settled
east of the Appalachian Mountains to move west to the territory acquired in
the Louisiana Purchase. Numerous treaties were negotiated by which the tribes
relinquished most of their land and eventually were forced to move west.

Initially the federal government used bargaining and negotiation to accomplish removal, but many tribes resisted (Prucha 1984). However, the election of
Andrew Jackson by a frontier constituency signaled the beginning of more force-
ful measures to accomplish removal. In 1830 Congress passed the Indian
Removal Act, which mandated the eventual removal of the eastern tribes to
points west of the Mississippi River, in an area which was to become the Indian
Territory and is now the state of Oklahoma. Dozens of tribes were forcibly
removed from the eastern half of the United States to the Indian Territory and
newly created reservations in the west, a long process ridden with conflict
and bloodshed.

As the nation expanded beyond the Mississippi River, tribes of the plains,
southwest, and west coast were forcibly settled and quarantined on isolated reservations. This was accompanied by the so-called Indian Wars—a bloody chapter
in the history of Indian-White relations (Prucha 1984; Utley 1963). This period
in American history is especially remarkable because the U.S. government was
responsible for what is unquestionably one of the largest forced migrations
in history.

The actual process of removal spanned more than a half-century and affected
nearly every tribe east of the Mississippi River. Removal often meant extreme
hardships for American Indians, and in some cases this hardship reached legend-
ary proportions. For example, the Cherokee removal has become known as the
"Trail of Tears." In 1838, nearly 17,000 Cherokees were ordered to leave their
homes and assemble in military stockades (Thornton 1987, p. 117). The march
to the Indian Territory began in October and continued through the winter
months. As many as 8,000 Cherokees died from cold weather and diseases such
as influenza (Thornton 1987, p. 118).

According to William Hagan (1979), removal also caused the Creeks to suffer dearly as their society underwent a profound disintegration. The contractors
who forcibly removed them from their homes refused to do anything for "the
large number who had nothing but a cotton garment to protect them from the
sleet storms and no shoes between them and the frozen ground of the last stages
of their hegira. About half of the Creek nation did not survive the migration and
the difficult early years in the West" (Hagan 1979, pp. 77–81). In the West, a
band of Nez Perce men, women, and children, under the leadership of Chief

Joseph, resisted resettlement in 1877. Heavily outnumbered, they were pursued by cavalry troops from the Wallowa valley in eastern Oregon and finally captured in Montana near the Canadian border. Although the Nez Perce were eventually captured and moved to the Indian Territory, and later to Idaho, their resistance to resettlement has been described by one historian as "one of the great military movements in history" (Prucha 1984, p. 541).

Assimilation

Near the end of the nineteenth century, the goal of isolating American Indians on reservations and the Indian Territory was finally achieved. The Indian population also was near extinction. Their numbers had declined steadily throughout the nineteenth century, leading most observers to predict their disappearance (Hoxie 1984). Reformers urged the federal government to adopt measures that would humanely ease American Indians into extinction. The federal government responded by creating boarding schools and the allotment acts—both were intended to "civilize" and assimilate American Indians into American society by Christianizing them, educating them, introducing them to private property, and making them into farmers. American Indian boarding schools sought to accomplish this task by indoctrinating Indian children with the belief that tribal culture was an inferior relic of the past and that Euro-American culture was vastly superior and preferable. Indian children were forbidden to wear their native attire, to eat their native foods, to speak their native language, or to practice their traditional religion. Instead, they were issued Euro-American clothes, and expected to speak English and become Christians. Indian children who did not relinquish their culture were punished by school authorities. The curriculum of these schools taught vocational arts along with "civilization" courses.

The impact of allotment policies is still evident today. The 1887 General Allotment Act (the Dawes Severalty Act) and subsequent legislation mandated that tribal lands were to be allotted to individual American Indians ... and the surplus lands left over from allotment were to be sold on the open market. Indians who received allotted tribal lands also received citizenship, farm implements, and encouragement from Indian agents to adopt farming as a livelihood (Hoxie 1984, Prucha 1984).

For a variety of reasons, Indian lands were not completely liquidated by allotment, many Indians did not receive allotments, and relatively few changed their lifestyles to become farmers. Nonetheless, the allotment era was a disaster because a significant number of allottees eventually lost their land. Through tax foreclosures, real estate fraud, and their own need for cash, many American Indians lost what for most of them was their last remaining asset (Hoxie 1984).

Allotment took a heavy toll on Indian lands. It caused about 90 million acres of Indian land to be lost, approximately two-thirds of the land that had belonged to tribes in 1887 (O'Brien 1990). This created another problem that continues to vex many reservations: "checkerboarding." Reservations that were subjected to allotment are typically a crazy quilt composed of tribal lands, privately owned "fee" land, and trust land belonging to individual Indian families.

Checkerboarding presents reservation officials with enormous administrative problems when trying to develop land use management plans, zoning ordinances, or economic development projects that require the construction of physical infrastructure such as roads or bridges.

The Indian New Deal

The Indian New Deal was short-lived but profoundly important. Implemented in the early 1930s along with the other New Deal programs of the Roosevelt administration, the Indian New Deal was important for at least three reasons. First, signaling the end of the disastrous allotment era as well as a new respect for American Indian tribal culture, the Indian New Deal repudiated allotment as a policy. Instead of continuing its futile efforts to detribalize American Indians, the federal government acknowledged that tribal culture was worthy of respect. Much of this change was due to John Collier, a long-time Indian rights advocate appointed by Franklin Roosevelt to serve as Commissioner of Indian Affairs (Prucha 1984).

Like other New Deal policies, the Indian New Deal also offered some relief from the Great Depression and brought essential infrastructure development to many reservations, such as projects to control soil erosion and to build hydroelectric dams, roads, and other public facilities. These projects created jobs in New Deal programs such as the Civilian Conservation Corps and the Works Progress Administration.

An especially important and enduring legacy of the Indian New Deal was the passage of the Indian Reorganization Act (IRA) of 1934. Until then, Indian self-government had been forbidden by law. This act allowed tribal governments, for the first time in decades, to reconstitute themselves for the purpose of overseeing their own affairs on the reservation. Critics charge that this law imposed an alien form of government, representative democracy, on traditional tribal authority. On some reservations, this has been an on-going source of conflict (O'Brien 1990). Some reservations rejected the IRA for this reason, but now have tribal governments authorized under different legislation.

Termination and Relocation

After World War II, the federal government moved to terminate its long-standing relationship with Indian tribes by settling the tribes' outstanding legal claims, by terminating the special status of reservations, and by helping reservation Indians relocate to urban areas (Fixico 1986). The Indian Claims Commission was a special tribunal created in 1946 to hasten the settlement of legal claims that tribes had brought against the federal government. In fact, the Indian Claims Commission became bogged down with prolonged cases, and in 1978 the commission was dissolved by Congress. At that time, there were 133 claims still unresolved out of an original 617 that were first heard by the commission three decades earlier (Fixico 1986, p. 186). The unresolved claims that were still pending were transferred to the Federal Court of Claims.

Congress also moved to terminate the federal government's relationship with Indian tribes. House Concurrent Resolution (HCR) 108, passed in 1953, called for steps that eventually would abolish all reservations and abolish all special programs serving American Indians. It also established a priority list of reservations slated for immediate termination. However, this bill and subsequent attempts to abolish reservations were vigorously opposed by Indian advocacy groups such as the National Congress of American Indians. Only two reservations were actually terminated, the Klamath in Oregon and the Menominee in Wisconsin. The Menominee reservation regained its trust status in 1975 and the Klamath reservation was restored in 1986.

The Bureau of Indian Affairs (BIA) also encouraged reservation Indians to relocate and seek work in urban job markets. This was prompted partly by the desperate economic prospects on most reservations, and partly because of the federal government's desire to "get out of the Indian business." The BIA's relocation programs aided reservation Indians in moving to designated cities, such as Los Angeles and Chicago, where they also assisted them in finding housing and employment. Between 1952 and 1972, the BIA relocated more than 100,000 American Indians (Sorkin 1978). However, many Indians returned to their reservations (Fixico 1986). For some American Indians, the return to the reservation was only temporary; for example, during periods when seasonal employment such as construction work was hard to find.

Self-Determination

Many of the policies enacted during the termination and relocation era were steadfastly opposed by American Indian leaders and their supporters. As these programs became stalled, critics attacked them for being harmful, ineffective, or both. By the mid-1960s, these policies had very little serious support. Perhaps inspired by the gains of the Civil Rights movement, American Indian leaders and their supporters made "self-determination" the first priority on their political agendas. For these activists, self-determination meant that Indian people would have the autonomy to control their own affairs, free from the paternalism of the federal government.

The idea of self-determination was well received by members of Congress sympathetic to American Indians. It also was consistent with the "New Federalism" of the Nixon administration. Thus, the policies of termination and relocation were repudiated in a process that culminated in 1975 with the passage of the American Indian Self-Determination and Education Assistance Act, a profound shift in federal Indian policy. For the first time since this nation's founding, American Indians were authorized to oversee the affairs of their own communities, free of federal intervention. In practice, the Self-Determination Act established measures that would allow tribal governments to assume a larger role in reservation administration of programs for welfare assistance, housing, job training, education, natural resource conservation, and the maintenance of reservation roads and bridges (Snipp and Summers 1991). Some reservations also have their own police forces and game wardens, and can issue licenses and levy taxes. The Onondaga tribe in upstate New York has taken their sovereignty one step further by issuing passports that are internationally recognized. Yet there is a great deal of variability in terms of how much autonomy tribes have over reservation affairs. Some tribes, especially those on large and well-organized reservations, have nearly complete control over their reservations, while smaller reservations with limited resources often depend heavily on BIA services....

CONCLUSION

Though small in number, American Indians have an enduring place in American society. Growing numbers of American Indians occupy reservation and other trust lands, and equally important has been the revitalization of tribal governments. Tribal governments now have a larger role in reservation affairs than ever in the past. Another significant development has been the urbanization of American Indians. Since 1950, the proportion of American Indians in cities has grown rapidly. These American Indians have in common with reservation Indians many of the same problems and disadvantages, but they also face other challenges unique to city life.

The challenges facing tribal governments are daunting. American Indians are among the poorest groups in the nation. Reservation Indians have substantial needs for improved housing, adequate health care, educational opportunities, and employment, as well as developing and maintaining reservation infrastructure. In the face of declining federal assistance, tribal governments are assuming an ever-larger burden. On a handful of reservations, tribal governments have assumed completely the tasks once performed by the BIA.

As tribes have taken greater responsibility for their communities, they also have struggled with the problems of raising revenues and providing economic opportunities for their people. Reservation land bases provide many reservations with resources for development. However, these resources are not always abundant, much less unlimited, and they have not always been well managed. It will be yet another challenge for tribes to explore ways of efficiently managing their existing resources. Legal challenges also face tribes seeking to exploit unconventional resources such as gambling revenues. Their success depends on many complicated legal and political contingencies.

Urban American Indians have few of the resources found on reservations, and they face other difficult problems. Preserving their culture and identity is an especially pressing concern. However, urban Indians have successfully adapted to city environments in ways that preserve valued customs and activities— powwows, for example, are an important event in all cities where there is a large Indian community. In addition, pan-Indianism has helped urban Indians set aside tribal differences and forge alliances for the betterment of urban Indian communities.

These alliances are essential, because unlike reservation Indians, urban American Indians do not have their own form of self-government. Tribal governments do not have jurisdiction over urban Indians. For this reason, urban Indians must depend on other strategies for ensuring that the needs of their community are met, especially for those new to city life. Coping with the transition to urban life poses a multitude of difficult challenges for many American Indians. Some succumb to these problems, especially the hardships of unemployment, economic deprivation, and related maladies such as substance abuse, crime, and violence. But most successfully overcome these difficulties, often with help from other members of the urban Indian community.

Perhaps the greatest strength of American Indians has been their ability to find creative ways for dealing with adversity, whether in cities or on reservations. In the past, this quality enabled them to survive centuries of oppression and persecution. Today this is reflected in the practice of cultural traditions that Indian people are proud to embrace. The resilience of American Indians is an abiding quality that will no doubt ensure that they will remain part of the ethnic mosaic of American society throughout the twenty-first century and beyond.

REFERENCES

Cornell, Stephen. 1988. *The Return of the Native: American Indian Political Resurgence.* New York: Oxford University Press.

Fixico, Donald L. 1986. *Termination and Relocation: Federal Indian Policy, 1945–1960.* Albuquerque, NM: University of New Mexico Press.

Hagan, William T. 1979. *American Indians.* Chicago, IL: University of Chicago Press.

Hall, Thomas D. 1989. *Social Change in the Southwest, 1350–1880.* Lawrence, KS: University Press of Kansas.

Hoxie, Frederick E. 1984. *A Final Promise: The Campaign to Assimilate the Indians, 1880–1920.* Lincoln, NE: University of Nebraska Press.

Nagel, Joanne. 1986. "American Indian Repertoires of Contention." Paper presented at the annual meeting of the American Sociological Association, San Francisco, CA.

O'Brien, Sharon. 1990. *American Indian Tribal Governments.* Norman, OK: University of Oklahoma Press.

Prucha, Francis Paul. 1984. *The Great Father.* Lincoln, NE: University of Nebraska Press.

Snipp, C. Matthew, and Gene F. Summers. 1991. "American Indian Development Policies," pp. 166–180 in *Rural Policies for the 1990s,* edited by Cornelia Flora and James A. Christenson. Boulder, CO: Westview Press.

Sorkin, Alan L. 1978. *The Urban American Indian.* Lexington, MA: Lexington Books.

Thornton, Russell. 1987. *American Indian Holocaust and Survival: A Population History since 1942.* Norman, OK: University of Oklahoma Press.

Utley, Robert M. 1963. *The Last Days of the Sioux Nation.* New Haven: Yale University Press.

Weibel-Orlando, Joan. 1991. *Indian Country, L.A.* Urbana, IL: University of Illinois Press.

38

"Is This a White Country, or What?"

LILLIAN B. RUBIN

"They're letting all these coloreds come in and soon there won't be any place left for white people," broods Tim Walsh, a thirty-three-year-old white construction worker. "It makes you wonder: Is this a white country, or what?"

It's a question that nags at white America, one perhaps that's articulated most often and most clearly by the men and women of the working class. For it's they who feel most vulnerable, who have suffered the economic contractions of recent decades most keenly, who see the new immigrants most clearly as direct competitors for their jobs.

It's not whites alone who stew about immigrants. Native-born blacks, too, fear the newcomers nearly as much as whites—and for the same economic reasons. But for whites the issue is compounded by race, by the fact that the newcomers are primarily people of color. For them, therefore, their economic anxieties have combined with the changing face of America to create a profound uneasiness about immigration—a theme that was sounded by nearly 90 percent of the whites I met, even by those who are themselves first-generation, albeit well-assimilated, immigrants.

Sometimes they spoke about this in response to my questions; equally often the subject of immigration arose spontaneously as people gave voice to their concerns. But because the new immigrants are dominantly people of color, the discourse was almost always cast in terms of race as well as immigration, with the talk slipping from immigration to race and back again as if these are not two separate phenomena. "If we keep letting all them foreigners in, pretty soon there'll be more of them than us and then what will this country be like?" Tim's wife, Mary Anne, frets. I mean, this is *our* country, but the way things are going, white people will be the minority in our own country. Now does that make any sense?"

Such fears are not new. Americans have always worried about the strangers who came to our shores, fearing that they would corrupt our society, dilute our culture, debase our values. So I remind Mary Anne, "When your ancestors came here, people also thought we were allowing too many foreigners into the country. Yet those earlier immigrants were successfully integrated into the American society. What's different now?"

SOURCE: Rubin, Lillian B. 1994. "Families on the Fault Line: America's Working Class Speaks about the Family, the Economy, Race, and Ethnicity," pp. 172–196. New York: HarperCollins Publishers. Copyright © 1994. Reprinted by permission of the author.

"Oh, it's different, all right," she replies without hesitation. "When my people came, the immigrants were all white. That makes a big difference."...

Listening to Mary Anne's words I was reminded again how little we Americans look to history for its lessons, how impoverished is our historical memory.

For, in fact, being white didn't make "a big difference" for many of those earlier immigrants. The dark-skinned Italians and the eastern European Jews who came in the late nineteenth and early twentieth centuries didn't look very white to the fair-skinned Americans who were here then. Indeed, the same people we now call white—Italians, Jews, Irish—were seen as another race at that time. Not black or Asian, it's true, but an alien other, a race apart, although one that didn't have a clearly defined name. Moreover, the racist fears and fantasies of native-born Americans were far less contained then than they are now, largely because there were few social constraints on their expression.

When, during the nineteenth century, for example, some Italians were taken for blacks and lynched in the South, the incidents passed virtually unnoticed. And if Mary Anne and Tim Walsh, both of Irish ancestry, had come to this country during the great Irish immigration of that period, they would have found themselves defined as an inferior race and described with the same language that was used to characterize blacks: "low-browed and savage, grovelling and bestial, lazy and wild, simian and sensual."[1] Not only during that period but for a long time afterward as well, the U.S. Census Bureau counted the Irish as a distinct and separate group, much as it does today with the category it labels "Hispanic."

But there are two important differences between then and now, differences that can be summed up in a few words: the economy and race. Then, a growing industrial economy meant that there were plenty of jobs for both immigrant and native workers, something that can't be said for the contracting economy in which we live today. True, the arrival of the immigrants, who were more readily exploitable than native workers, put Americans at a disadvantage and created discord between the two groups. Nevertheless, work was available for both.

Then, too, the immigrants—no matter how they were labeled, no matter how reviled they may have been—were ultimately assimilable, if for no other reason than that they were white. As they began to lose their alien ways, it became possible for Native Americans to see in the white ethnics of yesteryear a reflection of themselves. Once this shift in perception occurred, it was possible for the nation to incorporate them, to take them in, chew them up, digest them, and spit them out as Americans—with subcultural variations not always to the liking of those who hoped to control the manners and mores of the day, to be sure, but still recognizably white Americans.

Today's immigrants, however, are the racial other in a deep and profound way.... And integrating masses of people of color into a society where race consciousness lies at the very heart of our central nervous system raises a whole new set of anxieties and tensions....

The increased visibility of other racial groups has focused whites more self-consciously than ever on their own racial identification. Until the new immigration shifted the complexion of the land so perceptibly, whites didn't think of themselves as white in the same way that Chinese know they're Chinese and African-Americans know they're black. Being white was simply a fact of life, one that didn't require any public statement, since it was the definitive social value against which all others were measured. "It's like everything's changed and I don't know what happened," complains Marianne Bardolino. "All of a sudden you have to be thinking all the time about these race things. I don't remember growing up thinking about being white like I think about it now. I'm not saying I didn't know there was coloreds and whites; it's just that I didn't go along thinking, *Gee, I'm a white person.* I never thought about it at all. But now with all the different colored people around, you have to think about it because they're thinking about it all the time."

"You say you feel pushed now to think about being white, but I'm not sure I understand why. What's changed?" I ask.

"I told you," she replies quickly, a small smile covering her impatience with my question. "It's because they think about what they are, and they want things their way, so now I have to think about what I am and what's good for me and my kids." She pauses briefly to let her thoughts catch up with her tongue, then continues. "I mean, if somebody's always yelling at you about being black or Asian or something, then it makes you think about being white. Like, they want the kids in school to learn about their culture, so then I think about being white and being Italian and say: What about my culture? If they're going to teach about theirs, what about mine?"

To which America's racial minorities respond with bewilderment. "I don't understand what white people want," says Gwen Tomalson. "They say if black kids are going to learn about black culture in school, then white people want their kids to learn about white culture. I don't get it. What do they think kids have been learning about all these years? It's all about white people and how they live and what they accomplished. When I was in school you wouldn't have thought black people existed for all our books ever said about us."

As for the charge that they're "thinking about race all the time," as Marianne Bardolino complains, people of color insist that they're forced into it by a white world that never lets them forget. "If you're Chinese, you can't forget it, even if you want to, because there's always something that reminds you," Carol Kwan's husband, Andrew, remarks tartly. "I mean, if Chinese kids get good grades and get into the university, everybody's worried and you read about it in the papers."

While there's little doubt that racial anxieties are at the center of white concerns, our historic nativism also plays a part in escalating white alarm. The new immigrants bring with them a language and an ethnic culture that's vividly expressed wherever they congregate. And it's this also, the constant reminder of an alien presence from which whites are excluded, that's so troublesome to them.

The nativist impulse isn't, of course, given to the white working class alone. But for those in the upper reaches of the class and status hierarchy—those whose children go to private schools, whose closest contact with public transportation is the taxi cab—the immigrant population supplies a source of cheap labor, whether as nannies for their children, maids in their households, or workers in their businesses. They may grouse and complain that "nobody speaks English anymore," just as working-class people do. But for the people who use immigrant labor, legal or illegal, there's a payoff for the inconvenience—a payoff that doesn't exist for the families in this study but that sometimes costs them dearly. For while it may be true that American workers aren't eager for many of the jobs immigrants are willing to take, it's also true that the presence of a large immigrant population—especially those who come from developing countries where living standards are far below our own—helps to make these jobs undesirable by keeping wages depressed well below what most American workers are willing to accept....

It's not surprising, therefore, that working-class women and men speak so angrily about the recent influx of immigrants. They not only see their jobs and their way of life threatened, they feel bruised and assaulted by an environment that seems suddenly to have turned color and in which they feel like strangers in their own land. So they chafe and complain: "They come here to take advantage of us, but they don't really want to learn our ways," Beverly Sowell, a thirty-three-year old white electronics assembler, grumbles irritably. "They live different than us; it's like another world how they live. And they're so clannish. They keep to themselves, and they don't even *try* to learn English. You go on the bus these days and you might as well be in a foreign country; everybody's talking some other language, you know, Chinese or Spanish or something. Lots of them have been here a long time, too, but they don't care; they just want to take what they can get."

But their complaints reveal an interesting paradox, an illuminating glimpse into the contradictions that beset native-born Americans in their relations with those who seek refuge here. On the one hand, they scorn the immigrants; on the other, they protest because they "keep to themselves." It's the same contradiction that dominates black-white relations. Whites refuse to integrate blacks but are outraged when they stop knocking at the door, when they move to sustain the separation on their own terms' [*sic*]—in black theme houses on campuses, for example, or in the newly developing black middle-class suburbs.

I wondered, as I listened to Beverly Sowell and others like her, why the same people who find the life ways and languages of our foreign-born population offensive also care whether they "keep to themselves."

"Because like I said, they just shouldn't, that's all," Beverly says stubbornly. "If they're going to come here, they should be willing to learn our ways—you know what I mean, be real Americans. That's what my grandparents did, and that's what they should do."

"But your grandparents probably lived in an immigrant neighborhood when they first came here, too," I remind her.

"It was different," she insists. "I don't know why; it was. They wanted to be Americans; these here people now, I don't think they do. They just want to take advantage of this country["]....

"Everything's changed, and it doesn't make sense. Maybe you get it, but I don't. We can't take care of our own people and we keep bringing more and more foreigners in. Look at all the homeless. Why do we need more people here when our own people haven't got a place to sleep?"

"Why do we need more people here?"—a question Americans have asked for two centuries now. Historically, efforts to curb immigration have come during economic downturns, which suggests that when times are good, when American workers feel confident about their future, they're likely to be more generous in sharing their good fortune with foreigners. But when the economy falters, as it did in the 1990s, and workers worry about having to compete for jobs with people whose standard of living is well below their own, resistance to immigration rises. "Don't get me wrong; I've got nothing against these people," Tim Walsh demurs. "But they don't talk English, and they're used to a lot less, so they can work for less money than guys like me can. I see it all the time; they get hired and some white guy gets left out."

It's this confluence of forces—the racial and cultural diversity of our new immigrant population; the claims on the resources of the nation now being made by those minorities who, for generations, have called America their home; the failure of some of our basic institutions to serve the needs of our people; the contracting economy, which threatens the mobility aspirations of working-class families—all these have come together to leave white workers feeling as if everyone else is getting a piece of the action while they get nothing. "I feel like white people are left out in the cold," protests Diane Johnson, a twenty-eight-year-old white single mother who believes she lost a job as a bus driver to a black woman. "First it's the blacks; now it's all those other colored people, and it's like everything always goes their way. It seems like a white person doesn't have a chance anymore. It's like the squeaky wheel gets the grease, and they've been squeaking and we haven't," she concludes angrily.

Until recently, whites didn't need to think about having to "squeak"—at least not specifically as whites. They have, of course, organized and squeaked at various times in the past—sometimes as ethnic groups, sometimes as workers. But not as whites. As whites they have been the dominant group, the favored ones, the ones who could count on getting the job when people of color could not. Now suddenly there are others—not just individual others but identifiable groups, people who share a history, a language, a culture, even a color—who lay claim to some of the rights and privileges that formerly had been labeled for whites only." And whites react as if they've been betrayed, as if a sacred promise has been broken. They're white, aren't they? They're *real* Americans, aren't they? This is their country, isn't it?

The answers to these questions used to be relatively unambiguous. But not anymore. Being white no longer automatically assures dominance in the politics of a multiracial society. Ethnic group politics, however, has a long and fruitful history. As whites sought a social and political base on which to stand, therefore, it was natural and logical to reach back to their ethnic past. Then they, too, could be "something"; they also would belong to a group; they would have a name, a history, a culture, and a voice. "Why is it only the blacks or Mexicans or Jews that are 'something'?" asks Tim Walsh. "I'm Irish, isn't that something, too? Why doesn't that count?"

In reclaiming their ethnic roots, whites can recount with pride the tribulations and transcendence of their ancestors and insist that others take their place in the line from which they have only recently come. "My people had a rough time, too. But nobody gave us anything, so why do we owe them something? Let them pull their share like the rest of us had to do," says Al Riccardi, a twenty-nine-year-old white taxi driver.

From there it's only a short step to the conviction that those who don't progress up that line are hampered by nothing more than their own inadequacies or, worse yet, by their unwillingness to take advantage of the opportunities offered them. "Those people, they're hollering all the time about discrimination," Al continues, without defining who "those people" are. "Maybe once a long time ago that was true, but not now. The problem is that a lot of those people are lazy. There's plenty of opportunities, but you've got to be willing to work hard."

He stops a moment, as if listening to his own words, then continues, "Yeah, yeah, I know there's a recession on and lots of people don't have jobs. But it's different with some of those people. They don't really want to work, because if they did, there wouldn't be so many of them selling drugs and getting in all kinds of trouble."

"You keep talking about 'those people' without saying who you mean," I remark.

"Aw c'mon, you know who I'm talking about," he says, his body shifting uneasily in his chair. "It's mostly the black people, but the Spanish ones, too."

In reality, however, it's a no-win situation for America's people of color, whether immigrant or native born. For the industriousness of the Asians comes in for nearly as much criticism as the alleged laziness of other groups. When blacks don't make it, it's because, whites like Al Riccardi insist, their culture doesn't teach respect for family; because they're hedonistic, lazy, stupid, and/or criminally inclined. But when Asians demonstrate their ability to overcome the obstacles of an alien language and culture, when the Asian family seems to be the repository of our most highly regarded traditional values, white hostility doesn't disappear. It just changes its form. Then the accomplishments of Asians, the speed with which they move up the economic ladder, aren't credited to their superior culture, diligence, or intelligence—even when these are granted—but to the fact that they're "single minded," "untrustworthy," "clannish drones," "narrow people" who raise children who are insufficiently "well rounded."[2]

Not surprisingly, as competition increases, the various minority groups often are at war among themselves as they press their own particular claims, fight over turf, and compete for an ever-shrinking piece of the pie. In several African-American communities, where Korean shopkeepers have taken the place once held by Jews, the confrontations have been both wrenching and tragic. A Korean grocer in Los Angeles shoots and kills a fifteen-year-old black girl for allegedly trying to steal some trivial item from the store.[3] From New York City to Berkeley, California, African-Americans boycott Korean shop owners who, they charge, invade their neighborhoods, take their money, and treat them disrespectfully.[4] But painful as these incidents are for those involved, they are only symptoms of a deeper malaise in both communities—the contempt and distrust in which the Koreans hold their African-American neighbors, and the rage of blacks as they watch these new immigrants surpass them.

Latino-black conflict also makes headlines when, in the aftermath of the riots in South Central Los Angeles, the two groups fight over who will get the lion's share of the jobs to rebuild the neighborhood. Blacks, insisting that they're being discriminated against, shut down building projects that don't include them in satisfactory numbers. And indeed, many of the jobs that formerly went to African-Americans are now being taken by Latino workers. In an article entitled "Black vs. Brown," Jack Miles, an editorial writer for the *Los Angeles Times*, reports that janitorial firms serving downtown Los Angeles have almost entirely replaced their unionized black work force with non-unionized immigrants.[5]...

But the disagreements among America's racial minorities are of little interest or concern to most white working-class families. Instead of conflicting groups, they see one large mass of people of color, all of them making claims that endanger their own precarious place in the world. It's this perception that has led some white ethnics to believe that reclaiming their ethnicity alone is not enough, that so long as they remain in their separate and distinct groups, their power will be limited. United, however, they can become a formidable countervailing force, one that can stand fast against the threat posed by minority demands. But to come together solely as whites would diminish their impact and leave them open to the charge that their real purpose is simply to retain the privileges of whiteness. A dilemma that has been resolved, at least for some, by the birth of a new entity in the history of American ethnic groups—the "European-Americans."[6]...

At the University of California at Berkeley, for example, white students and their faculty supporters insisted that the recently adopted multicultural curriculum include a unit of study of European-Americans. At Queens College in New York City, where white ethnic groups retain a more distinct presence, Italian-American students launched a successful suit to win recognition as a disadvantaged minority and gain the entitlements accompanying that status, including special units of Italian-American studies.

White high school students, too, talk of feeling isolated and, being less sophisticated and wary than their older sisters and brothers, complain quite openly that there's no acceptable and legitimate way for them to acknowledge a white identity. "There's all these things for all the different ethnicities, you know, like clubs for black kids and Hispanic kids, but there's nothing for me and my friends to join," Lisa Marshall, a sixteen-year-old white high school student, explains with exasperation. "They won't let us have a white club because that's supposed to be racist. So we figured we'd just have to call it something else, you know, some ethnic thing, like Euro-Americans. Why not? They have African-American clubs."

Ethnicity, then, often becomes a cover for "white," not necessarily because these students are racist but because racial identity is now such a prominent feature of the discourse in our social world. In a society where racial consciousness is so high, how else can whites define themselves in ways that connect them to a community and, at the same time, allow them to deny their racial antagonisms?

Ethnicity and race—separate phenomena that are now inextricably entwined. Incorporating newcomers has never been easy, as our history of controversy and violence over immigration tells us.[7] But for the first time, the new immigrants are also people of color, which means that they tap both the nativist and racist impulses that are so deeply a part of American life. As in the past, however, the fear of foreigners, the revulsion against their strange customs and seemingly unruly ways, is only part of the reason for the anti-immigrant attitudes that are increasingly being expressed today. For whatever xenophobic suspicions may arise in modern America, economic issues play a critical role in stirring them up.

REFERENCES

Alba, Richard D. *Ethnic Identity*. New Haven: Yale University Press, 1990.

Roediger, David R. *The Wages of Whiteness*. New York: Verso, 1991.

NOTES

1. David R. Roediger, *The Wages of Whiteness* (New York: Verso, 1991), p. 133.

2. These were, and often still are, the commonly held stereotypes about Jews. Indeed, the Asian immigrants are often referred to as "the new Jews."

3. Soon Ja Du, the Korean grocer who killed fifteen-year-old Latasha Harlins, was found guilty of voluntary manslaughter and sentenced to four hundred hours of community service, a $500 fine, reimbursement of funeral costs to the Harlins family, and five years' probation.

4. The incident in Berkeley didn't happen in the black ghetto, as most of the others did. There, the Korean grocery store is near the University of California campus, and the woman involved in the incident is an African-American university student who was maced by the grocer after an argument over a penny.

5. Jack Miles, "Blacks vs. Browns," *Atlantic Monthly* (October 1992), pp. 41–68.

6. For an interesting analysis of what he calls "the transformation of ethnicity," see Richard D. Alba, *Ethnic Identity* (New Haven, CT: Yale University Press, 1990).

7. In the past, many of those who agitated for a halt to immigration were immigrants or native-born children of immigrants. The same often is true today. As anti-immigrant sentiment grows, at least some of those joining the fray are relatively recent arrivals. One man in this study, for example—a fifty-two-year-old immigrant from Hungary—is one of the leaders of an anti-immigration group in the city where he lives.

39

Are Asian Americans Becoming "White"?

MIN ZHOU

Are Asian Americans becoming "white"? For many public officials the answer must be yes, because they classify Asian-origin Americans with European-origin Americans for equal opportunity programs. But this classification is premature and based on false premises. Although Asian Americans as a group have attained the career and financial success equated with being white, and although many have moved next to or have even married whites, they still remain culturally distinct and suspect in a white society.

At issue is how to define Asian American and white. The term "Asian American" was coined by the late historian and activist Yuji Ichioka during the ethnic consciousness movements of the late 1960s. To adopt this identity was to reject the Western-imposed label of "Oriental." Today, "Asian American" is an umbrella category that includes both U.S. citizens and immigrants whose ancestors came from Asia, east of Iran. Although widely used in public discussions, most Asian-origin Americans are ambivalent about this label, reflecting the difficulty of being American and still keeping some ethnic identity: Is one, for example, Asian American or Japanese American?

Similarly, "white" is an arbitrary label having more to do with privilege than biology. In the United States, groups initially considered nonwhite, such as the Irish and Jews, have attained "white" membership by acquiring status and wealth. It is hardly surprising, then, that nonwhites would aspire to becoming "white" as a mark of and a tool for material success. However, becoming white can mean distancing oneself from "people of color" or disowning one's ethnicity. Pan-ethnic identities—Asian American, African American, Hispanic American—are one way the politically vocal in any group try to stem defections. But these group identities may restrain individual members' aspirations for personal advancement.

VARIETIES OF ASIAN AMERICANS

Privately, few Americans of Asian ancestry would spontaneously identify themselves as Asian, and fewer still as Asian American. They instead link their identities to specific countries of origin, such as China, Japan, Korea, the Philippines,

SOURCE: Zhou, Min. 2003. "Are Asian Americans Becoming White"? *Contexts*, 3(1): 29–37. Copyright © 2003 by the American Sociological Association. Reprinted by permission of SAGE Publications, Inc.

India or Vietnam. In a study of Vietnamese youth in San Diego, for example, 53 percent identified themselves as Vietnamese, 32 percent as Vietnamese American, and only 14 percent as Asian American. But they did not take these labels lightly; nearly 60 percent of these youth considered their chosen identity as very important to them.

Some Americans of Asian ancestry have family histories in the United States longer than many Americans of Eastern or Southern European origin. However, Asian-origin Americans became numerous only after 1970, rising from 1.4 million to 11.9 million (4 percent of the total U.S. population), in 2000. Before 1970, the Asian-origin population was largely made up of Japanese, Chinese and Filipinos. Now, Americans of Chinese and Filipino ancestries are the largest subgroups (at 2.8 million and 2.4 million, respectively), followed by Indians, Koreans, Vietnamese and Japanese (at more than one million). Some 20 other national-origin groups, such as Cambodians, Pakistanis, Laotians, Thai, Indonesians and Bangladeshis, were officially counted in government statistics only after 1980; together they amounted to more than two million Americans in 2000.

The sevenfold growth of the Asian-origin population in the span of 30-odd years is primarily due to accelerated immigration following the Hart-Celler Act of 1965, which ended the national origins quota system, and the historic resettlement of Southeast Asian refugees after the Vietnam War. Currently, about 60 percent of the Asian-origin population is foreign-born (the first generation), another 28 percent are U.S.-born of foreign-born parents (the second generation), and just 12 percent were born to U.S.-born parents (the third generation and beyond).

Unlike earlier immigrants from Asia or Europe, who were mostly low-skilled laborers looking for work, today's immigrants from Asia have more varied backgrounds and come for many reasons, such as to join their families, to invest their money in the U.S. economy, to fill the demand for highly skilled labor, or to escape war, political or religious persecution and economic hardship. For example, Chinese, Taiwanese, Indian, and Filipino Americans tend to be over-represented among scientists, engineers, physicians and other skilled professionals, but less educated, low-skilled workers are more common among Vietnamese, Cambodian, Laotian, and Hmong Americans, most of whom entered the United States as refugees. While middle-class immigrants are able to start their American lives with high-paying professional careers and comfortable suburban lives, low-skilled immigrants and refugees often have to endure low-paying menial jobs and live in inner-city ghettos.

Asian Americans tend to settle in large metropolitan areas and concentrate in the West. California is home to 35 percent of all Asian Americans. But recently, other states such as Texas, Minnesota and Wisconsin, which historically received few Asian immigrants, have become destinations for Asian American settlement. Traditional ethnic enclaves, such as Chinatown, Little Tokyo, Manilatown, Koreatown, Little Phnom Penh, and Thaitown, persist or have emerged in gateway cities, helping new arrivals to cope with cultural and linguistic difficulties. However, affluent and highly-skilled immigrants tend to bypass inner-city enclaves and settle in suburbs upon arrival, belying the stereotype of the "unacculturated" immigrant. Today, more than half of the Asian-origin population is spreading out in suburbs surrounding traditional gateway cities, as well as in new urban centers of Asian settlement across the country.

Differences in national origins, timing of immigration, affluence and settlement patterns profoundly inhibit the formation of a pan-ethnic identity. Recent arrivals are less likely than those born or raised in the United States to identify as Asian American. They are also so busy settling in that they have little time to think about being Asian or Asian American, or, for that matter, white. Their diverse origins include drastic differences in languages and dialects, religions, cuisines and customs. Many national groups also bring to America their histories of conflict (such as the Japanese colonization of Korea and Taiwan, Japanese attacks on China, and the Chinese invasion of Vietnam).

Immigrants who are predominantly middle-class professionals, such as the Taiwanese and Indians, or predominantly small business owners, such as the Koreans, share few of the same concerns and priorities as those who are predominantly uneducated, low-skilled refugees, such as Cambodians and Hmong. Finally, Asian-origin people living in San Francisco or Los Angeles among many other Asians and self-conscious Asian Americans develop a stronger ethnic identity than those living in predominantly Latin Miami or predominantly European Minneapolis. A politician might get away with calling Asians "Oriental" in Miami but get into big trouble in San Francisco. All of these differences create obstacles to fostering a cohesive pan-Asian solidarity. As Yen Le Espiritu shows, pan-Asianism is primarily a political ideology of U.S.-born, American-educated, middle-class Asians rather than of Asian immigrants, who are conscious of their national origins and overburdened with their daily struggles for survival.

UNDERNEATH THE MODEL MINORITY:

"WHITE" OR "OTHER"

The celebrated "model minority" image of Asian Americans appeared in the mid-1960s, at the peak of the civil rights and the ethnic consciousness movements, but before the rising waves of immigration and refugee influx from Asia. Two articles in 1966—"Success Story, Japanese-American Style," by William Petersen in the *New York Times Magazine*, and "Success of One Minority Group in U.S.," by the *US News & World Report* staff—marked a significant departure from how Asian immigrants and their descendants had been traditionally depicted in the media. Both articles congratulated Japanese and Chinese Americans on their persistence in overcoming extreme hardships and discrimination to achieve success, unmatched even by U.S.-born whites, with "their own almost totally unaided effort" and "no help from anyone else." (The implicit contrast to other minorities was clear.) The press attributed their winning wealth and respect in American society to hard work, family solidarity, discipline, delayed gratification, non-confrontation and eschewing welfare.

This "model minority" image remains largely unchanged even in the face of new and diverse waves of immigration. The 2000 U.S. Census shows that Asian Americans continue to score remarkable economic and educational achievements. Their median household income in 1999 was more than $55,000—the highest of all racial groups, including whites—and their poverty rate was under 11 percent, the lowest of all racial groups. Moreover, 44 percent of all Asian Americans over 25 years of age had at least a bachelor's degree, 18 percentage points more than any other racial group. Strikingly, young Asian Americans, including both the children of foreign-born physicians, scientists, and professionals and those of uneducated and penniless refugees, repeatedly appear as high school valedictorians and academic decathlon winners. They also enroll in the freshman classes of prestigious universities in disproportionately large numbers. In 1998, Asian Americans, just 4 percent of the nation's population, made up more than 20 percent of the undergraduates at universities such as Berkeley, Stanford, MIT and Cal Tech. Although some ethnic groups, such as Cambodians, Lao, and Hmong, still trail behind other East and South Asians in most indicators of achievement, they too show significant signs of upward mobility. Many in the media have dubbed Asian Americans the "new Jews." Like the second-generation Jews of the past, today's children of Asian immigrants are climbing up the ladder by way of extraordinary educational achievement.

One consequence of the model-minority stereotype is that it reinforces the myth that the United States is devoid of racism and accords equal opportunity to all, fostering the view that those who lag behind do so because of their own poor choices and inferior culture. Celebrating "model minorities" can help impede other racial minorities' demands for social justice by pitting minority groups against each other. It can also pit Asian Americans against whites. On the surface, Asian Americans seem to be on their way to becoming white, just like the offspring of earlier European immigrants. But the model-minority image implicitly casts Asian Americans as different from whites. By placing Asian Americans above whites, this image still sets them apart from other Americans, white or nonwhite, in the public mind.

There are two other less obvious effects. The model-minority stereotype holds Asian Americans to higher standards, distinguishing them from average Americans. "What's wrong with being a model minority?" a black student once asked, in a class I taught on race, "I'd rather be in the model minority than in the downtrodden minority that nobody respects." Whether people are in a model minority or a downtrodden minority, they are still judged by standards different from average Americans. Also, the model-minority stereotype places particular expectations on members of the group so labeled, channeling them to specific avenues of success, such as science and engineering. This, in turn, makes it harder for Asian Americans to pursue careers outside these designated fields. Falling into this trap, a Chinese immigrant father gets upset when his son tells him he has changed his major from engineering to English. Disregarding his son's talent for creative writing, such a father rationalizes his concern, "You have a 90 percent chance of getting a decent job with an engineering degree, but what chance would you have of earning income as a writer?" This thinking represents more than typical parental concern; it constitutes the self-fulfilling prophecy of a stereotype.

The celebration of Asian Americans rests on the perception that their success is unexpectedly high. The truth is that unusually many of them, particularly among the Chinese, Indians and Koreans, arrive as middle-class or upper middle-class immigrants. This makes it easier for them and their children to succeed and regain their middle-class status in their new homeland. The financial resources that these immigrants bring also subsidize ethnic businesses and services, such as private after-school programs. These, in turn, enable even the less fortunate members of the groups to move ahead more quickly than they would have otherwise.

NOT SO MUCH BEING "WHITE" AS BEING AMERICAN

Most Asian Americans seem to accept that "white" is mainstream, average and normal, and they look to whites as a frame of reference for attaining higher social position. Similarly, researchers often use non-Hispanic whites as the standard against which other groups are compared, even though there is great diversity among whites, too. Like most immigrants to the United States, Asian immigrants tend to believe in the American Dream and measure their achievements materially. As a Chinese immigrant said to me in an interview, "I hope to accomplish nothing but three things: to own a home, to be my own boss, and to send my children to the Ivy League." Those with sufficient education, job skills and money manage to move into white middle-class suburban neighborhoods immediately upon arrival, while others work intensively to accumulate enough savings to move their families up and out of inner-city ethnic enclaves. Consequently, many children of Asian ancestry have lived their entire childhood in white communities, made friends with mostly white peers, and grown up speaking only English. In fact, Asian Americans are the most acculturated non-European group in the United States. By the second generation, most have lost fluency in their parents' native languages (see "English-Only Triumphs, But the Costs are High," *Contexts*, Spring 2002). David Lopez finds that in Los Angeles, more than three-quarters of second-generation Asian Americans (as opposed to one-quarter of second-generation Mexicans) speak only English at home. Asian Americans also intermarry extensively with whites and with members of other minority groups. Jennifer Lee and Frank Bean find that more than one-quarter of married Asian Americans have a partner of a different racial background, and 87 percent of those marry whites; they also find that 12 percent of all Asian Americans claim a multiracial background, compared to 2 percent of whites and 4 percent of blacks.

Even though U.S.-born or U.S.-raised Asian Americans are relatively acculturated and often intermarry with whites, they may be more ambivalent about becoming white than their immigrant parents. Many only cynically agree that "white" is synonymous with "American." A Vietnamese high school student in New Orleans told me in an interview, "An American is white. You often hear people say, hey, so-and-so is dating an 'American.' You know she's dating a white boy. If he were black, then people would say he's black." But while they recognize whites as a frame of reference, some reject the idea of becoming white themselves: "It's not so much being white as being American," commented a Korean-American student in my class on the new second generation. This aversion to becoming white is particularly common among second-generation college students who have taken ethnic studies courses, and among Asian-American community activists. However, most of the second generation continues to strive for the privileged status associated with whiteness, just like their parents. For example, most U.S.-born or U.S.-raised Chinese-American youth end up studying engineering, medicine, or law in college, believing that these areas of study guarantee a middle-class life.

Second-generation Asian Americans are also more conscious of the disadvantages associated with being nonwhite than their parents, who as immigrants tend to be optimistic about overcoming the disadvantages of this status. As a Chinese-American woman points out from her own experience, "The truth is, no matter how American you think you are or try to be, if you have almond-shaped eyes, straight black hair, and a yellow complexion, you are a foreigner by default.... You can certainly be as good as or even better than whites, but you will never become accepted as white." This remark echoes a commonly-held frustration among second-generation, U.S.-born Asians who detest being treated as immigrants or foreigners. Their experience suggests that whitening has more to do with the beliefs of white America, than with the actual situation of Asian Americans. Speaking perfect English, adopting mainstream cultural values, and even intermarrying members of the dominant group may help reduce this "otherness" for particular individuals, but it has little effect on the group as a whole. New stereotypes can emerge and un-whiten Asian Americans, no matter how "successful" and "assimilated" they have become. For example, Congressman David Wu once was invited by the Asian-American employees of the U.S. Department of Energy to give a speech in celebration of Asian-American Heritage Month. Yet, he and his Asian-American staff were not allowed into the department building, even after presenting their congressional Identification, and were repeatedly asked about their citizenship and country of origin. They were told that this was standard procedure for the Department of Energy and that a congressional ID card was not a reliable document. The next day, a congressman of Italian descent was allowed to enter the same building with his congressional ID, no questions asked.

The stereotype of the "honorary white" or model minority goes hand-in-hand with that of the "forever foreigner." Today, globalization and U.S.-Asia relations, combined with continually high rates of immigration, affect how Asian Americans are perceived in American society. Many historical stereotypes, such as the "yellow peril" and "Fu Manchu" still exist in contemporary American life, as revealed in such highly publicized incidents as the murder of Vincent Chin, a Chinese American mistaken for Japanese and beaten to death by a disgruntled white auto worker in the 1980s; the trial of Wen Ho Lee, a nuclear scientist suspected of spying for the Chinese government in the mid-1990s; the 1996 presidential campaign finance scandal, which implicated Asian Americans in funneling foreign contributions to the Clinton campaign; and most recently, in 2001, the Abercrombie & Fitch t-shirts that depicted Asian cartoon characters in stereo-typically negative ways, with slanted eyes, thick glasses and heavy Asian accents. Ironically, the ambivalent, conditional nature of their acceptance by whites prompts many Asian Americans to organize pan-ethnically to fight back—which consequently heightens their racial distinctiveness. So becoming white or not is beside the point. The bottom line is: Americans of Asian ancestry still have to constantly prove that they truly are loyal Americans.

40

Feeling Like a Citizen, Living As a Denizen

Deportees' Sense of Belonging

TANYA GOLASH-BOZA

I met Victor in a barbershop near downtown Kingston. He walked, talked, and dressed like a young man from Brooklyn. Victor told me, with a heavy Brooklyn accent, "I'm from Brooklyn. I grew up in Brooklyn all my life." Although Victor considers himself to be from Brooklyn, he was born in Jamaica, in a hospital not too far from where we were sitting. When Victor was 4 years old, he and his mother took a plane from Kingston, Jamaica, to New York City.

Victor and his mother traveled to the United States as legal permanent residents (LPRs). An LPR is a foreign national who has been granted the privilege of residing permanently in the United States, and who qualifies for citizenship by naturalization after living in the United States for 3 to 5 years. Victor and his mother qualified for U.S. citizenship when Victor was 7 years old. Had Victor's mother become a U.S. citizen herself before Victor's 18th birthday, he could have become a U.S. citizen automatically. Victor's mother never went through the naturalization process. When Victor turned 18, he could have applied for naturalization. Yet he did not.

In 1996, when Victor was 24 years old, he was caught selling marijuana. He served 2½ years in prison and was deported to Jamaica because U.S. law requires deportation for non-U.S. citizens convicted of certain drug charges. In Jamaica, Victor has no friends or family, and finds it difficult to survive. He longs to return to New York where his mother and daughter live. Victor is one of many deportees I met who qualified for citizenship, yet who never became U.S. citizens. Had Victor gone through the naturalization process, he would not have been deported from the only land he calls home. Although he had been back in Jamaica for nearly a decade, Victor still considered Brooklyn "home." It seems almost a truism to say that we feel as if we belong in our homes. However, home invokes more of a sense of where *we* feel a sense of attachment, whereas belonging is more relational—it denotes both how *we* feel about a place as well as how *others* perceive us (Anthias, 2008; Olwig, 2002).

As a legal citizen of Jamaica and not of the United States, technically, Victor does not belong in the United States. For Victor, it seems perfectly natural that he belongs in the place he considers home. However, under U.S. law, his claims to belonging have little meaning. The mere fact of being born on U.S. territory provides you with legal citizenship, which guarantees you the inalienable right to remain in the United States, and to come and go as you see fit. In contrast, those persons born outside the United States are not automatically U.S. citizens unless they meet certain conditions, and can only become U.S. citizens if they qualify for *and* seek out naturalization.

Legal citizenship is "the legal correlate of territorial belonging. It signifies official recognition of a particularly close relationship between person and country" (Bhabha, 2009, p. 191). This official relationship entails the right to participate in politics as well as the right to not be banished from one's country of citizenship. This particular bundle of rights—the right to vote, hold public office, and not be deported—constitutes the totality of rights that U.S. citizens have that noncitizens do not.

In the United States, the Constitution guarantees a broad set of rights to all persons. Citizenship is not mentioned in the Constitution, and the U.S. Supreme Court has upheld the rights of noncitizens in this country to due process in criminal courts, to access to education, and to all of the rights guaranteed in the Constitution (Bosniak, 2006). A resident alien in the United States—officially referred to as a legal permanent resident—thus can live his or her life in the United States with the ability to exercise nearly all the rights of U.S. citizens. The line between citizen and noncitizen can thus seem blurry when we consider LPRs. The blurriness of this boundary, however, has come into focus in recent years as hundreds of thousands of LPRs have been deported. This article thus argues that legal citizenship has taken on renewed importance in the current U.S. context of mass deportation.

In the contemporary United States, millions of LPRs live as if they were citizens—they have access to nearly all of the rights of citizens. However, due to laws implemented in 1997, it has become remarkably easy for a LPR to be deported from the United States. I argue that the passage of these laws—which rendered LPRs deportable for fairly minor convictions—has given renewed meaning to the importance of territorial belonging for theories of citizenship. While the deservingness of citizens to remain within a nation's territory is hardly ever put into question, U.S. laws make it surprisingly easy to remove a noncitizen from its territory. The fact that noncitizens enjoy many of the same rights and sense of belonging as citizens makes their deportation all the harsher....

LPRs like Victor are denizens—people who have settled within the territory of a country yet lack citizenship.... Why don't denizens who qualify for citizenship apply? And, what does this tell us about the meanings of citizenship and alienage in the contemporary United States?

... Based on interviews with 30 deported Jamaicans who were once LPRs of the United States, I argue that denizens often feel "like citizens" based on their family and community ties to the United States, yet that their allegiance and sense of belonging is primarily to their family and community—not to the state. In this sense, there is a disconnect between the law—which privileges legal citizenship—and the daily lives of denizens—in which they can experience a profound sense of belonging in their communities.

LEGAL CITIZENSHIP, ALIENAGE, AND BELONGING

Legal citizenship signifies the official recognition of a particular relationship between a person and the government and guarantees territorial rights in the country of citizenship (Román, 2010). All persons residing on U.S. territory have some relationship with the government: We pay taxes, are subject to sanction if we violate U.S. laws, attend government-sponsored schools, and benefit from government services. Only those who have legal citizenship, however, have the chance to shape the laws that govern society, and only U.S. citizens are safe from expulsion from U.S. borders.

Citizenship and alienage define one another in a dialectical fashion—alienage is only important because citizenship is. The idea of citizenship renders possible the existence of aliens.... On the one hand, citizenship is inclusionary insofar as it brings together a group of people defined as citizens. On the other hand, this definition of some people as citizens and others as aliens is exclusionary. In the United States, LPRs are not citizens, even though they do have a pathway to citizenship. However, so long as they remain LPRs, they remain outside the borders of the political community.

Although only citizens of the United States are members of the political community (Román, 2010), the United States is more than a political community: It is also a cultural, social, and economic community. Many non-U.S. citizens who reside permanently in the United States consider themselves members of these communities, despite their formal exclusion from the political community. This reality points to a contradiction: Citizenship implies a political relationship with the government. However, what is most important in our everyday lives is not our relationship with our government but that with our families and communities. This formal relationship with the government, however, is the only way to ensure that non-U.S. citizens are not removed from their families and communities. Failure to formalize one's relationship with the United States by becoming an official member of the polity potentially has severe consequences, insofar as noncitizens can face deportation....

Some noncitizens in the United States can vote in local elections, and in the United States, not all citizens are eligible to vote. Notably, people convicted of felonies in many states cannot vote. As it turns out, in the United States, the sole clearly marked distinction between citizens and noncitizens is that U.S. citizens cannot be deported. As William Walters (2002) argues, deportation is a "technology of citizenship"—it actively creates a world that is divided into states. Without the possibility of deportation, states would have no authority over returning people to where they officially belong, and thus no control over who can become a member of the state's population. The possibility of deportation endows citizenship with meaning insofar as only noncitizens (or denaturalized citizens) can be deported (Gibney, 2013). A consideration of deportation allows us to think about how citizenship can function as a barrier to territorial rights—the right to live in a particular place (Parker, 2001)....

DEPORTATION OF LEGAL PERMANENT RESIDENTS
FROM THE UNITED STATES

If you lack citizenship in the United States, technically, you do not have the *right* to be in the United States; remaining within the United States is a *privilege* that can be revoked at any time. Deportation simply means revoking that privilege. Non-U.S. citizens can be deported without due process, and without consideration for their social, cultural, and family ties to the United States because, in U.S. law, deportation is not punishment. Although noncitizens in the United States are given the full spectrum of rights in criminal proceedings, they are denied many of these basic rights in immigration proceedings. It is in these proceedings that the importance of citizenship comes into sharp relief.

The idea that deportation is not punishment is based on a distinction between deportation and banishment, where banishment is punishment because it involves removing a person from a country where he belongs, yet deportation is the act of returning a person to where he belongs, and thus is not punishment. This legal idea of belonging is based exclusively on one's formal political status as a citizen or noncitizen, and leaves no room for the consideration of other forms of belonging to the polity.

Only non-U.S. citizens can be deported because U.S. citizens have territorial rights—the right to live within the borders of the United States. As U.S. citizens, this right cannot be revoked. To do so would be banishment, and banishment is not among the punishments the United States metes out to people convicted of crimes. In stark contrast, an LPR can be deported, even for minor infractions of the law. In many deportation cases, an LPR's social, cultural, and economic ties to the United States can be ignored.

In 1996, Congress passed two laws that profoundly changed the rights of all foreign-born people in the United States: the Anti-Terrorism and Effective Death Penalty Act and the Illegal Immigration Reform and Immigrant Responsibility Act of 1996. These laws were striking in that they eliminated judicial review of some deportation orders, required mandatory detention for many noncitizens, and introduced the potential for the use of secret evidence in certain cases.

One of the most pernicious consequences of these laws is related to the deportation of LPRs—noncitizens who have been granted legalization and have the right to remain in the United States on a permanent basis, so long as they do not violate provisions of the Immigration and Nationality Act. Prior to 1996, judges were permitted to exercise discretion in deportation cases. When deciding whether or not to deport a person who had been convicted of a crime, judges could consider the immigrant's rehabilitation, remorse, family support, and ties (or lack thereof) to their country of origin. The 1996 laws took away the judge's discretionary power in aggravated felony cases. Congress created the idea of an aggravated felony as part of the Anti-Drug Abuse Act of 1988 to provide harsh provisions for noncitizens convicted of murder and drugs and arms trafficking. The expansion of the definition of aggravated felonies in the 1996 laws meant that this category now includes any crime of violence or theft offense for which the term of imprisonment is at least 1 year, illicit drug offenses, as well as other violations (Kanstroom, 2000)....

CASE SELECTION

The present study is based on interviews with 30 Jamaican deportees—all of whom were LPRs in the United States. I chose Jamaica as the site for this study because the migration stream from Jamaica is long-standing, meaning that many Jamaicans are likely to have spent large portions of their lives in the United States, and the majority of Jamaicans in the United States are LPRs. Among Jamaican deportees, 95% are men; 28% arrived in the United States before age 16; and the average period of residence in the United States is 12 years. The majority of criminal deportees were expelled on drug charges, and very few had committed violent crimes (Headley, 2005). My sample reflects these demographics.

Jamaican LPRs are particularly likely to be deported. About 10% of LPR deportees between 1997 and 2006 were Jamaican, even though Jamaicans make up less than 2% of all LPRs in the United States. Most of the Jamaican deportees I interviewed had no intention of returning to their countries of origin prior to being deported. All of them left family members in the United States, including children, parents, and spouses. They had lived in the United States for extended periods of time: the shortest stay was 8 years and the longest 38 years.

METHODOLOGY AND SITE DESCRIPTION

In Kingston, I used snowball sampling and key informants to find interviewees. I employed two research assistants, both of whom were deportees, to help me find interviewees. As deportees, my two research assistants were ideal key informants and experts. Their insider status made it relatively easy for them to locate an otherwise hard-to-find population and gave me insight with regard to some of the deportees' experiences. In addition, having grown up in Kingston, my research assistants could explain the interviewees' backgrounds to me, such as the quality of schools they had attended and the class composition of the neighborhoods where they were raised. In addition, I interviewed three people in the Trenchtown neighborhood of Kingston through local personal contacts, and two in Ocho Rios through a deportee contact there. I also interviewed two people who introduced themselves to me because they could tell I was from the United States and they wanted to chat with an American. Using snowball sampling and a variety of points of entry, I was able to obtain a sample that closely resembles the overall deportee population in Jamaica. Although the deportee population in Jamaica is nearly all male, I interviewed one woman. I spent a total of 6 months in Jamaica between December 2008 and July 2009. The interviews ranged in length from 30 minutes to more than 2 hours, and were all audio-recorded. I had all of the audiotapes transcribed by a Jamaican university student to ensure the correct spelling of any patois or local slang. I checked each transcription for accuracy then coded the data for specific and emergent themes. For this article, I paid particular attention to the discussion of citizenship and belonging in the interviews.

THE CITIZENSHIP OF ALIENS

Interviews with Jamaican deportees render it clear that a sense of belonging to the United States is not limited to U.S. citizens. The Jamaican deportees I interviewed often felt as if they belonged in the United States. They built lives in the United States—many were married and most had children. Their parents and siblings became U.S. citizens. Many deportees protested their deportation on the grounds that their family members were all U.S. citizens. Another discursive strategy deportees used was to point out that the United States is their home because of the ties they built there. These deportees appealed to notions of *jus nexi*—where citizenship would be based on authentic connections to a society, developed through social, cultural, and economic participation in society.

Familial ties were a central reason deportees cited when arguing they belonged in the United States. Hazel was the only female former LPR I interviewed. Hazel, who had three children born in the United States, was deported after being convicted of a drug offense. Hazel told me,

> I deserve to go back. I didn't kill anybody. I didn't rob anybody. I did commit a crime, but I did my time. Allow me to at least travel and see my kids. They all are American; that makes me a citizen too because they were all born there; none of my children were born here [in Jamaica].

In Hazel's account, we can see how deportation makes national borders more meaningful. When borders separate a mother from her children, their rigidity becomes heightened. Hazel provides two reasons she should be permitted to return to the United States: (1) her crime [illicit drug possession] was not serious; and (2) her children are U.S. citizens. She believes her children's U.S. citizenship "makes [her] a citizen too."

Lamar also made similar claims. Lamar was an LPR. His grandmother, both of his parents, and his four children are all U.S. citizens. He explained,

> I grew up in the United States. Everything I learned was from over there. All my children, my mother and father, everyone is there. They should look into cases like that. And say, this man should be able to stay, maybe keep an eye on him or whatever, go to his parole officer or whatever. If you have your immediate family over there, you should be able to stay. Because, they are all citizens. I am like a citizen too because my parents are citizens.

Similar to Hazel, Lamar says that he is "like a citizen" because his family members are citizens. Lamar feels as though his deportation was unjust. He told me,

> The ones that deserve to stay over there should stay. You know … you grew up over there so why should you come back to where you were born? All of your roots are over there cause your mother and father are still over there. In my case, that's what it is.

Lamar believes he deserves to stay in the United States because of his family ties there. He uses a language of deservingness, rights, belonging, and even citizenship, even though he has no legal claim to territorial belonging. When Floya Anthias (2008) writes of how our sense of belonging is heightened through strategies of exclusion, she was not referring to the literal sense of exclusion that occurs in the context of deportation. However, we can see how Lamar's discourse of belonging became salient because of his deportation. What is notable in these interviews is that deportees use the language of citizenship to make these claims. Feeling like a citizen, however, is not sufficient grounds to contest deportation. The only surefire way to avoid deportation is to be a legal citizen.

Deportation was also very harsh for Victor, introduced in the beginning of the article. Victor told me he used to "sit in [his] room and stress the hell out" when he was deported. Arriving in Brooklyn at age 4 and spending his entire life there, Victor developed a deep sense of belonging to his community, even though he was not a citizen of the United States.

Victor believes he did not deserve to be deported. He feels he belongs in Brooklyn both because he grew up there and also because he perceives that the only people in the world who care about him live there. He explained,

> It is not good to live in a place for so many years, to live and have family that loves you, and make sure that you all right and then to come here. You are a person that used to work and you earn your own dollar. Then you're here.… What are you going to do? … I have a woman that loves me in America. I've been with her since she was 17 and I was 18. Now, I'm 31.… I have my daughter. I have my mother. I have my brothers. I have my sisters. I have my uncles. What the hell am I doing here?

Victor wants to return to America. He told me he would travel to the United States clandestinely if he could, even though that would mean he would live there without papers. For Victor, what is most important is that he is able to live near people who care about him. Victor claims he belongs in the United States because of his family and social ties there. Like many other deportees, he perceives that no one in Jamaica cares if he lives or dies.

These deportees made claims to the United States based on their relationship to their families, friends, and communities there. Their lack of similar ties to Jamaica makes them feel lost and out of place in the land of their birth. Their stories show that one can experience several types of belonging without formal citizenship status. Moreover, they show how important the right to territorial belonging can be for those who feel a strong sense of belonging. Their citizenship claims are based on their contributions to society and family ties to the United States. They made claims to legal citizenship on the basis of their sense of belonging....

THE SIGNIFICANCE OF CITIZENSHIP AND ALIENAGE

... The fact that LPRs felt a strong sense of belonging to the United States raises the question of why they did not seek out legal citizenship. I find that, somewhat ironically, they did not seek out citizenship because alienage was not a salient aspect of most of their lives. The fact that they felt a sense of belonging in the United States was a disincentive in terms of seeking out legal citizenship.

Never Got Around To It

Most deportees I spoke with told me that they thought about naturalization but never got around to it. Chris, for example, moved to the United States in 1969, when he was 16 years old. Barely out of high school, Chris married and had three children. They were living together in Brooklyn in 2006 when Chris had trouble with the law. A neighbor stole a stereo system from his apartment. Chris confronted him about the theft, and they got into an altercation. The neighbor pulled out a knife. Chris wrestled the knife from him, and stabbed him. Despite his claims to self-defense, Chris was convicted of assault and was sentenced to 1 year in jail. Chris's conviction of a violent crime that carried a sentence of 1 year in jail became an aggravated felony under immigration laws. Thus, after serving 8 months of his sentence, he was released from Riker's Island Jail, only to be taken directly to immigration detention. He spent 5 months in a detention facility in Texas and was deported.

When he was deported, Chris had been in the United States for 38 years, was married to a U.S. citizen, and had three U.S.-born children. Chris had not been back to visit Jamaica the entire time he was in the United States, and had not maintained ties with the land of his birth. Chris was an LPR in the United States and had been eligible for citizenship for decades. I asked Chris why he never applied for citizenship. He told me,

> I don't know. I don't even know what to say to you. Because many times, many, many a time, my wife, my friends, my cousins, my aunts, ... told me to go and apply for citizenship. And [they] just couldn't get me there honestly.

Chris watched his family members apply for and receive naturalization. His family members suggested to him that he should seek out naturalization. However, it was not important enough to him for him to find the time and money to do it. He now regrets not naturalizing—because his failure to naturalize means he must live apart from his wife and three children. For the 38 years he lived in the United States, however, his inability to vote or participate in the formal political polity held little importance to him. While living in the United States, citizenship was not important to Chris because of the rights it would grant him; it only became important when his lack of citizenship meant he would be separated from his home and family. As Chris led a law-abiding life, he never anticipated that he could be convicted of a crime and subsequently deported.

Chris's response was similar to many other deportees who had spent years in the United States. I asked Delroy, another LPR, why he did not naturalize. He told me his green card said "permanent resident," so he thought that was sufficient. He told me, "America is home.... I was born in Jamaica, but that's a long time ago." According to Delroy, having a green card led to "complacence" as he did not see a need to seek out citizenship. After living in the United States for 28 years, he felt "Americanized." Delroy was later deported after being convicted of domestic violence, which made him realize that although he felt "Americanized" after decades of living in the United States, technically, he belonged in Jamaica.

I asked Lamar, introduced above, why he never applied for citizenship if he felt so attached to the United States.

> At the time I thought it was okay being that I was [an LPR] 'cause I could travel. But if it were serious times like now—I would have applied for citizenship long ago. The opportunity never presented itself. I thought it was all right to be a permanent resident.

When Lamar lived in the United States, he thought it was sufficient to be an LPR. His deportation, however, made him see the importance of legal citizenship. Although he felt "like a citizen," he was not a U.S. citizen. Lamar and Hazel, like other Jamaican deportees, felt as if they belonged in the United States because of their family ties there. Lamar was deported in 2004, 31 years after he had moved to the United States and 7 years after the implementation of the 1996 laws. Nevertheless, he never made it a priority to seek out naturalization.

Other deportees gave similar reasons for never naturalizing. Elmer told me "It didn't dawn on me." Peter, like many others, was aware that you could be deported after committing a crime, but did not see himself getting on the wrong side of the law. As a law abiding "citizen," deportation seemed unlikely. As Peter said, "I worked and got along like any average person. I wasn't really worried about the fact that I might get deported one day for whatever reason. I never thought it would happen." Roy also never thought he would get deported, although, when I asked, he cited financial barriers as the primary reason he never sought citizenship.

> Well, that is a big question. I don't have a reason. I ask myself that big question all the time. You see, when you are caught up with work and work, it takes away some of your time. I was the only one working and bringing in the money. The money was tight. To apply for citizenship is costly. My wife was a stay at home mom. We were struggling financially. But, still I had enough time to get it, over 20 years.

The daily pressures of life superseded the perceived need to seek out citizenship for these men. Being a citizen of Jamaica and not the United States was not salient in their daily lives. It was only deportation that made them realize the importance of their failure to formalize their relationship with the U.S. government.

For these deported LPRs their lack of U.S. citizenship became an exclusion mechanism that prevented them from having territorial rights in the United States. This exclusion mechanism, however, was only made visible when they faced deportation. Prior to that moment, they rarely, if ever, thought about their alienage. Victor—introduced in the beginning of this article—told me, "I knew people that got deported, but I never thought I would get deported." Even Victor, who was involved in the underground economy, did not seem himself as deportable. Thus, it is not surprising that the other LPRs discussed here, who generally led law-abiding lives, did not think they were deportable.

One deportee, O'Ryan, applied for citizenship, but his application was too late. His case is instructive because it shows that applying is not enough. To avoid deportation, you need to go through the long and expensive process of naturalization. Legally, the question of citizenship is dichotomous. Either you are a U.S. citizen or you are not....

O'Ryan went to the United States as a small child. He applied for citizenship when his green card expired in 1996. His mother and cousin applied at the same time. His mother's citizenship went through, and then his cousin's. When he heard theirs had gone through, he went to check on his citizenship. The citizenship office told him he needed to redo his fingerprints. He finally received the letter saying he should go to the swearing-in ceremony in 2001, 5 years later.

However, O'Ryan had been arrested a few weeks earlier on a drug charge, and was in jail when his letter arrived. Thus, at the age of 25, O'Ryan was deported to a country he barely knew. This bureaucratic delay meant O'Ryan was deported from the land he considers home. O'Ryan told me: "That's where I grew up. That's where I know everybody. That's home to me." Before being deported, O'Ryan had never considered leaving New York City. He explained,

> I wasn't thinking about ever leaving New York.... This was like a
> wakeup call to me that nothing in life is guaranteed ... but I know that
> one thing is guaranteed that no matter where I go or what I do, I'm
> born in Jamaica. I am a Jamaican ... and I just gotta accept that. I keep
> hearing from my family that you're in Jamaica. You need to start
> thinking about Jamaica.

Being born in Jamaica makes O'Ryan Jamaican. This may seem obvious, but it took deportation for him to realize and accept his natural-born citizenship. And, it is not an easy realization. For O'Ryan, New York is home.

> I did everything there. I went to school there.... Everything that
> happened to me for the first time happened to me in New York. I have
> no experiences of Jamaica. I hate saying it, but all my experiences of
> Jamaica so far have basically been bad.... Every time I try to get myself
> started out here, I get shut down. Every time I try to open a door,
> somebody slams it in my face.

In New York, O'Ryan has his daughter, his former fiancé, his mother, his cousins, and his friends. Coming to terms with the fact that the land of his birth is where he officially belongs is no easy task. In the United States, O'Ryan had access to the cultural and social rights he needed to feel like a full member of society. He did not seek U.S. citizenship so that he could feel as if he belonged; he sought it to ensure he could stay in the United States. In Jamaica, O'Ryan has the political right to vote and the legal right to territorial belonging. However, he does not feel as if he belongs in Jamaica, where he has only weak family ties and feels out of place.

For most deportees, alienage was not significant in their daily lives in the United States; instead, they often feel out of place in their land of citizenship—Jamaica. As Lamar explained, they thought being an LPR was "all right." Despite their status as denizens of the United States and citizens of Jamaica, each of these deportees experienced cultural and social belonging to their communities in the United States, and a lack of belonging in Jamaica. A close look at their statements also reveals that their primary attachments are to their families and communities—not to the state. O'Ryan and Victor spoke about feeling home in New York or Brooklyn. Hazel spoke about wanting to be with her children. Their sense of belonging is highly localized.

DISCUSSION AND CONCLUSION

This article contributes to the scholarly discussion on the meanings of citizenship, alienage, and belonging through analyses of the narratives of Jamaican LPRs who have been deported. These stories render it evident that LPRs' sense of belonging to the United States is rooted in their families and communities. The strong sense of belonging that Jamaican deportees feel to the United States makes it clear that alienage was not a significant factor in most of their daily lives. Concomitantly, their lack of legal citizenship seemed to have little importance. Remarkably, the reason that alienage was not significant is because they had access to other forms of citizenship. They had access to social citizenship—the rights and responsibilities they had in terms of providing for their families. They had access to cultural citizenship—feelings of belonging to their communities and families. With access to cultural and social citizenship, these deportees did not perceive a pressing need to seek out formal, legal citizenship. They also had access to civil and legal rights. When they faced criminal charges in the United States, they were granted access to the same rights as U.S. citizens. It was only in deportation proceedings where they felt their rights were abrogated.

Undocumented migrants encounter obstacles to full inclusion that prevent them from feeling like citizens of the United States and make them more vulnerable to exploitation. Like undocumented migrants, LPRs are also deportable. However, because LPRs do not experience the barriers associated with illegality, neither their alienage nor their deportability are significant in their daily lives.

Deportation is an exercise of state power that can be enacted exclusively on noncitizens. For LPRs to be deported, they first have to be convicted of a crime. Many of these deportees saw themselves as law-abiding "citizens" and permanent residents of the United States. For them, alienage only was a category of exclusion in the most literal sense—their alienage enabled their deportation. Prior to that moment, alienage was generally not a salient part of their lives. Ironically, had alienage been more significant, had their alienage prevented their social inclusion, they may have been more inclined to seek out legal citizenship, which, in turn, could have prevented their deportation.

REFERENCES

Anthias, F. (2008). Thinking through the lens of translocational positionality: An intersectionality frame for understanding identity and belonging. *Translocations: Migration and Social Change, 4*(1), 5–20.

Bhabha, J. (2009). The "mere fortuity of birth"? Children, mothers, borders, and the meaning of citizenship. In S. Benhabib & J. Resnik (Eds.), *Migration and mobilities: Citizenship, borders, and gender* (pp. 197–227). New York: New York University Press.

Bosniak, L. (2006). *The citizen and the alien: Dilemmas of contemporary membership.* Princeton, NJ: Princeton University Press.

Gibney, M. (2013). Deportation, crime, and the changing character of membership in the United Kingdom. In K. Franko Aas & M. Bosworth (Eds.), *The borders of punishment: Migration, citizenship, and social exclusion* (pp. 218–236). Oxford, England: Oxford University Press.

Kanstroom, D. (2000). Deportation, social control, and punishment: Some thoughts about why hard laws make bad cases. *Harvard Law Review, 113*, 1890–1935.

Olwig, K. (2002). A wedding in the family: home making in a global kin network. *Global Networks, 2.3*, 205–218.

Parker, K. (2001). State, citizenship, and territory: The legal construction of immigrants in antebellum Massachusetts. *Law and History Review, 19*, 583–643.

Román, E. (2010). *Citizenship and its exclusions: A classical, constitutional and critical race critique.* New York: New York University Press.

Walters, W. (2002). Deportation, expulsion, and the international police of aliens. *Citizenship Studies, 6*, 265–292. doi:10.1080/1362102022000011612

41

Policed, Punished, Dehumanized
The Reality for Young Men of Color Living in America

VICTOR M. RIOS

Growing up, I experienced constant police harassment and brutality, making me normalize police violence in my community. I personally had my face stomped to the cement by police at age fifteen. My younger brother had been dragged out of a car through the window and beaten at age fourteen by a gang of notorious police officers who called themselves "the Riders." When we filed complaints with the Oakland Police Department or talked to lawyers for help we were ignored. It seemed then that there was nothing we could do about unsanctioned police violence, that no one else cared, and that there were no avenues for getting the word out; until now. The difference between 1997 and today is that a critical mass of young black and Latino men being killed by police and vigilantes has come to the national spotlight through social and mainstream media. This provides a new opportunity to expose the problem of state-sanctioned violence on marginalized populations and on cultivating a social movement that turns against the brutal punitive state.

The state-sanctioned police and vigilante violence on men of color like Oscar Grant (California), Trayvon Martin (Florida), Eric Garner (New York), Andy Lopez (California), and Michael Brown (Missouri)—to name a few high-profile cases—stems from an unchecked system of punitive social control that criminalizes young people of color from early ages. While some marginalized men of color are murdered by police and vigilantes, many more, a large number who live in poverty throughout the United States, experience a kind of social death or social incapacitation in which they are rendered as criminal suspects not just by police but by schools, community centers, social workers, merchants, community members, and even family members. By the time that these young men become young adults, their lives have been policed, punished, and dehumanized by various institutions. This hypercriminalization empowers and emboldens the criminal justice system, law enforcement, and vigilantes to harass, arrest, and shoot these young men at will. This youth control complex—the coalescing of various social institutions to punish, stigmatize, and dehumanize marginalized young people—provides the justification for law enforcement to render young black and Latino bodies as disposable. Society criminalizes marginalized young

SOURCE: Rios, Victor. 2015. From: *Deadly Injustice: Trayvon Martin, Race, and the Criminal Justice System*, edited by Devon Johnson, Patricia Y. Warren, and Amy Parrell. New York: New York University Press.

people so much that by the time a police officer or vigilante shoots them, they have already been rendered as deserving of draconian punishments, even death.

In the midst of this brutal treatment that marginalized young people experience in many communities throughout the United States, there is hope. In my twelve years of collecting data on young people who have been entangled in the criminal justice system, I have found that this very system of punitive social control that has socially incapacitated so many young people and killed many others has, paradoxically, become the most viable impetus for the marginalized masses to generate a viable social movement, one that dismantles the punitive state, in the twenty-first century. When calls are made to resist against police brutality, these very same individuals who have been rendered as apolitical and criminal by their communities become engaged and politicized. The marginalized masses will continue to resist, on their own terms, until a more just and dignifying system is created.... I discuss how the inner-city men I have followed over the years engage in both acts of survival and crimes of resistance. Sociologist Howard Becker (1963) finds that labeled youths resist by internalizing their label and committing more crime. My study finds a missing link in this analysis: the internalization of criminality is only one outcome in the labeling process; another outcome that young people who are labeled partake in is resistance—they internalize criminalization, flip it on its head, and generate action that seeks to change the very system that oppresses them. This is apparent in day-to-day contestations to hyperpolicing but also in recent mass protests around the country in the wake of multiple police killings of young black and Latino men.

PUNITIVE SOCIAL CONTROL AND SOCIAL INCAPACITATION

A black man is killed by police or vigilantes at least every twenty-eight hours in the United States. Many of these killings are dismissed by the legal system as legitimate. Even the U.S. Supreme Court has been involved in sanctioning police brutality. In a recent decision it ruled that even egregious police misconduct is not a violation of individual constitutional rights. This rampant state-sanctioned deadly force is only the tip of a larger iceberg: *punitive social control.* Young African American and Latino males suffer a plethora of police and vigilante harassment and brutality on the streets throughout the United States. Many scholars and activists have argued that to date little has been done by police departments, politicians, or the community to halt the everyday use of excessive force and racial profiling by police departments and private security on African American and Latino youths.

Vigilantes, those "everyday citizen" wannabe cops and security guards who are involved in criminalizing and punishing young males of color, have learned that the law places little value on the lives of the black population. This has emboldened individuals to take the law into their own hands. The punitive state is so powerful that it has now been embodied by those "everyday citizens" who seek to punish those considered criminal suspects. The killing of young men of color like Trayvon Martin is a consequence of an extremely punitive criminal justice system and negative public sentiment about inner-city minority boys. To prevent another brutal killing of yet another innocent minority teenager, the public and policy makers must urge police departments to change their differential treatment practices: to eliminate profiling, excessive force, and the vigilante mentality that many officers display.

Many of the young people I have followed, some for over a decade, have not been shot and killed by police but instead have experienced a slow but devastating process of *social death*, the systematic process by which individuals are denied their humanity. While individuals remain alive, they are socially isolated, violated, and prevented from engaging in social relations that affirm their humanity.... But beyond finding that incarceration produces a certain kind of social death, criminalization begins at a young age and injects young black and Latino boys with microdoses of social death. I refer to this "micro-aggression" form of social death as "social incapacitation." Social incapacitation is the process of dehumanization by which punitive social control becomes an instrument that prevents marginalized populations from becoming "productive" citizens who can feel dignity and affirmation from institutions of socialization....

As many young working-class black and Latino boys come of age, they realize that they are being systematically punished for being poor, young, black or Latino, and male. In the era of mass incarceration, when punitive social control has become a dominant form of governance, some young people are systematically targeted as criminal risks.... This process has created a generation of marginalized young people, who are prevented from engaging in a full affirmation of their humanity, let alone from gaining entry into roles that might give them social mobility. The logic and practice of punitive social control has prevented many marginalized young people from gaining acceptance in school, landing a job, or catching a break for minor transgressions from police and probation officers.

Criminalization does not only occur in the law; it crosses boundaries and follows the young people in this study across an array of institutions, including school, the neighborhood, the community center, and the family. In other words, the young men I have studied often found themselves in situations in which their deviant and nondeviant behaviors were constantly treated as threats, risks, and crimes by the various dominant institutions they attempted to navigate. I define this ubiquitous criminalization as the *youth control complex,* a system in which schools, police, probation officers, families, community centers, businesses, and other institutions collaborate to treat young people's everyday behaviors as criminal activity. Young people, who become pinballs within this youth control complex, experience what I refer to as *hypercriminalization,* the process by which an individual's nondeviant behavior and everyday interactions become treated as risk, threat, or crime and, in turn, have an impact on his or her perceptions, worldview, and life outcomes. The youth control complex creates an overarching system regulating the lives of young people, what I refer to as *punitive social control.* Criminalization results in punishment. Punishment, in this study, is understood as any outcome resulting from criminalization that makes young people feel stigmatized, outcast, shamed, defeated, or hopeless.

The youth control complex is not a new phenomenon. Poor and racialized populations have been criminalized in the United States since its inception. The black body has been one of the objects on which criminalization, punishment, social incapacitation, and social death have been executed and perfected. The transatlantic slave trade, savage whippings by slave owners, lynching, and police brutality have been a few of the many historical forms, often state-sanctioned if not state-imposed, of violent punishments executed on the black body. The state-sanctioned brutal police and vigilante murders of unarmed men of color in the twenty-first century can be understood as modern-day lynching by ballet. These killings serve the expressive function of sending a message to other men of color that they are to obey de facto and de jure rules of racial, gendered, and economic exclusion and subjugation.

Punishment of the brown body has been executed through the genocide of indigenous populations; appropriation of Mexican territory by the United States; and vigilante and police brutality against "bandidos," "immigrants," zoot-suiters, and gang youths. In an era of mass incarceration, developed over the past thirty years, punitive social control has fed an out-of-control minotaur, allowing it to expand its labyrinth by embedding itself into traditionally nurturing institutions, punishing young people at younger ages, and marking many for life. Criminalization is well disguised as a protective mechanism: zero-tolerance policies at schools are declared to provide the students who want to learn protection from bullies and disruptions; increased punitive policing is sold as protecting good citizens from violent gang members; longer incarceration sentences and adult sentencing appear to keep the bad guys from victimizing others and send a clear message to potential criminals; and so on.

The young men in this study compare encounters with police, probation officers, and prosecutors with interactions they have with school administrators and teachers who place them in detention rooms; community centers that attempt to exorcise their criminality; and even parents, who feel ashamed or dishonored and relinquish their relationship with their own children altogether. It seems, in the accounts of the boys, that various institutions collaborate to form a system that degrades and dishonors them on an everyday basis. As such, these young men's understanding of their environment as a punitive one, where they are not given a second chance, leads them to believe that they have no choice but to resist. These institutions, though independently operated with their own practices, policies, and logics, intersect with one another to provide a consistent flow of criminalization. The consequences of this formation are often brutal. Young Ronny, from Oakland, explains: "We are not trusted. Even if we try to change, it's us against the world. It's almost like they don't want us to change. Why they gotta send us to the ghetto alternative high school? We don't deserve to go to the same school down the street? When we try to apply for a job, we just get looked at like we crazy. If we do get an interview, the first question is, 'Have you been arrested before?'. . . We got little choice." Ronny understands his actions as responses to this system of punishment, which restricts his ability to survive, work, play, and learn. As such, he develops coping skills that are often seen as deviant and criminal by the system....

This web of punishment, the youth control complex, adds to young people's blocked opportunities but also generates creative responses, which allow the boys to feel dignified. Sometimes these responses even lead to informal and formal political resistance. However, some responses jeopardize the boys' ability to remain free and negatively impact their social relations with others.

... In an environment where there were few formal avenues for expressing dissent toward a system, which the boys believed to be extremely repressive, they developed forms of resistance that they believed could change, even if only temporarily, the outcome of their treatment. The boys believed they had gained redress for the punitive social control they had encountered by adopting a subculture of resistance based on fooling the system and by committing crimes of resistance, which made no sense to the system but were fully recognizable to those who had been misrecognized and criminalized.

Paradoxically, punitive social control also created a deep self-awareness of the class position in which these young people found themselves. These boys all demonstrated a clear understanding of the process of punishment. In addition, their deviant and delinquent actions, except when they were drunk or high, served as an attempt to act in their own rational interests. Marginalized young people have agency and political awareness; they have a clear understanding of the system of punitive social control that impacts their daily lives. While some of what the boys told me was one-sided, full of half-truths, and with a clear bias and misrecognition of their social conditions and the intentions of most social control institutions to genuinely help them, these young people could clearly articulate the mechanisms by which they ended up marked and tracked into the criminal justice system. Their actions, subcultures, and worldviews were developed in direct opposition to punitive social control. This resistance carried the seeds of redemption, self-determination, resilience, and desistance. Embracing the positive aspects of this resistance, teaching young people how to use it to navigate in mainstream institutions, and granting less damaging consequences for young people who break the law are all challenges that we must take on if we are to change the trajectory of punitive social control.

REFERENCE

Becker, Howard. 1963. *Outsiders: Studies in the Sociology of Deviance.* New York: Free Press.

42

The Myth of Immigrant Criminality and the Paradox of Assimilation

RUBÉN G. RUMBAUT AND WALTER EWING

Because many immigrants to the United States, especially Mexicans and Central Americans, are young men who arrive with very low levels of formal education, popular stereotypes tend to associate them with higher rates of crime and incarceration. The fact that many of these immigrants enter the country through unauthorized channels or overstay their visas often is framed as an assault against the "rule of law," thereby reinforcing the impression that immigration and criminality are linked. This association has flourished in a post-9/11 climate of fear and ignorance where terrorism and undocumented immigration often are mentioned in the same breath.

But anecdotal impression cannot substitute for scientific evidence. In fact, data from the census and other sources show that for every ethnic group without exception, incarceration rates among young men are lowest for immigrants, even those who are the least educated. This holds true especially for the Mexicans, Salvadorans, and Guatemalans who make up the bulk of the undocumented population. What is more, these patterns have been observed consistently over the last three decennial censuses, a period that spans the current era of mass immigration, and recall similar national-level findings reported by three major government commissions during the first three decades of the 20th century. The problem of crime in the United States is not "caused" or even aggravated by immigrants, regardless of their legal status. But the misperception that the opposite is true persists among policymakers, the media, and the general public, thereby undermining the development of reasoned public responses to both crime and immigration.

Among the findings in this report:

CRIME RATES HAVE DECLINED AS IMMIGRATION HAS INCREASED

- Even as the undocumented population has doubled to 12 million since 1994, the violent crime rate in the United States has declined 34.2 percent and the property crime rate has fallen 26.4 percent.

SOURCE: Rumbaut Ruben, and Ewing. 2007. The Myth of Immigrant Criminality and the Paradox of Assimilation. American Immigration Council, https://www.americanimmigrationcouncil.org/research/myth-immigrant-criminality-and-paradox-assimilation, p. 411.

- Cities with large immigrant populations such as Los Angeles, New York, Chicago, and Miami also have experienced declining crime rates during this period.

IMMIGRANTS HAVE LOWER INCARCERATION RATES THAN NATIVES

- Among men age 18–39 (who comprise the vast majority of the prison population), the 3.5 percent incarceration rate of the native-born in 2000 was 5 times higher than the 0.7 percent incarceration rate of the foreign-born.

- The foreign-born incarceration rate in 2000 was nearly two-and-a-half times less than the 1.7 percent rate for native-born non-Hispanic white men and almost 17 times less than the 11.6 percent rate for native-born black men.

- Native-born Hispanic men were nearly 7 times more likely to be in prison than foreign-born Hispanic men in 2000, while the incarceration rate of native-born non-Hispanic white men was almost 3 times higher than that of foreign-born white men.

- Foreign-born Mexicans had an incarceration rate of only 0.7 percent in 2000—more than 8 times lower than the 5.9 percent rate of native-born males of Mexican descent. Foreign-born Salvadoran and Guatemalan men had an incarceration rate of 0.5 percent, compared to 3.0 percent of native-born males of Salvadoran and Guatemalan descent.

- Foreign-born Chinese/Taiwanese men had an extremely low incarceration rate of 0.2 percent in 2000, which was three-and-a-half times lower than the 0.7 percent incarceration rate of native-born men of Chinese/Taiwanese descent.

- The incarceration rate of foreign-born Laotian and Cambodian men (0.9 percent) was the highest among Asian immigrant groups in 2000, but was more than 8 times lower than that of native-born men of Laotian and Cambodian descent (7.3 percent).

- With the exception of Laotians and Cambodians, foreign-born men from Asian countries had lower incarceration rates than those from Latin American countries, as did their native-born counterparts. This is not surprising given that immigrants from India, Taiwan, China, South Korea, and the Philippines are among the most educated groups in the United States, while immigrants from Cambodia, Laos, Mexico, and Central American countries are among the least educated.

IMMIGRANTS HAVE LOWER INCARCERATION RATES THAN NATIVES AMONG HIGH-SCHOOL DROPOUTS

- For all ethnic groups, the risk of imprisonment was highest for men who were high-school dropouts. But among the foreign-born, the incarceration gap by education was much narrower than for the native-born.

- The highest incarceration rate among U.S.-born men who had not finished high school was seen among non-Hispanic blacks, 22.3 percent of whom were imprisoned in 2000—more than triple the 7.1 percent incarceration rate among foreign-born black high-school dropouts.

- The incarceration rate of native-born Hispanic men without a high-school diploma in 2000 (12.4 percent) was more than 11 times higher than the 1.1 percent rate of foreign-born Hispanic high-school dropouts.

- Foreign-born Mexicans without a high-school diploma had an incarceration rate of 0.7 percent in 2000—more than 14 times less than the 10.1 percent of native-born male high-school dropouts of Mexican descent behind bars.

- Only 0.6 percent of foreign-born Salvadoran and Guatemalan high-school dropouts in 2000 were in prison, which was nearly 8 times lower than the 4.7 percent incarceration rate among native-born men of Salvadoran and Guatemalan descent who lacked high-school diplomas.

- The 0.9 percent incarceration rate of foreign-born Vietnamese high-school dropouts in 2000 was vastly lower than the 16.2 percent rate of native-born high-school dropouts of Vietnamese descent. The incarceration rate of native-born high-school dropouts of Indian descent (6.7 percent) was far greater than the 0.3 percent rate among foreign-born Indian high-school dropouts.

THE PARADOX OF ASSIMILATION

- The higher rate of imprisonment for native-born men than foreign-born men highlights a darker side to assimilation than is commonly recognized.

- The process of assimilation often involves the acquisition by immigrants and their descendants of English-language proficiency, higher levels of education, valuable new job skills, and other attributes that ease their entry into U.S. society and improve their chances of success in the U.S. economy.

- However, other aspects of assimilation are not as positive. For instance, immigrants, especially those from Latin America, have lower rates of adult and infant mortality and give birth to fewer underweight babies than natives despite higher poverty rates and greater barriers to health care. But their health status—and that of their children—worsens the longer they live in the United States and with increasing acculturation.

- The children and grandchildren of many immigrants—as well as many immigrants themselves the longer they live in the United States—become subject to economic and social forces, such as higher rates of family disintegration and drug and alcohol addiction, that increase the likelihood of criminal behavior among other natives.

The risk of incarceration is higher not only for the children of immigrants, but for immigrants themselves the longer they have resided in the United States. However, even immigrants who had resided in the United States for 16 plus years were far less likely to be incarcerated than their native-born counterparts.

43

Refugees, Race, and Gender
The Multiple Discrimination against Refugee Women

EILEEN PITTAWAY AND LINDA BARTOLOMEI

More than 80 percent of the world's refugees are women and their dependent children. Violence against women is rampant during armed conflict. It is manifested through involuntary relocation, as forced labor, torture, summary executions of women, forced deportation, and racist state policies denying or limiting public representation, health care, education, employment, and access to legal redress. Rape and other forms of sexual torture are now used routinely as strategies of war in order to shame and demoralize individuals, families, and communities. Resettlement policies actively discriminate against women on grounds of both race and gender. The gender blindness of the 1951 Refugee Convention and international law and domestic policy relating to refugee women has been recognized only relatively recently within the international system. The 1951 Refugee Convention does not recognize persecution based on grounds of gender as a claim for refugee status, nor is it clear that violence on grounds of gender can be considered as persecution. Rape has been recognized as a crime against humanity, a war crime, and an act of genocide in the Statutes of the International Criminal Court, but to date only thirty-two of the sixty nation states needed to ratify these statutes before they can become operational have done so.

RACISM AS A ROOT CAUSE OF REFUGEE GENERATION

... The Office of the United Nations High Commissioner for Refugees (UNHCR) estimates that there are some 21 million refugees and an additional 20 million internally displaced peoples across the world in more than forty countries.[1] Most wars are now intra-state rather than inter-state conflicts. Many of these civil wars are characterized by violence resulting from heightened ethnic tensions driven by economic goals.[2] These include disputes over access to natural resources and land, which intersect with goals of economic and ethnic supremacy.

SOURCE: Eileen Pittaway and Linda Bartolomei. 2001. "Refugees, Race, and Gender: The Multiple Discrimination against Refugee Women", in *Refuge*, Vol. 19, No 6. Reprinted with permission from Canada's Journal on Refugees.

... There are multiple manifestations of racism in the experience of refugees and other displaced peoples. Refugees are forced to leave their country or community of origin because of a well-founded fear of persecution for reasons of race, ethnicity, or nationality, religion, political opinion, or membership of a particular social group. Once the conflicts that caused them to flee are declared over, often following the intervention of superpowers, racism can preclude safe return and integration of refugees back into the communities from which they fled. Despite this knowledge, repatriation is often forced on refugee communities by host countries and UN agencies unable or unwilling to sustain the financial cost of the refugee population. Internal armed conflict, generating large numbers of internally displaced peoples, is most often institutionalized racism and must be recognized as such.

As the flow of uprooted peoples increases, many states are increasingly reluctant to host refugees. Narrow definition and interpretations of refugees, ... often leave those discriminated against on the grounds of minority or ethnic status unprotected. Refugees are routinely demonized by Western countries and the media as "illegal immigrants" and "economic migrants."[3] This is despite evidence that the majority of people seeking asylum have a genuine fear of persecution if returned to their home country, and despite the acknowledged contribution made by refugees to their host countries over the years.

THE GENDERED NATURE OF THE REFUGEE EXPERIENCE

... The experience and impact of racism during armed conflict is clearly a gendered experience: the majority of those who are killed or "disappeared" are men and male youths. This accounts for the refugee populations, who in the majority are women and their dependent children, who generally have been exposed to extreme physical violence.[4] Research has shown that the legal protections for women around the world, including refugee women who have experienced violence, are largely gender blind and do not address the reality of women's lives. Refugee women continue to be discriminated against in situations of armed conflict, in refugee determinations, and in resettlement because of their gender.

... THE INTERESECTIONALITY OF RACE AND GENDER

International awareness of the way in which multiple forms of discrimination intersect to inhibit the empowerment and advancement of women has its origins in 1975 at the UN First World Conference on Women, and subsequent women's conferences, the last of which, the Fourth World Conference on Women, was held in Beijing in 1995. The conference outcomes document, the Beijing Platform for Action (BPFA), was adopted by all member states.

It recognizes that factors such as age, disability, socioeconomic position, or membership in a particular ethnic or racial group could compound discrimination on the basis of sex, to create multiple barriers for women's empowerment and advancement. In documentation for the World Conference against Racism, the Committee to Eliminate Racial Discrimination noted that racial discrimination does not always affect women and men equally or in the same way: "There are circumstances in which racial discrimination only or primarily affects women, or affects women in a different way, or to a different degree than men."[5] ... The failure to address the "'differences' that characterise the problems of different groups of women can obscure or deny human rights protection due to all women."[6] Although all women are subject in some manner to discrimination based on gender, this distinction is compounded for some women when gender discrimination "intersects" with discrimination on other grounds, which may include, among other things, race, class, and color.

... Nongovernmental organizations (NGOs) around the world have documented the fact that the oppression women suffer because of their race, religion, caste, ethnicity, nationality, and other sociopolitical categories is aggravated by the discrimination they face because of their gender. As a result, women, more than men, are subjected to double or multiple manifestations of human rights violations.

THE INTERSECTIONALITY OF RACE AND GENDER IN REFUGEE SITUATIONS

During armed conflict, women can become the targets of "ethically motivated gender-specific"[7] forms of violence. Ideological frameworks developed by extreme forms of nationalism and fundamentalism that reify women's image as "bearers of the culture and values" have led to widespread sexual assaults against women as political acts of aggression. Such acts of sexual aggression are often fuelled by race- and gender-based propaganda.[8] An additional intersect of race and gender is the forcible impregnation of females from one ethnic group by males from another group as a form of genocide. Women bear the direct impact of these actions. Racism, racial discrimination, xenophobia, and related intolerance have increasingly been used to incite armed conflicts over resources and rights within and between countries around the world.

The "othering" of refugees—that is, regarding one or several sections of the community as "the other," or of intrinsically lesser value than the dominant culture or power holders—has increased, particularly in some countries in Europe where the concept of "fortress Europe" has fostered a climate of xenophobia and racism.

... Refugee women are actively discriminated against on the grounds of their ethnicity and their gender. They are often devalued or "othered" on grounds of their race, and this racial discrimination effectively removes any need by the aggressors to respect them by gender. This effectively "others" them twice and makes them prime targets for rape, systematic rape, and sexual torture for the purpose of shaming the men of their communities. Members themselves of patriarchal societies, women are also "othered" by their own communities, making this form of torture extremely effective, to the point where women are sometimes murdered in "honor killings" and are often rejected by their own communities because they have been "violated" by the aggressors.

Women are raped to humiliate their husbands and fathers, and for reasons of cultural genocide. They are forced to trade sex for food for their children. They are raped by the military, by border guards, and by the UN peacekeeping forces sent to protect them. Rape and sexual abuse are the most common forms of systematized torture used against women, and they range from gang rape by groups of soldiers to the brutal mutilation of women's genitalia. There is evidence of military training to commit these atrocities. In recent ethnic-based conflicts in Bosnia, Rwanda, Sierra Leone, and East Timor, rape and sexual violence have been used to target women of particular ethnic groups and as an instrument of genocide. Similar patterns are found in all armed conflict.

… Refugee women who have suffered rape and sexual abuse report keeping their trauma secret from determining (immigration) officers for fear of being labelled prostitutes and being denied refugee status or visas on moral grounds. Such fears are well documented by UNHCR, Amnesty International, and many aid agencies working with refugee women. A study conducted in Winnipeg, Canada, found that more than 50 percent of refugee women who had been raped, and 94 percent of other refugee sexual-assault victims, did not tell their refugee workers of their experience.[9] Far more sought help for psychosomatic symptoms related to the experience. Because the post-traumatic symptoms such as depression, loss of sleep, anger, fear of strangers, and feeling dirty are similar to those of other trauma, the root of the problem often goes unrecognized and untreated. There is still a conspiracy of silence surrounding the true extent of the problem, and until it is fully acknowledged women will not receive the services they deserve.

REFUGEE WOMEN AT RISK: A CASE STUDY

An examination of the Australian Women at Risk Program, illustrates the racism inherent in much refugee policy. This research, … highlights the gulf between policy and practice, and the gender blindness that has led to the ongoing discrimination against refugee women in international law and policy.

… The research project aimed to discover why the identification of women at risk was proving so difficult. Interviews were conducted with UNHCR officials, workers in refugee camps, and officials at Australian posts in Southeast Asia. Several implementation problems were identified, such as a lack of information and poor communication between levels of management, but these hurdles did not explain an apparent apathy toward the program.

A potential key to the problem became clear after it was noted that a total of seven out of twenty-two senior male officials in Australia, Thailand, and Hong Kong interviewed for the project had all used the same revealing phrase to describe the difficulties of identifying refugee women at risk.[10] They described the trauma that some women experienced as "only rape," implying that rape or the likelihood of being raped was insufficient grounds for considering a woman for the WaRP. These officials used the phrase when asked whether they considered rape and sexual abuse to be grounds for referring women to the WaRP.

Their argument was that if a woman was complaining of only rape and sexual abuse, she could not possibly be considered a woman at risk. As one man commented, "If only rape was the criterion, I could send you most of the women in this camp. It happens all the time, especially to the young single women, and we can't do much about it." A UNCHR official stated that rape was not grounds for refugee status, therefore it could not be grounds for the WaRP, and that to qualify for this program a woman had to be experiencing extreme forms of violence and not only rape. A third said rape was so common that it could not be seen as grounds for consideration and, anyway, that was how women got extra food (from the guards who raped them), and was therefore hardly likely to be classified as "extreme danger." The worst comment was that often what happened wasn't really rape anyway, because some women "exploited" their sexuality within the camp system in order to get favours from the guards. Another official commented that because it had often happened to women before they reached the camps, it was no longer an issue. And the final remark was that "it happens so often to these women that they get used to it, sort of expect it, and they don't see it as violence like being beaten up or tortured."

The interviewees were asked if anyone talked to the women about the rape and sexual abuse. Most acknowledged that such conversation did not occur because the women were too ashamed or shy to discuss such issues with male officers. It was apparent from the research that in the camps there was no treatment or support for women who had been raped or sexually abused prior to arrival, and that there was little protection within the camps. Interviews with women and service providers in Hong Kong indicated that often camp security staff perpetrated abuses within the camps. These comments highlighted not only insensitivity to gender but also racism, as they implied that refugee women were of lesser social standing and therefore of lesser value than those making the comments, who were mainly Anglo-Saxon. While it can never be proved, it can be hypothesized that they would not have made these comments about women from their own ethnic groups and class.

It is worth noting that the interviews conducted with refugee women in Australia and with the women in camps indicated that the rape of refugee women was not just the result of an opportunity that men seized when they found themselves in a position of power over vulnerable women. Much of the rape and sexual torture was planned and systematic. In camps it was institutionalized and a way of keeping control. These acts were undertaken with relative impunity. During conflicts, women were raped in an attempt to extract information, to shame communities, and to destroy the social fabric. The women were forcibly impregnated to destroy ethnic purity. They were often systematically tortured in a way that suggested that soldiers had been trained to do it; for example, the cutting of nipples with wire cutters after rape has been reported across Indochina and Indonesia. From Latin America come stories of genital mutilation with electric prods, with broken glass, and through the use of trained dogs.

Apparently, despite much rhetoric about protecting refugee women, many people in positions of influence were unwilling or unable to accept rape and sexual torture during an armed conflict as a major problem.

... If the needs of refugee women are to be recognized and addressed, there has to be change at an international level. The rape and sexual abuse of refugee women, during a conflict, in flight, or in refugee camps, has to be recognized as a war crime and be considered as persecution, and such a finding has to be reflected in international law and conventions. Without such recognition, domestic law and social policy designed to address the needs of these women, although grounded in international law, will constantly fail to fulfill their goals. ... Gender blindness, patriarchal values, and racism combine to ensure that the experiences of refugee women are not acknowledged or addressed.

REFUGEE WOMEN, RACISM, AND RESETTLEMENT

Racism is not only a cause of refugee movement, but it also continues in countries of settlement and resettlement. Gender discrimination is also entrenched in social structure. Refugee women, like many migrant workers, are frequently treated as second-class citizens in their countries of destination. Racist state policies of host countries in the West and the Asia-Pacific, particularly on labor and immigration, result in the exploitation of refugee and migrant women. They are discriminated against in terms of wages, job security, working conditions, job-related training, and the right to unionize. They are also subjected to physical and sexual abuse. When illegally employed, they have no access to labor laws. They are not given equal access to the law, nor are they treated equally under the law. Their employment opportunities are limited largely to domestic work or the sex industry, where their right to work, freedom of movement, reproductive rights, right to acquire, change, or retain their nationality, right to health and other basic human rights are violated. The result is that refugee women and their families are more vulnerable to religious, racial, and gender discrimination and exploitation.

Their stateless condition makes refugee women and children easy targets for traffickers. Trafficking has not been deterred by the imposition of restrictive and exclusionary immigration policies by host countries. On the contrary, such policies account for the increasing number of undocumented migrant female workers who have been trafficked or are most vulnerable to trafficking. Trafficking involves the recruitment, transportation, transfer, and harbouring of persons and is conducted by threat, use of force or other forms of coercion, abduction, fraud, and deception. The purposes of trafficking in persons include involuntary servitude—domestic, sexual, or reproductive—in forced or bonded labor in conditions akin to slavery. Refugee women, indigenous women, Dalit women, and women from ethnic minorities are some of the groups of women most vulnerable to trafficking. The extensive documentation of the exploitation of migrant and refugee women, especially from countries in the Asia–Pacific region, underscores the fact that migration and trafficking in women is a critical area of concern.

Racism directed at refugee populations in resettlement countries often causes refugee women to remain silent about their experiences of gender discrimination and violence within their own communities. Often racism within the broader community exacerbates the pressure on refugee women to maintain their traditional roles in order to keep their communities intact. The problems of many refugee women remain hidden in countries of resettlement. The racial barriers that men may face in access to employment and education are concerns more frequently aired in the public arena. As a result, the prevailing discourse in many resettlement countries among refugee advocates is that refugee men find resettlement far more difficult than do refugee women.

Refugees face systematic discrimination on the bases of race, ethnicity, and gender in the process of selection for resettlement in third countries—most often developed countries with predominantly white populations. Refugees are selected for resettlement from situations of refuge in first countries of asylum. There is a marked trend for resettlement countries to give first preference to refugees most likely to "blend" into the host country. Therefore, humanitarian response from countries of the North to refugee populations from the South is markedly different from response to refugees from the North.

… Racism experienced by many refugees in resettlement countries has multiple effects on women. Refugee men who are denied access to employment or decision making in the host country can attempt to retain their personal autonomy and power through controlling their wives and children, and the result is often an increase in domestic violence. Resettlement countries exhibit a strong preference for families with a male head, and do not often select single women with large families for resettlement, on the grounds that they will become an economic burden on the resettlement country. Resettlement services seldom acknowledge the experiences of refugee women and their need for services to be provided.

… The intersectionality of race and gender in refugee situations and the multiple forms of discrimination that it generates have been named and discussed. The issue will not go away.

NOTES

1. UNHCR, "Refugees by Numbers" (2000) on-line: United Nations High Commissioner for Refugees <http://www.unhcr.org/> (Accessed: 11 June 2001).

2. "When Is a Refugee Not a Refugee?" *Economist (US)*, 3 March 2001.

3. Ronald Kaye, "Defining the Agenda: British Refugee Policy and the Role of Parties," *Journal of Refugee Studies* 7, no. 2–3 (1994): 144–59; Guy S. Goodwin-Gill,

"International Law and Human Rights: Trends Concerning International Migrants and Refugees," *International Migration Review* 23, no. 3 (1989): 526–46. Theodor van Boven, "United Nations Strategies to Combat Racism and Racial Discrimination: Past Experiences and Present Perspectives," UN Doc. E/CN.4/1999/WG.1BP.7, para 56 (1999). UNHCR, "Asian Preparatory Meeting for the World Conference against Racism, Racial Discrimination, Xenophobia and Related Intolerance 2001: Declaration and Plan of Action" (2001) online WCAR Homepage <http://www.un.org/WCAR/docs.htm> (Accessed: 6 June 2001).

4. UNHCR, "Guidelines on Preventing and Responding to Sexual Violence against Refugee Women" (Geneva: UNHCR, 1995).

5. ICERD, "Gender Related Dimensions of Racial Discrimination" (2000) General Recommendation 25, International Committee on the Elimination of Racial Discrimination, UN Doc. ICERD/C/56/Misc.21/Rev.3.

6. UN, "Gender and Racial Discrimination," (2000) Report of the Expert Group Meeting 2000, online: United Nations <http://www.un.org/womenwatch/saw/csw/genrac/report.htm> (Accessed: 6 June 2001).

7. Ibid.

8. Ibid.

9. L. Pope, "Refugee Protection and Determination: Women Claimants" (summary of comments made at the CRDD Working Group on Women Refugee Claimants: Training Workshop for Members, Toronto, Canada, 1990).

10. Research data, WaR Project ANCCORW, interviews conducted by Pittaway in Hong Kong refugee camps in March 1991 and in Thai refugee camps in September 1991. Interviews in Australia were conducted by Pittaway and Sylvia Winton in September 1991.

44

The Intersectional Paradigm and Alternative Visions to Stopping Domestic Violence

What Poor Women, Women of Color, and Immigrant Women Are Teaching Us About Violence in the Family

NATALIE J. SOKOLOFF

This article focuses on how poor women and women of color and their allies ... are working against domestic violence in the United States in communities marginalized by race, class, gender, sexuality, and immigrant status....

In contrast to the earlier feminist approaches, the *intersectional* domestic violence approach challenges gender inequality as *the primary factor* explaining domestic violence: gender inequality is neither the most important nor the only factor that is needed to understand violence against many marginalized women in the home (Crenshaw, 1994). Gender inequality is only part of their marginalized and oppressed status....

In the intersectional domestic violence literature in the U.S., two distinct (and sometimes conflicting) objectives emerge: (1) *giving voice* to battered women from diverse social locations and cultural backgrounds (2) while still focusing on *socially structured inequalities* (e.g., race, gender, class, sexuality) that constrain and shape the lives of battered women, albeit in different ways. The first has been described as a *race/class/gender* (or *multicultural*) perspective, whose focus is on multiple, interlocking oppressions of individuals—i.e., on issues of *difference*; the second has been described as the *structural* perspective requiring analysis and criticism of existing systems of power, privilege and access to resources (see Andersen and Collins, 2001; Mann and Grimes, 2001). It is my position that an intersectional analysis must draw from *both* approaches because women's voices must be situated within the social structure of women's lives....

SOURCE: Sokoloff, Natalie J. 2008. "The Intersectional Paradigm and Alternative Visions to Stopping Domestic Violence." *International Journal of Sociology of the Family* 34(Autumn): 153–185.

... The major topic of this article ... [i]s to discuss some of the alternative models for paradigms that women marginalized by race, class, gender, sexuality, and immigrant status have begun using to combat violence against women in their homes and communities....

HOW AN INTERSECTIONAL ANALYSIS CAN HELP TO UNDERSTAND MARGINALIZED WOMEN'S EXPERIENCES OF DOMESTIC VIOLENCE

It is still common to hear that domestic violence cuts across all classes, races, and ethnic groups. While this is true, intersectional scholars challenge this uncritical view by arguing that poor women of color are the "most likely to be in both dangerous intimate relationships and dangerous social positions" (Richie, 2000, p. 1136). Richie argues that the anti-domestic violence movement's avoidance of a race, gender and class analysis of violence against women "seriously compromises the transgressive and transformative potential of the antiviolence movement's potential [to] radically critique various forms of social domination" (p. 1135). She concludes, failure to address the multiple oppressions of poor women of color jeopardizes the validity and legitimacy of the anti-domestic violence movement.

One dilemma is the problem of how to report race and class differences in domestic violence prevalence rates, especially in the U.S. where race is such a dominant part of the social agenda and class is typically ignored as a concept.... [T]here is tremendous diversity among women regarding the prevalence, nature, and impact of domestic violence—even within ethnic, racial, religious, socio-economic groups and sexual orientations (Hampton et al., 2005; West, 2005). Several studies indicate that Black women are severely abused (West, 2005) and murdered at significantly higher rates (Della-Giustina, 2005; Hampton et al., 2005; Websdale, 1997; West, 2005) than their representation in the population.

By itself, this information may serve little purpose but to reinforce negative stereotypes about African Americans and the Black community's alleged "culture of violence" in the U.S. That is one reason why Yasmin Jiwani (2001) focuses on structural issues—especially in immigrant of color communities—where culture is all too typically blamed for violence against women in their families. One solution to this problem of representation is to *contextualize* these findings within a *structural* framework—one that looks at socially organized systems of social inequality.... Research literature concludes that (1) Black women are *less* likely than white women to be battered when one controls for income and marital status (see Farmer and Tiefenthaler, 2003), (2) neighborhood context (a high percent of unpartnered parents, unemployment, poverty, and public assistance) is a more effective predictor of domestic violence than race/ethnicity (see Benson et al., 2003), and (3) the relationship between race and intimate partner violence may be spurious—"more likely, race is a proxy measure for neighborhood" (Potter, 2007: 370).

This being said, we must never forget the profound racism that exists in U.S. society, including the effects of living in racially segregated communities. Thus, for example, the degree of poverty is more intense in some African American communities. Whereas 75 percent of poor Blacks live in communities with other poor Blacks—and all its attendant disadvantages, only 25 percent of poor whites live in poor white communities. Instead, poor whites are more likely to live in communities with working class and some middle class white residents, which provide an immeasurable degree of resources available to that community (e.g., see Rusk, 1995). So comparing poor Blacks and poor whites is simply not "comparable." This must be taken into account in working toward lowering levels of domestic violence in Black, white and other communities of color.

Finally, recent research suggests that the level of collective efficacy in a community can be related to domestic violence (Almgren, 2005; Benson et al., 2003; Brown et al., 2005; Lauristen and White, 2001).... [W]omen who live in neighborhoods with greater collective efficacy are more likely to inform others of their abuse, thereby leading to greater levels of support for the women. Thus, it is not only socially structured inequalities but also the community's ability to feel empowered in relation to its many structured obstacles that lead to better opportunities for battered women to escape or challenge abusive situations at home. Lack of good jobs, decent education, livable housing, decent health care and good childcare, etc. can create conditions for communities and its members to not feel up to being able to protect their members, including the women in their homes.

HOW AN INTERSECTIONAL ANALYSIS CHALLENGES STEREOTYPES OF MARGINALIZED COMMUNITIES AND THEIR CULTURES

All too often, whites in the U.S., including feminists, have been quick to allocate blame to non-white cultures (especially Black and immigrant) for domestic violence....

According to Dasgupta (2005), "American mainstream society still likes to believe that woman abuse is limited to minority ethnic communities, lower socio-economic stratification, and individuals with dark skin colors" (212–213). This leads to stereotyping of battered women from "other" cultures. But it also fails to look at the strengths of non–dominant cultures and how they provide protective factors for battered women (Yoshioka and Choi, 2005)....

In terms of the negative stereotyping many women from immigrant and non-white communities experience, Uma Narayan (1997) describes how in the U.S., when we hear that women in India die "by fire" in dowry deaths (a man or some of his family members kill the woman because they are dissatisfied with her dowry to the paternal family line with whom she usually lives), the "culture" tends to be blamed. Thus, many people in the United States call dowry deaths horrendous—which of course they are, but then go on to explain their cause as less "enlightened" attitudes toward women and the "backwardness" of the South Asian culture. However, when women in the U.S. are killed by guns (at the same rate as dowry deaths in India), this is rarely, if ever, said to be due to the culture; rather it is usually blamed on the individual man's unstable personality at best and patriarchy at worst (Narayan, 1997). But the American culture is not said to be "backward" or to blame for the death....

In the U.S., safety for battered women and their children is typically premised on the idea that she will *separate from* or *leave* her abuser. We know that most women *do* leave their batterers: it takes on average 7 attempts before she is ultimately able to leave and be safe; and that the most violent and dangerous time for a woman in a battering relationship is just before, at the point of, or just after leaving (e.g., see Browne, 2003; Campbell et al., 2003). If she chooses to leave the batterer, she needs a strong shelter system or other support especially during the first 3 months, and then the first year. After that, the leave-taking process has a good chance of getting and keeping her out of an intolerable, abusive relationship (Campbell, 2008). But leaving their abusers might mean turning to "outsiders"—police, courts, doctors, domestic violence agencies, etc.—in the mainstream or dominant communities. Thus, while it may be true that she will face violence in her family or community, it is just as true that if she goes outside her home and her community, she may have to face a whole other set of hostilities. Moreover, leaving the batterer may mean leaving her entire community and its supportive aspects—often a not insignificant source of services and/or protections. [E.g., see the experience of an Orthodox Jewish woman in Baltimore who did all the "right" things to leave her abusive husband and whose community provided the kind of support that nurtured her through these terrible times and stigmatized her husband until his abuse stopped (Kay, 2006). Likewise, Holmes (2005) reports that certain Muslim Imams in Philadelphia shun men who abuse their wives].

The stereotyping of marginalized groups of battered women leads to major misunderstandings about what might help the women in their struggles to be free of violence: incarcerating the men or taking away the man's economic contribution to her and her children is counterproductive in many marginalized communities. Arguing that calling the police is the most important thing for them to do often backfires. Thus, as Garfield (2005, 2006) reminds us, when the violence against women movement changed from an advocacy to a state-based criminal justice operation, blame and punishment became the focus, *not social structural change* which is needed for violence against women to stop.

The discrimination in the criminal justice system, as in the larger U.S., is deep and profound. In the Black community, as many as one-third of all young African American men are in prison, jail, on probation or parole in the U.S. (Sokoloff, 2004). In some cities, like Baltimore and Washington D.C., it is over half (Donziger, 1996; Lotke and Zeidenberg, 2005)! Black women who are battered often do not feel the police and the criminal justice system will solve their problems; rather they may just intensify them (Richie, 2005). Native

American women have argued that violence against Native women has been part and parcel of the colonization of Native Americans in the U.S.; thus, violence by the state is intimately intertwined with violence by individual Native men against their partners....

Thus, many women of color argue, interpersonal violence and state violence against women and oppressed communities are intertwined. Moreover, many immigrants—both documented and undocumented—often fear the police, whether because of negative experiences in their home countries or because of the arrest and deportation of immigrants here in the U.S. which has only intensified since 9/11 2001.... As with African Americans and Native Americans, given the language, cultural, and structural barriers facing many battered immigrant women, the police, courts, and social services that are available often do not ensure their safety—in fact, may increase their vulnerability (Abraham, 2000; Bui, 2004; Dasgupta, 2005). In short, an intersectional analysis helps us see that the simplistic advice to either "leave" the abusive partner or to "call the police" when she does not have an adequate safety plan is often misguided. This may actually harm rather than help certain groups of women who must deal with violence in their homes as well as in and against themselves and their communities by larger outside forces in the U.S. (Coker, 2005)....

Community Based Models of Social Justice: Community Engagement/Community Accountability

In ethnically, racially, and economically marginalized and immigrant communities of color especially, an important movement has developed to reduce women's reliance on the criminal legal system.... Rather, the goal is to develop a community's capacity to mobilize and organize against, and assume accountability for, changing violence against women in that community. This is important in large part because most batterers are *never* seen by the criminal justice system—even under the best of circumstances. Those arguing for less reliance on the criminal legal system argue for a need to think outside the traditional criminal justice (or criminal legal) and social service models for addressing and eradicating violence. The degree to which these traditional models are incorporated into new paradigms varies considerably, but many see them as "adjuncts" or "back up" at best....

Smith (2005) suggests strategies that must deal simultaneously with the reality of structural as well as state violence, especially state violence within the U.S. criminal justice system. She argues marginalized women too often cannot look to the criminal justice system for safety because state violence is intimately connected to domestic violence. Thus, the "challenge women of color face in combating personal *and* state violence is to develop strategies for ending violence that *do* assure safety for survivors of sexual/domestic violence and *do not* strengthen our oppressive criminal justice apparatus"—a struggle that has been particularly powerful in the African American, Native American, and more recently Latino/a communities (Incite!, 2006). To understand the violence against women in their homes by intimate partners and by the state, one need only look at the ways in which women of color in particular have been brutalized both by their husbands/partners and by the police, prisons, etc. (see Ritchie, 2006; Faith, 2004). It is in large part because of the, at best, inadequate and, at worst, hostile treatment by police and courts in the U.S. that methods of engaging communities beyond the criminal justice system must be found to support poor, minority and immigrant women who are abused in their homes as well as in their communities....

Also, it is important to clarify that while culturally competent services ... are required to create social justice for marginalized battered women in the U.S., one should not be lured into thinking that these social services are adequate *by themselves* without large-scale structural changes. Domestic violence is part of the larger societal systems of violence and inequality (e.g., imperialism, racism, colonialism, patriarchy, etc.) and as such domestic violence must be attacked at its root causes: the socially structured systems of inequality—of race, class, gender, sexual orientation, immigrant status and the like.

In *Safety and Justice for All: Examining the Relationship between the Women's Anti-Violence Movement and the Criminal Legal System* (MsFoundation, 2003)... the MsFoundation elaborates on some of the community responses that are relevant to marginalized communities. Each one of these approaches must be seriously considered and thoroughly researched as well. They include: community squads to intervene with batterers; alternative 911 services that rush community residents to a crisis scene; community groups overseeing children's safety; alternative accountability such as restorative justice; popular education through alternative means like music and street theater; and men teaching men about domestic violence and how to challenge its destructive forces....

Empowering marginalized women to become their own advocates (e.g., see *Building Rhythms of Change*, 2001; Williams and Tibbs, 2002) and to organize against violence against women throughout their communities has been repeated in one version or another as a movement for social change in the African American, Latina, Native American, and immigrant domestic violence literature. Given that these actions are "home grown," the advocacy and organizing are "inherently culturally competent" and do not need to be "imported" from the outside (Dutton et al., 2000). As Alianza (2004) states in its analytic framework:

> We need to develop systems of support for victims/survivors within our communities. Latina survivors need to be recognized as *experts* in meeting these challenges; they must be involved in program design and service delivery at all levels. Programs need to include services that will give survivors better options and opportunities for becoming independent and more able to create relationships and homes free from violence.

Likewise, Mujeres Unidas y Activas, based in San Francisco and Oakland, California, sees itself as different from other domestic violence programs because it is "founded on the concept that immigrant women are uniquely equipped to find solutions to the problems that most directly affect their lives."... Thus battered immigrant women from the Latina community are used as peer mentors, group facilitators, community educators, and organizers (Mujeres, 2006).

Finally, Sista II Sista is a collective of young African American and Latina women, ages 13–25, that began in Bushwick, Brooklyn (NY) in 1995. These young women work in a non-hierarchical structure (using a flower instead of a hierarchy to express its organizational format) and sees as one of its main goals in the Freedom School (founded in 1996) the need to help young women of color see themselves as leaders in their communities. Leadership is encouraged through:

> identifying the issues that are central to their community and learning important ways to fight to shape their community positively around these issues…[like] creating concrete changes (culturally and institutionally) against violence against the women of color in their Bushwick, Brooklyn community (Sista II Sista, n.d.)…

A FINAL NOTE ON THE IMPORTANCE OF MICRO AND MACRO STRUCTURAL AND ECONOMIC CHANGE

Again, it is also important for us to see where we can learn from those in the U.S. and in other countries. At the micro level, the fact that criminal justice responses cannot protect women from re-victimization because they do not systematically address women's underlying economic, social, and political disadvantages is demonstrated by the work of Websdale and Johnson (2005) in rural Kentucky. They show how a more effective way to reduce woman battering is to empower battered women by providing the underlying structural conditions for independent housing, job training and opportunities, affordable childcare, and social services….

SUMMARY AND CONCLUSION

(1) What might it look life if communities had the resources to explore effective interventions that keep decision-making power within the community, and make it possible for women to stay?

(2) Where might we be if government accountability did not aim its efforts on criminal legal punishment, but instead centralized responsibility for basic needs [i.e., *human, economic, civil, political, and cultural rights of all human beings*] and human dignity and affirmed the human rights of all?

The content of this article, I hope, makes it very clear that we must go far beyond the inner sanctum of the mainstream traditional isolated nuclear family if we are to challenge violence against women in families, especially those living in communities that are marginalized by race, class, gender, ethnicity, immigrant status, and the like. Looking at violence against communities, families within communities, and against the women in those communities is key to understanding violence against women in their homes. And this cannot be done outside a framework of the socially structured conditions of inequality and oppression and movement toward greater structural change and social justice. Both micro- and macro-level analyses and social movements for structural and cultural change (for human, economic, civil, political, cultural, gender, sexual rights and human dignity for all) must be a part of the work that we do to eliminate violence against women in their homes....

... [T]o understand domestic violence generally, and particularly in the most marginalized communities, we must *contextualize* family violence within the larger systems of social, political, racial, gender, economic, and sexual inequality. Nothing less will do if we are to not only study but also work toward *changing* violence against women in the family....

REFERENCES

Abraham, Margaret (2000). *Speaking the Unspeakable: Marital Violence among South Asian Immigrants in the United States.* Piscataway, NJ: Rutgers University.

Alianza Latina Nacional para Erradicar la Violencia Domestica (2004). Nacional Directory of Domestic Violence Programs Offering Services in Spanish. Retrieved on January 5, 2006: http://www.dvalianza.org

Almgren (2005). The Ecological Context of Interpersonal Violence: From Culture to Collective Efficacy. *Journal of Interpersonal Violence*, 20: 218–224.

Andersen, Margaret and Patricia Hill Collins (2001). Introduction. In M. Andersen and P. H. Collins (Eds.), *Race, Class & Gender: An Anthology*, 4E (pp. 1–9). Belmont, CA: Wadsworth.

Benson, Michael, et al. (2003). The Correlation between Race and Domestic Violence is Confounded with Community Context. *Social Problems*, 51: 326–342.

Browne, Angela (2003). Fear and the Perception of Alternatives: Asking "Why Battered Women Don't Leave" Is the Wrong Question. In Barbara Raffel Price and Natalie J. Sokoloff (Eds.). *The Criminal Justice System and Women: Offenders, Prisoners, Victims, and Workers*, 3rd Ed. (pp. 343–359). New York: McGraw-Hill.

Bui, Hoan (2004). *In the Adopted Land: Abused Immigrant Women and the Criminal Justice System.* Westport, CT: Praeger.

Campbell, Jacqueline, et al. (2003). Risk Factors for Femicide in Abusive Relationships: Results of a Multisite Case Control Study. *American Journal of Public Health*, 93(7): 1089–1097.

Campbell, Jacqueline (2008). Danger Assessment: A Tool to Help Identify the Risk of Intimate Partner Homicide and Near Homicide as Part of Routine Mental Health Assessments. Jewish Women International National Training Institute Teleconference. (Online- January 17).

Coker, Donna (2005). Shifting Power for Battered Women: Law, Material Resources, and Poor Women of Color. In Natalie J. Sokoloff with Christina Pratt (Eds.). *Domestic Violence at the Margins: Readings on Race, Class, Gender & Culture.* NJ: Rutgers University, pp. 369–388.

Crenshaw, Kimberle (1994). Mapping the Margins: Intersectionality, Identity, and Politics. In Martha Albeitson (Ed.). *The Public Nature of Private Violence.*

Dasgupta, Shamita Das (2005). Women's Realities: Defining Violence against Women by Immigration, Race and Class. In Natalie J. Sokoloff (Ed.) (with Christina Pratt). *Domestic Violence at the Margins: Readings in Race, Class, Gender & Culture* (pp. 56–70). NJ: Rutgers University.

Della-Giustina, Jo-Ann (2005). *Gender, Race and Class as Predictors of Femicide Rates: A Path Analysis.* Ph.D. Dissertation, John Jay College of Criminal Justice and the Graduate Center, City University of New York.

Donziger, Steven (Ed.) (1996). *The Real War on Crime: The Report of the National Criminal Justice Commission.* NY: HarperPerennial.

Dutton, Mary Ann, Leslye Orloff, and Gail Aguilar Hass (2000). Summer. Characteristics of Help-Seeking Behaviors, Resources and Service Needs of Battered Immigrant Latinas: Legal and Policy Implications. *Georgetown Journal of Poverty Law and Policy,* 7(2): 245–305.

Faith, Karlene (2004). Progressive Rhetoric, Regressive Policies: Canadian Prisons for Women. In Barbara Raffel Price and Natalie J. Sokoloff (Eds.). *The Criminal Justice System and Women: Offenders, Prisoners, Victims, and Workers,* 3rd Ed., pp. 281–288.

Farmer, Amy and Jill Tiefenthaler (2003). Explaining the Recent Decline in Domestic Violence. *Contemporary Economic Policy,* 21: 158–172.

Garfield, Gail (2005). *Knowing What We Know: African American Women's Experiences of Violence and Violation.* Rutgers University.

Garfield, Gail (2006). Does the Violence Against Women Act Provide Justice for Battered Women? Unpub. Ms. John Jay College of Criminal Justice. New York City.

Hampton, Robert, et al. (2005). Domestic Violence in African American Communities. In Natalie J. Sokoloff (Ed.) (with Christina Pratt). *Domestic Violence at the Margins: Readings in Race, Class, Gender & Culture* (pp. 127–141). NJ: Rutgers University.

Holmes, Kristen (2005). Muslims in Philly Shun Men Who Abuse Wives. At www.CentreDaily.com (July 23).

Incite!/Critical Resistance Statement (2005). Gender Violence and the Prison Industrial Complex: Interpersonal and State Violence against Women of Color. In Natalie J. Sokoloff (Ed.) (with Christina Pratt). *Domestic Violence at the Margins: Readings in Race, Class, Gender & Culture* (pp. 102–114). NJ: Rutgers University.

Incite! Women of Color Against Violence (2006). *Color of Violence: The Incite! Anthology.* Boston: South End.

Jiwani, Yasmin (2001). *Intersecting Inequalities: Immigrant Women of Colour, Violence & Health Care.* At http://www.harbour.sfu.ca/freda/articles/hlth.htm

Kay, Liz (2006). A Woman's Plea for Closure: Orthodox Jewish Community Rallies Against Husband Who Denied a Religious Divorce. *The (Baltimore) Sun,* September 19, pp. IB + 2B.

Lauristen and White (2001). Putting Violence in Its Place: The Influence of Race, Ethnicity, Gender and Place on the Risk for Violence. *Criminology and Public Policy*, 1(1): 37–59.

Lotke, Jason and Eric Zeidenberg (2005). *Tipping Point: Maryland's Overuse of Incarceration and the Impact on Public Safety.* At: http://www.justicepolicy.org/images/upload/05-03_REP_MDTippingPoint_AC-MD.pdf

Mann, Susan and Michael Grimes (2001). Common Grounds: Marxism and Race, Gender and Class Analysis. *Race, Gender, Class*, 8 (2).

Mujeres Unidas y Activas. Retrieved on April 10, 2006. At: http://www.mujeresunidasnetwork.com/ncn/index.php?option=com_content&task=vie

Narayan, Uma (1997). *Dislocating Cultures Identities, Traditions, and Third World Feminisms.* NY: Routledge.

Potter, Hillary (2007). Reaction Essay: The Need for A Multi-faceted Response to Intimate Partner Abuse Perpetrated by African-Americans. *Criminology and Public Policy*, 6(2): 367–376.

Richie, Beth (2005). A Black Feminist Reflection on the Antiviolence Movement. In Natalie J. Sokoloff (Ed.) (with Christina Pratt). *Domestic Violence at the Margins: Readings in Race, Class, Gender & Culture* (pp. 50–55). NJ: Rutgers University.

Ritchie, Andrea (2006). Law Enforcement Violence against Women of Color. In Incite! The Color of Violence. *Color of Violence: The Incite! Anthology.* Boston: South End, pp. 138–156.

Rusk, David (1995). *Baltimore Unbound: A Strategy for Regional Renewal.* Balto.: Johns Hopkins University.

Sista II Sista-Social Justice Wiki. Got to http://socialjustice.ccnmtl.columbia.edu/index.php/Sista_II_Sista

Smith, Andrea (2005). Looking to the Future: Domestic Violence, Women of Color, the State and Social Change. In Natalie J. Sokoloff with Christina Pratt (Eds.), *Domestic Violence at the Margins: Readings in Race, Class, Gender, & Culture* (pp. 416–434). NJ: Rutgers University.

Sokoloff, Natalie J. (2004). Impact of the Prison Industrial Complex on African American Women *Souls: A Critical Journal of Black Politics, Culture, and Society*, 5(4): 31–46.

Websdale, Neil (1997). *Understanding Domestic Homicide.* Boston: Northeastern University.

West, Carolyn (2005). The "Political Gag Order" Has Been Lifted: Violence in Ethnically Diverse Families. In Natalie J. Sokoloff with Christina Pratt (Eds.). *Domestic Violence at the Margins: Readings in Race, Class, Gender & Culture* (pp. 157–173). NJ: Rutgers University.

Williams, Oliver J. and Carolyn Tibbs (2002). *Community Insights on Domestic Violence among African Americans: Conversations about Domestic Violence and Other Issues Affecting Their Community* (San Francisco and Oakland). Office of Justice Programs, Dept. of Justice, Office on Violence Against Women.

Yoshioka, Marianne and Deborah Choi (2005). Culture and Interpersonal Violence Research: Paradigm Shift to Create a Full Continuum of Domestic Violence Services. *Journal of Interpersonal Violence*, 20(4): 513–519.

PART IV

Intersectionality and Social Change

MARGARET L. ANDERSEN
AND PATRICIA HILL COLLINS

By the end of this book, students often want to know "What can I do?" We know there is not a simple answer to this question. People have a range of reactions to the social injustices brought about by race, class, and gender, once they know about them. We understand that developing and then acting on an intersectional analysis of race, class, and gender may be a lot to expect from many of our readers. For some, reading this book will be the first time they have used an intersectional lens to think about social inequality. Others will have experienced much of what is written about here, but not beyond the particulars of their own lives. Still others have already been working for change and find that the intersectional approach to race, class, and gender taken here provides new directions for their projects.

We know that developing an intersectional perspective and then deciding what to do about social inequality is a complex process—one without simple solutions or ways of thinking. Thus, we have developed this last section of the book to examine various ways that different people and groups use an intersectional perspective to think about social change and how they have acted to create change in the race, class, and gender systems we examine in this book.

In prior editions of this book, we encouraged readers to think inclusively about possibilities for social action and social change. We asked, is there something in your life that you care about so much that it would spur you to work for social justice? Is it your family? Something happening in your school or community? Something that affects your children, your friends, your faith, or your neighborhood? Your own experiences as a Black woman, or an affluent man, a queer-identified person, or a working-class college student? Perhaps for you there is a social issue that you are passionate about—violence against women, climate change, clean water, global poverty and hate crimes, for example. Generally, people think that people who work for social justice must be somehow extraordinary, like Martin Luther King Jr., or other heroic figures, but most people who engage in social activism are ordinary, everyday people who decide to take action about something that touches their lives. We invite readers to think about how race, class, and gender might shape their understandings of the need for social change and how they might go about working for it.

In preparing this edition, we see that many people are now working for social change around issues of race and gender equity, voting rights, gun violence, immigrant rights, and similar social justice projects. Many others are now aware of these issues and are considering what they want to do. Many are using an intersectional lens in thinking through their involvement and shaping their social action. Regardless of whether an individual is new to the social inequalities of race, class, and gender, or a seasoned veteran in trying to change them, an intersectional lens is an important resource for effective social change.

In Part IV, we present the media and social movements as two important sites where people, especially young people, have been especially visible in working for social change. These are also two sites where people are using the race, class, and gender framework developed here to think about and do something about social inequality. Both sites have changed dramatically over the past several decades.

Mass media and communications technologies such as the Internet, smart phones, and various social media have grown exponentially over the many editions of this book. In the 1990s, media was primarily a top-down endeavor that treated ordinary people primarily as consumers who received the ideas of experts. Journalists, academic experts, music companies, Hollywood studios, and celebrities created the music, books, films, television, and other products of popular culture. The dominant mass media was extremely powerful because misrepresented people had no speedy or effective way to rebut controlling images about Asians, Blacks, women, and Latinos. The images then could circulate unchecked.

Now, new technologies give ordinary people opportunities to be both consumers and producers of media. Access to new media has sparked a populist, bottom-up endeavor. People who have been disadvantaged within traditional mass media venues can increasingly produce and share their own media content. Women, people of color, young people, and working-class and poor people upload their own videos to YouTube, write blogs, share ideas on social media platforms like Facebook and Twitter, use their cell phones as citizen journalists to record newsworthy events in their lives, and create their own web sites rather than relying on corporate-sponsored content.

This more populist media has generated many ways for people to create their own content, perhaps using a national, even global, platform to distribute their ideas to others via cyberspace. Information is no longer rationed, nor is it only produced by the most powerful. People with access to these communications technologies can collaborate using email, text messaging, video chat technology, and other tools. This coexistence of traditional media and new social media platforms has fostered a sea change in how people engage media to work for social change.

New social movements have also changed over the past several decades. The contours of new social movements differ dramatically, in part, because of access to social media. Throughout this book we have shown how intersecting power relations of race, class, and gender take institutional forms in work, family, education, and other social institutions. Race, class, and gender affect societal patterns of social inequality that separate people into categories. Because the power relations of race, class, and gender segregate people from one another, it remains difficult for people to get to know one another in face-to-face encounters. The fact that people live such segregated lives has been and remains a challenge for organized political action.

Those who work for social change within the confines of their neighborhoods, regions, or nation-states have historically done so in isolation: They had no readily available way of communicating with people who shared similar interests. Working for social change across boundaries of segregated space created a different set of concerns. When people do not know one another and are dependent on biased or flawed media images, it is difficult to imagine what others are like. On the other hand, social media platforms can also work as "echo chambers." Consequently, people have to be careful not to isolate themselves among only those who think in the same way.

Currently, many people, both in the United States and globally, are finding ways to overcome these historical spatial barriers. Facebook, Twitter, Instagram, and similar social media platforms enable people, who may never meet in their daily lives, to communicate across social distances of city and suburbs, or physical distances across the globe. People who confront racism, sexism, poverty, homophobia, militarism and similar social problems can learn from others' struggles.

By themselves, new communications technologies and social media platforms have not catalyzed political activism—the very real social issues that affect people's lives do this—yet new communications technologies have provided a set of important tools for social action.

Young people have been at the forefront of these dramatic changes in media and in new social movements that aspire to bring about social change. Armed with intersectional perspectives, many young people are taking leadership in working for social change. Many know that they are crafting their own futures, and that the world that they inherited need not be the world that they leave behind.

MEDIA AND POPULAR CULTURE

The articles in the section "Media and Popular Culture" illustrate some of the ways that people use new communications technologies and popular culture for social change. Kishonna L. Gray ("Race, Gender, and Virtual Inequality: Exploring the Liberatory Potential of Black Cyberfeminist Theory") shows how Black women have used social media platforms to re-define a new political discourse.

Likewise, Jessica Vasquez-Tokos and Kathyrn Norton-Smith examine how Latino men at different phases of life resist the controlling images that are applied to them ("Talking Back to Controlling Images: Latinos' Changing Responses to Racism over the Life Course"). Their article shows that people do not blindly consume images nor live their lives according to the social scripts that are set forth for them. Instead, as their lives unfold, Latino men deepen their understanding of controlling images and make personal decisions of how they would deal with them.

Amy D. McDowell examines how people use creative ways to refute the controlling images that they confront ("'This Is for the Brown Kids!' Racialization and the Formation of 'Muslim' Punk Rock"). Muslim men confront contemporary images of them as terrorists; Muslim women, as oppressed by Muslim men. A common response to these images has been to assimilate as much as possible into American culture or to claim identity as Muslim as a valued religious identity. The youth in McDowell's study take a different path. They claim the popular culture area of punk, attracted to its rebellious stance toward mainstream culture. This article illustrates the power of art as a form of politics and the significance of popular culture in challenging mainstream ideas.

Peter A. Leavitt, Rebecca Covarrubius, Yvonne A. Perez, and Stephanie A. Fryberg ("Frozen in Time: The Impact of Native American Media Representations") show that the controlling images have been particularly damaging for Native people. Their work on dominant media stereotypes about American Indians documents that these are not mere images. Rather, they have an impact, and not a good one, on people's well-being. The importance of resisting them has, therefore, liberating potential, as each article in this section reveals. This is why resisting controlling images in the dominant media and using social media for criticism and political organization are so important.

SOCIAL MOVEMENTS AND ACTIVISM

Despite the significance of media and popular culture, for many people cyberactivism is not enough. Political mobilization around specific social issues is also an important site of change. Often drawing upon new communications technologies and platforms as organizing tools, some activists use grassroots approaches, while others combine media and face-to-face actions. Moreover, new social movements are increasingly drawing upon intersectional frameworks.

The articles in "Social Movements and Activism" illustrate how efforts to generate a more just society develop in many social contexts—schools, homes, communities, churches, and other locales. Change also occurs at many levels, ranging from personal change and small group-based change to institutional change and large-scale national and global social movements. Moreover, efforts to generate the structural changes of a more just society benefit when social movements draw upon intersecting frameworks. By seeing how relations of race, class, and gender affect their own projects, social activists get a better sense of what is possible and what may be needed to make change happen.

Whether social actors are located inside the institutions they wish to change or whether they stand outside its boundaries, the strategies they use reflect the opportunities and constraints of each specific site. As we discussed in Part III, social structures are not devoid of politics. Instead they are sites that organize inequalities of race, class, and gender. These, in turn, shape particular social issues. Where one works for social change is just as important as the kind of social change or one envisions.

Working from within corporations, schools, or other venues or can mean trying to change the institutional policies and practices that overtly discriminate based on race, class, and gender. It might also mean creating new policies and practices that serve people's needs better. Working for change outside formal social institutions brings a different set of political strategies. Boycotts, picketing, public demonstrations, leafleting, and other direct-action strategies long associated with social movements of all types typically constitute actions taken outside an institution. Activities such as these can be ignored by mainstream media, or misinterpreted when they are covered. Yet it is important to remember that direct action from outsider locations represents one important way to work for social justice. Interestingly, social media transcends these institutional boundaries, often linking people who work for change within organizations to those outside their borders.

The articles in this section examine varying aspects of how people draw upon intersectional frameworks to work for social change within social movement contexts. Building coalitions across race, class, and gender is an important challenge for any social movement, one that requires thinking about intersections. Hana Brown and Jennifer A. Jones ("Immigrant Rights Are Civil Rights") refute the perception that because African Americans and Latinos have competing interests, they do not engage in coalition politics. More accurately, black and brown people agree and disagree on specific issues, but within the United States, these differences occur within a common struggle for civil rights. Brown and Jones show how African Americans and Latinos in Mississippi worked to find their common issues. Their different experiences deepened their understanding of civil rights.

Veronica Terríquez's study ("Intersectional Mobilization, Social Movement Spillover, and Queer Youth Leadership in the Immigrant Rights Movement") of undocumented Latino youth shows how the civil rights movements and the queer movement use an intersectional lens to build coalitional politics. Terríquez details how working in one social movement enables young people to see connections with other movements. This article shows the working of intersectionality in action, one where race, class, gender, and sexuality work together to shape a movement for citizenship rights.

Dorothy Roberts and Sujatha Jesudason ("Movement Intersectionality: The Case of Race, Gender, Disability, and Genetic Technologies") examine how an intersectional perspective enables groups to collaborate when their differing vantage points might otherwise pull them apart. Looking at an organization called Generations Ahead, Roberts and Jesudason show how two different groups—women concerned about women's reproductive rights and disability rights activists—worked together despite their differences. Focusing on what connected them rather than what divided them, they formed a coalition for change, even though they were unlikely partners. Building such coalitions requires first acknowledging and understanding the different experiences that divide people, and then confronting those differences without one group asserting power over another. In the case that Roberts and Jesudason examine, women of color concerned about reproductive rights and disability rights advocates worked together through face-to-face discussions. They identified and articulated common values to construct bridging frameworks, thus cultivating a shared advocacy agenda.

Alfonso Morales ("Growing Food and Justice: Dismantling Racism through Sustainable Food Systems") analyzes an organization that is part of the sustainable food movement—the Growing Food and Justice for All Initiative (GFJI). This organization deliberately empowered a coalition of groups to expand the food movement to address the needs of low-income, racially and ethnically diverse communities. Morales shows how issues of race, class, and gender cannot be background variables, but must be central to social movements for social justice.

Just as we opened this book with Audre Lorde's classic article ("Age, Class, Race, and Sex: Women Redefining Difference") setting the stage for the emergence of intersectionality as a way of viewing the world, we close the volume with Sarah J. Jackson's ("(Re)Imagining Intersectional Democracy") discussion of intersectional democracy. Jackson reminds us that working for a social justice vision is never finished but is always under construction. Jackson brings many of the ideas in this volume full circle. Showing how the ideas of Black feminism have drawn upon and contributed to hashtag activism, Jackson reminds us where the intersectional framework began and how far it may need to go.

Collectively, the articles in this book show the different ways that people from diverse backgrounds have developed a vision for inclusion. Lorde's original message has traveled far beyond intersections of race, class, and gender to incorporate many different axes of power that cause social division. As these ideas continue to circulate, other systems of power may come into sharper view. All sites of change contain emancipatory possibilities, if we can only learn to imagine them. Whether working for individual empowerment or engaging in social activism, we must learn to see beyond the present to imagine the future. Thinking inclusively about race, class, and gender stimulates this vision of hope.

45

Race, Gender, and Virtual Inequality

Exploring the Liberatory Potential of Black Cyberfeminist Theory

KISHONNA L. GRAY

Black women have varied responses when employing Internet technologies for empowerment. New communication technologies have expanded the opportunities and potential for marginalized communities to mobilize in this context counter to the dominant, mainstream media. This growth reflects the mobilization of marginalized communities in virtual and real spaces, reflecting a systematic change in who controls the narrative. No longer are mainstream media the only disseminators of messages or producers of content. Everyday people have employed websites, blogs, and social media to voice their issues, concerns, and lives. Women, in particular, are employing social media to highlight issues that are often ignored in dominant discourse. However, access itself neither ensures power nor guarantees a shift in the dominant ideology. Many women recognize the potential of social media to improve their virtual and physical outcomes, but they also recognize the limits to which technologies can sustain a narrative counter to the current hegemonic structure. Regardless of how much content women create, the Internet will never have the power to dismantle society's dominant structures.

I argue that Black cyberfeminism may address the critique that traditional virtual feminist frameworks do not effectively grasp the reality of all women and may help theorize the digital and intersecting lives of women. Operating under the oppressive structures of masculinity and Whiteness that have manifested into digital spaces, women persevere and resist such hegemonic realities. Yet the conceptual frameworks intended to capture the virtual lives of women cannot deconstruct the structural inequalities of these spaces. Cyberfeminism, technofeminism, and other virtual feminisms may address women in Internet technologies, but they fail to capture race and other identifiers that must also be at the forefront of analysis.

Black cyberfeminism, as an extension of virtual feminisms and Black feminist thought, incorporates the tenets of interconnected identities, interconnected social forces, and distinct circumstances to better theorize women

SOURCE: Gray, Kishonna L. 2015. *Produsing Theory in a Digital World 2.0: The Intersection of Audiences and Production in Contemporary Theory*, edited by Rebecca Ann Lind, pp. 175–192. New York: Peter Lang Publishing.

operating within Internet technologies and to capture the uniqueness of marginalized women.

EXAMINING THE POSSIBILITIES AND LIMITATIONS OF CYBERFEMINISM

Broadly, cyberfeminism is a notion that the Internet has liberating qualities that can free us from the confines of our gendered bodies. The premise, however, has been criticized as both utopic and irrelevant to women's circumstances in new technologies. We cannot just forego our bodies in virtual spaces because much of our real-world selves are emitted into these spaces. The discussion must move beyond the confines of the digital and be reexamined for its potential to mobilize women in both digital and physical spaces. The virtual and physical selves are inseparable. We must critically engage with the recursive relationship between our physical environments and our virtual selves, and we must use the framework to improve women's lives.

Women of color have long recognized that self-determination is a critical component to moving beyond the parameters of hegemonic ideology. Black feminist thought in particular argues for self-definition, a reclaiming of identity, and empowerment for all women and other marginalized groups. In this essay, I build on cyberfeminism and Black feminist thought to articulate the utility of a Black cyberfeminist framework in examining the issues that continue to impede the progression of marginalized women in media, technology, virtuality, and physical spaces.

The Internet has been touted for its liberatory promise but the potential for such transformation could be thwarted by attacks on women in technology. For instance, #GamerGate, which began as an online movement concerned with ethics in game journalism, morphed into an attack on women and feminists. The continued sexism permeating gaming culture is part of a larger culture in technology that devalues women as full participants. This type of structural inequality is not adequately addressed by cyberfeminism. However, by incorporating a critical feminist stance, such systemic problems can be articulated while moving toward meaningful ends for women in these spaces.

How likely is it that Internet technologies can reach their liberatory potential? Many women remain on the periphery of Internet technology. Internet technologies and virtual communities are assumed to be White and masculine. These unequal power relations are accepted as legitimate and are embedded in the cultural practices of digital technology. But many women have resisted this perpetual state of second-class citizenship. Black feminists in particular have outlined a template for countering the hegemonic narrative often operating in technology. By blending cyberfeminism and Black feminist thought, I provide a frame to begin the discussion of allowing women to exist on their own terms and to craft their own narratives. This framework is not new, but it is distinct, given its purpose and intent. This approach details women's experiences and also provides meaningful solutions to combat inequitable power structures.

BLACK WOMEN, IDENTITY, MEDIA, AND CONTROL

Media portrayals offer singular visions of women's lives, their behaviors, and their roles. Women are consistently underrepresented and misrepresented across various media. Feminists are particularly concerned about the representations of women and femininity that promulgate unrealistic standards of physical appearance; girls and women evaluate themselves based on these idealized representations.

There are additional concerns for women of color. Television represents women of color as hypersexual, promiscuous, and immoral. Many media outlets rely on updated versions of minstrel-era stereotypes, such as the hot-tempered and loud-mouthed Sapphire, the domestic servant or Mammy, and the promiscuous Jezebel.

These images are in constant clash with women's reality. Women and girls face conflicting messages about who they are, who they should be, what they can become, and how they should act. Additionally, the racialized element inherent in mediated imagery further serves to perpetuate dominant ideology in the lives of women of color. Conflicting constructions of Black womanhood only serve to reify who is and who is not eligible for full inclusion into womanhood. Black women have long had their identities constructed by outside forces, by masculinity, and by other entities not valuing Black women's agency.

Black women and girls struggle for self-determination and self-definition against their ghettoized and distorted representations. Hegemonic ideologies dominate the narrative of female life in the public sphere; women must work hard to resist these destructive forces. Social media have provided a means to combat these oppressive narratives and allow women the ability to define their own realities. As cyberfeminists contend, Internet technologies are an effective means to resist repressive and oppressive gender regimes and enact equality. However, because Internet technologies still embody hegemonic ideologies and privilege Whiteness and masculinity, the potential to resist dominating structures of oppression may be slim.

This concept reflects a core component of Black feminist thought. As Lorde and Clark (2007) posited, the master's tools will never dismantle the master's house. This is fundamental reality that those with consciousness recognize: The oppressed will never be given full access to spaces, websites, blogs, social media, and other Internet technologies. Although technologies were never created with the intent to destroy the hegemonic structure, they can provide temporary or partial gains in countering the establishment. And because they provide empowerment to women who employ them, they are useful. But this compels one to ask whether the marginalized can ever truly be liberated from their oppressor. Using cyberfeminism as a starting point, it is necessary to critically examine the frameworks' limited ability to effect change.

BLACK FEMINIST THOUGHT IN THE DIGITAL ERA

Dealing with historical and contemporary oppression and marginalization, the lives of marginalized women in the digital era require an engagement with an emancipatory theoretical orientation, one that recognizes the distinctness of their shared and lived experiences. But even with the common threads woven into the patterns of women's lives, the ability to thwart the nature of dominant ideology proves daunting.

Collins (2000) outlined four perspectives unique to the standpoint of Black (and other marginalized) women: (a) self-definition and self-evaluation, (b) the interlocking nature of oppression, (c) the embrace of intellectual thought and political activism, and (d) the importance of culture. In what follows, I discuss these tenets and move toward a theoretical framework to understand the liberatory potential inherent in media and technology for women whose lived realities are reinforced through the intersecting nature of their ascribed identities.

SELF-DEFINITION AND SELF-EVALUATION

The oppressed have a unique standpoint in that they share particular social locations, such as gender, race, and/or class. Although damaging imagery of women of color permeates society, we can find evidence of contestation, resistance, and agency. These individuals share their meaningful experiences with one another, generating knowledge about the social world from their points of view. Despite this knowledge generation, oppressed populations lack the control needed to reframe and reconceptualize their realities. But particular advantages present themselves with the diffusion of information technologies: Women can create and control virtual spaces largely unregulated by the hegemonic elite. These spaces have the potential to foster the development of a group standpoint negating the impact of dominant ideology.

With these alternate spaces in place, women can begin to define their own identities and realities and influence perceptions of womanhood. They can resist the prevalence of controlling images (Collins, 2000), which reflect a system used to physically, economically, and socially control Black women. The power to manipulate images of Black women in such a way creates oppressive imagery that appears "natural, normal, [and] inevitable" (p. 5).

INTERLOCKING NATURE OF OPPRESSION

The ability for hegemonic imagery to influence perceptions exposes the ideological dimensions of women's oppression. Within this hegemonic domain of power, only ideological images characterizing marginalized women as less than human could advance and legitimate a system so fundamentally built on human degradation. This dynamic has existed since the arrival of colonists and since slavery. The images are merely recycled and remixed to further women's oppression.

Yet marginalized women consistently resist and rarely internalize these images (Collins, 2000); indeed, Black women struggle to indict the legitimacy of images such as mammies, matriarchs, welfare mothers, and so forth, as well as the integrity of those who circulate them. This process, which reveals the strong presence of a counterhegemonic consciousness, can be engaged via digital media. The relative ease with which digital spaces can be created presents oppressed groups with the ability to control and create positive content influencing our own images. For Black women, the Internet provides the potential space in which to thwart negative representations disseminated through the media.

Given the interlocking systems of oppression marginalized women experience, knowledge is especially prized for its functionality and intentionality and for its ability to help navigate and enhance one's life and community. Social networks, virtual communities, and other digital media are an extension of traditional communities, such as churches, families, and workplaces.

EMBRACING INTELLECTUAL THOUGHT
AND POLITICAL ACTIVISM

Throughout Black liberation movements, intellectualism has been at the core of the struggle. Intellectualism simply means the knowledge that one has about who one is. It is not rooted in educational attainment. This knowledge of self propels one to the realization of liberation. Black cyberfeminist communities seek collaborations and community building among all groups working to dismantle hegemonic structures, thus highlighting the expansion—the deconstruction—of the terms intellectual and activist. Many feminists adopt an either/or approach, assuming the role of either intellectual or activist, but Black feminists urge that the space must be open for all to take equal part, existing at the intersection of intellectualism and activism.

Action and thought are not at odds but complementary. According to Collins (2000), Black feminist thought leads to Black activism; a dialogical relationship suggests that changes in thinking may be accompanied by changed actions and that altered experiences may in turn stimulate a changed consciousness. For Black women as a collective, the struggle for a self-defined Black feminism occurs through an ongoing dialogue whereby action and thought inform one another.

EMPOWERMENT AND EMBRACING CULTURE

Empowerment is another important tenet of Black feminist theory. Women's power has the ability to produce transformation. Power in this sense is not about control, power over, or dominance. This power is practical; it exists in thoughts and emotions that may influence others, thus igniting interpersonal change and leading into a larger movement. Black women have historically used their power to empower others (Collins, 2000). One of the strongest movements occurring as I write this centers on #BlackLivesMatter. Having its roots in the death of young Black men at the hands of White police officers, the empowerment felt within the Black community and non-Black allies is directly linked to that interpersonal change.

The ability to be independent of the definitions set by the power structure and to produce what one wants to produce about oneself is a form of freedom.

Empowering one's self leads to embracing one's culture. During the Civil Rights Era of the 1960s, the call for Black power was about affirming Black humanity. White America often assumed it was the call for power and superiority over White society. Instead, it was the move to defend dignity, integrity, and institutions within Black culture.

IMAGINING BLACK CYBERFEMINISM

Black feminism can address concerns in the virtual lives of women leading toward a critical cyberfeminist framework. Here I modify the tenets of Black feminism to reflect women in digital realms. Specifically, Black cyberfeminism concerns itself with three major themes: (a) social structural oppression of technology and virtual spaces, (b) intersecting oppressions experienced in virtual spaces, and (c) the distinctness of the virtual feminist community.

SOCIAL STRUCTURAL OPPRESSION OF
TECHNOLOGY AND VIRTUAL SPACES

Ignoring the diverse lives of virtual inhabitants leads to the inability of marginalized bodies to define their own virtual realities. Marginalizing narratives perpetuated through the media reinforce limited conceptualizations of women. Black cyberfeminists urge women to regain control of hegemonic imagery, and Internet technologies allow for this. But as the limits of cyberfeminist and technofeminism illustrate, women need to ensure that they do not recreate oppressions. As I wrote this, I was reminded of an innovative form of Black cyberfeminist activism in the creation of the Twitter hashtag, #whitefeministrants. Online writer and blogger Mikki Kendall started the hashtag in response to the Whitening of feminist spaces where voices of color and otherwise marginalized women are excluded. The tweet, "My Feminism is More Important Than Your Anti-Racism: How to Properly Rank Oppression," immediately differentiates feminism by race and highlights the privileges and oppressions yet to be addressed within the feminist community. It also highlights that traditional spaces can be co-opted, allowing marginalized women to address their grievances. In this situation, marginalized women identified the power of social media to address the racialized distinctions within feminism.

INTERSECTING OPPRESSIONS IN VIRTUAL SPACES

The second theme of Black cyberfeminist theory is that women must confront and work to dismantle the overarching and interlocking structure of domination in terms of race, class, gender, and other intersecting oppressions. Because individuals experience oppression in different ways, we must not create a one-size-fits-all understanding of oppression. Black cyberfeminism requires understanding the diverse ways that oppression can manifest in the materiality of the body and how this translates into virtual spaces.

Black cyberfeminism also requires a recognition of the privileges that some marginalized bodies hold before we can begin dismantling these privileges and understanding the multitude of ways that intersectionality can manifest. Such an understanding might have prevented what was referred to as the feminist Twitter war and avoided claims that Black women and other women of color lead to toxicity in virtual spaces.

Black cyberfeminism, in the spirit of feminism, encourages a privileging of women's perspectives and ways of knowing because race, gender, class status, disability, sexuality, and a host of other identifiers generate knowledge about the world. Valuing these perspectives is the only way to liberate women from the confines of hegemonic notions deeming these identities unworthy.

Black cyberfeminism also recognizes that the lived experiences of women manifest in the virtual world as well. Women do not have the luxury of opting out of any aspect of their identity. By privileging these once marginalized identities, Black cyberfeminist spaces can begin to move women toward progressive and meaningful solutions to hegemonic notions about women.

Although all women share a common struggle, examining their intersecting realities reveals the distinctness of their lived experiences. Women may share sexual oppression, but it is not clear how this can unite all women whose lives, work, life expectancy, and family life are also structured by the hierarchies of racism, ethnicity, colonialism, or nationalism.

Power differences among women are so great that even the similar struggles against men are different. Women's struggle with technology is indirectly a struggle with masculinity, patriarchy, and male privilege; marginalized women also struggle with Whiteness. Cyberfeminists' inability to incorporate the structural nature of inequality results in a limited vision of liberation.

ACCEPTING THE DISTINCTNESS OF MARGINALIZED VIRTUAL FEMINISMS

Black cyberfeminism also addresses the distinct nature of how women utilize virtual technologies. Women have used social media for activism and change, as well as to advance contemporary feminism. The Internet has propelled activism and empowerment in that many individuals can take action on a single issue.

The tenets of Black cyberfeminism never detach the personal from the structural or the communal, which sets Black cyberfeminism apart. The key is in how marginalized women, specifically Black women, communicate and how Black women's Internet usage is a continuation of their offline selves.

Black women were once touted as poster children for the digital divide. What wasn't understood was the cultural and technical savvy that Black women incorporated to use technology on their terms and for their own purposes. A technology may have been created for one purpose, but Black women will employ it to fulfill their own needs, thus displacing the hegemonic establishment.

Black women engage in a variety of cultural forms beyond traditional virtual methods of blogging or tweeting. Black women employ music, poetry or spoken word, and other cultural art forms in their online lives. This direct extension of the physical into the digital acknowledges the accessibility and viability of these cultural artifacts to reproduce Black feminist thought.

Digital social media are important in that they represent, for women of color and other marginalized groups lacking resources, a path to a space where their voices are heard. The once voiceless can be heard, and that leads to empowerment. Twitter, Facebook, and other social networking sites have allowed women to empower themselves and mobilize their communities. As Black Twitter has illustrated, people of color have co-opted traditional virtual spaces for their own means to communicate and empower their communities. So by employing the cultural tradition of sygnifyin', marginalized bodies can express themselves with others without fear of retaliation or being othered within the spaces.

Black women's use of social media also reflects their incorporation of digital technologies and their continued efforts on the ground. Twitter and Facebook have been used to organize marches, highlight continued sexism on college campuses, and draw attention to any number of issues. Maybe, in fact, because of Black cyberfeminism's simultaneous engagement in the virtual and physical communities, the master's tools will be able to dismantle the master's house.

BLACK CYBERFEMINISM: FROM THE STREETS TO THE INFORMATION HIGHWAY

Black cyberfeminism, which represents the blending of multiple ideas into a cohesive analytical framework, simultaneously contributes to and widens the scope of cyberfeminism, technofeminism, and Black feminist thought. Although all three share many theoretical assumptions, values, and aims, their confluence is truly as distinct as the women who exist within Black cyberfeminism. Stemming from feminism's third wave, Black cyberfeminism represents a true engagement with the digital in the lives of wired women that encompasses a self-consciously critical stance toward the existing order with respect to the various ways that the digital affects women.

By bridging cyberfeminism and Black feminist thought, this framework is able to interrogate how women have understood their oppressed status, recognized the gendered and raced nature of the digital divide, and have made sense of their realities and experiences. Importantly, women are not passive bystanders in the information age waiting for their turn.

Black women are urged to recognize the distinctness of our cultures. It is this deep heritage that provides us with the energy and skills needed to resist and transform daily discrimination. As women, we must embrace the history of our oppression, understanding that this history informs how power relations pervade our lives. We can't simply adopt privileged points of view and expect significant change to occur in either the virtual or physical world. We must embrace one another's oppressions and even privileges, recognizing that we all come from distinct realities converging in virtual spaces. There must be an affirmative action to be inclusive of a variety of women and viewpoints. Women working together is the only way to achieve significant changes. We cannot adopt the exclusionary approach of previous generations of women. We must recognize our privileges—racial, heterosexual, lingual, and so forth—and move toward fairness and equality for all women. Digital spaces provide us with a significant opportunity to accomplish this feat.

REFERENCES

Collins, P. H. (2000). *Black feminist thought: Knowledge, consciousness, and the politics of empowerment.* New York, NY: Routledge.

Lorde, A., & Clarke, C. (2007). *Sister outsider: Essays and speeches.* Berkeley, CA: Crossing Press.

46

Talking Back to Controlling Images

Latinos' Changing Responses to Racism Over the Life Course

JESSICA VASQUEZ-TOKOS AND KATHRYN NORTON-SMITH

Five years after the 2010 United States Census reported that Latinos accounted for most of the nation's population growth (56 percent) from 2000 to 2010 (Passel, Cohn, and Lopez 2011), Donald Trump, a Republican presidential hopeful, stated in his presidential announcement, 'When Mexico sends its people, they're not sending the best. They're ... sending people that have lots of problems and they're bringing those problems. They're bringing drugs, they're bringing crime, [and] they're rapists.' This speech disseminated a controlling image, or deleterious representation, of Mexicans as criminals. 'Controlling images', as 'major instrument[s] of power', are ideological justifications of oppression that are central to the reproduction of racial, class, and gender inequality (Collins 1991, p. 68). The function of controlling images is 'to dehumanize and control' dominated groups (Collins 1986, S17). The social institutions that circulate controlling images do so to suppress less privileged groups, restricting minorities' access to upward mobility, self-efficacy, and power of self-definition (Collins 1991).

Controlling images are a hallmark of systemic racism, justifying 'the creation, development, and maintenance of white privilege, economic wealth, and socio-political power ... [rooted in] hierarchical interaction and dominance' (Feagin 2000, p. 14). Controlling images are conceptually different from prejudice because they are ideological collective representations, not psychological. Prejudice involves negative emotion and stereotypes (Quillian 2006) whereas controlling images are systemic and cultural, existing beyond the affective and cognitive. Controlling images, prejudice, and discrimination are related in that controlling images provide a 'strategy of action' (Swidler 1986) for prejudice and discrimination.... Controlling images are cultural tools that offer 'strategies of action,' thus bridging systemic racism, prejudice, and discrimination....

A cursory glance at media reveals that Latinos are typecast as docile menial labourers, unauthorized immigrants, criminals, gang members, rapists, seductresses, and athletes (Rodriguez 2008; Molina 2014). The controlling images most frequently reported to us by our largely middle-class sample of sixty-two

SOURCE: Vasquez-Tokos, Jessica and Kathryn Norton-Smith. 2017. "Talking Back to Controlling Images: Latinos' Changing Responses to Racism Over the Life Course," *Ethnic and Racial Studies*, 40(6): 912–930. Reprinted by permission of Taylor & Francis Group, LLC, a division of Informa plc (http://www.tandfonline.com). Permission conveyed through Copyright Clearence Center, Inc.

Latino men were that of gang members and sports athletes, both expressions of minority male virility and peripheral status. While popular imagination might more closely knit gang membership and athletics to blacks than Latinos, our empirical findings contradict this assumption, revealing how class status shapes racialized imagery of Latino manhood and how racialized treatment can transcend the black/brown divide (Rios 2011; Jones 2012). Honing in on the two most frequently reported controlling images of our sixty-two mostly middle-class Latino men respondents, we ask: How do controlling images of Latinos as gang members and sports athletes regulate opportunities, impose constraints, and channel emotions? How do Latinos respond to these controlling images? To what extent are these responses shaped by stage in life course?

... This article draws from sixty-two in-depth life history interviews with Latino men. The data come from two research projects conducted by the lead author in 2004–2005 and 2010. Both research projects focused on Latino families and asked questions concerning racial/ethnic identity, family, community, and race and gender issues. Latino men's relationship to gangs and sports was an emergent theme: the interview schedules did not address gangs or sports but nevertheless respondents repeatedly introduced the topics, reflecting their importance in their daily lives.

... The age breakdown for our respondents follows: 27 percent were teenagers to twenties, 51 percent were in their thirties to fifties, and 22 percent were in their sixties to eighties. For most respondents, the interviews were retrospective accounts that covered decades. Since we advance a life course argument that leverages life stories of adults that span decades, we do not compare age groups. Instead, in data collection and analysis we were attentive to stage in life course as respondents relayed encounters with racist controlling images. An advantage of life history in-depth interviews is that we can trace developments over time (such as forms of racism confronted and reactions) and theorize how these changes over the life course are patterned....

FINDINGS

This findings section first illustrates how people and institutions that circulate sports and gangs controlling images constrain Latino respondents and then theorizes resistance tactics. We find that age shapes resistance strategies, emotional tactics prevalent among youth and leadership techniques predominant among adults.

Forcibly Constrained

Effective controlling images regulate opportunities, impose constraints, and channel emotions. As Latino men interviewees reflected on their youth, they recalled feeling restricted by imagery that depicted them as athletic bodies lacking mental capacity worthy of an education. Schools teach formal curriculum as well as 'hidden curriculum', including modes of social control that reflect the dominant class's hegemony. Controlling images are one piece of hidden curriculum that indoctrinates minority students with lessons of racial subordination. The controlling image of the athlete curtailed Latino men's aspirations and diverted their educations onto sports fields. These men were rewarded only for their physicality, the athlete controlling image relying on a body/mind dichotomy established during slavery relative to black/white race relations that has had repercussions for other racial minority groups (Collins 1991, 2004; Molina 2014).

In his youth, the educational system diverted Harry Torres, now sixty-five, out of education and into athletics. Harry uses the phrases 'locked me out' and 'forced' to explain the institutional racism operating through controlling images that rerouted him toward sports:

> Since I was a straight-A student in grade school, they put me [in] all the AP [Advanced Placement] classes [in high school]. The teacher ... lock[ed] me out of it. This one teacher knocked me out of the highest English class: ... she sent me down to a lower English class. ... They forced me to take ... shop class ... I guess they figured this guy is going to pick up one of the manual skills. ... I didn't fight it. ... I went out for sports ... it wasn't as hard.

A strong student in elementary school, high school teachers used race as a proxy for mental ability and 'knocked [him] out' of advanced classes. Education is both a racialized and racializing institution (Vasquez 2011), and here we witness school teachers detouring Harry into menial labour classes. Authority figures racialized him as only fit for 'manual skills' that emphasize bodily rather than mental labour. Using 'inaction as an emotional strategy for preventing the emotional pain of racism' that struggle would entail (Evans and Moore 2015, p. 447), Harry 'didn't fight it' and instead 'went out for sports'. As a youth, Harry directed his attention to sports, succumbing to the controlling image and the system's expectations for him. With institutional racism propping up controlling imagery, Harry perceived resisting being typecast as a body without a brain as 'hard' whereas not 'fight[ing] it' translated to conforming to subpar expectations. The athlete controlling image naturalized sports participation and made bucking a system enforced by authority figures onerous to challenge.

Younger and older men, only some of whom were involved in gangs, complained about the difficulty of shirking the expectation of gang membership. Thirty-six-year-old Vincent Venegas, a former gang member, discusses the confinement of gang life: 'Since I was a gang member ... the possibilities are even smaller. ... It contained me.' Poignantly, 'contain' is a synonym for 'control', suggesting the efficacy of the controlling image which limited his access to legal jobs, self-expression, and contribution to society. Vincent's fifteen-year-old son, Pablo, 'fell into' the controlling image of the Latino gang member due to peer pressure: '*Cholos* [gangsters] were cool. They were cool because they were hip, they were fashion.' The controlling image required a particular display of masculinity: 'being a man, being cool, being ... gangster'. The gangster controlling image plus the controlling image of the low-achieving Mexican student combined to contain Pablo's academic success. Believing that he did 'not [need] to be educated because he already knew everything', Pablo and his peer group reserved academic achievement for Asian and White students: 'Oh, they want an education. They're going to college. Nah, they're too gay for us …. Let the white people get their education.' Controlling images threatened to derail his education until Pablo's uncle 'pounded some sense into [him]' saying: 'You are a minority. The blacks are a minority. The whites and Asians, they're getting their education. They're going to make more money …. They're the higher race…. You should be trying to get to the top.' Pablo's constraint stems from seductive controlling images that maintain the racial hierarchy. This narrative reveals the appeal of gang imagery, illustrating how in his multiracial urban context controlling images exert control by confining visions of the future – in Pablo's case, convincing him that being a gangster is a 'cool' aspiration until his uncle intervened.

These retrospective tales indicate that Latino men were more constrained by controlling images when they were young than once they entered adulthood. The controlling-ness of controlling images is stronger for youth who, by virtue of their age, are still learning how to navigate racist terrain. While some people can remain constrained by controlling images, our largely middle-class sample reported constraint in youth that transformed into resistance as they grew older.

Emotional Resistance: Gaining Esteem in Youth

Emotional resistance involves an internal, psychological, feelings- and attitude-centered struggle to recuperate self- and group-worth. There is a life course dimension to resistance styles: youth, who typically have lower levels of social power, expressed resistance emotionally whereas adults, who attain greater power with age, resisted through leadership activities related to their professions.

Twenty-eight-year-old Moises Ramos was emotionally committed to proving himself above the low bar of expectations that controlling images establish, sensing that his future prospects were at stake: 'People automatically judge you as being something you're not just because of the way you look.' In response, he adopted an '[I] don't give a fuck' attitude as a way to anesthetize the negativity embedded in controlling images. Moises explained his emotional resistance: 'just the attitude of "who cares what others think of me" and "that's not going to stop me from doing what I need to do to achieve my goals"' Emotional resistance can lead to behavioural resistance, as Moises notes: 'I've got to still prove myself because people are … always going to have this doubt about me … . So the more that I prove myself the better that I feel about myself.' A career counsellor at a city college, he challenges controlling images, using educated English to alter incorrect assumptions: 'If I open my mouth … [people] are … thrown off because supposedly I can't comprehend what they're saying.' A young adult, Moises employs two forms of resistance: he uses emotions and attitude to safeguard his sense of self and he uses speech acts to disrupt misperceptions.

Latino interviewees whose adolescence or young adulthood intersected with the social movements of the 1960s developed lasting emotional strategies of resistance. Raymond Talavera of California, now fifty-one, speaks to the importance of the Chicano Movement which asserted an unapologetic racial minority identity in rebutting controlling images. In his teenage years during the Movement, Raymond emotionally rejected controlling images: 'If I'm treated that way, it's *them*, not me ... . That was really the lesson of the Chicano Movement: "this is who we are, we don't have to be what the corporate leaders of America say we have to be."' Raymond conceptualizes emotional resistance as living beyond the restraints of controlling images:

> The Chicano Movement ... said, '... We are who we are, this is who we are. I know who I am. I know I don't have to be Joe Smith, Jr. of the Junior League ... . I am Raymond Talavera of the Hispanic Chamber. ... Accept us for who we are.'

The call to 'accept us as who we are' is poignant since freedom equates to living outside of controlling images. Social movements like the Chicano Movement can teach emotional resistance strategies which can last a lifetime and be converted into other forms of resistance as people age....

Leadership as Resistance: Giving Back to the Community in Adulthood

Resistance does not stop with emotions but can convert into action as people age and their social power increases. When younger respondents reported behavioural resistance, it was limited to sports contests or fighting, a physical manifestation of emotional turmoil. Among adult respondents, most of whom were professionals, resistance materialized as 'giving back' to their community. Encounters with controlling images inspired economically stable Latino men to contribute to their racial group through leadership efforts that teach equality, cultivate civic improvement, and model paths of mobility.

Most adult Latino respondents who discussed action-oriented resistance to controlling images referenced education.... Latino respondents, who as men were socialized to be active in the public sphere, were vocal about contributing positively to schools as a mentor, teacher, or administrator. These men formally expand their students' knowledge base and informally serve as models of professionals. By investing in the younger generation, these Latino educators 'give back' by broadening students' vision and demonstrating that they can unlock the shackles of negative imagery....

Not all who resisted controlling images through leadership did so through education, some staging resistance through other professions that serve the public, such as sixty-two-year-old architect Gilbert Ornales. As a native of Los Angeles, Gilbert's awareness of racial segregation peaked when he witnessed the creation of Dodger Stadium through the relocation of Latino residents of Chavez Ravine: 'They were just going into the *barrios* [neighborhoods] and moving people. ... Things were being planned for Chicano neighborhoods without no input [sic] or without sensitivity to the people living there. That's how we decided ... we're gonna ... be spokesmen architects for the people.' Gilbert incorporates community representation in his public architectural projects. He explains how he sought feedback from gang members when redesigning a park that now bears the name of a local gang:

> I designed [the park] with the input of the gang leaders. They wanted a Mexican park .... I designed ... a stage and ... stairs ... [that] simulate ... the mountain in Mexico City, Popocatepetl ... . They wanted a piñata. So I designed a piñata pole ... . One of the gang guys went back to school and wrote a paper on the process.

By including gang members in community redevelopment and highlighting their community-member and student statuses, Gilbert troubles the dichotomy between gang and student identities.

By enacting resistance through leadership in professional realms, these Latino men convert frustration with controlling images into constructive action. Emotions remain present in that they undergird and motivate action. By undertaking leadership activities, these men weaken controlling images by promoting life pathways beyond the prescription of controlling images. By giving back to the community through teaching, mentorship, and community involvement, these men invest in their racial group and challenge restrictive imagery....

CONCLUSION

Institutions and people that circulate controlling images of Latinos aim to constrain their opportunities, self-images, and futures. Controlling images organize experience, channel emotions, and shape expectations and aspirations. A tool of systemic racism, portrayals of Latinos as sports athletes and gang members can be effective, forcibly constraining Latino men's futures. Yet resistance occurs, younger men enacting emotional forms of resistance and older men activating leadership tactics in their professions.

Life course stage shapes antiracist strategies. Adults possess more social power than youth, people transforming the emotional resistance of youth into action-oriented resistance in adulthood. While these categories may overlap, the data reveal that age conditions form of resistance to controlling images. By investigating *reactions* to racialized depictions, we give agency to the oppressed and showcase the process of social change.

REFERENCES

Collins, Patricia Hill. 1991. *Black Feminist Thought.* New York: Routledge.

Collins, Patricia Hill. 2004. *Black Sexual Politics.* New York: Routledge.

Evans, Louwanda, and Wendy Leo Moore. 2015. "Impossible Burdens: White Institutions, Emotional Labor, and Micro-Resistance." *Social Problems* 62(3): 439–454.

Messner, Michael A. 1992. *Power at Play.* Boston, MA: Beacon Press.

Molina, Natalia. 2014. *How Race is Made in America*. Berkeley: University of California Press.

Quillian, Lincoln. 2006. "New Approaches to Understanding Racial Prejudice and Discrimination." *Annual Review of Sociology* 32(1): 299–328.

Rios, Victor M. 2011. *Punished: Policing the Lives of Black and Latino Boys*. New York: New York University Press.

Romero II, Tom I. 2004. "Wearing the Red, White, and Blue Trunks of Aztlán: Rodolfo 'Corky' Gonzales and the Convergence of American and Chicano Nationalism." *Aztlan* 29(1): 83–117.

Saraswati, L. Ayu. 2013. *Seeing Beauty, Sensing Race in Transnational Indonesia*. Honolulu, HI: University of Hawai'i Press.

47

"This is for the Brown Kids!"

Racialization and the Formation of "Muslim" Punk Rock

AMY D. MCDOWELL

"The darkie ones are terrorists—How simple can it be?" is a lyric from "Post-9/11 Blues," a poppy hip-hop track by Riz MC about life as the Other in the wake of anti-Muslim racism. With a peppy, tongue-in-cheek beat, the song underscores how non-Muslims "read" brown bodies through an Islamophobic lens.... Although Muslims and non-Muslims suffer from the "post-9/11 blues," we know surprisingly little about how marginalized groups respond to anti-Muslim racism in nonreligious ways and venues....

This research uses the case of Taqwacore punk rock to examine how panethnic youth collectively respond to anti-Muslim racism in culturally provocative, nonreligious ways. The word *Taqwacore* is a hybrid term composed of *taqwa*, which means "God consciousness" in Arabic, and *core* from *hardcore punk*. Taqwacore is a geographically dispersed cultural community of bands, filmmakers, authors, and artists, mostly in the United States, who began socializing and collaborating after the release of Michael Muhammad Knight's (2004) novel about an imagined Muslim punk house in Buffalo, New York, titled *The Taqwacores* (Fiscella 2015; Murthy 2010). Since its inception, Taqwacore has garnered a substantial amount of media attention from mainstream outlets including *Rolling Stone*, the *New York Times*, National Public Radio, *Newsweek, The Guardian*, and *Time*. This influx of media attention from mainstream outlets offers free publicity to Taqwacore bands, films, and authors. But this media also frequently misrepresents Taqwacores as a monolithic group of rebellious "Muslim punks" who, as the *Times* put it, "balance morning prayers with sex, drugs, and rock 'n' roll" (Dalton 2010). In fact, Taqwacore is not entirely made up of Muslims or restricted to a mash-up of Islam and punk rock. Taqwacore is composed of secular Muslim, spiritual Muslim, and non-Muslim youth who call themselves "brown kids" or "*for* brown kids." For Taqwacores, *brown* signifies a strategic and positional punk rock stance against the stereotype that Muslims and brown-bodied Others are a monolithic group defined by religious conservatism or extremism.

Punk rock culture is strongly associated with social nonconformity and independent cultural production, both of which Taqwacores draw upon to create a

SOURCE: McDowell, Amy D. 2017. "This is for the Brown Kids!": Racialization and the Formation of "Muslim" Punk Rock. *Sociology of Race and Ethnicity*, 3(2): 159–171. Copyright © 2016 by American Sociological Association. Reprinted by Permission of SAGE Publications, Inc.

resistant racial identity as "brown kids." From their perspective, punk rock is the bold act of embracing their status as the Other in a society that rejects and stereotypes them on the basis of race and/or religion. As one Taqwacore interviewee remarked, "Punk allows me to say, 'Hey, I'm a Muslim ... I already know you can't accept me so here's a big fuck you to you!'" His use of punk fits well with other marginalized anti-oppression punk movements such as Afropunk, Chicano-punk, and Queercore punk, which transform punk rock culture into a platform for nurturing and promoting peripheral spaces of belonging and identity. In line with this philosophy of punk rock, Taqwacores consider artists, musicians, and independent writers who rally against anti-Muslim racism part of their punk movement not because the sound or style fits standard accounts of punk but because the *message* is for "brown kids."...

In this article, I show how Taqwacores create a racial identity as "brown kids" that is panethnic and opposed to anti-Muslim racism along two primary axes: (1) by calling out whiteness in punk and keeping it out of their punk and (2) by redefining punk in favor of racial and religious outsiders....

RACIALIZATION AND RESISTANCE

This research extends a growing body of research on anti-Muslim racism and the responses to it in the aftermath of September 11, 2001. The main claim in this literature is that "race" is not strictly confined to physical characteristics such as skin tone. Cultural and religious characteristics such as a Muslim name, a hijab, clothing, or an accent can also be "raced." A wide range of people are grouped into the "Muslim" category and subjected to Islamophobia, a hostile attitude toward Islam and Muslims based on the idea that Islam and Muslims represent a threat to the West. The fact that non-Muslims are misidentified as Muslim proves that Muslims are understood on racial, rather than solely religious grounds (Tyrer 2013).

The process by which Muslims become "raced" is captured by the concept of racialization, which entails grouping people together on the basis of physical or cultural traits and then ascribing a set of characteristics to that group....

Taqwacore punk affirms that newly formed groups change, alter, and transform ascribed racial categories. Indeed, the ubiquity of anti-Muslim racism in American society is what stimulates Taqwacores to "relax and widen their boundaries" (Okamoto 2014:2) in pursuit of an umbrella panethnic identity. By forming and asserting this new group identity, they both disrupt common misconceptions of what it means to be Muslim in American society and establish new modes of cultural membership and belonging....

Taqwacores deploy an essentialist "brown" identity in punk to band people together against anti-Muslim racism. But Taqwacores do not merely *adapt* an ascribed identity. Instead, they *create* a "brown" identity that calls conventional religious, racial, and ethnic identities into question. Taqwacores form this new "brown" identity by participating in "reflexive racialisation," a concept that highlights how racially and ethnically marginalized individuals learn about their common struggles as "Others" and develop shared understandings of social inequality through self-generated, self-policed media.... Taqwacores take "reflexive racialisation" a step further: rather than engaging in talk about who they are, they talk about who they are *not*: they are not white and they are not mainstream.

To understand how Taqwacores create a broadly defined "brown" punk identity, I show how these youth define themselves against whiteness and take pride in being the Other....

I use three sources of qualitative data 65 hours of participant observations, 20 in-depth interviews, and Taqwacore artifacts such as Taqwacore music, films, and the Taqwacore webzine. The majority of these data were collected in 2009, 2010, and 2011 in the United States, when Taqwacore music, films, and public discussions about Taqwacore were at their peak. By the time I was wrapping up data collection in 2012, Taqwacore had "essentially evaporated" (Fiscella 2015:101), and many of the "original" Taqwacores were starting to claim, "Taqwacore is dead."...

Keeping Whiteness Out

Taqwacores create a punk identity as "brown kids" by claiming that whiteness is a problem in punk and by keeping whiteness out of their punk. In this community, whiteness is not solely a matter of skin tone. Whiteness stands for racially dominant groups and ideas as well as Western imperialism, Christian supremacy, and everyday practices of racial and religious exclusion. For this reason, some whites, such as white Muslims or those who advocate antiracist politics, are accepted as part of Taqwacore punk; they embody whiteness only to a degree.

Interviewees claim that white punks regularly test Taqwacores about their knowledge of punk rock history, insinuating that people of color are not genuine members of the subculture. P.C., a non-Muslim Taqwacore, explains that she never felt fully included in her local U.S. underground music scene. As a "brown girl," she says that she was quizzed by white punks about her capacity to appreciate "good" music and implicitly labeled a "poser." Angered by the racism, P.C. exclaims, "I'm as American as macaroni and cheese! I don't understand why I wouldn't like this music!" Sam, a Muslim Taqwacore who wears the hijab, says that she has come under attack in white punk spaces. She recalls an interaction she had with a white punk at a show in Sydney, Australia:

> I remember one time someone came up to me to introduce themselves. I thought, "Oh, this is cool. It's just someone wanting to get to know me." But when we sat down later on [to talk], he was just like so, "You believe in Islam, right?" and I was like "Yes." And he was like "I don't know, I can't say I agree with it." And I said, "That's cool, I suppose everyone has their own thing going on." And he was like, "I don't know, I don't think you can be a punk *and* a Muslim."

... Bearing in mind that Taqwacores feel that white punks are insensitive to racial and religious oppression, many Taqwacores are apprehensive about whites getting involved in their community. Sam feels that Taqwacore is supposed to be a "tight-knit" group of people who share similar beliefs and experiences "so it's a little bit odd," she says, "when a white person comes in and says that they want to embrace this culture." Similarly, when I asked Zahira, a Taqwacore blogger, about who belongs to Taqwacore, she responded,

> I mean it has been a little weird that there are people who have been identifying as Taqwacore who are not Muslim and not *really* punk. Um, and, I think it's kind of toes the line of exotifying this movement. And that part makes me uncomfortable. The exotification of this space. I really do think it should be a safe space for people who identify in this way versus the exotification of people in this community and then trying to like fit in.

When I pressed Zahira to specify the people who exotify Taqwacore, she answered in a direct tone, "They are white and they are not Muslim and they don't really listen to punk music." She "takes issue" with white, non-Muslim people calling themselves Taqwacore: "We're doing this for ourselves. We're not doing this to teach white folks," she adds....

Although some Taqwacores warn that whites dilute the political edge of punk for "brown kids," others argue that whiteness is an elusive category that cannot be tacked onto all light-skinned individuals in the same way and to the same degree. Accordingly, the Taqwacores who feel that whites can belong argue that some whites fight against racism or experience other forms of oppression....

Taqwacores make punk for "brown kids" by vocalizing their opposition to whiteness in punk and keeping it out. Yet Taqwacores do not simply use skin tone to determine whiteness; instead, whiteness is viewed as a practice of racial and religious privilege, bigotry, and exclusion. Accordingly, Taqwacores feel that the whites who contribute to the creation and dissemination of Taqwacore by playing in bands, writing books and blogs, and reviewing its music, are "punk as fuck," and Taqwacore punk to a degree because they stand against anti-Muslim racism. To the contrary, whites who downplay racial and religious injustices are not really punk because they maintain exclusive white punk rock spaces that are unwelcoming to "brown kids."

Making Punk for Brown Kids

The other primary way Taqwacores create punk for "brown kids" is by redefining punk in favor of racial and religious Others. They use *punk* as a synonym for "brown kids" who feel triply marginalized by white-dominated punk rock, mainstream American society, and traditional ethnic and religious communities. Jehangir, a punk rock character who plays the part of an unorthodox spiritual guide in *The Taqwacores* (Zahra 2010), epitomizes this stance when he claims that Taqwacore is where "all the crazy rejects and fuckups of the community come together ... nobody likes them. Muslims say they're not really Muslims. The punks say they're not really punks." As Jehangir's remark that "nobody likes them" suggests, Taqwacores take pride in being the true punk rock outsiders. Being a "reject" and a "fuck-up" is social assurance that Taqwacores do not obey social conventions, even punk rock conventions....

Another way Taqwacores make punk for "brown kids" is by embracing stigmatized racialized signifiers, such as the *keffiyeh* (traditional Arab headdress) in punk rock spaces. Members of The Kominas (which translates to "scumbag" or "scoundrel" in Punjabi) played loud Bollywood-esque punk riffs while sporting traditional Arab garments at a live show in Cambridge. In the context of a majority white performance bill, the lead singer's turban, *thawb* (an ankle-length garment with long sleeves), and sandals marked an unconventional space for the performance of Muslim identity. In doing so, he also interrupted the racial homogeneity of the larger, mostly white, punk rock milieu....

Making Muslim identity visible in punk rock venues not only solidifies the boundaries of Taqwacore punk for "brown kids"; doing so also disrupts racial stereotypes that Muslims are a monolithic group defined by uniform beliefs, practices, and values....

CONCLUSION

This research expands new scholarship on collective responses to anti-Muslim racism. It shows how Taqwacore punks collectively resist anti-Muslim racism in decidedly nonreligious ways. They do this by creating a racial identity as "brown kids" that stands against whiteness within the context of punk rock, a music culture rooted in antiestablishment identities and politics.

Taqwacores make the case that although punk is supposed to provide refuge for social misfits, racist and religious bigotry still finds expression in white-controlled punk rock spaces. Some white punkers claim to have authority on what punk means and often discount the racial struggles that brown-bodied and Muslim youth face in Western societies today. Some even go so far as to claim that a person cannot be punk *and* Muslim, insinuating that all Muslims are first and foremost defined by Islam, and nothing more. By calling out whiteness in punk and larger U.S. society, Taqwacores create their own punk rock space, one that is set apart and critical of white "mainstream" punk. In the process, they establish racism as something more complex than the traditional American white versus black divide. As they do this, they create a racial identity as "brown kids" that refuses to be defined by skin tone or a shared religious practice or ethnic background, but rather by an appreciation for *difference* that both magnifies and vilifies white supremacy.

REFERENCES

Dalton, Stephen. 2010. "Never Mind the Burkas, Here's the Islamic Punks." *The Times.* Retrieved November 15, 2010 (http://www.thetimes.co.uk/tto/arts/music/article2552007.ece).

Fiscella, Anthony. 2015. "Universal Burdens: Stories of (Un)freedom from the Unitarian Universalist Association, The MOVE Organization, and Taqwacore." PhD dissertation, Centre for Theology and Religious Studies, Lund University, Sweden.

Knight, Michael Muhammad. 2004. *The Taqwacores*. Brooklyn, NY: Soft Skull.

Murthy, Dhiraj. 2010. "Muslim Punks Online: A Diasporic Pakistani Music Subculture on the Internet." *South Asian Popular Culture* 8(2):181–94.

Okamoto, Dina. 2014. *Redefining Race: Asian American Panethnicity and Shifting Ethnic Boundaries*. New York: Russell Sage.

Read, Jen'nan Ghazal. 2008. "Muslims in America." *Contexts* 7(4):39–43.

Zahra, Eyad [Director]. 2010. *The Taqwacores*. Runmanni Filmworks.

48

"Frozen in Time"

The Impact of Native American Media Representations on Identity and Self-Understanding

PETER A. LEAVITT, REBECCA COVARRUBIAS, YVONNE A. PEREZ, AND STEPHANIE A. FRYBERG

Mass media messages are a nearly inescapable feature of modern life. Media displays and perpetuates shared ideas and images, or social representations, of the social world. For example, mass media offers an array of characterizations that associate different identity groups with different possibilities for how to be a person (i.e., how to act or behave) in society. These representations typically reflect and reify stereotypes of groups (e.g., African Americans as athletes and musicians, women as sexualized beings) that vary in quality (e.g., accuracy and valence—positive or negative representations) and quantity (e.g., number and breadth). For some social identity groups, such as White, middle-class individuals, the media provides an abundance of positive, varied representations, whereas for others, such as working-class and racial-ethnic minority individuals, it provides a limited number of predominantly negative and narrow representations. The purpose of this article is to examine how the quality and quantity of media representations influence identity and self-understanding, particularly when a group, such as Native Americans, is greatly underrepresented.

According to the theory of invisibility (Fryberg and Townsend, 2008), when a group is underrepresented in the media, members of that group are deprived of messages or strategies for how to be a person. Although media effects are typically small, the ability of media to shape how individuals experience and understand various-groups, contexts, or domains, is well documented (see Mastro, 2009). Notably, it is not merely the quality of media characterizations of groups that contribute to identity and shared understanding (e.g., public perceptions about the defining characteristics and behaviors of the group and about norms for how to treat the group), but the quantity of portrayals (e.g., the sheer number of portrayals) also communicates a message about the group's vitality in society. Accordingly, the limited representations associated with minority groups in the media, in terms of both quantity and quality, are likely to convey to group members that they do not belong and cannot be successful in a number of achievement-related fields (e.g., education, business) where minority groups are scarcely (if ever) seen in the media.

SOURCE: Leavitt, Peter A., Rebecca Covarrubius, Yvonne A. Perez, and Stephanie A. Fryberg. 2015. "'Frozen in Time': The Impact of Native American Media Representations on Identity and Self-Understanding." *Journal of Social Issues* 71(1): 39–53.

The invisibility of Native Americans in mass media provides a unique vantage point for examining how media representations impact both identity and self-understanding. Native Americans are typically depicted as 18th and 19th century figures (i.e., as teepee dwelling, buckskin and feather wearing, horse riding people) and, in the rare cases in which they are shown as contemporary people, they are negatively stereotyped as poor, uneducated and prone to addictions. This type of limited and negative representation of Native Americans is referred to as relative invisibility (Fryberg and Townsend, 2008). Many groups experience relative invisibility (e.g., Latino Americans, gay and lesbian, and working class individuals), but what differentiates Native Americans is that they uniquely experience absolute invisibility in many domains of American life. Specifically, they are rarely (if ever) seen as contemporary figures in the media, which means they are absent from depictions of mainstream public spaces, such as schools and hospitals, and from many professional positions, such as teachers, professors, doctors, and lawyers. In this way, Native Americans, more than other social groups, are seen and learn to see themselves through the lens of negative stereotypes or they look to the messages projected about the contemporary world and simply do not see themselves represented. In the remainder of this article, we will provide an overview of the available media representations of Native Americans and highlight the impact of these representations on Native American identities and self-understanding. First, we will examine the pervasiveness and the influence of media content in American society for different social groups. Second, we will review the quality and quantity of Native American representations in the media. Finally, we will discuss the psychological consequences of Native American invisibility in the media on identification and self-understanding.

PERVASIVENESS OF MEDIA CONTENT AND INFLUENCE

Many of the impressions people form of diverse individuals and groups are the result of vicarious or indirect experience through media rather than direct, in-person contact. To illustrate, 98.9% of American households have a television set and 92.6% of Americans watch television regularly. In the past two decades, new media technologies have also become central and influential in American life. For instance, 80.9% of households have personal computers, 78.7% of households use the Internet regularly, and 78% of adult Internet users read the news online. Eighty-five percent of adults own a cell phone, 63% of cell phone owners use the Internet on their phone, and more than 40% of Americans play video games regularly. These media vehicles offer messages or representations about different groups and about how to think about or understand the social world.

These social messages or representations reflect the widely shared, yet taken-for-granted, ideas, practices, and policies that individuals use to understand or orient themselves within their everyday social contexts and to communicate with one another. These representations convey information about the good or right way to be a person, including how individuals represent or think about themselves in the past, present, and future. Social representations communicate, for example, that this is how a certain kind of person talks and behaves, this is how to interact with this kind of person, and this is what this kind of person can achieve.

Due to its pervasiveness, mass media is a potent channel by which social representations are created and maintained in mainstream society. They provide a surrogate representation for real-world exposure in cases where interpersonal contact between majority and minority group members is limited and/or nonexistent. Popular media is, in many cases, the only exposure some people have to members of other groups. This is problematic when the media conveys inaccurate or stereotypical representations about social groups, or when the media fails to provide a representation at all (i.e., a group is invisible).

MEDIA REPRESENTATIONS OF NATIVE AMERICANS

Media is not an "equal-opportunity self-schema afforder"; that is, it does not provide equal social representations of how to be a person for all groups. Some groups are represented less often and in more negative ways than others. We contend that this inequality puts groups, such as Native Americans, at a psychological disadvantage compared to groups who are abundantly and positively represented. Close examination of the population statistics and media portrayals of Native Americans reveals that they are largely invisible in contemporary American life.

In the United States, individuals who identify in the census as Native Americans constitute 1.6% of the population, whereas individuals who report being Native American and some other racial-ethnic group(s) constitute 4.1% of the population (DeVoe and Darling-Churchill, 2008). In contrast, content analyses of primetime television and popular films reveal that the inclusion of Native American characters ranges from no representation to 0.4% of characters being Native American. Similarly, less than 1% of children's cartoon characters and 0.09% of video game characters are Native American. Taken together, whereas Native Americans make up a small portion of the population, they are considerably more underrepresented in the media. In fact, they are often invisible in the media.

The representational issue, however, is not simply that Native Americans are numerically underrepresented, but that the quality of representations is also constrained. For instance, whether Native Americans are depicted as sports team mascots (e.g., Washington Redskins) or Hollywood film characters (e.g., Pocahontas), they are typically portrayed as 18th and 19th century figures. Furthermore, these representations not only locate Native Americans as historical figures, they also depict them as particular types of Native Americans (e.g., Sioux, Apache, Navajo). Considering the diversity of Native American groups, these narrow representations not only define Native Americans as a homogeneous group "frozen in time," but also render invisible hundreds of diverse tribal cultures.

Moreover, the advent of the Internet has allowed these types of portrayals to reach a wide audience. A simple Internet image search for Native Americans via Google and Bing—the two most widely used search engines—directs the searcher to the same historical imagery. For the purpose of illustration, we examined the first 100 image results for each of the terms "Native American" and "American Indian" returning 200 images total from both search engines. We found that 95.5% of Google ($n = 191$) and 99% of Bing ($n = 198$) images were historical representations. These search results highlight the extent to which media consumers are inundated with a narrow set of historical images of Native Americans.

Finally, as noted above, the absence or misrepresentation of contemporary Native American media representations is amplified when Americans have no direct, daily contact with Native Americans. According to the U.S. Census Bureau, only 14 U.S. states have Native populations greater than 100,000 and nearly one fourth of Native Americans live on reservations. Hence, the likelihood Americans would have direct in-person contact with Native Americans that could counter the misrepresentation or invisibility of Native Americans in the media is quite small.

PSYCHOLOGICAL CONSEQUENCES OF NATIVE AMERICAN INVISIBILITY

In this section, we will outline the psychological implications of invisibility in the media on how both Native Americans and non-Natives understand what it means to be Native American in contemporary society. Specifically, we will discuss how invisibility contributes to the homogenization of identity, development of identity prototypes, and deindividuation and self-stereotyping among contemporary Native Americans.

Homogenization of Native American identities

In mainstream media, the limited and narrow depictions homogenize Native American identities. The result being that it stifles self-understanding such that it limits perceptions of how Native Americans should appear and behave. The homogenization of Native American identities inhibits the ability of Native Americans to see their group or to imagine themselves as anything other than the limited media portrayals. Moreover, in the absence of direct, in-person contact, the homogenizing of Native American identities creates a reference point around which Native Americans must orient themselves as they negotiate their identities.

Fryberg et al. (2008), for example, tested how homogeneous media portrayals of Native Americans impact self-understanding and perceptions of potential. Using the most common media portrayals, Native American students were exposed to either the Cleveland Indian mascot, Disney's Pocahontas, negative stereotypes (dropout rates, rates of alcohol abuse, and depression rates), or no media image (control), and then answered questions about self-understanding (e.g., self-esteem, community worth) or potential (e.g., achievement-related possible selves). Compared to the control group, exposure to prominent media portrayals led Native American high school and college students to have more negative feelings about their self (i.e., decreased self-esteem) and community (i.e., decreased community worth), and depressed academic future possibilities (i.e., diminished achievement- related possible selves).

By creating a homogeneous identity reference group, media portrayals of Native Americans also constrain self-understanding for Native Americans; inhibiting the opportunity to explore a variety of atypical identities. The limited representations convey to Native Americans that they do not belong and cannot be successful in atypical domains. Interestingly, these representations need not be negative to have pernicious effects on self-understanding. Homogeneous positive stereotypes, such as "Asians are good at math and science," also undermine self-understanding and performance because they also deny group members a variety of atypical identities.

Prototypes of Native American identities

Invisibility also contributes to the development and perpetuation of prototypes or socially agreed upon "best examples" of what it means to be "Native American" by non-Native Americans in contemporary society. In the absence of direct in-person contact or other pertinent sources of information, media representations emerge as prototypes that establish the quality and quantity of characteristics people associate with different groups and, thus, influence the psychological resources afforded to individual group members.

When individuals are seen as prototypical of groups with more privileged media profiles—greater quantity and more favorable quality—they are afforded status, esteem, and identity benefits. Conversely, when individuals identify with groups who are not afforded privileged media profiles (i.e., lesser quality and/or unfavorable quality), they are (1) viewed as prototypical of their group and thus associated with the less privileged media profile or (2) viewed as non-prototypical of their group and thus not recognized as members of their group. Being prototypical or not is related to access to various psychological resources.

Overall, being prototypical of one's group is associated with higher in-group status, suggesting additional incentives to be seen as prototypical. In fact, nonprototypical group members are not only viewed as lower status, but they also experience greater degrees of invisibility, greater insecurity about fitting in with their group, and less positive feelings about their group. Of course, when the group prototype is negative, being seen as highly prototypical can also have harmful effects. Eberhardt et al. (2006), for example, demonstrated that when African

American criminal defendants were seen as more prototypical of their group (i.e., appeared more stereotypically African American), they received harsher sentences. The issue is not simply that the prototype influences equitable sentencing, but that these prototypes are so tacit and invisible that they influence individuals who may otherwise believe that they hold egalitarian values.

The effects are not limited to the criminal justice system. Prototypicality in education, for example, impacts who is and who is not identified with academic representations. Take the prototype for a "good student." When people think about this prototype, African Americans, Latino Americans, and Native Americans are rarely included. As a result, it is difficult for individual members of these groups to identify with the prototype and thus to reap the benefits of enhanced feelings of self-worth and belonging that identification with this prototype provides.

In fact, for groups who are not typically associated with desired prototypes, research reveals that actively creating associations (e.g., creating in-group "good student" representation yields psychological benefits. In education, for example, enhancing the self-relevance of the "good student" prototype has been shown to alleviate performance decrements for underrepresented students. Moreover, when groups who experience stereotypes about their academic abilities (e.g., women in math, Black students and intelligence) think about self-relevant role models who demonstrate competence and success, the performance-inhibiting effects of negative stereotypes are diminished. Similarly, reading about or identifying self-relevant role models increases school motivation and belonging.

One limitation to fostering these self-relevant associations in the media is that high achieving nonprototypical individuals (e.g., John Herrington, the first Native American astronaut, Charles Curtis, the first Native American Vice President of the United States, or Wilma Mankiller, the first woman Chief of the Cherokee Nation) may be seen as an "exception to the rule." Research on "superstar role models" reveals that although exposure to such notable figures offers momentary cognitive and emotional benefits, it yields no long-term motivational effects. In other words, these high achieving individuals are so extraordinary that their success is also seen as atypical (i.e., nonprototypical) of Native Americans and thus unattainable for Native American students. This is not to say that such superstar role models are bad, but that they are not enough. Realistic, attainable, and plentiful positive role models are needed to yield lasting changes in self-understanding and potential.

Deindividuation and Self-Stereotyping

Alongside prompting homogenous and prototypical representations of groups, media invisibility also influences how individuals contend with and are impacted by these representations—specifically deindividuation and self-stereotyping. Deindividuation refers to the point at which an individual sees oneself as interchangeable with other members of the group. For individuals who are highly identified with their social group, deindividuation is a source of enhanced self-esteem and belonging. These psychological consequences are contingent on two assumptions:

(1) that group members choose to belong or identify with the group and (2) that the group is valued in other's eyes. In other words, when individuals deindividuate with their group, they are perceived by other individuals as having selected from a variety of positive options. For some groups, however, these assumptions are incorrect. Their homogeneous, prototypic media representations are limited (or invisible) and/or largely negative, such that the choice in terms of self-understanding and identity is much more constrained.

Assibey-Mensah (1997), for example, demonstrated that the limited, yet widely held stereotypic and negative representations of African Americans influence how African American youth identify with various role models. That is, African American youth do not identify with positive African American academic role models, but instead disproportionately identify with the publicly available stereotypic role models (e.g., athletes and film stars). The quandary for African American youth is that the choice to deindividuate to their group representation may yield immediate psychological benefits in terms of self-esteem and belonging, but also constrains individual potential by rendering invisible more viable personal and professional pathways.

The act of "choosing" to deindividuate with these group representations is referred to as self-stereotyping. What undergirds this process is a belief that if one wants to be regarded as a member of their group then they must identify, or self-stereotype, with the associated representations. For example, identification with negative representations of Native Americans is one way to justify and cope with failure (self-handicapping; "doing what everyone expects of me"). What self-stereotyping demonstrates is that members of underrepresented groups may be motivated to identify with any available representation simply because one representation is better than no representation (i.e., absolute invisibility). The one representation, no matter how unfavorable or inaccurate, provides answers to the "Who am I?" questions that people are motivated to answer and provides a reference point around which to negotiate one's identity with others.

CONCLUSION

Media invisibility has notable consequences for identity and self-understanding. By promoting limited, homogeneous prototypes of Native Americans, the media inhibits the development of characteristics or abilities beyond those supported by these Native American prototypes and inadvertently promotes maladaptive self-strategies (e.g., deindividuation and self-stereotyping) that undermine individual potential. Media invisibility also highlights the fact that the representation and identification process is not necessarily a conscious and agentic endeavor. Native Americans, like most Americans, make the choice to consume mass media, but the psychological consequences and, in particular, the consequences of being depicted by narrow and limited representations are rarely publicly shared or discussed. Moreover, Native Americans have little control over how or when their group is portrayed in the media. In other words, they are not making an active choice to be represented in these negative and limiting ways and although they can contest or reject these representations, they cannot control the impact of these representations on how other people think about Native Americans. By highlighting the impact of absolute and relative invisibility, this article brings to the foreground the process by which Native Americans and non-Natives develop expectations about how Native Americans should look and behave and how these expectations influence self-understanding, identity development, and inter group relations.

... Given the pervasiveness of media in everyday life and the impact that media representations or a lack of representation can have on psychological well-being, the ideas presented here have widespread implications for public policy in general and policy makers in particular. First, media outlets have tremendous potential to either harm (by way of fostering negative stereotypes) or to help (by way of fostering new identities and new future possibilities) Native Americans. Given the inherent inequality in how different groups are represented, it may behoove policy-makers to create policies that require media outlets to attend to how and when they represent diverse groups. Moreover, groups that are underrepresented, such as Native Americans, often have little direct influence on the way they are portrayed in mass media. Policy-makers can play a critical role by ensuring that Native Americans are represented and included in decisions that represent and affect their communities. Take the issue of Native American mascots as an example. There is evidence that these mascots harm Native American students and influence intergroup relationships in college settings. Policy-makers should advocate for school environments that are free from limiting and negative representations that influence the future potential of Native American students.

Creating widespread and large-scale change in the way society portrays and thinks about Native Americans is no easy task. Such an endeavor will require the cooperation of many people in many different domains (e.g., media, education) of society. Fortunately, there is a precedent for this work. Policy-makers and educators have long been involved in changing the way minority groups are represented. Continued support from policy-makers and educators can and will go a long way in helping historically underrepresented and underserved groups, like Native Americans, be seen and understood as the unique, diverse, and contemporary people that they are.

REFERENCES

Assibey-Mensah. G. O. (1997). Role models and youth development: Evidence and lessons from the perceptions of African-American male youth. *The Western Journal of Black Studies, 21*(4), 242–252.

DeVoe, J. F., & Darling-Churchill, K. E. (2008). *Status and trends in the education of American Indians and Alaska Natives: 2008* (NCES 2008–084). Washington, DC: National Center for Education Statistics, Institute of Education Sciences, U.S. Department of Education.

Eberhardt, J. L., Davies, P. G., Purdie-Vaughns, V. J., & Johnson, S. L. (2006). Looking deathworthy: Perceived stereotypicality of black defendants predicts capital-sentencing outcomes. *Psychological Science, 17*(5), 383–386. doi: 10.1111/j.1467–9280.2006.01716.x.

Fryberg, S. A., & Townsend, S. M. (2008). The psychology of invisibility. In G. Adams, M. Biernat, N. R. Branscombe, C. S. Crandall & L. S. Wrightsman (Eds.), *Commemorating Brown: The social psychology of racism and discrimination.* (pp. 173–193). Washington, DC, US: American Psychological Association.

Mastro, D. (2009). Racial/ethnic stereotyping and the media. In R. Nabi & M. B. Oliver (Eds.), *The Sage handbook of mass media effects* (pp. 377–391). Thousand Oaks, CA: Sage Publications.

49

Immigrant Rights are Civil Rights

HANA BROWN AND JENNIFER A. JONES

It was a brilliantly sunny morning in Jackson when a group of activists amassed at the offices of the Mississippi Immigrant Rights Alliance (MIRA). They were there for the organization's annual Civic Engagement Day, and the hallways were lined with coolers of drinks, pots of beans, and stacks of tamales—fuel to sustain the activists during the afternoon of lobbying and marching that lay ahead. It was January 2015, and the state legislature was poised to begin debate on 11 bills designed to restrict the rights of undocumented immigrants.

MIRA's offices were packed with Latino immigrants, mostly of Mexican origin, who had raised money for a coach bus to travel the three hours from the Gulf Coast to Jackson for the events. State police regularly pull people over for "driving while brown" so traveling the interstate in cars would have been risky for the undocumented in the group. Arranging and paying for a chartered bus took tremendous commitment and coordination from the activists. But their resolve was strong and their goals were clear: convince legislators to defeat the eleven anti-immigrant bills and begin the push for in-state tuition benefits for undocumented students.

Before the group began its walk to the capitol, Bill Chandler, MIRA's director, introduced many of the faces in the room and then quickly got down to business. A veteran labor and civil rights organizer, Chandler has spent the last ten-plus years organizing rallies, filing lawsuits, and lobbying Mississippi's political leaders to support immigrant rights. On this day, he handed out worksheets that explained how a bill gets processed in the state legislature. Speaking of MIRA's challenges and triumphs, he energized the group for the upcoming march, a press conference on the capitol steps, and an afternoon of lobbying sessions with legislators. Shortly thereafter, the crowd headed out the door.

When the activists donned their red, white, and blue caps and gowns and grabbed their placards and buttons, their chances of success seemed slim. They lobbied and marched against the backdrop of Mississippi's grizzly history of slavery, sharecropping, and civil rights repression. The recent and rapid growth of Mississippi's Latino immigrant population and the backlash against that growth are but the latest chapters in the state's long and tumultuous history of racial inequality and political conservatism. But when the legislative session came to a close, two months after MIRA's coordinated lobbying and advocacy efforts, all

SOURCE: Brown, H. Jones, J.A. 2016. "Immigrant Rights are Civil Rights." *Contexts*, 15(2), 34–39. Copyright © 2016 by the American Sociological Association. Reprinted by permission of SAGE Publications, Inc.

of the so-called "bad bills" were dead. Not a single anti-immigrant bill even reached the floor for debate. It was a significant victory, but it was hardly the first of its kind for MIRA. Since 2005, and during a time when anti-immigrant measures have spread like wildfire in other states, MIRA and its allies have spearheaded the defeat of 282 anti-immigrant bills in Mississippi. Only one such bill has passed in the state in over a decade. No funding was ever appropriated for the measure, and it has never been enforced.

When we first arrived in Mississippi, we wanted to understand the array of policy approaches that "new immigrant destination" states have taken to their changing demographics. Some states, like Alabama and South Carolina, have captured national attention with their sweeping anti-immigrant measures, but the less told tale is of the places that, despite conservative racial and political histories, are witnessing increasingly successful resistance to this punitive turn. To understand these patterns, we combed through archival materials, marched with activists, and interviewed key stakeholders involved in immigration politics in Mississippi. We found that, even more surprising than the state's rejected anti-immigrant bills, was the coalition behind these defeats. MIRA counts among their friends and allies not only the Black Legislative Caucus, but also every major civil rights and labor organization in the state. Over the course of our research, informed by our previous work on race and public policy in North Carolina and Georgia, we came to understand that this Black-Latino coalition was not only influencing the sphere of immigration politics in Mississippi, it was part of a strategic political vision slowly being realized across the South.

NATURAL ENEMIES?

The powerful Black-Latino coalition that has emerged in Mississippi flies in the face of what social scientists generally assume about Black-Latino relations. Researchers and the media have long contended that the two groups have a contentious and adversarial relationship. Political scientists Paula McClain and Claudine Gay, for example, claim that Blacks and Latinos are in near constant competition for resources. This competition supposedly breeds contempt between the groups and leads Blacks and Latinos to adopt stereotypes and negative attitudes toward each other. The result, scholars claim, is that Black-Latino political coalitions are unlikely because the groups' political interests are fundamentally incompatible. So widespread is the presumption of Black-Latino conflict that media outlets across the country—from Los Angeles to Miami, from Charlotte to Dallas—have published stories about ongoing tensions brewing between the groups. Stories of coalitions and cooperation are few and far between.

Armed with this knowledge, we were fascinated, in the early stages of our research, to see evidence of a seemingly cohesive and politically influential Black-Brown alliance in Mississippi. As we combed through more than a decade's worth of news articles from Mississippi's largest newspapers, it became increasingly clear how central this alliance was to the state's political scene.

Mississippi's pro-immigrant movement, we learned, is run by a close-knit team of immigrant rights workers, civil rights movement veterans, and labor leaders, most of whom are Black or Latino. The staunchest allies of Latino immigrants in the legislature are members of the Black Legislative Caucus. So strong is their support that immigration is widely known as a "Black issue" in the state.

We saw this cooperation in action when we accompanied MIRA on its march and rally that January day in 2015. Latino immigrant activists marched with Black representatives from the Southern Christian Leadership Conference, the ACLU, and local unions, and these Black leaders joined Latino youth in chants of "Si Se Puede!" and echoed with cries of "Yes We Can!"

The deeper we dug into the Mississippi story, the more interested we became in this alliance and its effects on state politics. We learned that the alliance was not accidental, but strategic—and years in the making. MIRA, journalist David Bacon writes, has built its well-oiled advocacy machine around the belief that "Blacks plus immigrants plus unions equals power." They have worked diligently over the last decade to build a broad interracial coalition strong enough to resist the force of the state's violent racial past and the racial conservatism of the present.

IMMIGRANT RIGHTS ARE CIVIL RIGHTS

As its name suggests, MIRA's central focus is on immigration issues in Mississippi. But one look at the organization's archives makes clear the decades of organizing and activism that have given shape to the organization. A room with binder-lined walls tells the story of how the contemporary Mississippi immigrant rights movement grew out of earlier struggles for social justice in the 1960s and 1970s. Virtually all of the major players in MIRA are veteran activists and religious leaders who understand, much like social movement scholars do, the complex, strategic, and painstaking work of creating social change.

The biographies of MIRA's leaders make this clear. Take Bill Chandler, a White Californian who looks a bit like Kenny Rogers. Chandler moved permanently to Mississippi in 1989 as a labor organizer, following long stints in Texas and Detroit where he registered voters and organized nurses and other skilled workers. Earlier in his career, he had organized Latino grape workers in California with such luminaries as Dolores Huerta and Cesar Chavez, before moving eastward to work on Charles Evers' campaign for Mississippi governor. By the time Chandler returned to Mississippi, his organizing work had instilled in him a belief that the future of progressive politics depended on a strong Black-Latino coalition. This was particularly true in Mississippi, where African Americans accounted for over one-third of the population (more than any other state) and the Latino population was beginning to grow.

This knowledge weighed in Chandler's mind when casinos began booming on Mississippi's Gulf Coast in the 1990s. The gaming industry aggressively recruited undocumented Latino workers who began arriving in the state in numbers for the first time. Later, the casinos contracted out their hiring to recruiters who, in turn, abused the workers. Rather than go underground, several of the Latino casino workers began to organize. They presented a number of demands regarding working conditions and wages. Management met with them, agreed to the demands—and, the next day, had the immigrant workers arrested and put into deportation proceedings.

The veteran labor organizer, Chandler, went to the coast immediately to support the largely Black union leadership responding to the incident. Just as unions in the state began to embrace immigration as an issue, rumors were spreading that Latino immigrants were being denied enrollment in public schools—a clear violation of federal law. Given the state's legacy of Jim Crow segregation and racially unequal schooling, many Black leaders saw these practices as a direct affront to their decades-long civil rights agenda. Chandler and a small group of religious leaders seized the opportunity to recruit SCLC leader and ALF-CIO organizer, Jim Evans, to the fight for immigrant rights. Soon after, they also recruited former Civil Rights leader (and soon to be Jackson city Mayor) Chokwe Lumumba to the cause. Chandler had marched in Detroit with Lumumba, who had recently returned to Mississippi to serve as a public defender. Together, the group built an organization that, from its inception, was intentionally composed of civil rights, labor, faith-based, and community organizations from around the state, all working alongside and on behalf of Latino immigrants. Chandler captained the ship, but it was the state's Black leadership who embraced immigration as the new civil rights issue of their time.

Doug McAdam and other movements scholars have shown that participants in the Civil Rights movement often moved on to participate in other social justice campaigns, leveraging their activist networks and organizing experience to build new movements. The ability of Lumbumba, Evans, and Chandler to erect MIRA and build a Black-Latino alliance rested on exactly these kinds of ties from movements past. When MIRA's leaders approached the state's Black legislative caucus for support on immigration issues, they knew how to frame immigration in a way that would resonate with the group's commitment to racial justice. "When we took migration issues to the Black caucus," Chandler recounted, "they jumped on board. It didn't take much for them to adjust their thinking and see that [immigrant rights and civil rights] were the same issues." Unanimous supporters of immigrant rights, Black caucus members have leveraged their seniority and control over key legislative committees to prevent most anti-immigrant bills from ever reaching the floor for a vote.

By 2005, MIRA had established its bona fides, marshalling enough support in the legislature to kill several anti-immigrant bills and launching a media campaign designed to spread the message that Blacks and Latinos were not enemies, but allies. When Hurricane Katrina ravaged the Mississippi Gulf Coast in August of that year, MIRA took on new relevance. After the storm, an estimated 350,000 Latino construction workers, mostly undocumented, arrived in Mississippi. They labored in hazardous conditions to clean up the area and rebuild devastated properties, only to have their wages stolen by ruthless contractors. At the same time, many emergency shelters refused to admit established

Latino residents who had lost their homes and often forcibly evicted those who managed to gain entry. MIRA and its team of largely Black and Latino legal advisers soon became the main lifeline for Latino immigrants in the region, forcing shelters to change their practices and, more impressively, winning over one million dollars in back wages for immigrant workers cheated by their employers.

In the coming years, anti-immigrant policy efforts swelled in states across the nation and federal immigration enforcement escalated. In 2008, federal immigration officials undertook what remains the largest immigration raid in U.S. history. Descending up on a Howard Industries plant in Laurel, MS, they arrested and deported over 600 undocumented Latino workers. The raid left a scar on the community, particularly the African American employees of the plant, many of whom watched their coworkers and fellow union members get taken away from their jobs and families. In the coming weeks and months they worked with MIRA to find the families of those arrested and deported and to raise funds to support them.

MIRA not only fostered social ties between Blacks and Latinos on the ground, but they also doubled down on their legislative work. As they did in so many other states, national organizations like the Federation for American Immigration Reform (FAIR) and the American Legislative Exchange Council (ALEC) had partnered with state Tea Party representatives to craft a spate of anti-immigrant bills. In 2009 alone, thanks largely to MIRA and the Black Caucus, 77 anti-immigrant bills died in committee. By 2015, the group had killed nearly 300 bills, leaving Mississippi the only state in the Deep South without any enforceable anti-immigrant bills.

These efforts were facilitated by MIRA's continued bootson-the-ground work of building a strong Black-Latino alliance. Over the last decade MIRA has regularly held joint press conferences with the Black caucus and engaged in lobbying days and protests alongside Black legislators. Nearly every year, Latino immigrants and Black leaders hold unity conferences, designed to build ties between the groups. Equally important, immigrant rights activists now serve as allies to labor unions and civil rights organizers, joining workers on the picket lines, calling on Latinos to support investment in Black communities, and participating in affiliate events.

MIRA's key players also established strong ties with local media outlets, giving them a key forum to advance the idea that Blacks and Latinos share a common experience of discrimination and a common goal for change. When we analyzed the language Mississippi newspapers used to discuss Black-Brown relations, the effects of MIRA's strategic meaning work were clear. Up until 2006, news media in the state characterized Black-Brown relations as rife with conflict and indifference. Yet by 2009, nearly 100% of stories about Black-Brown relations in Mississippi discussed similarity and cooperation rather than conflict, a trend that persists today. The sense of linked fate is so prominent in the political arena that Mississippi NAACP president Derrick Johnson has publicly announced, "Any legislation that discriminates against Latinos, discriminates against African Americans." In Mississippi, Black-Latino alliances have produced not only legislative successes, but have changed the way many in the state think and talk about racial politics.

A NEW POLITICS FOR A NEW SOUTH

MIRA's leaders attribute their success to coalition-style politics and its rootedness in the social justice battles of earlier eras. "I think what helps us," Chandler explains, "is that we have relationships we've built up over the years, we have people coming from different struggles. That is the power of the South I think. And we saw right away the potential for significant political change."

He is not alone in this assessment. Few states managed to build an influential Black-Latino coalition as early as MIRA, but alliances are emerging now across the region. After the passage of sweeping anti-immigrant bills in South Carolina, Alabama, and Georgia in recent years, local organizers have galvanized efforts to build precisely the kind of Black-Brown networks that have fueled MIRA's success. In fact, immigration advocates in many states have looked directly to MIRA and Mississippi to replicate their efforts. In Alabama, the Hispanic Interest Coalition of Alabama (HICA) has recruited African American leaders to its board, taking MIRA's leadership on as advisers in the process. A newer coalition, the Alabama Coalition for Immigrant Justice (ACIJ), includes Black-Brown unity as one of its core tenets and is a lifetime member of the Birmingham NAACP chapter.

MIRA was founded on the principle that "all immigration laws from the beginning are about two things: racism and managing labor." In treating these issues as interwoven, the coalition has established a political campaign and style premised on the idea that Black and Latino issues are inextricably linked, that there are "power connections" between racism against African American and discrimination against immigrants. As the group's leaders are well aware, the implications of this alliance are vast. Mississippi's Black-Latino coalition and the emergent alliances in other parts of the South aim not only to protect the rights of immigrants but to spark a new "power politics" that spans beyond immigration to other causes. Rather than shore up racial divides, these coalitions have their sights set on forging a new progressive political alliance that can reshape the region's political landscape for years to come.

50

Intersectional Mobilization, Social Movement Spillover, and Queer Youth Leadership in the Immigrant Rights Movement

VERONICA TERRIQUEZ

Undocuqueer [undocumented and queer] leaders have been very visible in the immigrant rights movement in recent years. They have been at the forefront of many of the major protests and actions for the DREAM Act. I believe they have played a critical role in making sure that the public becomes aware of our situation and how we are fighting for our rights in this country.

—Jaime, 22-year-old, straight-identified, undocumented youth activist from Los Angeles

Jaime's claims above echo those of other participants in and observers of the youth-led arm of the contemporary immigrant rights movement—that is, queer youth activists were highly visible in the movement in the early 2010s. At the time, these undocumented youth activists, both queer and straight, were often referred to as the DREAMers because they had been leading the charge to gain congressional support for the federal Development, Relief, and Education for Alien Minors (DREAM) Act. The aim of this proposed legislation was to provide a pathway to citizenship for eligible undocumented youth.

SOURCE: Veronica Terriquez. 2015. "Intersectional Mobilization, Social Movement Spillover, and Queer Youth Leadership in the Immigrant Rights Movement." *Social Problems*, 62: 343–362. Reprinted by permission from Oxford University Press.

Borrowing from the gay and lesbian movement's "coming out of the closet" narrative, the DREAMers organized "Coming Out of the Shadows" campaigns in 2010 through 2012. Activists publicly declared their undocumented status in order to combat the stigma associated with their precarious legal situation and humanize their experiences in the eyes of broader audiences. Among some activists, these declarations of a legal status were accompanied by the disclosure of a queer (here meaning lesbian, gay, bisexual, transgender, or other non-heterosexual) identity in online venues, public demonstrations, and mainstream media.

LGBTQ visibility in the immigrant youth movement presents an interesting puzzle for at least a few reasons. First, LGBTQ people of color and immigrants encounter a suite of challenges to publicly disclosing a queer identity. Second, this movement's primary focus was on a pathway to citizenship for the undocumented, not on mainstream LGBTQ issues; as such, an LGBTQ identity was not central to advancing the movement's goals. And third, while queer individuals historically have been active in movements focused on the rights of immigrants and people of color, their LGBTQ identities have often remained invisible. These prior movements often made salient a single and unifying identity, and did not attend to how multiple and overlapping "intersectional" identities shaped activists' experiences.

How, then, can movements mobilize their members based on their intersectional identities? How can they generate high levels of participation among multiply marginalized subgroups within their broader constituencies? To address these questions, I examine the identity formation processes within the DREAM movement in California, the state with the largest undocumented immigrant youth population in the United States. My empirical analysis relies on data from 410 web surveys and 50 semi-structured, in-person interviews collected from Latino and Asian-Pacific Islander activists. My findings suggest that queer-identified youth comprised a significant proportion of movement participants and that they were more civically engaged than their straight-identified peers.

I argue that LGBTQ representation and visibility can, in large part, be attributed to identity formation processes within the DREAM movement, rather than simply the self-selection of LGBTQ-identified individuals into this movement. I assert that the recognition and activation of multiply marginalized identities at various levels of identity formation—at the broader movement, organizational, and individual levels—catalyzed *intersectional mobilization*, that is, high levels of activism and commitment among movement participants who represent a disadvantaged subgroup within a broader marginalized constituency. I also demonstrate how the "social movement spillover" (Meyer and Whittier 1994) of the "coming out" narrative from the LGBTQ to the DREAM movement not only empowered immigrant youth to publicly disclose their legal status, it also enabled them to declare a stigmatized sexual identity. As such, I argue that social movement spillover can produce a *boomerang effect*. That is, it can reinforce the aims of the movement from which the strategy originated. This study contributes to theorizations of how social movement identity processes foster active participation among multiply marginalized groups....

INTERSECTIONALITY AND BARRIERS TO THE PUBLIC DISPLAY OF AN LGBTQ IDENTITY

The intersectionality literature offers a useful starting point for understanding LGBTQ activists' prominence in the DREAM movement. Intersectionality theory indicates that each individual has multiple identifiers that are linked to broader structures of inequality (Collins 1986, 2000; Crenshaw 1989). Determining an individual's social location, these identities include a person's race, gender, class, sexual orientation, and legal status. Varying in relevance and salience across social contexts (Hancock 2011; Townsend-Bell 2011), these identities shape how individuals experience unequal power relations and make sense of their surroundings.

An intersectional theoretical framework suggests that non-White youth from low-income and undocumented backgrounds might experience a number of challenges to openly disclosing a queer identity. For example, Latino and Asian immigrant youth may lack a strong identification with a larger white LGBTQ population that can sometimes hold racist attitudes toward people of color. Family ties, traditional gender roles, conservative religious values, and widespread homophobia can prevent LGBTQ individuals, including immigrants, from openly disclosing their sexual orientation. Financial dependence may prompt some LGBTQ immigrants to keep their sexual orientations "tacit" (understood, but not discussed) because they do not want to risk losing their family's financial support.

Moreover, being undocumented potentially can add to the risks of coming out as a sexual minority. Undocumented immigrants often live "in the shadows," hiding the fact that they lack authorization to live in the United States. They face the risk of deportation, must cope with the stigma of being labeled "illegal," and frequently endure significant economic and social hardships related to their legal status. Because they rely heavily on family networks for support, queer undocumented immigrants may face legal consequences and further social marginalization if their families and communities reject them because of their sexual orientation. This research offers empirical evidence that undocumented youths' experiences with multiple layers of marginalization pose barriers to coming out of the closet....

Semi-Structured Interview Findings

... I use semi-structured interview data to assess the role of immigrant families and communities in shaping the disclosure of an LGBTQ identity among undocumented youth activists. I then examine how collective identity formation processes facilitated the intersectional mobilization of LGBTQ activists within the DREAM movement.

Barriers to the Public Display of an LGBTQ Identity among the Undocumented

Interview data rule out the possibility that queer representation in the DREAM movement can be explained by immigrant family and community environments that are supportive of the public disclosure of an LGBTQ identity. Rather, the evidence indicates that activists often encountered obstacles to coming out of the closet within the contexts of their families or broader immigrant communities. For most queer DREAMers in this study, coming out as LGBTQ was a painful experience, and many still struggled with being open about their sexual orientation. In fact, even though almost all queer activists reported that their fellow straight peers knew about their sexual orientation, two-fifths had still not come out of the closet within the context of their families or larger communities at the time they were interviewed for this study. As such, the experiences of immigrant youth activists reflect those of other LGBTQ individuals who strategically disclose their sexual orientation in some social contexts, but not others.

With a handful of exceptions, most queer activists claimed that homophobia within their families or broader immigrant communities presented a barrier to coming out of the closet. The story of Alberto (age 26) provides an illustration of the homophobia some confronted. Born in Guadalajara, Mexico, Alberto grew up in a socially conservative family. Alberto recalled an experience as a child growing up in Southern California that taught him early on that being gay was not acceptable.

> There was a boy in elementary school who was really, really feminine. I mean, he took it to the ballpark—you know, like home run femininity. He was fabulous. But I remember my mom saying, "Oh, my God, I would be devastated if I was his mother." And that really hit me. It stuck with me.

Like several other youth, Alberto also noted that his parents' religious beliefs made coming out especially challenging. He explained:

> My dad belonged to one of the most conservative organizations within the Catholic Church, and he was very vocal about his distaste for certain things, including homosexuality. So I thought it was wrong to be gay because I grew up in that environment.

Consequently, for Alberto coming to terms with his sexual orientation was a very painful process that created a lot of family conflict, especially since his parents actively sought a "cure" for his gayness.

In addition to risking the disapproval of their families, several interviewees worried about potential negative material consequences that might result if their families learned about their sexual orientation. Most DREAMers, regardless of their sexual orientation, relied heavily on family and extended networks for their economic survival and other resources. Loss of family financial support was a real concern for some LGBTQ activists, as 24-year-old Samir asserted:

> We can't [legally] have jobs, so we already have these financial issues. Once you come out to your family—and if they don't respond very well—then there is that chance of losing your bed, a place to sleep. There's a lot more you can lose because you can't really take care of yourself financially when you're undocumented.

Indeed, a few queer undocumented youth encountered significant financial vulnerabilities because their families were unaccepting. For example, after being kicked out of his family's household for being gay, one young man resorted to living in his car; lacking regular employment, some days he went hungry. Another young woman's family ceased helping her pay for college after she came out as a lesbian. She was ineligible for government financial aid at the time and was forced to withdraw from school when she ran out of funds.

Coming out to unsupportive family members could also have legal consequences. This was the case for 24-year-old Nacho who first met his estranged father as a teenager. Initially impressed by his academic achievements, his father, a U.S. citizen, introduced Nacho to his extended family and promised to fix his papers. "They were all super supportive of me," Nacho recounted. "They were going to help me pay for college. They were going to help me get my legal status." This changed, however, when he came out as gay. "They pretty much cut me off, and it felt horrible … It was really heartbreaking. It hurt so much." His family's homophobia not only took an emotional toll on Nacho, it also cost him a potential opportunity to become a legal permanent resident.

In addition to noting family related barriers to coming out, a few queer DREAMers also expressed concerns about disclosing their identities in the context of their community activism. The immigrant rights movement as a whole seeks to gain support from mainstream and religious audiences not always accepting of LGBTQ individuals. For example, Zaira, a 24-year-old activist, explained:

> I never say that I'm queer. I deal with a lot of religious people, whether it's in the immigrant rights movement or out in the community. If I come out as queer to them, I think they'll judge me before they see the work that I'm doing. Coming out as queer might be a hindrance to my work.

Zaira believed that the broader Latino immigrant community and other potential allied communities would not be as receptive to her political work if they learned she lived with a female partner. Like other activists, Zaira reported "covering" or minimizing her queer identity in what she perceived to be a homophobic broader community.

While most queer DREAMers encountered challenges to being out within their families and communities, there were several who openly shared their queer identities in almost all social settings. This minority had gained the acceptance of their families, and this familial support appeared to afford these young people a greater ability to disclose their queer identities in a wide range of social settings.

THE INTERSECTIONAL MOBILIZATION OF UNDOCUQUEERS WITHIN THE DREAM MOVEMENT

Movement-Level Collective Identity Formation and the Spillover of the Coming Out Strategy

Evidence suggests that the DREAM movement's Coming Out of the Shadows identity strategy—initiated in 2010 by LGBTQ undocumented activists in

Chicago—played an important role across the movement nationally in helping some undocumented LGBTQ activists to come out of the closet and become politicized around both their legal status and sexual orientation. Unlike in the fat acceptance movement, in which several activists first came out as queer and then came out as fat (Saguy and Ward 2011), many DREAMers' experiences of coming out of the shadows as undocumented inspired them also to come out as LGBTQ. In other words, *out* LGBTQ individuals did not disproportionately self-select into the movement, as many struggled with publicly declaring their sexual orientation within their immigrant families or communities. Rather, the case of LGBTQ DREAMers demonstrates how social movement spillover produced a boomerang effect whereby the coming out strategy migrated from the gay rights movement to the immigrant rights movement, where, in turn, its effectiveness emboldened immigrant activists to support LGBTQ inclusivity. This boomerang effect fueled the intersectional mobilization of undocuqueers.

Notably, DREAMers' Coming Out of the Shadows national strategy explicitly acknowledged the intersectional identities of activists at the outset. For example, the online "Coming out of the Shadows—A How to Guide," posted by the national organization Dream Activist, states: "In the same way the LGBTQ community has historically come out, undocumented youth, some of whom are also part of the LGBTQ community, have decided to speak openly about their status" (see Dream Activist 2010). In line with this pronouncement, activists across the country used social media to publicly counter the double stigma associated with their undocumented and LGBTQ identities (Zimmerman 2012).

Mateo's experience serves as an example of the impact of the DREAM movement's Coming Out of the Shadows campaign on queer identity development and disclosure among activists. As a college freshman, Mateo became an active member of his campus organization, but he initially kept his LGBTQ identity a secret. As Mateo attested, this changed as a result of the movement's Coming Out of the Shadows campaign:

> It wasn't always okay to say that you were undocumented, but I guess when I first started college it was taboo, completely. I think now it's okay to say, "I'm undocumented ..." For me, I first got comfortable saying that I'm undocumented before I started saying "I'm queer" or "I'm gay" or "I'm LGBTQ." I guess once you go through saying you're undocumented, it's easy to accept a queer identity and be able to talk about it with other people. Maybe it's because you already have this sense of self-empowerment based on your undocumented status, that you can use that for your queer identity.

The evidence therefore suggests that the spillover of the coming out identity strategy at the movement level contributed to intersectional mobilization by facilitating the widespread public declarations of two marginalized identities— undocumented and queer. Spillover, in this case, was facilitated by overlapping memberships between the LGBTQ and immigrant rights movements. However, other activists became inspired to disclose a queer identity in connection to coming out of the shadows. Likely contributing to the significant representation of out LGBTQ members among the ranks of the DREAMers (as suggested by survey findings), this case of spillover resulted in a boomerang effect through the destigmatization of LGBTQ identities among immigrant youth and perhaps the broader publics they sought to reach. This boomerang effect also contributed to the LGBTQ inclusivity of DREAM organizations....

DISCUSSION

... Notably, the DREAM movement emerged in an era of increasing LGBTQ activism and social acceptance, which likely facilitated the participation of queer members. After all, the historical context in which an organization is established has an enduring impact on the cultural assumptions that shape members' norms and practices. The youthfulness of the DREAM movement also likely contributed to its receptivity toward LGBTQ members, as young people tend to be less homophobic than older populations. Yet, neither the recent vintage of DREAM organizations nor the age composition of their members can entirely explain the high levels of visibility of queer activists.

Findings from this study indicate that queer youths' representation and visibility in this movement can, in part, be attributed to their intersectional mobilization or the recognition and activation of undocumented and queer identities in the broader movement, within DREAM organizations, and among individual activists. At the broader movement level, the Coming Out of the Shadows identity strategy empowered some members to disclose publicly not only their legal status, but also their sexual orientation. Data point to the possibility that this case of spillover boosted the number of visible queer activists within this movement. As such, the migration of the coming out narrative produced a boomerang effect by reinforcing LGBTQ rights among immigrant youth. Meanwhile, at the organizational level, multi-identity work consisting of deliberate efforts to combat homophobia contributed to organizational environments in which queer members felt welcome and safe. Finally, queer activists' intersectional consciousness further intensified their own activism as they sought to address LGBTQ issues within the immigrant rights movement and build bridges with the gay rights movement. This study of the DREAM movement shows how undocumented immigrant youth exercise agency in collectively shaping their own social and political incorporation....

REFERENCES

Collins, Patricia H. 1986. "Learning from the Outsider Within: The Sociological Significance of Black Feminist Thought." *Social Problems* 33:14–32.

Collins, Patricia Hill. 2000. *Black Feminist Thought: Knowledge, Consciousness, and the Politics of Empowerment.* New York: Routledge.

Crenshaw, Kimberle Williams. 1989. "Demarginalizing the Intersection of Race and Sex: A Black Feminist Critique of Antidiscrimination Doctrine, Feminist Theory, and Antiracist Politics." *University of Chicago Legal Forum* 139–68.

Dream Activist. 2010. "Coming Out of the Shadows – A How To Guide." Retrieved June 7, 2015 (www.nysylc.org/2010/03/coming-out-of-the-shadows-week-how-to-guide/).

Hancock, Ange-Marie. 2011. *Solidarity Politics for Millennials: A Guide to Ending the Oppression Olympics.* New York: Palgrave Macmillan.

Meyer, David S. and Nancy Whittier. 1994. "Social Movement Spillover." *Social Problems* 41:277–98.

Townsend-Bell, Erica. 2011. "What is Relevance? Defining Intersectional Praxis in Uruguay." *Political Research Quarterly* 64:187–99.

Vargas, Jose Antonio 2012. "Not Legal Not Leaving." *Time*, June 25. Retrieved January 7, 2013 (http://time.com/2987974/jose-vargas-detained-time-cover-story/).

Wang, Dan J. and Sarah A. Soule. 2012. "Social Movement Organizational Collaboration: Networks of Learning and the Diffusion of Protest Tactics, 1960-1965." *American Journal of Sociology* 117:1674–722.

Ward, Jane. 2008. "Diversity Discourse and Multi-Identity Work in Lesbian and Gay Organizations." Pp. 233–55 in *Identity Work in Social Movements*, vol. 30, edited by J. Reger, D. J. Myers, and R. L. Einwohner. Minneapolis, MN: University of Minnesota Press.

Whittier, Nancy. 2012. "The Politics of Coming Out: Visibility and Identity in Activism against Child Sexual Abuse." Pp. 145–69 in *Strategies for Social Change*, vol. 37, edited by G. M. Maney, R. V. Kutz-Flamenbaum, and J. Goodwin. Minneapolis, MN: University of Minnesota Press.

Yoshino, Kenji. 2007. *Covering: The Hidden Assault On Our Civil Rights.* New York: Random House.

Yukich, Grace. 2013. "Constructing the Model Immigrant: Movement Strategy and Immigrant Deservingness in the New Sanctuary Movement." *Social Problems* 60:302–20.

Zimmerman, Arely. 2012. "Documenting Dreams: New Media, Undocumented Youth, and the Immigrant Rights Movement." Case Study Working Paper, Media, Activism, and Participatory Politics Project, University of Southern California, Los Angeles, CA.

51

Movement Intersectionality

The Case of Race, Gender, Disability, and Genetic Technologies

DOROTHY ROBERTS AND SUJATHA JESUDASON

INTRODUCTION

Intersectional analysis does not apply only to the ways identity categories or systems of power intersect in individuals' lives. Nor must an intersectional approach focus solely on differences within or between identity-based groups. It can also be a powerful tool to build more effective alliances between movements to make them more effective at organizing for social change. Using intersectionality for cross movement mobilization reveals that, contrary to criticism for being divisive, attention to intersecting identities has the potential to create solidarity and cohesion. In this article, we elaborate this argument with a case study of the intersection of race, gender, and disability in genetic technologies as well as in organizing to promote a social justice approach to the use of these technologies. We show how organizing based on an intersectional analysis can help forge alliances between reproductive justice, racial justice, women's rights, and disability rights activists to develop strategies to address reproductive genetic technologies. We use the work of Generations Ahead to illuminate how intersectionality applied at the movement-building level can identify genuine common ground, create authentic alliances, and more effectively advocate for shared policy priorities.

Founded in 2008, Generations Ahead is a social justice organization that brings diverse communities together to expand the public debate on genetic technologies and promote policies that protect human rights and affirm a shared humanity. Dorothy Roberts is one of the founding board members of Generations Ahead, and Sujatha Jesudason is the Executive Director.

Since its inception, Generations Ahead has utilized an intersectional analysis approach to its social justice organizing on reproductive genetics. Throughout 2008–2010, the organization conducted a series of meetings among reproductive

SOURCE: Roberts, Dorothy, and Sujatha Jesudason. 2013. "Movement Intersectionality." *Du Bois Review* (Fall): 313–328. Copyright © 2013 W.E.B. Du Bois Institute for African and African American Research. Reprinted with the permission of Cambridge University Press.

justice, women's rights, and disability rights advocates to develop a shared analysis of genetic technologies across movements with the goals of creating common ground and advancing coordinated solutions and strategies. This cross-movement relationship- and analysis-building effort laid the foundation for successfully resisting historical divisions between reproductive rights, racial justice, and disability rights issues in several important campaigns. In examining the ways in which the theory and practice of intersectionality are used here we hope to demonstrate the kinds of new alliances that now become possible—alliances that can be both more inclusive and effective in the long term.

FROM DIFFERENCE TO RADICAL RELATEDNESS

In her classic article, "Demarginalizing the Intersection of Race and Sex," Kimberlé Crenshaw (1989) focused on Black women to show that the "single-axis" framework of discrimination analysis not only ignores the way in which identities intersect in people's lives, but also erases the experiences of some people. As a result, she argued, "[b]lack women are sometimes excluded from feminist theory and antiracist policy discourse because both are predicated on a discrete set of experiences that often does not accurately reflect the interaction of race and gender" (p. 140). The intersectional framework revealed that Black women suffer the combined effects of racism and sexism and therefore have experiences that are different from those of both White women and Black men, experiences which were neglected by dominant antidiscrimination doctrine (Crenshaw 1989). Extending from the example of Black women, an intersectional perspective enables us to analyze how structures of privilege and disadvantage, such as gender, race, and class, interact in the lives of all people, depending on their particular identities and social positions.... Furthermore, intersectionality analyzes the ways in which these structures of power inextricably connect with and shape each other to create a system of interlocking oppressions, which Patricia Hill Collins (2000) termed a "matrix of domination" (p. 18).

The value of intersectional analysis, however, is not confined to understanding individual experiences or the ways systems of power intersect in individuals' lives. Over the last two decades, feminist scholars have discussed and debated the potential applications of intersectionality. As a "framework of analysis" or "analytic paradigm," intersectionality has been applied to theory, empirical research, and political activism....

By highlighting the differences in experiences among women, it might seem that an intersectional approach would make coalition building harder. Some scholars have criticized its attention to identity categories for hindering both intra- and cross-movement mobilization by splintering groups, such as women, into smaller categories, and accentuating the significance of separate identities (Brown 1997). As Andrea Canaan (1983) observed in *This Bridge Called My Back*, the singular focus on identity can lead us to "close off avenues of communication and vision so that individual and communal trust, responsibility, loving and knowing are impossible" (p. 236).

Yet intersectionality presents an exciting paradox: attending to categorical differences *enhances* the potential to build coalitions between movements and makes them more effective at organizing for social change.

How can illuminating differences build solidarity? First, it is only by acknowledging the lived experiences and power differentials that keep us apart that we can effectively grapple with the "matrix of domination" and develop strategies to eliminate power inequities. This is not a matter of *transcending* differences. To the contrary, activists interested in coalition building must confront their differences openly and honestly. "Our goal is not to use differences to separate us from others, but neither is it to gloss over them," writes Gloria Anzaldua and AnaLouise Keaton (2002, p. 3). Intersectionality avoids the trap of downplaying differences to reach a false universalism and superficial consensus—a ploy that always benefits the most privileged within the group and erases the needs, interests, and perspectives of others. An intersectional approach should not create "homogenous 'safe spaces' " where we are cordoned off from others according to our separate identities (Cole 2008, p. 443). Rather, it can force us into a risky place of radical self-reflection, willingness to relinquish privilege, engagement with others, and movement toward change.

Second, once differences are acknowledged, an intersectional framework enables discussion among groups that illuminates their similarities and common values.... Commonality is not the same thing as sameness. Searching for and creating commonalities among people with differing identities through active engagement with each other is one of intersectionality's most important methodologies not only for feminist theorizing but also for political activism.

Third, analysis of our commonalities reveals ways in which structures of oppression are related and therefore highlights the notion that our struggles are linked. Despite our distinct social positions, we discover that "we are all in the same boat" (Morales 1983, p. 93). Not only does intersectionality apply to everyone in the sense that all human beings live within the matrix of power inequities, but also that the specific intersections of multiple oppressions affect each and every one of us.

Of course, these intersecting systems affect individuals differently, depending on the specific context and their specific political positions. This is why engagement between groups with differing perspectives is critical to understanding the dynamics of inequality and to organizing for social change. Rather than erasing our identities for the sake of coalition, we learn from each other's perspective to understand how systems of privilege and disadvantage operate together and, therefore, to be better equipped to dismantle them....

Far from building walls around identity categories, then, intersectionality forces us to break through these categories to examine how they are related to each other and how they make certain identities invisible. This shift from seeing our differences to seeing our relatedness requires that we understand identity categories in terms of matrices of power that are connected rather than solely as features of individuals that separate us (Cole 2008; Dhamoon 2011).... "While analytically we must carefully examine the structures that differentiate us, politically we must fight the segmentation of oppression into categories such as 'racial issues,' 'feminist issues,' and 'class issues," writes Bonnie Thornton Dill (1983, p. 148). Indeed, our radical *interrelatedness* is equally as important as our differences. To us, the radical potential for intersectionality lies in moving beyond its recognition of difference to build political coalitions based on the recognition of connections among systems of oppression as well as on a shared vision of social justice. The process of grappling with differences, discovering and creating commonalities, and revealing interactive mechanisms of oppression itself provides a model for alternative social relationships.

AN INTERSECTIONAL ANALYSIS OF RACE, GENDER, DISABILITY, AND REPRODUCTIVE GENETIC TECHNOLOGIES

... Reproductive justice is a prime example of applying an intersectional frame to both political theorizing and political action. Women of color developed a reproductive justice theory and movement to challenge the barriers to their reproductive freedom stemming from sex, race, and class inequalities (Nelson 2003; Roberts 1997, 2004; Silliman et al., 2004). Reproductive justice addresses the inadequacies of the dominant reproductive rights discourse espoused by organizations led by White women that was based on the concept of choice and on the experiences of the most privileged women. Thus, women of color contributed to the understanding of and advocacy for reproductive freedom by recognizing the intersection of race, class, and gender in the social control of women's bodies.

What if we complicated the matrix even more by including disability as an identity and political category in theorizing and organizing by women of color? Far from being a marginal social division because it affects fewer people, disability helps to shape reproductive and genetic technologies and policies that affect everyone.... Like intersectionality's central claim that "representations of gender that are 'race-less' are not by that fact alone more universal than those that are race-specific" (Crenshaw 2011, p. 224), representations of race and gender that neglect disability are no more universal than those that are based solely on able bodies.... It was only when we engaged with disability rights activists that we began to grapple with their perspectives on reproductive politics and changed our own perspectives in concrete ways.

Just as the dominant conception of discrimination imposed by courts erases Black women, organizing for social change along certain categories can obscure the importance of other perspectives and opportunities for building coalitions to achieve common social justice goals. Disability rights discourse largely has failed to encompass racism, and anti-racism discourse largely has failed to encompass disability. The disability rights and civil rights movements are often compared as two separate struggles that run parallel to each other, rather than struggles that have constituents and issues in common, ... even as both people of color and people with disabilities share a similar experience of marginalization and "othering" and even though there are people of color with disabilities (Pokempner and Roberts, 2001).

Race, gender, and disability do not simply intersect in the identities of women of color with disabilities, however. Rather, racism, sexism, and ableism work together in reproductive politics to maintain a reproductive hierarchy and enlist support for policies that perpetuate it (Roberts 2009, 2011). In her past work, Roberts (1997) has contrasted policies that punish poor women of color for bearing children with advanced technologies that assist mainly middle- and upper-class White women not only to have genetically-related children, but to also have children with preferred genetic traits. While welfare reform laws aim to deter women receiving public assistance from having even one additional healthy baby, largely unregulated fertility clinics regularly implant privileged women with multiple embryos, knowing the high risk multiple births pose for premature delivery and low birth weight that requires a fortune in publicly-supported hospital care. Rather than place these policies in opposition, however, Roberts argued in "Privatization and Punishment in the New Age of Reprogenetics" (2005) and "Race, Gender, and Genetic Technologies: A New Reproductive Dystopia?" (2009) that they are tied together. Policies supporting both population control programs and genetic selection technologies reinforce biological explanations for social problems and place reproductive duties on women that privatize remedies for illness and social inequities.

Advances in reproduction-assisting technologies that create embryos in a laboratory have converged with advances in genetic testing to produce increasingly sophisticated methods to select for preferred genetic traits, and de-select for disability. Liberal notions of reproductive choice obscure the potential for genetic selection technologies to intensify both discrimination against disabled people and the regulation of women's childbearing decisions. These technologies stem from a medical model that attributes problems caused by the social inequities of disability to each individual's genetic make up and that holds individuals, rather than the public, responsible for fixing these inequities. Disability rights activists have pointed out that prenatal and pre-implantation genetic diagnosis reinforce the view that "disability itself, not societal discrimination against people with disabilities, is the problem to be solved" (Parens and Asch, 1999, p. s13). This medicalized approach to disability assumes that difficulties experienced by disabled people are caused by physiological limitations that prevent them from functioning normally in society, rather than the physical and social limitation enforced by society on individuals with disabilities (Saxton 2007). Although disabilities cause various degrees of impairment, the main hardship experienced by most people with disabilities stems from pervasive discrimination and the unwillingness to accept and embrace differing needs to function fully in society.

Locating the problem inside the disabled body rather than in the social oppression of disabled people leads to the elimination of these bodies becoming the chief solution to impairment. By selecting out disabling traits, these technologies can divert attention away from social arrangements, government policies, and cultural norms that help to define disability and make having disabled children undesirable (Wendell 1996). Genetic selection is also discriminatory in that it reduces individual children to certain genetic traits that by themselves are deemed sufficient reasons to terminate an otherwise wanted pregnancy or discard an embryo that might otherwise have been implanted (Asch 2007).

The expectation of genetic self-regulation may fall especially harshly on Black and Latina women, who are stereotypically defined as hyperfertile and lacking the capacity for self-control (Gutierrez 2008; Roberts 1997). In an ironic twist, it may be poor women of color, not affluent White women, who are most compelled to use prenatal genetic screening technologies. This paradox is revealed only by a political analysis that examines the interlocking systems of inequity based on gender, race, and disability that work together to support policies that rely on women's management of genetic risk rather than social change. This intersectional analysis also reveals that reproductive justice, women's rights, and disability rights activists share a common interest in challenging unjust repro-genetics policies and in forging an alternative vision of social welfare.

THE DYNAMICS OF INTER- AND INTRA-MOVEMENT MOBILIZATION ROOTED IN AN INTERSECTIONAL FRAMEWORK

... An intersectional approach provides a method for overcoming ... barriers to collaboration and even using differences between identity categories and causes as a tool for more effective strategizing and action....

In the last several years, as scholars and activists, working with the staff and board of Generations Ahead, we have used an intersectional framework as an integral part of our organizing work. Intersectionality has been an essential tool in shaping the mission, vision, and work of the organization, in deepening our understanding of the social and ethical implications of genetic technologies, and in building unlikely partners to advance a social justice agenda....

At the heart of Generations Ahead's method of cross-movement alliance-building are three main elements: honestly and openly discussing in face-to-face conversations key areas of conflict among movements; articulating common values upon which bridging frameworks could be constructed; and cultivating a shared advocacy agenda, followed by joint strategizing and collective action, to address specific issues. These elements put in practice the key theoretical insight of an intersectional analysis ... —that uncovering how dominant discourses and systems marginalize certain groups in intersecting ways and at specific sites can be a basis for solidarity. By acknowledging differences, not transcending them, activists can more effectively grapple with the "matrix of domination" because an intersectional analysis ultimately reveals how structures of oppression are related and therefore our struggles are linked. To be successful, this process required building trust by learning about each other's movements and concretely demonstrating solidarity for each other's issues, for example, by co-sponsoring and attending each other's events (Generations Ahead 2009).

Based on this model, Generations Ahead organized a series of meetings among reproductive rights and justice, women of color and Indigenous women, and disability rights advocates to dig deeper into the areas of tension between movements and to develop a shared analysis of genetic technologies across movements, with the goals of creating common ground and advancing coordinated solutions and strategies....

In order to openly and honestly identify the distinctive ways in which reproductive and genetic technologies affected different constituencies, the participants were asked to divide themselves up into self-identified constituency groups. It was clearly acknowledged that participants were not being asked to privilege or prioritize any one identity over others, but rather that they were being asked to share the unique and distinguishing perspectives of different constituencies. The twenty-one participants divided up into the following groups: Indigenous women, Asian women, women of African descent, women (of color) with disabilities, and Latinas living in the United States. Queer identified people agreed to raise their specific concerns within all of the other groups. Each group's members then spent time identifying the particular benefits and concerns genetic technologies raised for their group, and the values that they wanted to see integrated into any advocacy on this issue.

Rather than starting the discussion about the benefits and risks of genetic technologies based on a universal and generic human being, these constituency groups were able to do several interesting things simultaneously. First, when asked to consider these technologies from the standpoint of their identity-specific perspective, these issues became more relevant for all participants. None of the participants were users of these technologies, and, up until that moment, most felt that they were not relevant to their lives and social justice advocacy. But once they were able to connect what felt like an abstract, futuristic, and privileged issue to their lives and communities, their investment in the issue shifted. Most participants were now able to reflect on and attach genetic technologies to issues that they deeply cared about: sex selection and son preference for Asian women; genetic determinism and eugenics for women of African descent; prenatal disability de-selection for women with disabilities; blood quantum and tribal identity for Indigenous women; and family formation and fertility for Latinas. By the end of the discussion, all participants were able to understand the issues raised by genetic technologies as an extension of their existing social justice commitments and concerns (Generations Ahead 2008).

Second, the participants were able to make these linkages as a part of a larger, shared "matrix of domination," rather than as a hierarchical analysis of oppression. Because everybody was able to speak to the intersections with their lived experiences, and since all identities were equally valued, the discussion quickly and easily transitioned to shared struggles and solidarity, rather than a debate over who was more or less oppressed or privileged in the development and use of genetic technologies. Shared concerns were quickly visible in the similar histories of reproductive oppression and genetic determinism, and the ways in which biology, bodies, and reproduction have been historically categorized, regulated, stigmatized, and controlled for some groups.

In addition, participants in each group discussed other intersecting identities that clearly cut across all groups, such as immigration status, class, sexual orientation, gender identities, and age. Acknowledging these other intersections prevented any one individual or one group to claim the "most oppressed" or "most victimized" identity. It meant that everyone in the room enjoyed privilege in at least one, if not more, of their identities. Since no one in the room could be either pure victim or pure oppressor, participants were more willing and comfortable acknowledging their own privilege and less attached to any presumed victim status. This led to, as Gloria Anzaldua (2002) noted, more thorough self-reflection and openness to learning from and engaging with others. Everybody felt like they belonged together because of, not in spite of, their differences.

And finally, owing to the sense of "we are all in this together" and newly recognized links between genetic technologies and their existing social justice commitments, the whole group was able to identify and articulate a shared set of values and perspectives that they wanted to promote in any analysis of the social implications of genetic technologies. They pinpointed values that they felt were important to help guide work in this area, values such as: start with an intersectional analysis of power and inequities at the center of any analysis of benefits and risks, include community in identifying solutions, and make sure to address the underlying factors that cause unequal outcomes and don't just blame it on the technology per se (Generations Ahead 2008).

Participants then worked together to develop a condensed list of shared values that everyone could take back to their organizations and continue to use to inform any shared or individual advocacy in this area. The group collectively affirmed values such as: put human welfare, not profit, at the center of the use of these technologies; recognize that individuals, families, and communities are socially, culturally, and politically determined, not solely biologically or genetically; include those most impacted by these technologies to be a part of the decision-making about their use; and acknowledge the intersectionality of diverse lived experiences and advocate for long-term, holistic solutions (Generations Ahead 2008).

This convening laid the groundwork for future, more challenging conversations and collaborations between reproductive justice and disability rights leaders. The lessons and praxis of using an intersectional approach were then applied to a series of five roundtable conversations between two groups that have a long history of tension, mistrust, and aversion to working together—reproductive rights and disability rights advocates. These roundtable discussions started with the most difficult area of disagreement between these two movements—their differing approaches to genetic testing technologies and abortion....

These discussions were started with an open acknowledgment of this third rail of disagreement, and recognition that there was a mutual history of hurt and fear, where each movement felt that the other did not appreciate its perspective or deep concerns about the other movement's perspective....

As a result of their engagement over conflicts and common values, the advocates were able to agree on a shared alternative paradigm for addressing genetic technologies based on "long-term, comprehensive, intersectional policies that create structural changes in social inequality" (Generations Ahead 2009, p. 6).

Instead of these two groups being at loggerheads over whether to regulate abortion and prenatal screening to prevent the de-selection of people with disabilities or allow unfettered reproductive freedom that could lead to the eugenic elimination of disability, participants were able to define a set of shared values. These include:

- Reproductive autonomy should include support for people making the choice to have children, including children with disabilities, and support to raise their children with dignity.

- All women who choose to parent should be valued as parents and all children should be valued as human beings, including children with disabilities.

- Policy advocacy should focus on providing social and material supports to women, families, and communities, not on when life begins, whose life is more valued, or who can be a parent.

- Both movements should broaden their agendas to fight to improve the social, political, physical, and economic contexts within which women and people with disabilities make decisions about their lives. The focus should be on changing society, not on individual decision-making (Generations Ahead 2009, p. 2).

Through these shared values all participants were able to affirm women's self-determination and the value of people with disabilities, so that one was not pitted against the other. And they were able to include an analysis that encompassed concerns about race, class, immigration, and sexual orientation....

CONCLUSION

As the work of Generations Ahead illustrates, the radical potential for intersectionality lies in moving beyond its acknowledgement of categorical differences to build political coalitions based on the recognition of connections among systems of oppression as well as on a shared vision of social justice.... In the process we have learned several important lessons for how to "do" intersectionality in organizing and advocacy.

First, a good process for radical relationship- and alliance-building requires forthrightly acknowledging the multiple intersecting lived experiences of all participants. Radical alliances can only be built on the basis of being honest about differences and disagreements. This honesty is what creates the potential for new solidarities based on shared but different experiences. Second, trust must be developed through the process. Alliance building is a progressive, developmental process where trust is built through repeated contact, connection, conversation, and collective action. Identifying multiple and intersecting interests is crucial to creating repeated opportunities for collaboration. The third lesson is related to a willingness on the part of all participants to change their perspectives and politics. An intersectional framework is a critical tool for disrupting oppressed-oppressor binaries, and opening up the possibilities for discovering values and experiences in common. And the final lesson is to keep the focus on shared values. While scholars and advocates for social change might disagree on general strategy and tactics, they can more easily agree on shared values that can form the basis for a common vision, as well as for joint action on specific campaigns. Here again, an intersectional approach is useful in deconstructing disagreements and reconstructing similar experiences and hopes....

In acknowledging that all of us have multiple identities and by including all of those identities in the organizing process, intersectionality in practice can be a powerful tool for grappling with differences and uncovering shared values and bridging frameworks. This process provides a basis for collective action and a model for alternative social relationships rooted in our common humanity. Instead of separating groups, as some have argued, using an intersectional framework can create new and authentic alliances even among historically oppositional groups that can lead to more inclusive, focused, and effective efforts for social change. Intersectionality as a theory and practice for social change can, and should, be used as a critical tool in struggles for social justice that seek to include us all.

REFERENCES

Anzaldua, Gloria and AnaLouise Keating (Eds.) (2002). *This Bridge We Call Home: Radical Visions for Transformation*. New York: Routledge.

Asch, Adrienne (2007). Why I Haven't Changed by Mind about Prenatal Diagnosis: Reflections and Refinements. In Erik Parens and Adrienne Asch (Eds.), *Prenatal Testing and Disability Rights*, pp. 234–258. Washington, DC: Georgetown University Press.

Brown, Wendy (1997). The Impossibility of Women's Studies. *Difference: A Journal of Feminist Cultural Studies*, 9(3): 79–101.

Canaan, Andrea (1983). Brownness. In Cherrie Moraga and Gloria Anzaldua (Eds.), *This Bridge Called My Back: Writings by Radical Women of Color*, pp. 232–237. New York: Kitchen Table, Women of Color Press.

Cole, Elizabeth (2008). Coalitions as a Model for Intersectionality: From Practice to Theory. *Sex Roles*, 59: 443–453.

Crenshaw, Kimberlé (1989). Demarginalizing the Intersection of Race and Sex: A Black Feminist Critique of Antidiscrimination Doctrine, Feminist Theory and Antiracist Politics. *University of Chicago Legal Forum*, pp. 139–167.

Crenshaw, Kimberlé (2011). Postscript. In Helma Lutz, Maria Teresa Herrera Vivar, and Linda Supik (Eds.), *Framing Intersectionality: Debates on a Multi-Faceted Concept in Gender Studies*, pp. 221–233. Surrey, England: Ashgate Publishing Limited.

Dhamoon, Rita Kaur (2011). Considerations on Mainstreaming Intersectionality. *Political Research Quarterly*, 64(1): 230–243.

Generations Ahead (2008). A Reproductive Justice Analysis of Genetic Technologies: Report on A National Convening of Women of Color and Indigenous Women. <http://www.generations-ahead.org/files-for-download/articles/GenAheadReport_ReproductiveJustice.pdf> (accessed May 17, 2012).

Generations Ahead (2009). Bridging the Divide: Disability Rights and Reproductive Rights and Justice Advocates Discussing Genetic Technologies, convened by Generations Ahead 2007–2008. <http://www.generations-ahead.org/files-for-download/articles/GenAheadReport_BridgingTheDivide.pdf> (accessed May 21, 2012).

Generations Ahead (2010). Robert Edwards, Virginia Ironside, and the Unnecessary Opposition of Rights. <http://www.generations-ahead.org/resources/the-unnccessary-opposition-of-rights> (accessed May 17, 2012).

Gutierrez, Elena R. (2008). *Fertile Matters: The Politics of Mexican Origin Women's Reproduction*. Austin, TX: University of Texas Press.

Hill Collins, Patricia (2000). *Black Feminist Thought: Knowledge, Consciousness and the Politics of Empowerment*, 2ed. New York: Routledge.

Morales, Rosario (1983). We're All in The Same Boat. In Cherrie Moraga and Gloria Anzaldua (Eds.), *This Bridge Called My Back: Writings by Radical Women of Color*, pp. 91–93. New York: Kitchen Table, Women of Color Press.

Nelson, Jennifer (2003). *Women of Color and the Reproductive Rights Movement*. New York: NYU Press.

Parens, Erik and Adrienne Asch (1999). The Disability Rights Critique of Prenatal Genetic Testing: Reflections and Recommendations. *The Hastings Center Report*, 29(5): s1–s22. The Hastings Center.

Pokempner, Jennifer and Dorothy E. Roberts (2001). Poverty, Welfare Reform, and the Meaning of Disability. *Ohio State Law Journal*, 62: 425–463.

The Prenatally and Postnatally Diagnosed Conditions Awareness Act Fact Sheet. <http://www.generations-ahead.org/files-for-download/success-stories/InfoSheetBrownbackKennedy Legislation_final.pdf> (accessed May 17, 2012).

Roberts, Dorothy (1997). *Killing the Black Body: Race, Reproduction, and The Meaning of Liberty*. New York: Pantheon.

Roberts, Dorothy E. (2004). Women of Color and the Reproductive Rights Movement. *Journal of the History of Sexuality*, 13: 535–539.

Roberts, Dorothy E. (2005). Privatization and Punishment in the New Age of Reprogenetics. *Emory Law Journal*, 54: 1343–1360.

Roberts, Dorothy (2009). Race, Gender, and Genetic Technologies: A New Reproductive Dystopia? *Signs*, 34: 783–804.

Roberts, Dorothy (2011). *Fatal Invention: How Science, Politics, and Big Business Re-create Race in the Twenty-first Century*. New York: The New Press.

Saxton, Marsha (2007). Why Members of the Disability Community Oppose Prenatal Diagnosis and Selective Abortion. In Erik Parens and Adrienne Asch (Eds.), *Prenatal Testing and Disability Rights*, pp. 147–164. Washington, DC: Georgetown University Press.

Silliman, Jael, Marlene Gerber Fried, Loretta Ross, and Elena R. Guttierez (2004). *Undivided Rights; Women of Color Organize for Reproductive Justice*. Boston, MA: South End Press.

Thornton Dill, Bonnie (1983). Race, Class, and Gender: Prospects for an All-Inclusive Sisterhood. *Feminist Studies*, 9(1): 131–150.

Wendell, Susan (1996). *The Rejected Body: Feminist Philosophical Reflections on Disability*. New York: Routledge.

52

Growing Food and Justice

Dismantling Racism through Sustainable Food Systems

ALFONSO MORALES

M any Americans, particularly low-income people and people of color,
... are overweight yet malnourished. They face an overwhelming variety
of processed foods, but are unable to procure a well-balanced diet from the
liquor stores and mini-marts that dominate their neighborhoods.

These groups are food insecure, but furthermore, they are victims of food
injustice.... For the last twenty years there has been a kind of "call and response"
that has produced a web of relationships among government, scholars, nonprofit
organizations, and foundations all interested in understanding food insecurity and
food injustice. Definitions of important concepts like food security have been
developed, organizations like the Community Food Security Coalition have
grown up, foundations, universities, and government have developed programs
to fund food research and practice, and nascent food justice organizations have
emerged and are now populating communities around the country.

In this article I describe one of the newest threads in this web of activity,
that of the Growing Food and Justice for All Initiative (GFJI), a loose coalition of organizations developed under the auspices of Growing Power, Inc., a
food justice organization based in Milwaukee, Wisconsin, with offices in
Chicago, Illinois, and a loose coalition of regional affiliates. Food justice organizations borrow from most every strand in the web of interrelated organizations and ideas, but they focus on issues of racial inequality in the food system
by incorporating explicit antiracist messages and strategies into their work. This
article chronicles in part how GFJI developed in response to the relative
absence of people of color in the food system.... I show how food justice
organizations have responded to GFJI in different ways and how they are
weaving together various threads from the larger web into their own activities
and toward their own goals as they develop their own approaches to food
justice.

SOURCE: Morales, Alfonso. 2011. "Growing Food and Justice: Dismantling Racism
through Sustainable Food Systems." Pp. 149–176 in *Cultivating Food Justice: Race, Class,
and Sustainability*, edited by Alison Hope Alkon and Julian Agyeman. Cambridge, MA:
Massachusetts Institute of Technology Press. © 2011 Massachusetts Institute of
Technology, by permission of The MIT Press.

FOOD JUSTICE IN HISTORICAL AND
CONTEMPORARY ECONOMIC CONTEXT

... Over the last fifty years major grocery chains have sought suburban locations to accommodate larger stores, more parking spaces, and higher profits (USDA 2009). Eisenhauer (2001) refers to this trend as "supermarket redlining," or the process by which corporations avoid low-profit areas. Consider the impact on food access of such decision making. In 1914 American cities had fifty neighborhood grocery stores per square mile, an average of one for every street corner (Zelchenko 2006). Mayo (1993) documents store design and industry changes that transformed groceries from small neighborhood operations to large chains. Just as some were rediscovering healthy food in the 1960s, grocers began following the migration of the middle class from the city to the suburbs. The Business Enterprise Trust indicated the attitude, "It makes no sense to serve distressed areas when profits in the serene suburbs come so easily" (qtd. in Eisenhauer 2001). For instance, between 1968 and 1984, Hartford, Connecticut, lost eleven out of its thirteen grocery chains, and between 1978 and 1984 Safeway closed more than 600 inner city stores around the country (Eisenhauer 2001).

This mass departure reduced food access for low-income and minority people. Morland's multistate study (2002) found four times as many grocery stores in predominantly white neighborhoods as predominantly black ones, and other studies have noted that inner-city supermarkets have higher prices and a smaller selection of the fresh, wholegrain, nutritious foods (Sloane 2004).... When taken with a general retreat from "hunger" by the USDA ... market-driven relocation of groceries to the suburbs left behind the conditions for a public health disaster.

Food, and poor nutrition in particular, is a risk factor in four of the six leading causes of death in the United States—heart disease, stroke, diabetes, and cancer.... We know that race and class inequalities produce insufficient nutrition and increase food-related disease. We know that what people eat and how they eat contributes significantly to mortality, morbidity, and increasing health care costs. By contrast, we know how food relates to good health (Institute of Medicine 2002). And when we think of access to fresh, whole foods we typically think it is dependent on income and on where one lives. Thus, decision making on locating grocery stores created "food deserts," and public health problems, but also germinated a new food coalition: the Community Food Security Coalition....

FOOD JUSTICE AND THE GROWING FOOD AND
JUSTICE FOR ALL INITIATIVE

The effort to reconstruct the foodscape for people of color has augmented the discussion of food security with organizing around the concept of food justice. This idea grew from racial inequality in food access and its accompanying public health problems. In the same way that the civil rights movement grew from racial inequalities in housing, voting, transportation, and the like, new voices are naming the racism in food, but they are not alone. Tom Vilsack expressed these racial inequalities, in his first major speech as the U.S. secretary of agriculture under President Obama. Vilsack wondered aloud what the founder of the USDA, Abraham Lincoln, would find if he walked into the USDA building today and asked, "How are we doing?" "And he'd be told," said Vilsack, "Mr. President, some folks refer to the USDA as the last plantation.' And he'd say, 'What do you mean by that?' 'Well it's got a pretty poor history when it comes to taking care of folks of color. It's discriminated against them in programming and it's made it somewhat more difficult for some people of color to be hired and promoted. It's not a very good history, Mr. President" (Federation of Southern Cooperatives 2009)....

While racially motivated food justice has scant and scattered organizational infrastructure, its current manifestation does have a name and a place of origin.... The GFJI was established under the auspices of Growing Power, Inc., the Milwaukee-based organization founded by MacArthur Award winner Will Allen, Erika Allen's father. Since 1993, Will, Erika, and Growing Power staff have worked with diverse local communities to develop community food systems responsive to the circumstances of people of color.... Will Allen started a two-acre urban farm in a food-insecure Milwaukee community. The produce is sold in the community at affordable prices. But Growing Power is much more than an urban farm: it sponsors national and international workshops on food security, maintains flourishing aquaponics and vermicomposting programs, helps teach leadership skills, and provides on-site training in sustainable food production. The organization embodies the community-based, systemic approach ... by reconnecting vulnerable populations to healthy food and by developing empowered individuals in economically and socially marginalized communities....

By choosing to focus explicitly on racism and sustainable food systems, GFJI created space for a diverse community to join together and support one another in the eradication of racism and the growth of sustainable food systems. As Pothukuchi points out, "in the 1990s, the community food security concept was devised as a framework for integrating solutions to the problems faced by poor households (such as hunger, limited access to healthy food, and obesity), and those faced by farmers (such as low farm-gate prices, pressures toward consolidation, and competition from overseas)" (2007, 7). By adding the additional concern of racism, GFJI has effectively tightened the connections already implicit in the concept of community food security: racially diverse households and farmers are some of the most at-risk groups in the communities targeted by USDA Community Food Projects. GFJI demonstrates the widespread appeal of an organization that combines the challenging topics of racism, sustainable agriculture, and community food security. But as Will Allen points out, these problems cannot be untangled, and must be tackled simultaneously. He notes, "We are all responsible for dismantling racism and ensuring more sustainable communities, which is impossible without food security."

For its member organizations, GFJI acts as a coordinating body, a source of emotional and spiritual sustenance, and a site for germinating and sharing fresh ideas. Each month one or more "germinators" convene a conference call, on a subject of interest to member organizations. Topics range widely, from the impact of the Obama administration on food-related problems to power sharing; and from developing effective multicultural leadership to sharing strategies for getting food to low-income communities. These monthly calls, which sustain the initiative without having to support an organizational infrastructure, are an important and ongoing source of ideas and support for these organizations....

Often led by people of color, food justice organizations see dismantling racism as part of food security. By taking an explicitly racialized approach, the food justice movement moves away from the colorblind perspective.... The food justice approach aligns movement organizations explicitly with the interests of communities and organizations whose leaders have felt marginalized by white-dominated organizations and communities. By creating a space explicitly intended for the exploration of the particular challenges facing communities of color, the food justice movement has encouraged these communities to get the help and the support that they require to continue their work. The GFJI provides some logistical support, an annual conference, and networking opportunities, but perhaps most important, a sense of community, continuity, and connection among colorful communities working to improve their food security.

By bringing the individual organizations together in a cohesive body, GFJI relates racial social critique to antiracist tools of sustainable agriculture in the service of creating a nationwide network dedicated to implementing food justice strategies. In addition, GFJI provides its members with the important sense of belonging to an organization whose goals are directly aligned with their own: although many sustainable agriculture and CFS (community food security) organizations have fostered the ambitions of people and communities of color over the years, they have not always shared the antiracist agenda. GFJI places racism front and center in the context of food and agriculture, allowing its members to feel confident that they are engaged in a community that shares their ideals and aspirations, and creating an infrastructure through which these organizations can support each other. This is a particularly important point in light of the overwhelming challenges involved in challenging the deeply entrenched and richly funded agribusiness industry, in addition to government agencies with deeply racist histories, such as the USDA. In the first coalition conference—GFJI I—organizations from around the country presented their work across race and the food system, and not surprisingly responded to the agenda in different ways, making clear two things: first, the variety of approaches and practices there are among antiracist food system organizations; and second, how the GFJI fulfilled its purpose by becoming an opportunity for each organization to articulate its unique approach, learn from the nuances others described, and even find that the GFJI might not be what they need....

REFERENCES

Eisenhauer, Elizabeth. 2001. In Poor Health: Supermarket Redlining and Urban Nutrition. *Geojournal* 53:125–133.

Federation of Southern Cooperatives. 2009. Transcript of U.S. Agriculture Secretary Tom Vilsack's speech to Federation of Southern Cooperatives/Land Assistance Fund's Georgia Annual Farmer's Conference, January 21, <http://www.federationsoutherncoop.com/albany/Vilsackspeech.pdf>.

Institute of Medicine. 2002. *The Future of the Public's Health.* Washington, DC: National Academy Press.

Mayo, James M. 1993. *The American Grocery Store.* Santa Barbara, CA: Greenwood Publishing Group.

Morland, Kimberly, Steve Wing, Ana Diez Roux, and Charles Poole. 2002. Neighborhood Characteristics Associated with the Location of Food Stores and Food Service Places. *American Journal of Preventive Medicine*, 22(1):23–29.

Pothukuchi, Kami. 2007. *Building Community Food Security: Lessons from Community Food Projects*, 1999–2003. Community Food Security Coalition, <http://www.food-security.org/BuildingCommunityHoodSecurity.pdf>.

Sloane, David C. 2004. Bad Meat and Brown Bananas: Building a Legacy of Health by Confronting Health Disparities around Food. *Planners Network* (Winter): 49–50.

United States Department of Agriculture (USDA). 2009. *Access to Affordable and Nutritious Food: Measuring and Understanding Food Deserts and Their Consequences.* Washington, DC: Economic Research Service.

Zelchenko, Peter. 2006. Supermarket Gentrification: Part of the "Dietary Divide." *Progress in Planning* 168(1):11–15.

53

(Re)Imagining Intersectional Democracy from Black Feminism to Hashtag Activism

SARAH J. JACKSON

To say that Black lives matter[1] has become both a technological and cultural phenomenon in the United States is an understatement. The hashtag and those discursively linked to it have been used more than 100 million times, and the visibility and persistence of Black lives matter activists—from highway shutdowns in America's largest cities to the takeover of presidential candidates' political rallies—have led to widespread social and political debate about what has been dubbed "the new civil rights movement" (Freelon et al. 2016; Jackson and Foucault Welles, 2015). Yet there seems to be considerable consternation among academics, journalists, and politicians about how to incorporate the standpoints of a new generation of activists into our national politics. In this essay I discuss how these activists have manifested Black feminist impulses through social media and beyond, and suggest it is the responsibility of those invested in (re)imagining a more democratic process to closely consider the radically intersectional lessons of the current movement.

The Black lives matter movement can be traced to the legacy of the larger Black freedom movement, but also more recently to the work of millennial Black activist organizations like the Dream Defenders and the Black Youth Project 100 (Cohen and Jackson 2016). Members of these organizations and the young people who align themselves with their work have come of age in a country overwhelmingly celebratory of its racial progress but silent on the lasting impact of its racial sins. Eduardo Bonilla-Silva, among others, has detailed how neoliberal color-blind politics have dangerously entrenched the American impulse to reject explicit political critiques of white supremacy as unnecessary or outmoded. Similarly, millennials have borne witness to an America claiming postfeminism alongside seemingly constant attacks on women's bodies and bodily autonomy, and a country simultaneously moving toward greater lesbian,

SOURCE: Jackson, Sarah J. 2016. "(Re)Imagining Intersectional Democracy from Black Feminism to Hashtag Activism." *Women's Studies in Communication*, 39(4): 375–379. DOI: 10.1080/07491409.2016.1226654. Copyright © The Organization for Research on Women and Communication (ORWAC), www.orwac.org. Reprinted by permission of Taylor & Francis Ltd, http://www.tandfonline.com on behalf of The Organization for Research on Women and Communication (ORWAC).

gay, bisexual, transgender, and queer (LGBTQ) inclusion while silencing queer critiques and bodies that do not mold to the mainstream (McRobbie 2004; Walters 2016). Today's racial justice activists have responded to these political contradictions with discourse and tactics both familiar and unfamiliar to members of the old guard. In particular, millennial activists have rejected the respectability politics that guided much of the civil rights movement of the 1950s and 1960s and have turned to new technologies as tools for the promulgation and solidification of messages, nurturing a counterpublic community that centers the voices of those most often at the margins.

Black Lives Matter's organizational founders, and other members of the larger Movement for Black Lives collective, have insisted on discourses of intersectionality that value and center all Black lives, including, among others, Black women, femmes, and queer and trans folk.[2] This stands in contrast to the impulses of the historical Black freedom movement, which chose to center cisgender Black men in racial justice struggles, intentionally and strategically marginalizing cisgender Black women and queer folks who might be seen as unworthy of rights within the logics of capitalist heteropatriarchy (Smith 2006). As Alicia Garza explains in "Herstory of the #BlackLivesMatter Movement" (among other public interviews and writing), the phrase and hashtag were first publicly used in 2013 as a response to the acquittal of George Zimmerman for the 2012 murder of teenager Trayvon Martin, but represent an idea much larger than that singular case or moment.

Along with other Black Lives Matters founders Opal Tometi and Patrisse Cullors, Garza has long insisted on the radical intersectionality of the phrase, writing:

> Black Lives Matter is a unique contribution that goes beyond extrajudicial killings of Black people by police and vigilantes. It goes beyond the narrow nationalism that can be prevalent within some Black communities, which merely call on Black people to love Black, live Black and buy Black, keeping straight cis Black men in the front of the movement while our sisters, queer and trans and disabled folk take up roles in the background or not at all. Black Lives Matter affirms the lives of Black queer and trans folks, disabled folks, Black-undocumented folks, folks with records, women and all Black lives along the gender spectrum (Garza 2016).

It matters that Garza, Tometi, and Cullors are women, Garza and Cullors are queer, and Tometi is the daughter of immigrants; their activism prior to and since the founding of Black Lives Matter includes organizing domestic workers, immigration activism, criminal justice reform, advocating for victims of domestic violence, and health care rights. Thus, the founders of Black Lives Matter not only live intersectional lives but also have long been engaged with intersectional activism. For Garza, Cullors, and Tometi, the personal and political are truly intertwined and the political is unquestionably intersectional.

Of course, contemporary activists have not succeeded in centering intersectional concerns from within a void. Decades of work—particularly by Black and queer feminist academics and activists like Barbara Smith, Audre Lorde, Kimberlé Crenshaw, bell hooks, and Patricia Hill Collins, plus the painfully slow but increasing inclusion of Americans who experience gender, race, and/or class oppression in institutions of higher education—have exposed a growing number of young Americans to academic theories of intersectionality early in their politicization and education. Further, grassroots community-organizing traditions that have been practiced and refined for decades by Black women like Fannie Lou Hamer, Ella Baker, and Marsha P. Johnson have empowered new generations of Black women to demand their experiences be made central to community concerns. Thus, the aspiration of the new radical Black politic is for Black teenage mothers, homeless Black trans people, and queer Black androgynes to matter in our national political consciousness as much as anyone else.

The visibility of these efforts is unsurprisingly mired in an evolving ideological battle between the margins, where intersectional concerns are centered, and the mainstream, where politics simultaneously make minute accommodations of intersectional frameworks while silencing larger critiques. Pushback against the contemporary politics of radical intersectionality comes from many sites, some painfully close. Primarily, mainstream institutions and elites, through discursive absences, have erased intersectional concerns from the most visible accounts of and responses to contemporary activism. This erasure has been perpetuated by those the movement would like to consider allies, like President Barack Obama and sympathetic journalists who often exclude women, girls, and trans and queer folk on the still too rare occasions they directly address racialized state violence; and by demagogues on the right who have proudly declared themselves enemies of the movement they denigrate using old narratives of Black pathology and new ones of a mythical color blindness.

Further, some old-guard members of the civil rights movement, many of whom represent a middle-class, churchgoing, Black elite, have themselves erased the intersectional concerns of contemporary activists by clinging to values of respectability, and members of both the feminist and LGBTQ establishment have suggested the tactics, politics, and style of Black lives matter activists are divisive. And finally, the cultural impulse toward centering male activists has created some rupture in the movement over the rising celebrity of figures like DeRay McKesson instead of women organizers like Garza, Tometi, and Cullors (Cobb).

Yet I suggest it is twenty-first-century activists, as exemplified by the Black lives matter movement, who have finally succeeded in making intersectional issues of racial oppression visible to mainstream America. Their tools have been both the networked counterpublics which nourish them and the physical shedding of respectability in how they interrupt and take up space in the political and cultural places that have too long included them only if they willingly remain at the margins. The latter has been illustrated repeatedly in 2015 and 2016 as Black lives matter activists have insisted on the right to space, speech, and critique at Hillary Clinton and Bernie Sanders events. As I write, queer Black lives matter activists have gained national attention by withdrawing from the San Francisco Pride parade to protest increased police presence and are holding a sit-in at

Toronto Pride, demanding the exclusion of police and the inclusion of resources for queer disabled people and queer Black youth.

The hashtags that have arisen alongside and in collusion with #BlackLivesMatter are similarly reflective of this unrepentantly intersectional Black liberation politic. In online counterpublic networks, Black women—cis and trans, some everyday citizens with little access to institutional power and others media personalities, some queer, some straight, some poor and disabled—have played an outsized role in shaping recent national conversations about everything from police brutality to gender identity to popular culture, with the creation of hashtags like #BlackLivesMatter, #GirlsLikeUs, and #OscarsSoWhite (Bailey and Gossett 2010; Jackson and Foucault Welles 2015). Among others, Black women have created and popularized hashtags that observe the raced aspects of street harassment, center the experiences of incarcerated Black trans women, offer interventions to mainstream white feminist erasures of Black women's experiences, and critique the appropriation and commodification of Black women's styles.

Certainly there are limits to the democratic possibilities of online engagement. The mainstream logics of scholarly research, journalism, and politics have sometimes reinterpreted the counterpublic discourses created online in ways that strip them of their meanings and make their context and creators invisible (Kingston Mann 2014). Further, the internet itself, including the technological architecture of social media spaces, has complex ties to projects of militarism, advertising, and surveillance that tip the scales of power against counterpublics (Foster and McChesney 2014). Yet hashtags arising from a Black feminist politic take advantage of this architecture to perform the two basic functions of counterpublic discourse: reflect the experiences and needs of a marginalized community and call on mainstream politics to listen and respond.

In June 2016, for example, the hashtag #GirlIGuessImWithHer spread quickly across the Twittersphere as voters disenchanted with the limited possibilities of the current democratic process resigned themselves to supporting Hillary Clinton. Like the other hashtags noted here, #GirlIGuessImWithHer was started by a Black woman in what both scholars and the media have come to refer to as "Black Twitter" (Sharma 2013; Florini 2014). In its very language, the underlying ability of "Girl I Guess" to laugh, snark, and sigh at being in an always-exhausting position of not having one's full interests represented in politics reflects a unique form of communication typified by Black women's speech (Bailey and Gumbs 2010; Houston 1992). The hashtag, which almost immediately was picked up by a diversity of Twitter users, intentionally rearticulated the Clinton political slogan #ImWithHer, generating discussion and debate with those engaged in mainstream politics. Those using #GirlIGuessImWithHer cited intersectional concerns regarding Hillary Clinton's role in the 1994 Crime Bill, welfare reform, and more recent policy initiatives they believe have had a deleterious impact on women and people of color globally.

The impulse to dismiss communication that arises from a counterpublic centering millennial Black feminist thought can be seen in how media outlets initially reported on #GirlIGuessImWithHer as simply a disgruntled eruption from Sanders supporters, despite the complexity of political frameworks motivating those who started it (Rigueur 2016). In considering hashtags like #GirlIGuessImWithHer, scholars face the challenge of not reproducing limited constructions of democracy and democratic engagement that have legitimized our less than representative—and certainly *not* radically inclusive—academic institutions and national politics. Ultimately, hashtags and other forms of situated knowledge arising from networked counterpublics and embraced by a new generation of Black activists should be treated as important contributions to the democratic process. They are a call to include nuanced issues of identity in activist spaces and in national political conversations. They are a call to recognize the value of political thought that arises from the margins and that just might transform our democracy, if we let it.

NOTES

1. In this essay I use the lowercase "Black lives matter movement" to refer to the contemporary iteration of the Black freedom movement, I capitalize Black Lives Matter when referring to the formal organization of the same name or in the context of the hashtag #BlackLivesMatter, in which it is generally capitalized for ease of reading.
2. The platform and policy demands of the Movement for Black Lives can be found at https://policy.m4bl.org/platform/.

WORKS CITED

Bailey, Moya, and Reina Gossett. "Analog Girls in Digital Worlds: Dismantling Binaries for Digital Humanists Who Research Social Media." *Routledge Companion to Media Studies and Digital Humanities*. Ed. Jentery Sayers. New York: Routledge, 2016. Print.

Bailey, Moya, and Alexis Pauline Gumbs. "We Are the Ones We've Been Waiting For: Young Black Feminists Take Their Research and Activism Online." *Ms. Magazine* Winter 2010: 41–42. Print.

Bonilla-Silva, Eduardo. *Racism Without Racists: Color-Blind Racism and the Persistence of Racial Inequality in the United States*. Lanham, MD: Rowman & Littlefield, 2006. Print.

Cobb, Jelani. "The Matter of Black Lives." *The New Yorker*. 14 March 2016. Web. 5 July 2016.

Cohen, Cathy J., and Sarah J. Jackson. "Ask a Feminist: A Conversation with Cathy J. Cohen on Black Lives Matter, Feminism, and Contemporary Activism." *Signs: Journal of Women in Culture and Society* 41.4 (2016): 775–92. Print.

Florini, Sarah. "Tweets, Tweeps, and Signifyin': Communication and Cultural Performance on 'Black Twitter.'" *Television and New Media* 15.3 (2014): 223–37. Print.

Foster, John Bellamy, and Robert W. McChesney. "Surveillance Capitalism: Monopoly-Finance Capital, the Military-Industrial Complex, and the Digital Age." *Monthly Review* 66.3 (2014): 1–1. Print.

Freelon, Deen Goodwin, Charlton D. McIlwain, and Meredith D. Clark. *Beyond the Hashtags: #Ferguson, #BlackLivesMatter, and the Online Struggle for Offline Justice.* Washington, DC: Center for Median and Social Impact, School of Communication, American University, 2016. Print.

Garza, Alicia. "A Herstory of the #BlackLivesMatter Movement." *Feminist Wire.* 7 October 2014. Web. 5 July 2016.

Houston, Marsha. "The Politics of Difference: Race, Class, and Women's Communication." *Women Making Meaning: New Feminist Directions in Communication.* Ed. Lana F. Rakow. New York: Routledge, 1992. 45–59. Print.

Jackson, Sarah J., and Brooke Foucault Welles. "#Ferguson Is Everywhere: Initiators in Emerging Counterpublic Networks." *Information, Communication, and Society* 19.3 (2016): 397–418. Print.

Jackson, Sarah J., and Brooke Foucault Welles. "Hijacking #myNYPD: Social Media Dissent and Networked Counterpublics." *Journal of Communication* 65.6 (2015): 932–52. Print.

Kingston Mann, Larisa."What Can Feminism Learn from New Media?" *Communication and Critical/Cultural Studies* 11.3 (2014): 293–97. Print.

McRobbie, Angela. "Post-Feminism and Popular Culture." *Feminist Media Studies* 4.3 (2004): 255–64. Print.

Rigueur, Leah Wright. "Young Black People Are Radically Reimagining What Political Activism Can Be." *The New York Times.* 1 March 2016. Web. 6 July 2016.

Sharma, Sanjay. "Black Twitter? Racial Hashtags, Networks, and Contagion." *New Formations* 78.1 (2013): 46–64. Print.

Smith, Andrea. "Heteropatriarchy and the Three Pillars of White Supremacy: Rethinking Women of Color Organizing." *Color of Violence: The Incite! Anthology.* Ed. Incite! Women of Color Against Violence. Cambridge, MA: South End Press, 2006. 66–73. Print.

Walters, Suzanna Danuta. *The Tolerance Trap: How God, Genes, and Good Intentions Are Sabotaging Gay Equality.* New York: New York UP, 2016. Print.

Index

Note: Page numbers followed by "f" indicate figures.